Proceedings

— ISADS 97 —

THIRD INTERNATIONAL SYMPOSIUM ON AUTONOMOUS DECENTRALIZED SYSTEMS

ISADS 97

Proceedings

— ISADS 97 —

THIRD INTERNATIONAL SYMPOSIUM ON AUTONOMOUS DECENTRALIZED SYSTEMS

April 9–11, 1997 ▪ Berlin, Germany

Sponsored by

IEEE Computer Society
Information Processing Society of Japan
The Society of Instrument and Control Engineers of Japan

In cooperation with

International Federation for Information Processing
International Federation of Automatic Control
Gessellschaft fuer Informatik

Supported by

Hitachi ▪ DeTeBerkom ▪ NEC ▪ Digital
GMD FOKUS ▪ Hewlett Packard ▪ IBM

IEEE Computer Society Press
Los Alamitos, California

Washington ▪ Brussels ▪ Tokyo

IEEE Computer Society Press
10662 Los Vaqueros Circle
P.O. Box 3014
Los Alamitos, CA 90720-1264

IEEE Computer Society Press Order Number PR07783
IEEE Order Plan Catalog Number 97TB100111
Library of Congress Number 96-80464
ISBN 0-8186-7783-X
Microfiche ISBN 0-8186-7785-6

Additional copies may be ordered from:

IEEE Computer Society Press	IEEE Service Center	IEEE Computer Society	IEEE Computer Society
Customer Service Center	445 Hoes Lane	13, Avenue de l'Aquilon	Ooshima Building
10662 Los Vaqueros Circle	P.O. Box 1331	B-1200 Brussels	2-19-1 Minami-Aoyama
P.O. Box 3014	Piscataway, NJ 08855-1331	BELGIUM	Minato-ku, Tokyo 107
Los Alamitos, CA 90720-1314	Tel: +1-908-981-1393	Tel: +32-2-770-2198	JAPAN
Tel: +1-714-821-8380	Fax: +1-908-981-9667	Fax: +32-2-770-8505	Tel: +81-3-3408-3118
Fax: +1-714-821-4641	misc.custserve@computer.org	euro.ofc@computer.org	Fax: +81-3-3408-3553
Email: cs.books@computer.org			tokyo.ofc@computer.org

Editorial production by Penny Storms
Cover by Joe Daigle
Printed in the United States of America by Braun-Brumfield, Inc.

 The Institute of Electrical and Electronic Engineers, Inc.

Table of Contents

Keynote Address

Session 1A: Multiagent Systems I

Session 1B: Telecommunication Services

Session 1C: Distributed Algorithms

Session 5C: Applications: Building Control

Chair: Edgar Nett, GMD SET, Germany

Panel Session P3

   Chair: Jürgen Nehmer, University of Kaiserslautern, Germany

General Chair's Message

Welcome to Berlin and the Third International Symposium on Autonomous Decentralized Systems 97. This is the third event in a series of symposia, of which the first two took place in Japan and the United States, respectively. Within the last year, the field of autonomous decentralized systems has become most interesting as we try to understand and provide solutions for systems in which computer and telecommunication techniques and technologies melt into a global information infrastructure.

Object-oriented paradigms, information market systems, mobile agent technologies, broadband networks, and multimedia are technologies that will have a dramatic impact on the future systems used by our information society. I hope that you will find the program stimulating and that you will take advantage of the opportunity to meet informally with your colleagues from around the world to engage in social as well as technical dialogues.

An international symposium of this size and diversity requires the support of many people and organizations. We would like to give special recognition to Hitachi, DeTeBerkom, NEC, Digital, Hewlett Packard, IBM, and GMD FOKUS for sponsoring this event.

J. Nehmer did a splendid job as program chair and has presented you with an excellent technical program in the local organization.

For their dedicated work in the local organization, I would like to thank Barbara Intelmann, Petra Hoepner, Volker Tschammer, Eckhard Moeller, and Roland Schwarz.

Finally, my personal thanks to the keynote speakers and to all authors, panel and session chairs, panelists, and attendees who, I am confident, will make ISADS 97 a success.

Again, welcome to Berlin and ISADS 97.

Radu Popescu-Zeletin
General Chair

GMD FOKUS
Hardenbergplatz 2
10623 Berlin, Germany
E-mail: zeletin@fokus.gmd.de

Program Chair's Message

ISADS 97—like its predecessors ISADS 95 and ISADS 93—focuses on novel network applications and supporting hardware/software technologies in the network and computer field to make those types of applications possible. Globally distributed control systems such as railroad control or air traffic control systems are subject to extreme requirements in terms of the timeliness, reliability, performance, and security of the various subsystems involved. These stringent requirements call for an integrated approach to future distributed applications that combines computing and telecommunication aspects with a unified development and run-time support environment.

We believe we have composed an attractive program with contributions from industry and research that were selected from more than 100 submitted papers. These papers provide excellent insight into actual research and development projects and the planning stage for the future. In three companion panel sessions, special issues related to the topic of network computing and standardization efforts in the ADS field are highlighted.

I hope ISADS 97 will be a great success.

J. Nehmer
Program Chair

Universitaet Kaiserslautern
Fachbereich Informatik
Postfach 3049
67653 Kaiserslautern, Germany
E-mail: nehmer@informatik.uni-kl.de

Symposium Committees

General Chair

Radu Popescu-Zeletin, GMD FOKUS/Technical University of Berlin, Germany

Program Committee

Chair: Juergen Nehmer, *Univ. Kaiserslautern,* Germany
Vice-Chair: Richard Soley, *OMG, USA*
Vice-Chair: Yuji Inoue, *NTT, Japan*

Ricardo Baeza-Yates, *Univ. de Chile, Chile*
Farokh B. Bastani, *Univ. of Houston, USA*
Hendrik Berndt, *Global One, USA*
Sujeet Chand, *Rockwell International Corp., USA*
Jung W. Cho, *KAIST, Korea*
Partha Dasgupta, *Arizona State Univ., USA*
Domenico Ferrari, *Univ. Cattolica, Piacenza, Italy*
Hiroyuki Fujita, *Tokyo Univ., Japan*
Michel Gien, *Chorus Systems, France*
Deb Guha, *TINA-C, USA*
V. Jagannathan, *West Virginia Univ., USA*
Charles Jung, *IBM, USA*
Tohru Kikuno, *Osaka Univ., Japan*
Kane Kim, *Univ. of California, Irvine, USA*
Shinzo Kitamura, *Kobe Univ., Japan*
Hermann Kopetz, *T. Univ., Vienna, Austria*
Guy Leduc, *Univ. de Liege, Belgium*

Jeffrey N. Magee, *Imperial College, UK*
Bernd Mahr, *T. Univ., Berlin, Germany*
Manuel d. J. Mendes, *Pont. Univ. Catolica, Brazil*
Kinji Mori, *Hitachi, Japan*
Max Muehlhaeuser, *Univ. of Linz, Austria*
C. V. Ramamoorthy, *Univ. of California, Berkeley, USA*
Gerd Schuermann, *GMD FOKUS, Germany*
Norio Shiratori, *Tohoku Univ., Japan*
Ralf Steinmetz, *T. Univ., Darmstadt, Germany*
Liba Svobodova, *IBM Research, Switzerland*
Makoto Takizawa, *Tokyo Denki Univ., Japan*
Ahmed Tantawy, *IBM Research, USA*
Volker Tschammer, *GMD FOKUS, Germany*
Paulo Verissimo, *F. C. U. L., Portugal*
Feng-Jian Wang, *National Chiao-Tung Univ., Taiwan*
Roberto Zicari, *Univ. of Frankfurt, Germany*

Operations Committee

Silke Cords, *GMD FOKUS, Germany*
Petra Hoepner, *GMD FOKUS, Germany*
Barbara Intelmann, *GMD FOKUS, Germany*
Katsumi Kawano, *Hitachi, Japan*
Eckhard Moeller, *GMD FOKUS, Germany*
Roland Schwarz, *GMD FOKUS, Germany*
Volker Tschammer, *GMD FOKUS, Germany*

Advisory Committee

Juergen Kanzow, *DeTeBerkom, Germany*
Nobuhiko Koike, *NEC, Germany*

Steering Committee

Chair: Stephen S. Yau, *Arizona State Univ., USA (IEEE-CS)*
Yoshikazu Nishikawa, *Kyoto Univ., Japan (SICE)*
Masanori Ozeki, *RTRI, Japan (IPSJ)*

Reviewers

Bruno Achauer
Osamu Akashi
Yahya Y. Al-Salqan
Terauchi Atsushi
Luiz Bacellar
Ricardo Baeza-Yates
Rogerio Barra
Farokh B. Bastani
Rida A. Bazzi
Hendrik Berndt
Reinhard Bertram
Olivier Bonaventure
Jan Borchers
Stephen Brady
Frank Buddrus
Sujeet Chand
M.-F. Chen
K. Joseph Cleetus
J. Felix Costa
Partha Dasgupta
Juergen Dittrich
Christian Ebener
Klaus-Peter Eckert
Klaus-D. Engel
Fabrizio Ferrandina
Domenico Ferrari
Kerstin Fischer
Emmerich Fuchs
Hiroyi Fujita
Michel Gien
Fernando Gomide
Reinhard Gotzhein
Hermann Haertig
Takeo Hamada
Philipp Harrschar
Jean-Charles Henrion
Hiroaki Higaki
Shigeki Hirasawa
Yigal Hoffner

C.-H. Hu
K.-C. Huang
Yuji Inoue
Ikhyeon Jang
Euihoon Jung
Young-Jae Kang
Shigeki Kawano
Tohru Kikuno
Ji-Yun Kim
Jin-Suk Kim
Yu-Seok Kim
Tetsuo Kinoshita
Shinzo Kitamura
Eckhart Koerner
Hermann Kopetz
Andreas Krueger
Thomas Kunkelmann
Boseop Kwon
Sven-Eric Lautemann
Guy Leduc
Luc Leonard
Jack Liang
S.-C. Liao
Waldomiro Loyolla
Klaus Madlener
Mauricio Magalhaes
Jeffrey N. Magee
Bernd Mahr
Friedemann Mattern
Manuel J. Mendes
Markus Michalek
Dietmar Millinger
Marie-Luise Moschgath
Kenji Moriyasu
Max Muehlhaeuser
Shin'ichi Nagano
Juergen Nehmer
Roman Nossal
Masayuki Orimo

Giovanni Pacifici
Jaehyung Park
Charles Pecheur
Chotipat Pornavalai
Peter Puschner
Freeman Rawson
Sumitra Reddy
Rimbert Rudisch
Oliver Scheck
Christoph Schommer
Gerd Schuermann
Dimitrios Serpanos
Eltefaat Shokri
Richard Soley
Hyojung Song
Chittur Subbaraman
Liba Svobodova
Takayuki Tachikawa
Kaoru Takahashi
Katsuya Tanaka
Ahmed N. Tantawy
Christopher Temple
Daisuke Teratani
Junichi Toyouchi
Volker Tschammer
Tatsuhiro Tsuchiya
Keith Vanderveen
Paulo Verissimo
Hartmut Vogler
Feng-Jian Wang
Lars Wolf
J.-Y. Yang
I-Ling Yen
Hongtao Yin
Dong-ho Yoo
Akira Yoshizawa
Gerhard Zimmermann
Martina Zitterbart

Keynote Address

Network Evolution:
Convergence of Decentralized Information
Technology and Telecommunications

Speaker

Hagen Hultzsch

Network evolution - convergence of decentralised information technology and telecommunications

Dr. Hagen Hultzsch

Member of Deutsche Telekom AG's Board of Management

1. Introduction

For the past few years, the term "information society" has been ever-present in the media. And what is more: programmes involving gigantic investments have been launched everywhere in the USA-Europe-Eastern Asia triad in order to develop the required infrastructure.

It is often overlooked that this process does not actually involve the creation of an information society but of a telematic society. For, the characteristic element is not so much the surge in the amount of information that is available, but the revolutionary increase in performance both in the field of information technology and in telecommunications. Only the combined effects of these two developments can lead to the "information highway". So it is information technology and telecommunications which have paved the way for this telematic society. They offer such huge potential that together with entertainment electronics, they are referred to as the "convergence industry" and at the same time as the "last frontier" since they are considered to be the last discernible growth field.

There is hardly any other example which confirms the quotation of Le Corbusier more forcefully: "One does not stage a revolution by rebelling, but by delivering the solution!" The combined effects of information technology and telecommunications are leading to whole new developments and economic structures, which can indeed be called revolutionary. Many people have not yet become aware of the consequences which these developments will have. We are experiencing the transition from the production society to the innovation production society, which no longer fo-

cuses on cost-effective production but on the production of innovations. And it was the convergence industry which made this reversal in trend possible in the first place. Whereas progress made in information technology enables us to cope with tasks which are becoming increasingly complex, telecommunications is eliminating dependance on distance and time from the value-added process. Both together are leading the economic process into a completely new dimension.

2. The market volume of the convergence market

What is the real potential of this development into an innovation production society, which is facilitated by telematics? The forecasts are just as numerous as varying. If one assumes that annual investments in Europe made into the development of innovative products, from the design of chairs to the Ariane spaceship, amount to ECU 4 trillion, a mere 5 percent of this amount adds up to ECU 200 billion. To put it in different terms: a process acceleration of 5 percent attributable to the use of innovative telematics corresponds to ECU 200 billion - so the convergence market is in fact the last frontier.

Apart from this perhaps somewhat abstract consideration, concrete forecasts have also been made by market researchers for certain submarkets of the convergence market. There are, for instance, estimates predicting that the global market volume for multimedia will grow from DM 15 billion in 1993 to DM 140 billion by the year 2000, which represents a growth rate of 38 percent. Other estimates are based on even larger figures; this is probably attributable to the fact that there is no conclusive definition of the term "multimedia". Nonetheless, all forecasts basically agree on one thing, namely that the growth rates are enormous.

3. Prospects of new telematic services: telecooperation, telemedicine, videoconferencing

The prospects are extremely promising - but how do we turn visions into reality? Actually, many applications which characterise this innovation production society are already in the first stages of development and will achieve full market maturity in the next few years.

Example: Teleworking
It has long been clear that telework is an instrument which enhances not only productivity, but also the quality of life. The basic operational costs for the classical teleworker only amount to a few hundred dollars per month, even if he or she works 5 hours a day online. However, many people are working as teleworkers today without even realizing it - the employee at the call centre whose work depends totally on the telephone, or the manager who can work just as effectively in his automobile, in a hotel or at home using telematics. Actually, the way in which managers work presents the ideal application field for telework.

Example: Telemedicine
Because of the general conditions prevailing in society, the medical field is an ideal breeding ground for telematic services. The way people handle the resource "information" and the way in which it is exchanged between those involved in the health market is lagging way behind the already existing facilities. To this very day and in many cases, X-rays, diagnoses and invoices are still not being exchanged at high speed and digitally, but are sent through slow and manual channels. Atoms are moved instead of bits. Yet the changing age structure of the population and the increase in the demand for medical services can only be managed effectively if much more efficient processes are found - all this can be accomplished with telemedicine. Initial applications are meanwhile available on the market in this field.

Example: Videoconferencing
The market for videoconferencing equipment has been experiencing a boom for several years now, recording growth rates of several hundred percent each year. Observers anticipate that the mass market for PC videoconferencing including joint editing of documents will be achieved by 1997 at the latest due to the drop in prices. More sophis-

ticated equipment with near TV quality and dimensions will most likely follow suit.

4. What needs to be done?

These few examples show quite clearly what potential lies at our fingertips with telematic applications. The crucial question must therefore be: if the requirements are so high and the market prospects are so good - how can we develop these markets? This can only be accomplished if we solve several key tasks, which are not just of a technical nature and which are explained under the three following headings.

4.1 Bandwidth

Of course, we must solve the bandwidth problem, in other words, we must bring more bandwidth into people's homes and offices at more affordable prices. Even in Germany, where Deutsche Telekom has installed the most dense optical fiber network in the world comprising 125,000 cable kilometers, the "network capacity bottleneck" is a familiar term. It is caused by the explosive surge in networking in recent years, particularly in the residential sector. As everywhere else in the world, Germany is frantically trying to find solutions, namely additional broadband networks for online services and better transmission systems. However, Germany has an excellent starting position with its nationwide ISDN network, which provides the basis for digital, cost-effective medium-capacity transmission systems. This is crucial for the networking of PCs in the residential sector too and therefore for achieving the breakthrough into the mass market.

ISDN, by contrast, is only the preliminary stage to ATM, which will come into our living rooms and offices in the form of xDSL technologies much quicker than we can currently imagine. ATM is already available in Germany - the only country where it can be activated by the user - in all the conurbations and way beyond them. The further prospects of the cost-effective bandwidth evolution associated with ATM and xDSL is determined by the rapidly progressing convergence of the telephone, TV and PC to form an integrated, multifunctional, multimedia and interactive system which gives access to different services. Perhaps discussions about the NC will be enriched by a new facet: the NC with ATM connection. The development history of the NC is a typical example which shows how rapidly visions can be turned into products in this dynamic environment, products that are available in all the department stores.

4.2 Ergonomic telematic services

However, it is much more important to develop telematic products and services which meet market requirements than to improve the already efficient infrastructure. A lot of people still show clear signs of hesitance, which can be attributed to the fact that products tend to ignore ergonomic aspects. This is not so much a technical problem, as a lack of customer orientation in product development. The gap between developers and users has become smaller in the past few years, but the convergence industry is still characterised more strongly by thick manuals and awkward installation procedures than any other industry - even though the term "plug and play" is starting to define a new standard of quality. And it is absolutely vital that a new standard of quality be found: the acceptance of telematic services can only be enhanced with ergonomics and sensor techniques, or to be more precise: with communications ergonomics and communications sensor techniques. The infostructure can only generate economic growth, personal benefit, entertainment, a standard of living, in other words a comprehensive surge of activity, if applications are user-friendly.

4.3 More speed

The development of the innovation production society is associated with a change in paradigm; instead of what used to be said in the past about "big things swallow small things", the motto of the present day and the future has become much more appropriately "fast things swallow slow things". This motto describes the competition which companies face. But it also describes the demands made on staff: a large company needs a large number of staff - a fast company needs staff who work fast! And speed is only possible through innovation - the future belongs to innovative companies. The history of numerous companies from the Internet environment, which is probably the best-known product of the convergence industry, proves that this thesis is true. The will to embrace innovation on an ongoing basis must be implanted in our brains and become the motivating force of our actions.

Yet speed also means changing the focus of product development. It will, of course, still be necessary to develop completely new, innovative products in costly and lengthy development processes and this will be done in laboratories. But the acceptance of the telematic society can only be achieved through "iterative innovation". For, experience shows that most people prefer evolutionary developments to a revolution. The dynamics of technical developments must also be reflected in the products of the convergence industry, as speed has become a feature of the telematic society.

5. Task to be performed by network operators

What concrete tasks can be deduced from this for network operators? Apart from the problems associated with bandwidth, which is their very field of activity, they are particularly needed when it comes to the generation of new services. For, global liberalisation creates multi-infrastructures, which implies that owning one's own infrastructure no longer gives anyone the decisive competitive edge. In the long run, telecommunications infrastructures alone will not be sufficient for those who want to survive in a competitive environment. According to calculations made by Time magazine, the cost of a telephone call has dropped to one thousandth of the former value over the past 40 years. The provision of new telematic services will establish itself rapidly as a quality feature of successful telecommunications providers.

Moreover, the installation of a relevant platform will also gain significance. The catchwords harmonisation and standardisation are extremely important in this context. They are the basic prerequisites for growth in a competitive environment and therefore for the growth of markets.

One example is the global association of telecommunications and information technology companies, TINA-C (Telecommunication Information Networking Architecture Consortium), in which the significance of software for telecommunications manifests itself. Success has been achieved in the individual fields: standards such as ISDN, GSM, DVB or MPEG for image transmission have established themselves in telecommunications the same way in which object orientation has become the speed-promoting factor in information processing. And in the Internet, JAVA has become as successful as its triumphal march has been fast - neither of them has come as a surprise if we regard the Internet technologies and these utilisation structures, which are characterised by universality as a product of the telematic society with the features mentioned in the foregoing.

So the challenge is therefore: to harmonise the infrastructures and to combine them with contents as a means of developing new business segments. So Deutsche Telekom will not be the only network operator which intends to evolve from being a telecommunications provider into a telematics provider. The spectrum of activities will inevitably expand from network infrastructure through network access systems and terminals to contents.

6. Conclusion

The convergence industry is the growth market today, it is the proverbial "last frontier". In the society that is emerging, it will be quick to take on the same role which the electrical or mobility industry took on in the past. Nowadays, 12 percent of the value of an automobile consists of electronics - by the year 2000, it will be 20 percent!

A decisive prerequisite for its development can be attributed to the information highway - characterised by economics, ergonomic, iterative and therefore rapid innovation - which allows information technology and tele-communications to converge equally on the basis of global standards.

However, the "mental acceptance" of the changes which this will bring about in the business sector is equally important:

- The production society will be replaced by the innovation production society
- Project-oriented attitudes and working methods will be introduced as well as
- Flexible organisations and processes.

Companies with numerous sites, having their development department in India, sales and marketing division in the USA and controlling/finance in Germany, are already a reality today and are contributing towards enhancing productivity in the business sector, as well as towards enhancing the quality of life in the residential sector.

Each and every one of us is helping to shape this development. We are already creating a platform today that will reflect the quality of life of future generations. Our task is to play an active role in helping our society to evolve into an innovation production society and to seize any opportunities which may present themselves. Our goal must be to create a platform of prosperity and quality of life by vigorously making use of the convergence of information technology and communications technology to create the innovation production society.

Session 1A

Multiagent Systems I

Chair

Guenter Karjoth

Towards a Multi-Agent System for the Supervision of Dynamic Systems

M.-K. Allouche
allouche@emse.fr

C. Sayettat
sayettat@emse.fr

O. Boissier
boissier@emse.fr

L.S.C. - SIMADE - E.M.S.E.
158 cours Fauriel,
42023 Saint-Etienne cedex 02 France

Abstract

The supervision of dynamic systems is essential in industrial applications. The decentralization of such systems make their supervision more difficult. In this article, we tackle the supervision problem by using a temporal scenario recognition approach. The supervision is performed by a society of agents where an agent is considered as a watching process responsible for a subset of possible scenarios of the functioning of the system. The cooperation among agents is based on interaction protocols as well as dependence networks.

1 Introduction

The increasing complexity of industrial systems emphasizes their problems with *supervision*. Many plants as power stations, electrical networks, manufacturing systems produce observations that can be used to predict and diagnose their functioning.

Our aim is to build up a system that is able to supervise the evolution of an industrial system using a pre-established set of scenarios. These *scenarios* [13] represent the possible operating modes of the system by identifying first its possible states and second, the events that make it change from one state to another. Once the scenarios are made, they are connected with events continuously emitted by the industrial system, in order to be recognized. The scenario recognition process is described in the next section.

In the majority of the current approaches, the scenario recognition is centralized when systems are often composed of several physically distributed components. Due to abundance of captured information and its heterogeneous nature, coherence and congestion problems may arise.

In this article, we present a decentralized system where the supervision is performed by a society of cooperative agents (STARS [3], [2]). An agent is in charge of watching over a part of the supervised system using the emitted events in order to recognize the corresponding scenarios. The supervision of the industrial system is achieved by the definition of explicit dependence networks and interaction protocols among agents. Similar approaches to control industrial processes are also described in [19].

In section 2 we present the scenarios model as well as their recognition process. In section 3 we present the proposed agents' models allowing the recognition of scenarios. The section 4 describes the dependence network among agents, an interaction language as well as the necessary interaction protocols to enable the agents to recognize scenarios. In section 5, we illustrate the agents' functioning in order to perform the task of supervision. We conclude in section 6 and give some perspectives on our work.

2 The scenarios recognition

The *supervision* as mentioned in [18] may be accomplished by using temporal probabilistic networks [20], by event based approaches [9] or by using scenario recognition approaches. In the first two approaches, the whole graph of possible states of the supervised system is needed. This assumption is not applicable in real industrial contexts. We have prefered one in the last class of approaches which makes use of graph comparisons techniques [13] instead of the recognition of motifs in strings of characters [16] or constraints propagation techniques [14][11][12]. In this approach, only parts of the graph of states are considered : each part corresponds to a specific behavior or a functionning mode of the industrial system and can be modelled by a scenario (see examples given below). The supervision can be performed using an off-line or an on-line mode. The definition of the agents' models in the proposed system is based on the analysis of all the performed tasks during this recognition process. The recognition process consists in capturing events emitted by the supervised system and using them to activate certain scenarios that may be recognized later.

2.1 Scenario/Sub-scenario

A *scenario* is composed of a graph of temporal constraints and a set of preconditions allowing its activation. The graph of constraints is defined by a set of *states* and *events*. A state is made up from the combination of a non-temporal proposition and an interval indicating a period of time during which the proposition holds. An event marks the startpoint or the endpoint of a state. For example, the state "*machine.state = busy*" and its interval $[t_0, t_1]$ correspond to two events having respectively as occurrence timepoints t_0 and t_1 (see example given below).

Scenario. To make clearer the structure of a scenario, we will model the operating mode of a machine in a production line by the scenario "machine". This scenario is built up from a "story" given by experts in the field. Temporal constraints that exist among states and events of the scenario are indexed by (n°) :
- "The machine is free"
- "Then a part to be tooled is present in front of the machine. The machine is activated and becomes busy" (1)
- "The normal time of machining a part must not exceed 2mn" (2)
- "Once the part is finished, the machine becomes free again" (3)
- "During all the machining time, the motor of the machine must operate perfectly" (4)

The possible states of the machine are given in the following table (a timepoint variable t_i corresponds to the occurrence of an event) :

State	Interval
machine.state = busy	$I = [t_0, t_1]$
machine.function = machining	$J = [t_2, t_3]$
machine.motor-state = ok	$K = [t_4, t_5]$

The temporal constraints set among events are based on comparison types introduced by Allen[1]. They are set between timepoints or intervals (Start(I) and End(I) accede to bounds of interval I) :

Start(I)	{equal}	Start(J)	(1)
Start(J)	{[0, 120]before}	End(J)	(2)
End(I)	{equal}	End(J)	(3)
J	{during}	K	(4)

From this table, we construct a graph of constraints where nodes are events and edges are constraints set between two events. This graph is called the "*scenario body*". In our example, the graph is represented in figure 1(upper part). Constraints (5) and (6) are implicit constraints stating for example, that the beginning of a state comes before its ending.

Sub-scenario. The basic notion of scenario introduced above can be extended so that scenarios can admit one or several sub-scenarios. A *sub-scenario* is represented as an event in the body of the scenario it belongs to. As an example, we take the scenario "machine". We consider that the starting event of the abstract state "*machine.motor-state = ok*" corresponds to the recognition of another scenario "motor" that is a sub-scenario of the scenario "machine". The ending event corresponds to the rejection or the exit of the scenario "motor". This scenario is built up from the following story given by experts in the field :
- "The motor is started"
- "Then the starter of the motor is switched on" (1).
- "The water temperature must reach at least 40°C within 30s after the motor starting" (2). "It must remain higher than 40°C for as long as the motor is functioning" (3).
- "The oil temperature must reach at least 85°C within 35s after the motor starting" (4). "It must remain higher than 85°C for as long as the motor is functioning" (5).
- "The motor speed must lie between 1000 rpm and 1500 rpm within 50s after the motor starts" (6), "moreover, oil and water temperatures must be respected" (7),(8). "Furthermore, the starter must be switched off 10 to 30s before the motor speed has reached its authorized value" (9). "The motor speed must remain lower than 1500 rpm as long as the motor is functioning" (10).

The possible states are represented in the following table :

State	Interval
motor.mode = start	$I' = [i_0, i_1]$
motor.starter = on	$J' = [i_2, i_3]$
water.temperature $\geq$ 40	$K' = [i_4, i_5]$
oil.temperature $\geq$ 85	$L' = [i_6, i_7]$
1000 $\leq$ motor.speed $\leq$ 1500	$M' = [i_8, i_9]$

Temporal constraints indexed by [n°] in the story are listed below :

Start(I')	{before}	Start(J')	(1)
Start(K')	{[0,30]after}	Start(I')	(2)
K'	{finish,overlapped by}	I'	(3)
Start(L')	{[0,35]after}	Start(I')	(4)
L'	{finish,overlapped by}	I'	(5)
Start(M')	{[0,50]after}	Start(I')	(6)
M'	{during}	K'	(7)
M'	{during}	L'	(8)
J'	{[10,30]before}	M'	(9)
M'	{finish,overlapped by}	I'	(10)

The figure 1 shows the constraints graph(lower part). Note that constraints (11), (12), (13) and (14) are implicit constraints.

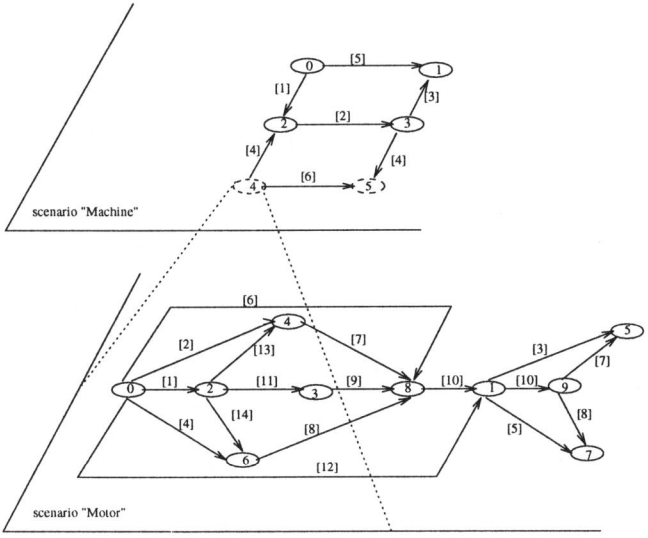

Figure 1: Graph of constraints of scenarios "machine/motor".

2.2 The recognition process

The recognition process consists in performing algorithmic operations on graphs of temporal constraints as well as in performing decision tasks.

Graph minimalisation. A graph of temporal constraints may contain incoherencies. To detect these incoherencies, an operation of *minimalisation* allows the construction of a minimal graph from an ordinary initial graph. An *incoherency* is detected if the minimal graph contains empty constraints.

Recognition by graphs comparisons. During its functioning, an industrial system emits information in

the form of events. From their occurrence times it's possible to construct as for scenarios, a graph of temporal constraints. This graph may evolve with time as a result of new incoming information. This graph is called the *"body of the current session"*.

The recognition process consists in (i) *activating* a set of scenarios having satisfied preconditions, (ii) *comparing* the body of the current session with the bodies of selected scenarios according to different *recognition modes* between a session and a scenario. These modes range from *partial compatibility* to *total satisfaction*. For example a session is compatible with a scenario if and only if the result of merging their graphs of constraints is a coherent graph. Other recognition modes are also described in [13]. All comparison operations have cubic complexity ($O(n^3)$).

3 Agents models

As described above, we can distinguish two types of tasks within the recognition process : operations on graphs of constraints (minimalisation, merging, intersection, ...) and decision operations relating to selection of a recognition mode, activation and scenarios recognition. To perform these tasks, we define two agents' models : the execution agents *"E-Agents"* and the decision agents *"D-Agents"*. Each D-Agent is responsible for watching over a subset of scenarios. The attribution of scenarios to agents may be based on several criteria. It may be based either on the physical distribution of components that the supervised system is composed of, or on its hierarchical structure into sub-systems. In both of these cases an agent receives only information coming from specific components or sub-systems that are under its responsability. In this way, he has only to preserve locally the coherence of the information he has [17]. It's easier to do so, because the information being smaller, homogeneity increases and congestion problems decrease. In this work, the distribution of scenarios among agents is based on the second criterion : an agent is responsible for the recognition of a set of scenarios determined by the hierarchical structure of the system functioning.

3.1 D-Agent model

In order to describe the D-Agents' role in performing the task of supervision, we will proceed according to two points of view [10]. The first one corresponds to a micro-level where a D-Agent is considered as an autonomous entity having its own set of scenarios to recognize using its own capacity of reasoning, of decision and of planning. The second one is a macro-level where the supervision of the system is made by a society of agents (see section 4).

D-Agents have capacity of reasoning and can make decisions. During the task of supervision, they select the best candidate among activated scenarios using events coming from the supervised system and messages coming from other agents. When a D-Agent chooses to recognize a scenario, he chooses first the *recognition mode* according to the semantical importance confered to the scenario. Second, he constructs a plan of recognition actions which will be executed with cooperation with other E-Agents and/or D-Agents.

A D-Agent has a *mental state* in which is represented information about him and about the agent's activities in the system :

• *internal description* of the agent : its goals, beliefs as well as facts about the supervised system. Goals deals with actions to be executed. Each action can be decomposed in a plan of basic actions. Beliefs concern the current session and the incoming events. Facts indicate for example, on the one hand if a scenario describes a good or a bad functioning and on the other hand which recognition mode is required or recommended for this scenario.

• *description of others* including their goals and beliefs in order to enable cooperation with other agents. (see section 4).

• *dependence network* allowing a D-Agent to know whom he depends on and conversly. It's also a way for D-Agents to define their *acquaintances*, i.e. the agents with which to cooperate.

Interactions allow D-Agents to update their goals, plans, beliefs as well as the description they have of others (see section 4).

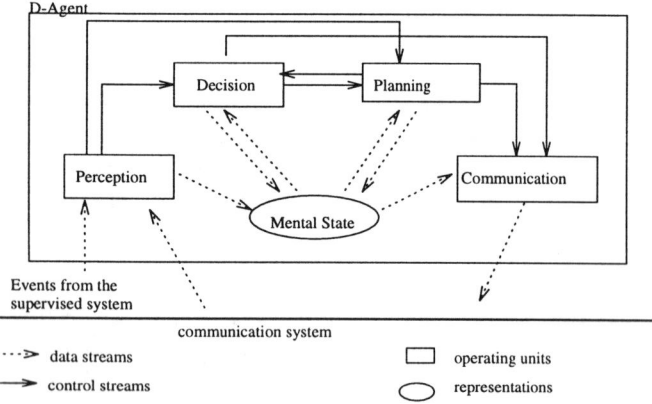

Figure 2: D-Agent Architecture

A D-Agent is composed of 4 modules (see Fig. 2)

that can consult and modify his mental state :

• *perception* : a D-Agent receives messages from both D-Agents and E-Agents. He's also in charge of capturing events comming from the supervised system;

• *decision* : this module is in charge of making decisions regarding a scenario recognition when a message or an event is perceived by the perception module. This module is also in charge of informing others about a recognized scenario, requesting information and finally cancelling a task if an occurred event makes changes in the agent's mental state, especially in his set of goals;

• *planning* : when a D-Agent decides that it's essential to recognize a scenario, the planning module generates a plan of all recognition tasks for sub-scenarios. It is also in charge of locally watching over their execution. Once all tasks are executed, the planning module informs the decision module of the ending of the high-level task it was responsible for.

• *communication* : this module manages communication with other agents. The system in its current state uses interaction protocols and an interaction language whose syntax is described in section 4.2.

3.2 E-Agent model

Each E-Agent is in charge of one basic task which is an operation on a graph of temporal constraints (see section 2.1). Basic tasks are executed during comparisons and incoherencies detection (see section 2.2). An E-Agent has thus limited capacities of reasoning and its major role is to satisfy requests sent by D-Agents.

An E-Agent is also composed of four modules :

• *perception* : this module allows an E-Agent to receive messages from D-Agents. Messages are requests to execute the basic task an E-Agent is dedicated to;

• *communication* : this module allows an E-Agent to send to a D-Agent the result of a basic task execution;

• *execution* : this module executes the basic task that the E-Agent is able to perform;

• *control* : when a request is perceived the information is transmitted to the control module which duplicates the execution module in order to satisfy the request. The result of an execution is sent to the control module which sends a message via the communication module to the D-Agent having sent the request. Duplication of the execution module allows satisfying several requests coming from different D-Agents in a parallel way.

4 The society of agents

To perform the task of supervision by a society of agents based on the models described above, agents must respect a certain organization. Agents' *organization* is often based on the *dependence* notion [7][22], or on the global task assigned to a group of agents

[24]. Building up an organization necessitates a study of the individual behavior as well as the group behavior within the society of agents [6]. In the first steps of our study, organization is based on the notion of dependence networks and the society of agents keeps the same organization during all the supervision period i.e. agents can not be reorganized dynamically [15].

4.1 Definition and Construction of a dependence network

A dependence network can be seen as a tool allowing agents to determine with which they will cooperate in order to perform the task of supervision (*acquaintances*). An agent depending on another agent forms a part of its acquaintance and vice versa. In the following sections, we show how dependences among agents strictly depend on distribution of scenarios among them.

Dependence network definition The distribution of scenarios among agents leads to the definition of dependences among them : agent x depends on agent y if x wishes to recognize a scenario b having a sub-scenario a which is under agent y's responsability. We define then a dependence operator D between two agents x and y for two scenarios b and a as the following :

$$x \; D_a^b \; y \equiv a \in d(y) \text{ and } b \in d(x) \text{ and } a \in s(b)$$

where : $d(x)$ is the set of scenarios forming agent x's reponsability, and $s(S)$ is the set of sub-scenarios of the scenario S.

Dependence network construction From the applications d and s, each agent constructs its own dependence network using the following algorithm. Agent reponsible for scenario S is designated as $r(S)$. The set of scenarios whose S is a sub-scenario (several scenarios may have the same sub-scenario) is given by the application $f(S)$.

```
For each a ∈ d(x) do
    For each b ∈ s(a) and b ∉ d(x) do x D_b^a r(b)
    For each b ∈ f(a) and b ∉ d(x) do r(b) D_a^b x
EndFor
```

A scenario being under the reponsibility of only one agent, r is an application. Therefore agents don't have common resources and then don't have resource dependences but only functional ones [22].

4.2 Interaction protocols

The communication is based on sending direct messages among agents. We describe the syntax of messages in the Backus Naur's form below. A message is composed of three fields, a field giving information about how the message must be routed in the communication system, a field describing the content of the message and a field defining a conversation context.

⟨Msg⟩::=⟨Communication⟩ ⟨Content⟩ ⟨context⟩
• the field ⟨Communication⟩ includes the message emission date and identities of the sender and the receiver of the message.

⟨Communication⟩::=⟨date⟩ ⟨sender⟩ ⟨receiver⟩
• the ⟨Content⟩ field takes the following syntax :

⟨Content⟩ ::= REQUEST ⟨Info-Rec⟩ ⟨scenario⟩ |
      RESPOND ⟨Answer⟩ |
      INFORM ⟨Rec⟩ |
      REFRAIN
⟨Rec⟩ ::= ⟨Rec-Mod⟩ ⟨scenario⟩ |
      ⟨scenario⟩ '<>' ⟨scenario⟩
⟨Info-Rec⟩ ::= RECOGNIZE | EXPLAIN
⟨Answer⟩ ::= ⟨Rec-Mod⟩ | EVT | SUB-SC | REQ
⟨Rec-Mod⟩ ::= COMP | SAT | RE

The field ⟨scenario⟩ is a symbol representing a scenario. The field ⟨context⟩ is the unique identification in the system for a conversation between two agents.

Exchanged messages among E-Agents and D-Agents follow the same syntax as the previous messages. Only the filed ⟨Content⟩ is modified : it's the expression of one or two graphs of temporal constraints depending on the cardinality of the operation the E-Agent is able to perform.

We have defined basic interaction protocols [5] [4] to control the interactions among agents :
• *Protocol of request* : A D-Agent x decides to recognize a scenario S having a sub-scenario s. Consulting its dependence network which contains $x D_s^S y$ (s is under the responsability of agent y), he sends a request to agent y asking him to recognize the sub-scenario s :

(d1 x y)(REQUEST RECOGNIZE s)(nxy)

As D-Agents can work in parallel, the sub-scenario s may have been already recognized by agent y. In this case, agent y sends a response to agent x informing him of the result of the recognition of s (for example that the current session is compatible (COMP) with s) :

(d2 y x)(RESPOND COMP)(nxy)

If the sub-scenario s is not recognized yet, y adds the recognition of s as a new goal to be achieved :

(goal s x nxy)

13

This goal contains the symbol of the scenario to be recognized (s), the agent concerned in the result of recognizing s (x) and the context of conversation to be precised to agent x when the result will be communicated (nxy). This goal will persist until the correspondant task is executed (recognition or rejection) or until x cancels his request.

• *Protocol of refraining a task* : An agent's decision regarding the recognition of a scenario is based either on captured events coming from the supervised system or on explicit demand of another agent. In the first case, the agent may make an erroneous decision due to lack of information. Having received additional information, an agent x may be obliged to change some of its goals and then be obliged to cancel tasks to be performed by other agents to satisfy his request.

To cancel a task, x sends a message of this type :

(d x y)(REFRAIN)(nxy)

• *Protocol of information* : Agent y having recognized the scenario s, notices the existence of the dependence $x D_s^S y$ in his dependence network. He can then inform x about this recognition. In fact, agent y considers that the recognition of s may be crucial for the recognition of S. In this case, y sends a message of the type :

(d y x)(INFORM COMP s)(nxy)

These protocols are used to build complex interaction protocols in order to control the negociation among agents, for example [8]. A detailed description of the negociation aspects can be found in [3]. The need of such a protocol arises from the incoherency that can appear during the supervision process accomplished by the society of agents because of the distributed functioning of the agents. Indeed, the exchanged information is limited and decisions are local to each agent. For this reason, two agents may have incoherency between their goals : they can decide to recognize two scenarios having contradictory semantics. Through communications, an agent detecting an incoherency between its goals and another agent's goals, starts a negociation phase with the corresponding agent. This negociation may also involve other agents. It consists in request/inform protocol phases among the agents in order to make an agent locally change its goals.

5   Example of resolution : Recognition of scenarios "machine" and "motor"

To illustrate the use of a dependence network and the protocols we have just defined above, we give an example showing the cooperation between two agents in order to recognize a scenario and its sub-scenario.

Consider two agents X and Y respectively responsible for the scenarios "machine" and "motor" presented in section 2.1.

The dependence networks of agents X and Y contain both the dependence $X D_{motor}^{machine} Y$. Suppose that X decides to recognize the scenario "machine", its planning module will put the communication module in charge of sending a list of messages using the request protocol. Messages exchanged among agents are represented in figure 3.

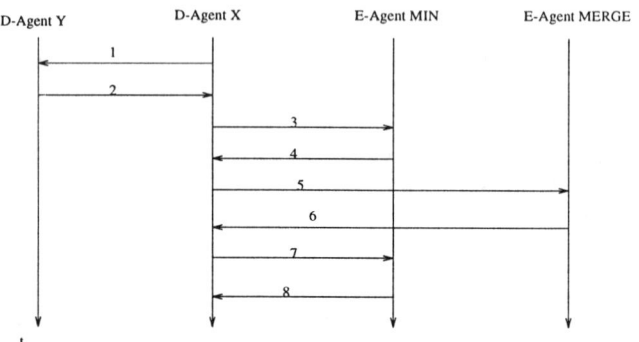

Figure 3: Recognition of scenarios "machine" and "motor"

1 : X asks Y to recognize the scenario "motor". The conversation context is "1xy"

(d1 X Y) (REQUEST RECOGNIZE *motor*) (1xy)

2 : Y responds informing X that its current session is compatible with the scenario "motor". To avoid overloading the figure, we didn't represent the exchanged messages during the recognition

(d2 Y X) (RESPOND COMP) (1xy)

3 : X sends the body of its current session to the E-Agent of minimalisation MIN

(d3 X MIN) (*machine-session*) (1xmin)

4: After minimalisation the E-Agent MIN sends to X the body of its current session in its minimal form

(d4 MIN X) (*machine-session*) (1xmin)

5: In order to compare the body of its current session with the body of the scenario "machine", X sends the correspondant graphs to the E-Agent "MERGE", which can merge two graphs of temporal constraints

(d5 X MERGE) (*machine-session machine*) (1xmerge)

6 : After merging, the resulting graph is sent to agent X

(d6 MERGE X) (*merge*) (1xmerge)

7 : To check the graph coherence, X asks its minimal form to the E-Agent MIN

(d7 X MIN) (*merge*) (2xmin)

8 : Receiving a graph without empty constraints, agent X deduce that its current session is compatible with the scenario "machine".

$$(d8 \text{ MIN } X) \ (merge\text{-}min) \ (2x\text{min})$$

This example shows how two agents cooperate in order to perform the task of supervision. This example can be easily extended to several agents.

6 Conclusion and perspectives

In this article, we have presented a model of agents dedicated to supervise industrial systems. Each agent locally handles the supervision by recognizing scenarios and interacts with other agents to insure the supervision of the complete industrial system. The agents' organization is defined as a dependence network which is determined from the distribution of scenarios among agents.

The introduction of the sub-scenario notion is well-adapted to model behaviors of systems having a hierarchical structure and aims to reduce the impact of the complexity of recognition algorithms ($O(n^3)$). The recognition of scenarios and their sub-scenarios must be performed simultaneously and require cooperation among entities that are responsible for these scenarios. Thus, a society of agents is in charge of performing the task of the supervision in a distributed way and this will contribute to eliminate two types of problems posed in classic supervision approaches where information is collected from all the system components and is sent to a control center to be analysed:

• when events coming from the supervised system are abundant, congestion problems may arise. In our approach, the supervison is local in each agent's domain and problems of this type are avoided.

• since all occurred events must be *sent* to the control center, some of them may be lost and this may hide other problems that will remain undetected. In our approach, dividing the supervised system into several domains allows having independant problems and limits the impact of these problems.

The cooperation is based on interaction protocols permitting also the detection of incoherencies in the system. Although it's simple, the interaction language is sufficient in the current state of this work. The independent functioning of agents increases the system flexibility.

A prototype of this system has been developed in C++ on Sparc-Stations. Presently, it is tested to supervise a production line in an industrial plant.

This work emphasizes for a fact the importance of the temporal factor in supervising industrial systems. Our future work, consists in taking it into account the agents' decisions making and interactions.

Thus, agents will be able to reason about the past and the future [21] and [23] in order to provide predicative diagnosis. Using dependence networks deduced from distribution of scenarios among agents gives new perspectives on the scenarios recognition, especially the introduction of new types of semantical relations among scenarios, as opposition and complementarity.

Acknowledgments

The authors wish to thank Christopher Yukna for his constructive suggestions concerning the structure of this article.

References

[1] J.-F. Allen, "Towards a General Theory of Action and Time", Artificial Intelligence 23, pp 123-154, 1984.

[2] M.-K. Allouche, C. Sayettat, "A Multi-Agent Architecture for Supervising Dynamic Systems", Proceedings of the First International Workshop on Decentralized Intelligent and Multi-Agent Systems, Krakow, Poland, 1995.

[3] M.-K. Allouche, O. Boissier, C. Sayettat, "Society of Temporal Agents : a Recognition Process Supersion", Internal report n. 9612, Ecole des Mines de Saint-Etienne, France (in french), 1996.

[4] M. Barbuceanu, M.S. Fox, "COOL - A language for Describing Coordination in Multi-Agent Systems", in Proceedings of First Internation Conference on Multi-Agent Systems, AAAI Press/The MIT Press, pp 17–24, 1994.

[5] O. Boissier, Y. Demazeau, "ASIC : An Architecture for Social and Individual Control and its Application to Computer Vision", Lecture Notes in Artificial Intelligence, 1069, Springer Verlag, Proceedings of the MAAMAW Workshop, Odense, DK, August 1994, 1996.

[6] P. Carle, A. Collinot, K. Zeghal, "Concevoir des organisations : la méthode Cassiopée", Journée "Systèmes Multi-agents" organized by PRC-IA, (in french), Paris, 1994.

[7] C. Castelfranchi, R. Conte, "Mind is not enough : Pre-cognitive bases of social interaction", Proc. Simulating societies symposium, university of Surrey, Guildford, UK, 1992.

[8] M. Chang, C. Woo, "SANP : a communication level protocol for negotiations", Decentralized AI III, Werner & Demazeau ed, Elsevier Science Publishers B.V, 1992.

[9] M.-O. Cordier, S. Thiébaux, "Event-based diagnosis for evolutive systems", Proc. Workshop on diagnosis from first principles (DX-94), New Paltz (long version in : rapport IRISA n° 819, 1994.

[10] Y. Demazeau, J.-P. Muller, "Decentralized Artificial Intelligence", in Decentralized A.I. Elseiver Science Publishers B.V. (North-Holland), 1990.

[11] C. Dousson, M. Ghallab, "Suivi de reconnaissance de chroniques", Revue d'Intelligence Artificielle, 8(1):29-61, (in french) 1994.

[12] C. Dousson, "Suivi d'Evolutions et Reconnaissance de Chroniques", Thèse de doctorat, Rapport LAAS N°94000, LAAS 7, av. du Colonel Roche 31077 Toulouse-France, (in french) 1994.

[13] D. Fontaine, "Reconnaissance de Scénarii Temporels", Rapport Interne. Université de Technologie de Compiègne, France, (in french) 1993.

[14] M. Ghallab, A. Mounir-Alaoui, "Managing Efficiently Temporal Relations through Indexed Spanning Trees", Proc. 11th IJCAI, 1297-1303, Detroit, 1989.

[15] F. Guichard, J. Ayel, "Logical reorganisation of dai systems", ECAI 94, Workshop on Agents Theories, Architectures and Languages, 1994.

[16] F. Lévy, "Recognising scenarios, a study", Proc. Workshop on diagnosis from first principles (DX-94), New Paltz, 1994.

[17] B. Malheiro, N.-R. Jennings, E. Oliveira, "Belief Revision in Multi-Agent Systems", ECAI 94. 11th European Conference on Artificial Intelligence. Edited by A.Cohn, Published in 1994 by John Wiley & Sons, Ltd, 1994.

[18] Collectif du projet inter-PRC, "Gestion de l'évolutif et de l'incertain dans une base de connaissances", 5èmes Journées Nationales PRC-GDR IA, Intelligence Artificielle, Nancy-France, (in french) 1995.

[19] C. Roda, N.-R. Jennings, E.-H. Mamdani, "ARCHON: A Cooperation Framework for Industriel Process Control", in Cooperating Knowledge Based Systems (ed DEEN, S.M.), pp 95-112, Springer Verlag, 1990.

[20] B. Séroussi, J.-L. Golmard, "An algorithm for finding the K most probable configurations in Baysian networks", Int J Approx Reasoning, 1:205-233, 1994.

[21] Y. Shoham, "Agent-oriented programming", In Artificial Intelligence 60, pp 51–92, 1993.

[22] J.-S. Sichman, Y. Demazeau, "Exploiting Social Reasoning to Deal with Agency Level Inconsistency", in Proceedings of First International Conference on Multi-Agent Systems, AAAI Press/The MIT Press, pp 352–359, 1994.

[23] M.-P. Singh, "Multiagent Systems, A theoretical Framework for Intentions, Know-How, and Communications", Lecture Notes in Artificial Intelligence 799, 1993.

[24] E. Le Strugeon, R. Mandiau, G. Libert, "Proposition d'organisation dynamique d'un groupe d'agent en fonction de la tâche", Actes des Premières Journées francophones en IAD-SMA, Toulouse, (in french) 1993.

A Constructing Scheme for Autonomous Distributed Control Systems with Multi-Agent Society

S. Y. Huang   Y. Umetani   Y. Yamada

Intelligent System Laboratory, Toyota Technological Institute

2-12-1 Hisakata, Tempaku, Nagoya 468, Japan

E-mail: {huang, umetani, yamada}@toyota-ti.ac.jp

Abstract

An autonomous distributed control (ADC), as a promising approach in dealing with complex control problems, is capturing the attentions of automation. Yet, there is still no available platform supporting its industrial applications. This paper introduce an ongoing project toward this goal carried out in our laboratory. In this work, we use agents and an agent society to model an ADC system. We propose a new agent architecture for intelligent machines, and a supervisor agent for human to provide a supervision control for agents. To facilitate the coordinating behaviors of agents, an agent communication language CCL is developed, and an object-oriented implementation of the agent system proposed is presented too, in this paper. Furthermore, the support system under development is introduced in simple. A case study concerning the shape control of a variable geometry truss is also given to show the availability of the agent system.

Key Words: *Autonomous distributed control, agent, intelligent machine, intention model, reaction model, supervisor, multi-agent society*

1. Introduction

Along with the applying of more complicated control systems in modern industrial automation, there is an increasing trend in decentralization of control systems. Whereas several merits such as adaptability, expandability, robustness, etc. of systems can be obtained with the aid of such decentralization, a crucial problem, that is how to make these distributed sub-systems work cooperatively, is also confronted. The fact is that, handling a decentralized system has gone beyond the view point for the system control to date.

Regarding this problem, the issue is that, needed is a new control scheme, for which the main research interests have been focused on the control mechanism of organs in living systems, or animals in colonies or groups [1]. The results reported of these investigations until now all indicate such a fact, that is either a living system or a living colony performs its control upon the autonomous activities of distributed organs or individuals, and in turn the autonomous activities are coordinated through a series of activating signals. These signals are emitted from a central system and transported through a communication system independent of the physical systems of the autonomous organs or individuals. Thus, an innovative control paradigm, which is defined to be an autonomous distributed control (ADC) [2], was triggered,.

It is obvious that a communication system plays a key issue in ADC systems, and a computer network is thought of a promising environment. In [7], a cooperating autonomous decentralized system has been developed, providing an open architecture to support the dynamic expansion of systems as well as vogue communications that occur frequently in ADC systems and is supported in broadcast or multicast. However, whereas a system architecture upon the data exchanging level is proposed to model ADC systems[7], the individual involved an ADC systems is not dealt with. Unlike it however, in our ongoing project, autonomous individuals and ADC systems are equally emphasized, and modeled with agents and an agent society separately.

Agents have been broadly studied in depth in AI community, pursuing a distributed problem solving. Several models have been raised out: reactive or cognitive models for agents, and network models for multi-agent societies [4, 5, 6]. But, between these studies and ADC systems still exists quite a large gap. Hereby briefly we conclude the main requirements of ADCs as follows:

- the synchronization of autonomous activities and the process control are bottleneck problems to those agents corresponding to automatic machines;
- one could not expect a perfect ADC system without human's supervision;
- an agent system should be able to model various control patterns, including complete autonomous control, central control as well as the hybrid of them;

17

- the controlling codes developed off-line should be inherited.

Taking into account these requirements, we construct an agent 'IM_Agent' for automatic machines, and introduces 'supervisor' agent for human. The supervisor agent provides a supervision control for an IM_Agent group. Thus, with the different combinations of IM_Agents and a supervisor, several control patterns ranging from central control to autonomous distributed control can be achieved. Tightly bound with these two agents categories, an agent communication language CCL (Cooperative Command Language) is also developed. Concerning process control of agents, our issue goes to a chronology or a traffic control for broadcast protocols flowing among agents. Saying alternatively, the broadcast communication is achieved through a consecutive delivery of protocols among agents, and the transport contract can be made in CCL options. As to the existing codes, with the aid of the abstraction of action primitives, these codes can be integrated into IM_Agent.

To date, several different architectures for autonomous agents have been proposed, among which two fundamental classes can be distinguished: horizontally layered architecture and vertically layered architecture[5]. In our work, we choose a vertical architecture for the IM_Agent, due to that it is more simple in design and easy in implement[5]. However, as an abstractive framework to meet various applications of the IM_Agent in industry, a need combining these layers freely might be required. Hence, different from the architecture in [5], we employ a separate mechanism to govern the control change of a layer output. Moreover, we take cooperation behaviors no longer as an independent layer. Rather, cooperative behaviors of an agent are considered involved implicitly in both the intention model and the reaction model, being that almost each action of an agent is constrained by other agents'.

Concerning multi-agent systems, similar to other works [6], also we employ a network model. However, instead of a yellow page management on an agent network[6], we prefer to an autonomous network that has no central controller. The agent network is separated into a variety of agent groups to bound the broadcast range. Finally, in order to illustrate the availability of the agent system proposed, a case study is shown considering the shape control of a highly redundant variable geometry truss.

2. Agent Architecture

Let us make a declaration about the word "Intelligent Machine" before we begin to explain the architecture of IM_Agent. In this study, we define intelligent machine to

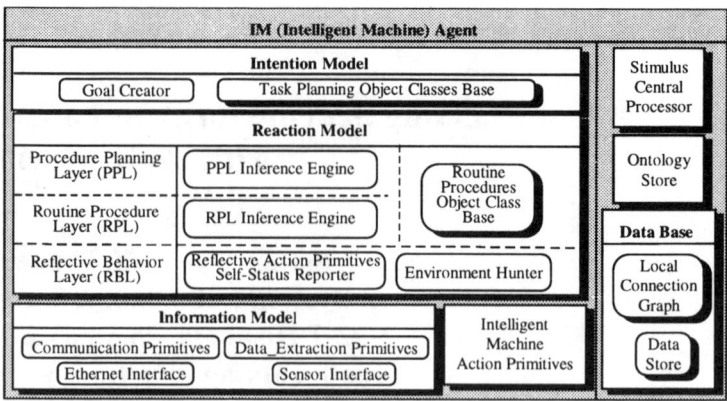

Figure 1. Agent Architecture for Intelligent Machine

such an ad hoc mechanical system that is controlled by computers, where a computer implies an entity encompassing CPU, long-term memory, network interface and OS. It should be noted that we do not limit the concept intelligent machine to robots. As a matter of fact, an intelligent machine, as an abstracted concept, is used to cover NC machines as well as even assembly lines, provided they are controllable with computers. Here, computer is a key issue in order to make intelligent machines conjunctive to agents.

An IM_Agent is constructed upon an assumption that the controllable activities of an intelligent machine can be abstracted into a series of action primitives. Its architecture is illustrated by Figure 1, where three models can be found. These three models are layered vertically, denoting a control and knowledge hierarchy. The Stimulus Central Processor is provided to achieve coherence of the three models.

2.1 Information Model

The information model in the IM_Agent is built to provide agents with a central information source about the outside environment. It consists of a flat communication platform and a sensor interface. The communication platform is built upon Ethernet, using CCL protocol that is constructed upon TCP/IP protocol, and working both as a server and a client. A sensor interface embedded in the IM_Agent separates agents from particular sensor types through defining a set of data extraction primitives.

2.3 Reaction Model

A reaction model is of core in the IM_Agent architecture, reflecting an "activation ⇒ response" behavioral pattern of agents. The activation that may come either from the external or the intention model, is transported from the Stimulus Central Processor (SCP).

18

The reaction model consists of three layers:
- Reflective Behavior Layer (RBL)

This layer shows the highest priority responses of the IM_Agent, and behaves concurrently with other layers. On this layer, various reflective actions can be excited when an agent (itself or the others related) drops into hazardous situations, in which case, an agent must act for its self-protection or in coordination with other agents' actions. The activation may come from other agents or the Environment Hunter, a background process recognizing the environmental situation around an IM_Agent. The Self-Status Reporter refers to a fast reply concerning its current status when inquired.

- Routine Procedure Layer (RPL)

The RPL embodies an inference engine and a collection of well-trained routine procedures constructed upon the action primitives. It always works in response to activation that is in the form of backward chaining expression and results in a backward inference. Activation is transported from the SCP. In general, routine procedures are separated into a set of group by CCL protocols, and takes the form as follows:

$procedure\_pattern\_name_i$: $inciting\_switch_i$ {
LOOP
   $action\_primitive_l$: $condition\_statement_l$;
       ...;
   break : $condition\_statement_{k+1}$;
ENDLOOP
}

These patterns are coded into objects and handled in an object base; where, $inciting\_switch_i$ is a condition block keeping all possible cases that incite a procedure; $condition\_statemen_j$ ($j \in (1, k+1)$) is a relational expression and returns value **true** or **false**; '$action\_primitive_j$' would get into work when the value of the expression is **true**.

- Procedure Planning Layer (PPL)

The PPL begins to work in the case of that activation is a vogue cooperating requests together with a set of antecedents. Such activation usually results in a forward chaining inference using the forward rules embodied in an agent, and finally leads to an implement of a routine procedure that is thought of the most suitable response. Except an inference engine, the PRL shares with the RPL a common routine procedure base.

2.3 Intention Model

The intention model shows goal-driven behaviors of agents. With this model, an agent runs actively on its own, and meanwhile asks for other agents' cooperation related. The activity pattern concerning this model can be drawn to be "Goal $\Rightarrow$ Activation $\Rightarrow$ Action". The core of this model is the Goal Creator, which sets up a set of control tasks for a particular goal, and in turn calls the SCP to create a series of activation to its reaction model as well as other agents asking for their cooperation. The Task Planning Object Class gathers all task planning algorithms that are coded into objects. In general, the intention model is initiated by men through a supervisor agent. So far, it is of interest to note the distinction between the PPL and the intention model. Saying briefly, the PPL just gives a self acting plan merely for an agent under the constraints in coordination to other agents' actions, working for a short duration. Whereas the intention model serves for long-term tasks during which, an agent would actively ask for other agents' cooperation toward its goal. In this situation, the agent working with its intention model performs just like a pivot of an agent group.

2.4 SCP, Ontology Store and Data Base

The SCP performs just like a central neuron in the IM_Agent. It deals with the protocols filleted from the RBL, and in turn executes the RPL, or the PPL when the RPL fails to give a suitable response. Also, the SCP is responsible for transporting act requests from the intention model to the reaction model. In addition, the SCP can also interrupt the execution of the intention model when needed. The data base contains a data store, which keeps the data used frequently by an agent, and a graph concerning the topological connections of the local network. In fact, both the data store and the graph just work as working buffers for some database server or agent name server. As to the ontology store, it is also a working buffer for some ontology server and keeps the contents used frequently.

3. Multi-Agent Society

When a society is referred, implied are always a heterogeneous entity configuration together with a variety of relationships. In order to model an ADC system, we need again to construct an agent society. To this end, two types of agent except for the IM_Agent are further introduced.

3.1 Supervisor and Observer

Although what we are seeking for with the aid of agents is an ADC system, one could hardly expect a unmanned working scene. So, an interface for men is thought of being necessary in ADC systems. In this work, a supervisor agent is built up to corresponding to men (Figure 2). As depicted, a supervisor consists of three models. The Information Model, similar to an IM_Agent's, has an network interface together with a set of communication primitives. But it does not need to hold

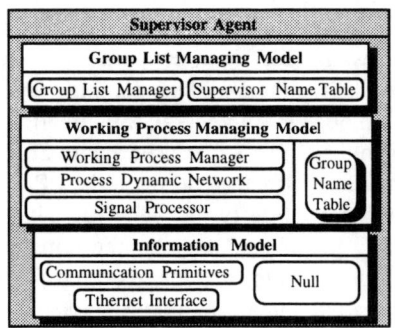

Figure 2. Architecture of Supervisor

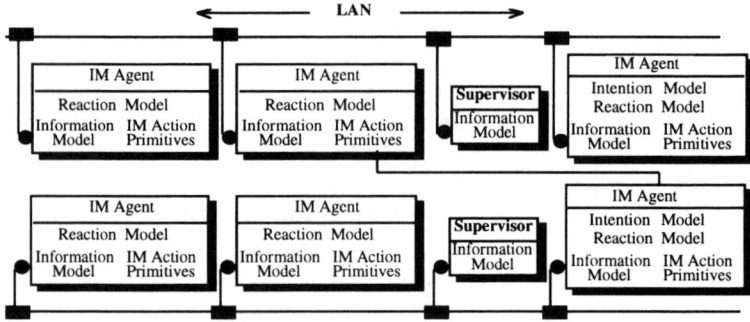

Figure 3. Agent Society

some sensor interface. The Working Process Managing Model provides men with an entry to enter an IM_Agent working group. It holds a Signal Processor for treating protocols received; a Process Dynamic Network built upon Petri Net for providing an abstraction that models the dynamic processes of an IM_Agent group to a supervisor; and a Working Process Manager for offering an interactive environment with men. With this interface, men can render manipulations on IM_Agents. In general, a supervisor can direct more than one working group simultaneously through its Group List Manager. With a Supervisor Table, a supervisor could avoid the conflict with others.

The observer agent is prepared to deal with various feedback systems involved in control. To date, it is still left as an interface for the development in future.

3.2 Hierarchy of IM_Agents

Some static, permanent relationships exist among IM_Agents, and forms the cornerstone of an agent society. Decomposing the working ability of an agent, that is a direct manner in relating agents to each other permanently, would result in a hierarchical organization of agents. As for an IM_Agent, its IM Action Primitives may be separated into a set of control tasks of other agent. Seeing that a control task usually requires a cooperation of agent group, a hierarchical configuration of agents is thereafter embraced.

3.3 Agent Working Group

So far, let us address the concept of agent working group that has been used several times above. An agent working group is composed of more than one heterogeneous agent, working for a common goal. Here, heterogeneous agents imply that at least two categories of agents, namely a number of IM_Agents and a supervisor are involved. Within a working group, only one agent is permitted working with its intention model, which means

that the others have to run on their reaction models. Our argument is that such combining constraints of intention model and reaction model is a key in avoiding conflicts and achieving coordination of agent activities. The case that no agent works with its intention model in a group may be rational. In this case a central control pattern is met, and a supervisor is needed to perform as a central commander.

Concerning agent working groups, an issue should be asserted, that is all IM_Agents within a group work non-hierarchically. The hierarchical activities of an agent result in a new agent group in hierarchy. Then an agent society is formed based on these hierarchical working groups (Figure 3).

3.4 Agent Identifier

Similar to the human society, an agent should have its identifier for recognizing it uniquely. However, what is often met is the case that, agents have to be picked up according to some vogue working status. So, a better idea is to construct an agent identifier keeping both the static location of an agent in the network and the dynamic status. To this end, we compose an identifier Agent_ID consisting of Static_ID and Dynamic_ID.

Static_ID locates an agent uniquely in a LAN or the Internet, indicating a peer-to-peer communication. It consists of a name and a domain name of the Internet:

agent_name@local.group.site

Thus, in a UNIX environment, for example, an agent can be located to be an object class aggregate within *local.group.site/home/agent_ name/agent_name.class*.

Dynamic_ID is composed of {**group**, **status**}; where '**group**' contains {*group_id, inter_id*}; '**status**' is an application extension exhibiting the dynamic features

4. Cooperative Command Language

Communication is the most significant medium to link distributed agents into a society. In view of that intelligent

20

control is the underlying approach to supporting ADC, a communication manner for exchanging data structure[7] is not enough. Instead, a set of protocols is required to support a variety of message types and communicative acts. Consequently, we issue go to an agent communication language (ACL)[8].

Within an ADC system, we can distinguish three basic communicative categories: (1) data/knowledge query or manipulation, which is carried out while an agent interacts with some ontology server or database server; (2) cooperative negotiation or act, which occurs among agents within same working groups; (3) and on-line training of an IM_Agents by a supervisor through transporting new procedural knowledge to the agent. Accordingly, we take such a strategy to build our communicative platform: using KQML[11] in the future as a universal language to cover the interactions between IM_Agents and other data/ knowledge systems, and developing a Cooperative Command Language (CCL) as a complement to KQML, to assist in the speech acts of (2) and (3) above.

CCL is currently under development. Below, we provide an explanation to it in, with regard to its present state, no more than natural language.

(1) **stop** the actions of other agents in a working group;
(2) **negotiate** with other agents for a cooperative goal;
(3) ask for other agents' physical **act**s;
(4) **inquire** other agents' status;
(5) **issue** some information to other agents;
(6) **teach** an IM_Agent how to work through transporting new procedures from a supervisor to it.

Basically, these fundamental protocols except for **teach** are achieved in an asynchronous manner, and implemented on UDP/IP protocol. '**teach**' is realized upon TCP protocol. Below given is a BNF expression for CCL.

CCL	::= ⟨CCL_head⟩ ⟨command_context⟩
	[⟨priority⟩] [⟨dispatch_pattern⟩]
⟨CCL_head⟩	::= ⟨receiver⟩ ⟨sender⟩ [⟨title⟩] [**in-reply-to:**
	⟨f-title⟩]
⟨receiver⟩	::= ⟨ Agent_ID ⟩
⟨sender⟩	::= ⟨ Agent_ID ⟩
⟨title⟩, ⟨f-title⟩	::= **text**
⟨ Agent_ID ⟩	::= [⟨Static_ID⟩] [⟨Dynamic_ID⟩]
⟨Static_ID⟩	::= *agent_name@domain_name*
⟨Dynamic_ID⟩	::= ⟨group⟩ [⟨status⟩]
⟨group⟩	::= (*group_id*, [*inter_id*])
⟨ status ⟩	::= application extension
⟨command_context⟩	::= ⟨primitive⟩ ⟨contents⟩ [⟨attachment⟩]

⟨primitive⟩	::= **Stop**	**Negotiate**	**Act**	**Inquire**	**Issue**
		Teach
⟨contents⟩	::= applications extension
⟨attachment⟩	::= ⟨data⟩	⟨data_bulk⟩	*object.class*
⟨data_bulk⟩	::= ⟨ data⟩*
⟨data⟩	::= **text**	**vector**	**matrix**	**real**	**integer**
⟨priority⟩	::= integer
⟨dispatch_pattern⟩	::= (to be extended)

It is easy to note from the CCL syntax above that an Agent_ID can be assigned incompletely. With such an Agent_ID, an agent cannot be determined uniquely. As a matter of fact, an incomplete Agent_ID indicates implicitly a broadcast or multicast communication. Besides, the option ⟨dispatch_pattern⟩ is prepared in particular to handle the transferring sequence of stimulus among agents. In this way, the chronology of agents' acting processes can be controlled.

5. Object-Oriented Implementation of the Agent System

At present, a development concerning a support environment for ADC systems is ongoing upon object-orientation. Java is selected as its developing platform, so that a dynamic system structure for the IM_Agent could be obtained. Getting a dynamic program structure, in our opinion, is a key issue in supporting on-line expansion of programs when an IM_Agents are trained by a supervisor. Furthermore, Java makes it possible to construct agents available on LANs, regardless their particular platforms. Below, we outline the implementation of the system.

5.1 Object Models

An agent does not correspond to an object merely. Instead, a group of objects is aggregated within an agent. Especially, in order to make an agent network autonomous so as to facilitate its management, an agent, named Com_Agent, is separated from the IM_Agent, or the supervisor agent responsible for agent communication. Then, the IM_Agent, or the supervisor agent is connected to a Com_Agent in inheritance, making use of its communicating capacity. Figure 4 shows the object model of the Com_Agent, using the notations proposed in [10]. We can easily note that there are two servers working concurrently. In this sense, a Com_Agent comprises a generalized client-server system.

The IM_Agent, which is thought of the most complicated, is mainly described. As shown in Figure 5, the IM_Agent consists of a group of objects. The object Reaction RB Generator corresponds to the reflective layer

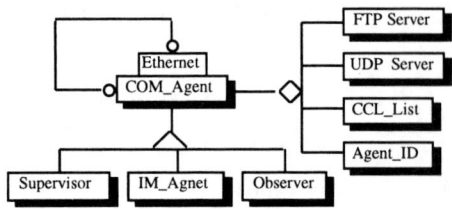

Figure 4. Object Model — Com_Agent

of the reaction model, running as an independent thread with the highest priority. The Data_Extraction Primitives is encapsulated with the object Sensor Data Extractor. The intention model of the IM_Agent is abstracted into two objects. The object Goal Creator is in charge of choosing a set of task planner from the object Intention TP Base, which keeps all the planners. In response to a supervisor's demand in CCL, the Goal Creator would return a control goal along with a set of planners to the object Intention TP Executor for building a task planner queue. The object Reaction PP Agency works as a PPL inference engine based upon a rule base, namely the object Reaction PP Base, and returns a call for a suitable routine procedure held by the object Reaction RP Base to the object Reaction RP Executor. The key role of the IM_Agent is achieved by the object Stimulus Central Processor (SCP). It processes requests received from other agents or its own Intention TP Executor, and in turn transports these requests to the object Reaction RP Executor which would result in the execution of a routine procedure, or delivers the request in CCL to other agents for possible cooperation. Be care that, the object Intention TP Executor just generates either an act pattern for the object Reaction RP Executor, or a cooperative request to other agents.

5.2 Dynamic Models

Rather the dynamic model proposed in [10], here just a simplified process flows running within the combination (IM_Agent, Com_Agent) is given out. As shown in Figure

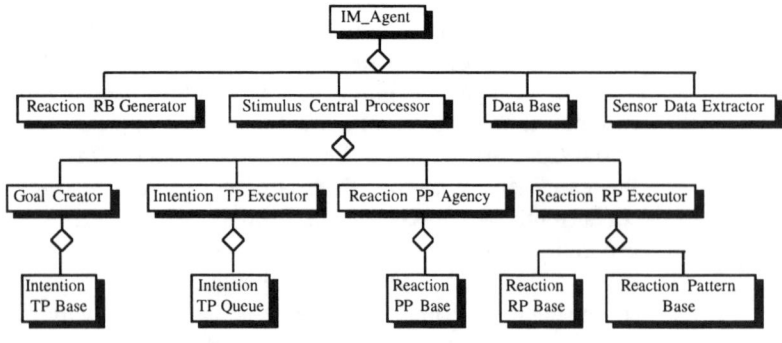

Figure 5. Object Model — IM_Agent

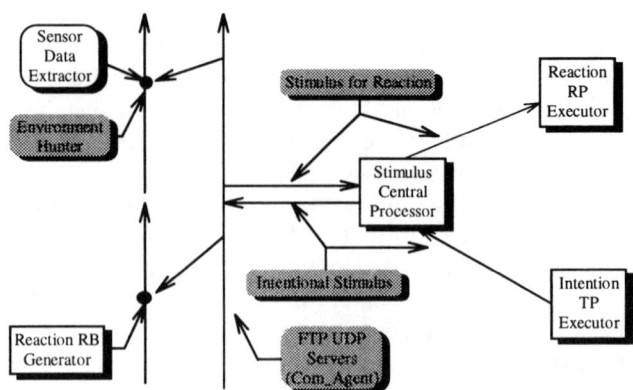

Figure 6. Parallel Activities in IM_Agent

6, there are four fundamental processes working concurrently. FTP/UDP servers, Environment Hunter and Reaction RB Generator act as daemon processes, and the SCP then forms the main process. While an IM_Agent works with its reaction model, the processing flow takes the form of protocol $\Rightarrow$ SCP $\Rightarrow$ Reaction RP Executor. Otherwise, an IM_Agent works with its intention model running in the pattern Intention RP Executor $\Rightarrow$ SCP $\Rightarrow$ (other agents & Reaction RP Executor).

5.3 Network

Since an Agent network is implemented on Internet, and Static_ID of Agent_ID is arranged in the name space of DNS (Domain Name System), we can use Internet name servers and routes to search agent hosts. Concerning the network management, we refer to a strategy, that is to let each agent handle its local connections with a Local Connection Graph (see Figure 1).

6. Support Platform of Multi-Agent System for Autonomous Distributed Control

To advance the application in ADC of the multi-agents proposed in this work, we are also engaging with the development of a supporting platform for multi-agent systems (Figure 7). The system is divided into server application and client applications, and consists of four components: (1) several agent frameworks in relation to different agent categories such IM_Agent, supervisor, etc.; (2) an agent manager client; (3) an agent design client; (4) ANS server.

The agent manager client provides two services: (1) agent network design, namely determining the host of a new agent on LANs; (2) agent network operations including **start** and **stop**, which make agents active or inactive

22

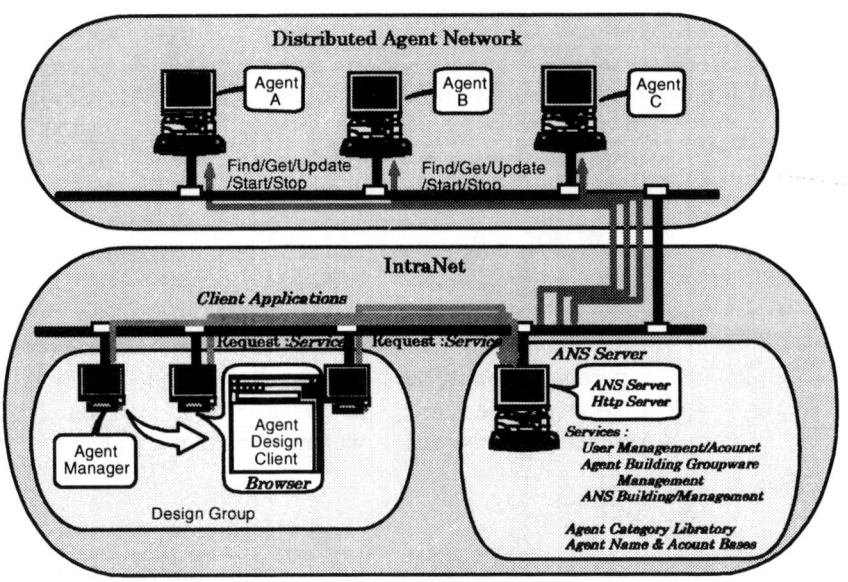

Figure 7. Agent-Oriented Network Operating System

respectively. The agent design client refers to the design of contents based upon some agent framework. ANS server is the core of the entire system. It holds an agent category library, where various agent frameworks such as IM_Agent, supervisor, etc. can be found. As the design server of agent system, it supports design acts of groupware, and performs the actual update operations on old agent programs when requested by a design client. On the other hand, as an agent manager server, it is in charge of the installation, maintenance and operations of agents on heterogeneous LANs, corresponding to the commands, namely **search**, **host**, **move**, **erase, access** and **get,** from an agent manager client. Meanwhile, it also offers a transaction management on agent programs automatically. Figure 7 shows such a system which is under development at present.

7. Case Study: Shape Control of a Variable Geometry Truss

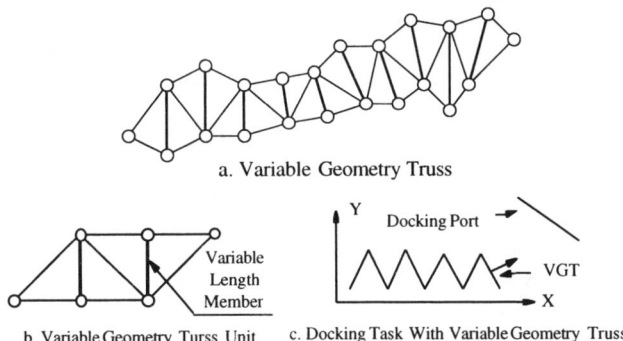

a. Variable Geometry Truss

b. Variable Geometry Turss Unit   c. Docking Task With Variable Geometry Truss

Figure 8. Docking Task with a Variable Geometry Truss

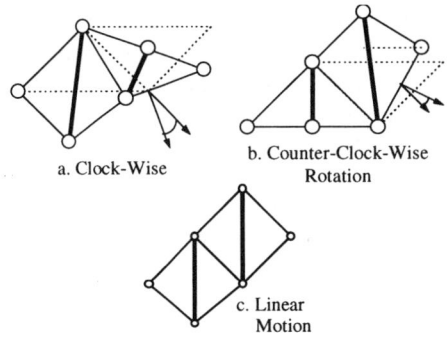

a. Clock-Wise   b. Counter-Clock-Wise Rotation

c. Linear Motion

Figure 9. Three Motion Patterns of a Variable Geometry Truss Unit

An autonomous distributed control for a variable geometry truss (VGT)[3] has been dealt with as a case study. A VGT (Figure 8a), applied mainly in the space missions, is usually equipped with highly redundant actuators[3] in adaptation to the complex environment in the space,. A VGT unit is shown by Figure 8b. It is easy to find that, through repeating VGT units along the longitudinal direction, a VGT can be formed. Therefore, a VGT unit can be taken as an autonomous individual and modeled into an IM_Agent.

Regarding a VGT unit, its shape change is made available with the two variable length members. Subsequently, the shape change of a VGT unit can be summarized into three fundamental patterns (Figure 9). Geometrically, any shape of a VGT can be achieved with combinations of the three shape patterns of overall units. Saying alternatively, entailing the autonomous acts of each agent toward the control goal of the system, the entire shape required for a VGT can be achieved, provided that synchronization of each agent's acts is coordinated effectively. Below, a coordinating algorithm concerning a docking task (Figure 8c) is given to illustrate the availability of the control scheme issued above. Let the docking unit work with its intention model, and the other agents run with their reaction models.

First, let us expand the item ⟨status⟩ of Dynamic_ID in the way:

⟨status⟩	::= [⟨role⟩] [⟨state⟩] [⟨distance⟩]
⟨role⟩	::= **master** I **cooperator** I **idler**
⟨state⟩	::= **moving** I **waiting** I **sleeping**
⟨distance⟩	::= **furthest** I **closest** I **middle**

Moreover, set ⟨contents⟩ of ⟨command_context⟩ be objective *obj*, that is a key word, of the command. A function '*violate*' is also provided upon a Dynamic_ID to check the rationality of a status change required by other agents. Usually, a unit working with its reaction model is a 'cooperator', or else a 'master'. Below, the reaction model without a PP Layer is established for a unit, where the first bold item is a command primitive with parameters omitted, and the subsequent expression is a condition for executing the command.

Reflective Layer {

   Pause : $(primitive = \textbf{stop}) \vee (notifyby(Environment_Hunter) = $ 'danger');

   Avoid : $notifyby(Environment_Hunter) = $ 'obstacle';

   Issue:: $(primitive = \textbf{Inquire}) \wedge (obj \in Data\_Base)$;

   Alter_State: $(primitive=\textbf{Negotiate}) \wedge .NOT.(violate(obj, $ Dynamic_ID)

}

Routine_Layer {

   Liner_motion: $(primitive = \textbf{Act}) \wedge (match(obj)= $ 'linear') $\wedge$ work_capable();

   Clock_wise_rot:$(primitive = \textbf{Act}) \wedge (match(obj)= $ 'clock_wise') $\wedge$ work_capable();

   C_clock_wise_rot:$(primitive=\textbf{Act}) \wedge (match(obj)= $ 'c_clock_wise') $\wedge$ work_capable();

   Forward: $((dispatch\_pattern=$left-order) :left ?right);

}

Where, function 'match' checks whether an *obj* received satisfies a routine pattern; *work_capable* conforms the working capacity a unit keeps. Reflective Layer is incited when an obstacle is met, a stop is required by other agents, the current state is inquired, or a state change is asked for. Routine Layer has three routine patterns, as well as an act for passing a protocol to other agents according to the indication of *dispatch_pattern*.

Below, we give the docking algorithm, which is separated into two tasks: one is let the docking unit enter the nearby of the docking port through asking for other agents' cooperation.

Docking_Task {

   LOOP $(Goal\_Generator() \Rightarrow task) \neq$null}

     IF (task = 'Close to Target') THEN

       LOOP $(notifyby(Environment\_Hunter) \neq$'complete')

         Send_CCL (dest_ID(state$\Leftarrow$sleeping $\vee$ waiting,

           distance$\Leftarrow$(old=furthest):closest?furthest), self_ID,

           context(**Act** $(action\_mode() \Rightarrow obj)$),

           dispatch_pattern((distance=closest) :left-order ?right-order));

       ENDLOOP;

     ELSE IF (task = 'docking') dock ();

   ENDLOOP

}

Where, function *action_mode*() returns an act pattern for other agents according to the relative position between the docking agent and the port; item 'left-order' indicates a transport mode of broadcast, that is forwarding the protocol to its left agent; and vice versa for **'right-order'**.

8. Conclusions

With the aim of providing a support platform to an autonomous distributed control (ADC), a multi-agent system is introduced. A new agent architecture is proposed to model intelligent machines, and the concept of a supervisor agent is employed to establish a human's supervision control for intelligent machines. With an agent society model based upon these two agent categories, a constructing scheme for ADC systems can be obtained. Moreover, an object-oriented implementation of the agent system suggested, that is carried out on LANs, is described in simple. In order to facilitate the applications the multi-agent system proposed, an support environment being developed is presented too. Finally, a case study considering the shape control of a variable geometry truss is carried out based upon the agent system.

References

1. Inou and Y. Umetani , 1993, 4, "Small Intestinal Movement in Vcivo and the Neuro-Mechanical Control Mechanisms," Inter. Sym. on Autonomous Decentralized Systems (IDSAS 93), pp. 407-413.

2. Ito Masami, et al Eds., 1994, "Autonomous Distributed System," Report for the Research Project Grant in Aid for Scientific Research on Priority Areas by the Ministry of Education Science and Culture.

3. Natori, S.Y. Huang, Y. Umetani et al., "Basic Concepts on Object-Oriented Architecture of Intelligent Adaptive Structures." *Proc. of the 5th Inter. Conf. on Adapt. Struct. (Sendai, Japan)*. 1994, 12, pp. 34-50.

4. Wooldridge and N. R. Jennings, 1994, "Agent Theories, Architectures, and Languages: A Survey," Intelligent Agents: ECAI-94 Workshop on Agent Theories, Architecture, and Languages (M. Wooldridge et al. Eds), Amsterdam, The Netherlands, pp. 1-39.

5. Muller, M. Pischel, M. Thiel, 1994, "Modeling Reactive Behavior in Vertically Layered Agent Architectures," Intelligent Agents: ECAI-94 Workshop on Agent Theories, Architecture, and Languages (M. Wooldridge, N. R. Jennings Eds), Amsterdam, pp. 261-276.

6. Iglesias, J. C. Gonzalez and J. R. Velasco, 1995, "MIX: A General Purpose Multi-Agent Architecture," Intelligent Agents II: Agent Theories, Architecture, and Languages (M. Wooldridge, et al. Eds.), Montreal, pp. 251-266.

7. H. Wataya, K. Kawano and K. Hayashi, 1995, "The Cooperating Autonomous Decentralized System Architecture," Inter. Sym. on Autonomous Decentralized Systems (IDSAS 95), pp. 40-47.

8. Mayfield, Y. Labrou and T. Finin, 1995, "Evaluation of KQML as an Agent Communication Language," Intelligent Agents II: Agent Theories, Architecture, and Languages (M. Wooldridge et al. Eds.), Montreal, pp.347-360.

9. James Rmbaugh, et al., *Object-Oriented Modeling And Design*, Prentice-Hall International Editions, 1991.

10. ARPA Knowledge Sharing Initiative. Specification of the KQML Agent-Communication Language. ARPA Knowledge Sharing Initiative, External Interface Working Group Working Paper, Dec. 1992.

A Hybrid Agent Model: a Reactive and Cognitive Behavior

Zahia Guessoum

Equipe RTCD, LAFORIA-IBP, Université Paris 6, Boîte 169, 4 Place de Jussieu, F-75252 PARIS

guessoum@laforia.ibp.fr

http://www-laforia.ibp.fr/~guessoum

Abstract: *To model complex systems, software agents need to combine cognitive abilities to reason about complex situations, and reactive abilities to meet hard deadlines. We propose an operational hybrid agent model which mixes well known paradigms (objects, actors, production rules and ATN) and real-time performances. This agent model combines cognitive and reactive modules. The cognitive module uses a rule-based system and is provided with a synchronization mechanism to avoid the possible inconsistencies of the asynchronous execution of several rule bases. The reactive asynchronous perception and communication modules allow the agent to dynamically adapt its behavior to changes in the environment. To manage the interactions between these fundamentally different modules (reactive and cognitive), we use an ATN-based module.*

Key Words: *Agents, Reactive, Cognitive, Hybrid, Object-Oriented, Actors, Production Rules, ATN.*

1 Introduction

Several agent models have been proposed. Two main approaches can be distinguished: *cognitive* and *reactive* (see [20, 7]). In the cognitive approach, each agent contains a symbolic model of the outside world, about which it develops plans and makes decisions in the traditional (symbolic) AI way. In the reactive approach, on the other hand, simple-minded agents react rapidly to asynchronous events without using complex reasoning.

Neither a completely reactive nor a completely cognitive approach is suitable for building complete solutions for real-life applications. Hybrid models [8, 3] have been proposed to combine the advantages of both reactive and cognitive models. In these models, agents are decomposed in a set of modules which can in turn be of a reactive or cognitive nature. However, the problem with such models is that of controlling the interactions between these fundamentally different modules: reactive and cognitive modules, so to say, do not live with the same time scale, which makes it difficult to integrate the different temporal sequences.

Our hybrid agent model relies on a first layer made up of interactive modules that can be either reactive or cognitive, and interact asynchronously and concurrently. A higher level *supervision* module schedules these interactions via an ATN so as to impose a global temporal sequencing. As we shall see, this time scale is based on the firing of individual rules as its basic unit.

In our model, each reasoning module owns a rule base, which represents the bulk of its cognitive abilities. Control of the reasoning process is achieved by a set of metarules (called a metabase) also owned by the module. These metarules operate on objects that are also accessed by the supervisor's ATN, thus establishing the supervision link at the level of the individual rule firing.

The purpose of this paper is to present our hybrid agent model and its implementation. Section 2 presents the main *Smalltalk*-based tools of our implementation: *Actalk* and *NéOpus*. Section 3 describes our hybrid agent model. It presents the individual modules, the control mechanisms which are used and the real-time characteristics of this model. In Section 4, we report on the implementation with a concurrent object language. In Section 5, we report on the implementation with a concurrent object language. In Section 5, we report on the application of our model to patient monitoring and to economic agents evolution modeling. Finally, we discuss the advantages of our model to design and implement multi-agent systems (MASs).

25

2 Technical Context

We use *Smalltalk-80* as our base language. This enables our implementation to make use of various software components (alias frameworks) such as the *Smalltalk* Discrete Event Simulation Package (see [11]) for observing the temporal behavior of the system under simulated real-time conditions. We have used *RPC-talk* [19], which provides *Smalltalk-80* with RPC (Remote Procedure Call) facilities, to put any number of machines at the service of our MASs. Our actors are built with *Actalk* [2] a generic platform for implementing various actor models in *Smalltalk-80*. *Actalk* may be seen as the foundation stone of our model. The rule bases and metabases of the reasoning modules use *NéOpus* [18], a first order inference engine completely embedded in *Smalltalk-80*. We have added a few features to customize these components to meet the specific needs of our model.

2.1 Actalk

Actalk allows to transform ordinary *Smalltalk* objects into actors. Asynchronism, a basic principle of actor languages, is implemented by enqueuing the received messages into the mail box, thus dissociating message reception from its interpretation. In *Actalk*, an actor is composed of three objects:

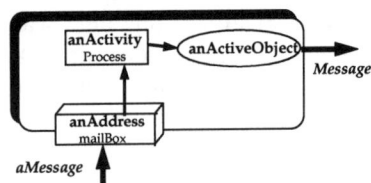

Fig. 1 Actor structure in Actalk

- An instance of class *Address* represents the mail box of the actor. It defines the way messages will be queued for later interpretation;
- An instance of class *Activity* represents the internal actor activity. It provides autonomy to the actor. It owns a Smalltalk process which continuously removes messages from the mail box and launches their interpretation by the actor;
- An instance of class *ActiveObject* represents the behavior of the actor, i.e. the way individual messages will be interpreted.

  To build an agent with *Actalk*, all one has to do is to create the three components (call them *ad*, *act* and *actObj* respectively) and put them together as an actor by sending the message *active* to *actObj*. Customizing *Actalk* therefore means defining subclasses of *Address*, *Activity* and *ActiveObject*.

2.2 NéOpus

NéOpus realizes a neat integration of rule-based programming with *Smalltalk-80*. It uses the Rete algorithm to compile rule bases. One of its prominent features is declarative specification of control with metarules [17]. Metarules are similar to rules and operate on so-called control objects. With each rule base there may be associated a metabase which controls the firing of its rules (see Fig.2).

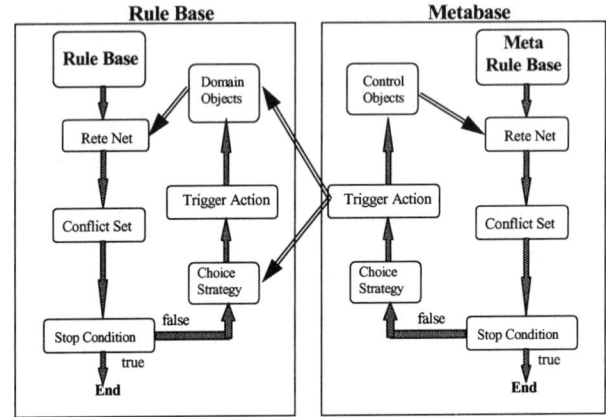

Fig. 2 Control with metarules in NéOpus.

A number of metabases have been designed by *NéOpus* users to define standard types of control. In particular, specific metabases have been built to reason on critical situations [5]. We have adapted some of them to our needs by changing a few individual metarules.

3 A Hybrid Agent Model

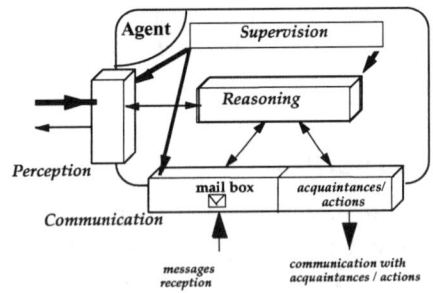

Fig. 3 Agent model

As underlined in the introduction, our model relies on two layers made up of modules. The first layer modules describe the agent activities. The second layer includes only one module: the supervision module. The latter manages the interaction between the first layer modules.

In this section, we describe the modules of the two layers, the mechanisms control we used and the real-time characteristics of this model.

3.1 The first layer modules

In this section, we describe three examples of modules of the first layer: the *perception*, the *reasoning* and the *communication/action* modules.

The Perception Module manages the interactions between the agent and its environment. It monitors sensors, translates and filters sensed data in accordance with the instructions of the reasoning or supervision modules. It may package information concerning the same phenomenon to facilitate interpretation. The data set obtained is used mainly by the reasoning module.

The Reasoning Module is responsible for generating adequate responses to the messages transmitted by the communication module, or to the changes detected by the perception module. To do this it relies on two kinds of capacities: operative, represented by the standard behavior of the associated *Smalltalk* objects (procedures, alias *methods* in the *Smalltalk* terminology), and cognitive, embodied in a *NéOpus*-based asynchronous production system [10]. This production system mainly comprises: (1) a rule base which includes objects describing the agent's environment and rules representing suitable operations over these objects; (2) an inference engine which includes the dependency and anti-interference mechanisms; and (3) a metabase which provides a declarative representation of the control of reasoning.

The Communication/Action Module allows the agent to receive and to send messages asynchronously. It filters the received messages, determines their priority (LIFO, FIFO,...) and the type of treatment to give them. It owns the list of the agent's acquaintances. It sends them messages with various modes corresponding to specific protocols (*urgent, answer needed, ...*). The messages and their modes are provided by the reasoning module.

This module also effects the direct actions (modification of the environment via effectors).

3.2 The Supervision Module

This module allows the agent to schedule its internal activities according to its world state. It synchronizes the execution of concurrent actions of the other modules. It relies on two notions: *states* and *transitions. States* qualify the context as perceived by the other modules. Changes in the context are reflected as *transitions* between states. Now, these states and transitions naturally build up an ATN. The ATN is a synthetic and deterministic representation of the agent's behavior.

Different states correspond to each module. The combination of these states defines the global agent's state. Each transition links an input state with an output state. The various signals received by the agent's modules represent the conditions of transition (reasoning terminated, has urgent message, ...) and the actions of transition (activate reasoning, terminate reasoning, ...) change the state of the various modules. When these conditions are verified, the transition actions are executed and the agent's state is modified.

3.3 Control Mechanisms

Intelligent agents with high degree of autonomy should perform well complex tasks in a dynamic environment for extended periods of time. Here we propose a control mechanisms at two levels. At the agent level, we propose a self-control to achieve autonomy and to coordinate its own activities. At the multi-agent level, we propose a coordination mechanism to avoid redundant actions and conflicts between agent.

3.3.1 Self-control

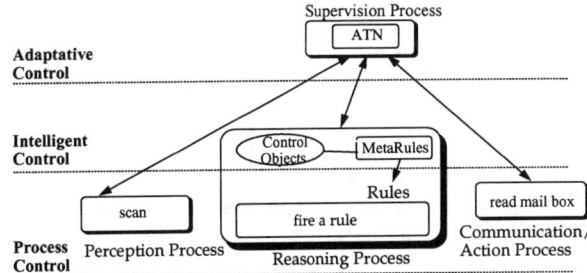

Fig.4 Self-control architecture

The self- control is managed at three levels (see Fig. 4): 1) at the supervision module level, an ATN allows to specify the agents behavior in accordance to the observed internal states; 2) at the reasoning module level, a metalevel architecture specifies declaractively the agent reasoning strategy and 3) at the internal parallelism level, a management of the processes associated to the asynchronous concurrent modules allows to specify the availabe-ressources allocation.

3.3.1.1 Supervision Module level: Adaptative Control

The supervision module allows the agent to dynamically adapt its behavior the its universe changes. The adaptation corresponds to a new organization mode of the agent in its decisional relations, interactions with its environment or relations with other agents. The agent is able to control its abilities (communicational, behavioral, ...) according to its current environment.

3.3.1.2 Reasoning Module level: Intelligent Control

One of our main hypothesis is: The reasoning control

strategies can be separated from the domain knowledge, but they must be supervised by the agent itself. Unlike blackboard approaches [16] our agents encapsulate the two kinds of knowledge: domain and control knowledge.

The system *NéOpus* we use to represent the agent reasoning capacities provides a declarative control architecture. The latter is founded on the following reflexivity principle: The control of a rule-base execution is fully supervised by an other metalevel rule base called metabase. It provides a declarative specification of control with metarules. With each rule base there may be associated a metabase which controls the firing of its rules. As system complexity increases, such control mechanism will need, for instances, state estimators. Thus, control objects are used to give reasoning these states estimators. They are used by the metarules and can be also be accessed by the supervisor's ATN, thus establishing link at the reasoning process.

3.3.1.3 Internal Parallelism level: Process Management

Each agent is mapped to a process. The execution model of all agents is closely related to the process management by *Smalltalk-80*. In the latter, a scheduler (Processor) does not stop an active process i.e. it is not preemptive. Therefore, each agent must of its own free the processor to give a chance to other agents by inserting the expression *Processor yield*.

In order to accommodate several agents and/or modules on a single processor, we must simulate parallelism. To do so we choose a process allocation strategy at two levels: at the supervision module level (simulation of parallelism between agents) and at the perception, reasoning and communication modules level (simulation of the internal parallelism of the agent).

At the supervision module level, the agent suspends its activity after each transition. At the reasoning module level, process control is performed after each rule firing. At the communication and the perception modules, the process control is effected after each mail box reading and at the end of the method scan.

3.3.2 Coordination mechanism

We are interested in situations where an agent shares a collection of resources with other agents. Thus, agents must adapt themselves to take advantages of resources as needed, but must coordinate their actions to avoid inconsistencies. Researchers have evolved a range of approaches to coordinate a collection of autonomous entities. [4, 9] describe the mechanisms that improve the network coherence: organization, exchanging metalevel information, local and multi-agent planning, and explicit analysis and synchronization.

With a synchronization mechanism as proposed by Ishida in [15], each agent protects itself against conflicts and redundant actions concerning the other agents, at a cost of (1) reduced concurrency, and (2) synchronization overhead. However, if the level of dependency is low, and the granularity of actions is high, this mechanism can provide useful coordination, as observed by L. Gasser. This is the case for our model. The granularity of our agents is high, each agent possesses its own rule base and the agent's rules are fired sequentially. Therefore dependency and interference between individual rules of the same agent do not have to be considered. Thus the dependency between agents is low.

There remains to avoid inconsistencies between different agents. To do so, we implement Ishida's dependency mechanism with a technique which improves its overall performance via a better anti-interference mechanism.

3.3.2.1 The Dependency Graph

Following Ishida's ideas, we distinguish two kinds of objects: local objects used by a single agent, and global objects used by several agents. The principle of the dependency mechanism may be defined as follows:
- each agent is provided with a dependency graph (Ishida' terminology) giving for each global object the list of other agents which use it;
- the dependency graph is used by the communication module of the agent to inform the other agents about the modification, creation or removal of global objects by the reasoning module. It is updated gradually when rules are triggered: if a global object is removed by rule actions, it is also automatically removed from the dependency graph and a message is sent to the other agents to update their graphs.

Inter agents conflicts resulting from access to global objects are thus avoided.

3.3.2.2 Anti-Interference Mechanism

Interference exists among two rules if there is a global object that both rules access and at least one modifies. The principle of our anti-interference mechanism may be defined as follows:
- each agent is provided with a list of those global objects that it is currently modifying (objects-in-use);
- each agent is provided with the collection of the objects-in-use lists of the other agents;
- we add two steps in the inference engine cycle:
  * *test*: before the firing of the selected rule, the agent verifies that no global object that is modified by the selected rule is also being modified by some other agent (as is apparent from its objects-in-use

list collection). In this case, the global objects that are modified by the selected rule are added to its own objects-in-use list and the rule is triggered, otherwise another fireable rule is selected;

* *updateList*: after the firing of the rule, the agent removes from its own *objects-in-use* list those global objects that were added to its in the *test* phase, i.e. the global objects that were modified by the rule.

The proposed solution prevents conflicts and redundant actions. It avoids also the synchronization messages used by Ishida.

3.4 Real-Time Characteristics of the proposed Agent Model

The proposed hybrid agent model has the main characteristics of a real-time system: 1) Reactivity and adaptability and 2) Anytime reasoning to guarantee a solution in a bounded time.

3.4.1 Reactivity and Adaptability

Reactivity and adaptability are crucial characteristics in the complex process control such artificial ventilation monitoring. The main problem to solve in such systems is the real-time reactivity to asynchronous changes of the environment and the adaptability of the agent behavior to the state changes provoked by these changes.

The characteristics of the proposed model which gives it reactivity and adaptability are the following:

- the use of asynchronous perception module allows to supervise in real-time a large information sources of the agent environment,
- the use of asynchronous communication module allows to respond in real-time to the other agents requests,
- the use of an ATN-based supervision module allows the agent to observe and to adapt its behavior to its universe changes as indicated by its asynchronous modules.

These characteristics give limited time reactivity to our agents. As regards real-time response, our model is adequate provided a time granularity of single rule firing. This seems to be acceptable in many industrial applications [1].

3.4.2 Anytime Reasoning

The anytime reasoning is a promising solution based on the anytime algorithms [21]. Thus, we have provided our agents with anytime reasoning abilities to guarantee a fixed time-consuming. This reasoning assumes that there

are several solutions with different qualities each of which represents approximate view of the final solution [16].

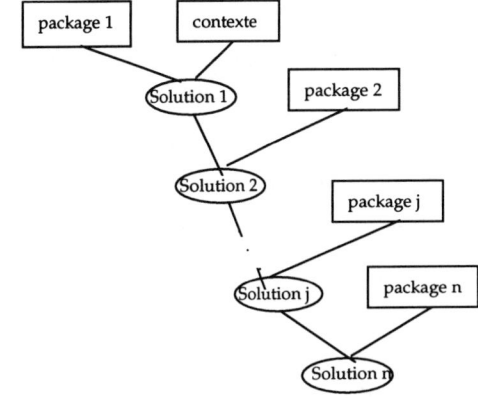

Fig. 5 Schematic structure of the progressive reasoning

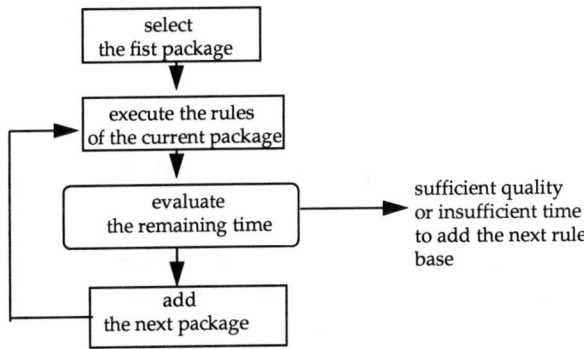

Fig. 6 A single execution cycle of the real-time metabase

The proposed technique is founded on the anytime reasoning on the one hand and on two main features of the rule-based system we use on the other hand. These main features are: 1) the declarative reasoning control and 2) the inheritance of rule bases. We propose to decompose each agent rule base in several packages. The first package provides fast solution but often of worst quality. The addition of the other packages and the context enrichment improves the solution quality (see Fig. 5).

To implement this progressive reasoning, we have defined and implemented a metabase which uses several deliberation stages where the solution quality is improved progressively (see Fig. 6).

To update the deadline, we have redefined the rule firing method. The new one allows to update, at each cycle, the remaining time.

4 From Concurrent Objects to Agents

To implement the proposed autonomous agent model,

we use *Actalk* In the realized environment, an agent is an actor composed by the following objects (see Fig. 7):

- An object (instance of the class *AgentActivity* subclass of *Activity*) which describes the agent activity.
- An object (instance of the class *BasicAgent* subclass of *ActiveObject*) which describe the agent ATN structure.
- a set of objects which describe the different modules describing the agent behavior.

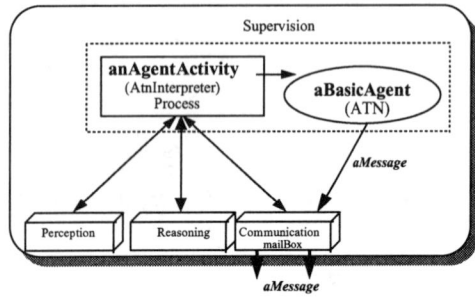

Fig. 7 Different objects of the implemented agent

The two first objects implement the supervision module of the agent and the last subset implements its first layer. The class describing the communication module derives from the class Address of *Actalk*.

In *Actalk*, an objet *Activity* manages the messages which are interpreted by an other object *ActiveObject*. In our model, the agent activity is described by an ATN which schedules the different activities (perception, reasoning and communication/action). So, we have redefined the instance method *body* used by *createProcess* which creates a process to take out continuously the messages present in the actor mailBox.

```
! Activity methodsFor: 'activity setting'!
body
[true] whileTrue: [self acceptNextMessage]

createProcess
^[self body] newProcess
```

The new function of this process is to interpret the agent ATN.

```
!AgentActivity      methodsFor:      'activity
setting'!
body
[true] whileTrue: [self atnInterpreter]
```

The use of an object-oriented concurrent language allows to benefice of the inheritance mechanism. To realize a social agents hierarchy, we have reused this mechanism. The implemented agents can have a simple behavior (reactive) or a complex one (cognitive or

hybrid). The agents have the same general structures, but they differ in:

- Their sensor-driving layer, they don't have all a perception module and a communication module.
- The behavior, the know-how, the domain and control knowledge they own.

Several agent classes belonging to different complexity levels have been defined (see Fig. 8).

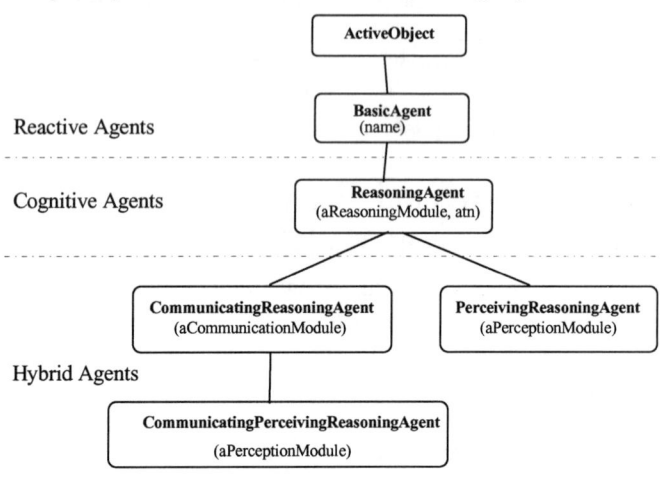

Fig. 8 Classes describing agent structures

At each class is associated an ATN to define the behavior type. Each class implements one or several agent creation functions.

5 Applications

To validate the operational environment (*DIMA*) which is based on the proposed hybrid agent model, we have implemented three applications: 1) manufacturing process simulator [11]; 2) *NéoGanDi* [12]: a multi-agent system to control mechanical ventilation and 3) *Meveco* [13]: a multi-agent system to model economic agents evolution.

In this section, we briefly describe *NéoGanDi* and *Meveco*.

5.1 NéoGanDi

The system deals with patients suffering from respiratory insufficiency, assisted with mechanical ventilation. The problem is to monitor in real-time various ventilation signals (tidal volume (Vt), respiratory rate (RR) and expired-CO_2 pressure (PCO_2)), in order to diagnose the patient current state and to adapt the mechanical assistance accordingly. To perform this task, it is necessary to develop a complex temporal reasoning to diagnose the time-course of the patient's status [6]. In alarming situations such as hypoventilation or apnea the current therapy must be modified quickly (1 second). A first system, *NéoGanesh*, is in use at the hospital Henri Mondor (Créteil near Paris) [6]. The extension based on a distributed architecture using our agent model, aims at increasing the system reactivity and incorporating additional distributed medical expertise.

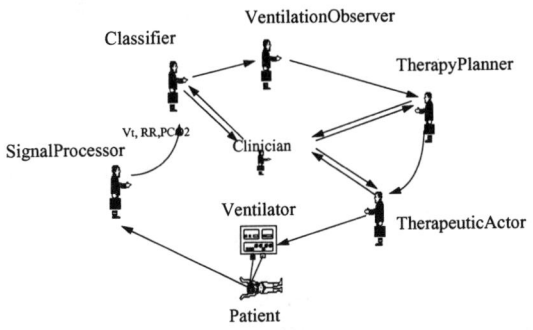

Fig. 9 Overview of the mechanical ventilation agents

NéoGanDi is composed by 7 agents (see Fig. 9). All these "intelligent" agents have the same general structure but they differ in 1) their sensor-driving layer: perception and communication/action modules; 2) their behavior: the know-how, the domain and control knowledge. For example, agent *SignalProcessor* has only a simple behavior to process data acquisition. Whereas, agent *Classifier* exhibits a complex behavior to appreciate the time-course of the patient's ventilation.

This experimental application reuses the whole of the operational *NéoGanesh* system. We plan to run it in the same medical environment as *NéoGanesh*, and thus to obtain real-life performance measurements. Note that recent works (which rely, for the most part, on blackboards [14]), so far lacks in clinical experiments.

In order to evaluate this new version, we have realized 1000 experimentations (on a station SUN SPARC 10). The results show that the total response time of the expert system is better than the total response time of a not distributed system (*NéoGanDi* implemented on only one machine). This is principally due to the use of asynchronous messages (use of a mail box).

	Response Time (Ordinary situation)	Response Time (Alarming situation)
NéoGanesh	4232 ms s	4067 ms
NéoGanDi (1 machine)	4738 ms	3475 ms
NéoGanDi (2 machines)	4240 ms	2156 ms

Tab. 1 : Comparison of NéoGanesh and the two versions of NéoGanDi

However, the total response time of *NéoGanesh* in alarming situation is improved of ~57 % when using the distributed version of *NéoGanDi*. This improvement is not only due to the use of two machines, it is also due to the use of our model. For example, the use of an ATN has significantly improved the time reactivity of the system.

5.2 Meveco

Standard economic literature is centered around the paradigm of homogeneity. Investors, consumers or traders are supposed to be identical in beliefs. Such homogeneity deprives economic agents of their essential features: these agents have different interpretations of the surrounding world.

Meveco is a multi-agent system to model the evolution of economic agents which are modeled as autonomous, cognitive, adaptive and heterogeneous agents. Unlike mathematical and expert systems approaches, ours allows a comparative and incremental evaluation of their relevance to the observed phenomena and their validity. It allows also to determine the most relevant strategy.

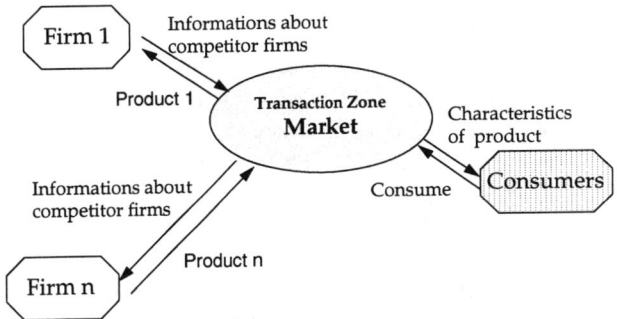

Fig. 10 Overview of the economic agents

These agents combine cognitive abilities to reason about competition, and reactive abilities to interpret the surrounding world. Each agent owns a model of the other agents. It scans the market to have some data (for example price and quality of the different products) and builds the model of the others.

The implemented system have provided results which have allowed to demonstrate the importance of the existence of heterogeneous, intelligent economic agents for the robustness of the system to abrupt external perturbations.

6 Conclusion

As we saw earlier, realistic agent models require the combination of reactive and cognitive abilities. Using such hybrid models for real-life applications needs to allow interactions between cognitive modules and reactive modules which have different time scales. The solution we propose has two main characteristics. First, we introduce a supervision module based on an ATN to control the interactions between modules that can be reactive or cognitive and to impose a global temporal sequencing. Second, the parallelism granularity of the time scale is based on the firing of a rule.

We have chosen an environment which combines actors, objects, production rules and processes. We benefit from the well known mechanisms of objects-oriented programming. Thus, by using the inheritance mechanism, the model can be easily extended. For example, to study communication between agents, we may define new subclasses to integrate a specific module based on speech acts .

Finally, in this paper, we have limited our study of real-time aspects to the agent level. Real-time agents are necessary to most real-life applications but they are not sufficient to real-time applications. It seems very interesting to study how the agents society cooperate to solve a global problem in real-time.

7 References

[1] F. Barachini and R. Granec. Productions systems for process control: advances and experiences. Applied Artificial Intelligence 7: 301-316, 1993.

[2] J-P. Briot. Modélisation et classification de langages de programmation concurrente à objets: l'expérience Actalk. Proc. of LMO'94, Grenoble, 1994.

[3] S. Bussmann and Y. Demazeau. An agent model combining reactive and cognitive capabilities. Proc. of IEEE International Conference on Intelligent Robots and Systems - IROS' 94, München, 1994.

[4] E. H. Durfee, V. Lesser, D. D. Corkill. Coherent Cooperation Among Communicating Problem Solvers. IEEE Transactions on Computers 36(11): 1275-1291, 1987.

[5] M. Dojat and F. Pachet. NéoGanesh: an extensible Knowledge-Based System for the Control of Mechanical Ventilation. 14th IEEE-EMBS, Paris, pp. 920-921, 1992.

[6] M. Dojat M. Harf, D. Touchard M. Laforest, H. Lemaire et L. Brochard. Clinical evaluation of a knowledge-based system providing ventilatory management and decision for exhibition during weaning from mechanical ventilation. American Journal of Respiratory and Critical Care Medicine, vol. 153, pp. 997-1004, 1995.

[7] J. Ferber. Les systèmes multi-agents, vers une intelligence collective. InterEditions, Paris, 1995.

[8] I. A. Ferguson. TouringMachines: An Architecture for Dynamic, Rational, Mobile Agents. PhD thesis, Clare Hall, University of Cambridge, UK.

[9] L. Gasser. An Overview of DAI. In Distributed Artificial Intelligence. N. M. Avouris and L. Gasser (eds.), Klewer Academic Publisher, Boston, 1992.

[10] Z. Guessoum. Systèmes asynchrones de production. Proc. of 2ème Journées IADSMA'94, Voiron, pp. 169-180, pp. 9-11 may, 1994.

[11] Z. Guessoum and Deguenon. A Multi-Agent Approach for distributed Discrete Event Simulation. Proc. of DIMAS'95, Poland, pp. 183-190, November, 1995.

[12] Z. Guessoum and M. Dojat. *A real-time agent model in an asynchronous object environment.* MAAMAW'96, in Lecture Notes in Artificial Intelligence, Agents Breaking Away, Walter Van de Velde and John Perram (eds.), Netherlands, January 1996.

[13] Z. Guessoum et R. Durand. *Des agents intelligents pour modéliser l'évolution des entreprises.* Intelligence Artificielle Distribuée, J-P. Müller et J. Quinqueton (eds.), pp. 47-58, Paris, 1996

[14] B. Hayes-Roth. Blackboard architecture for control. Artificial Intelligence 26: 251-321, 1985.

[15] T. Ishida. Methods and effectiveness of parallel rule firing. IEEE Conf. on Artificial Intelligence Applications, Washington, pp. 116-122, 1990.

[16] A. Mouaddib and Shlomo Zilberstain Knowledge-Based Anytime Computation. IJCAI'95, pp. 775-781, Montreal, 1995.

[17] F. Pachet and J-F. Perrot. Rule Firing with Metarules. Proc. of SEKE'94, Jurmala, Latvia, p. 322-329, 1994.

[18] F. Pachet. On the Embeddability of Production Rules in Object-Oriented Languages. Journal of Object-Oriented Programming 8(4): 19-24, 1995.

[19] F. Wolinski. RPC-Talk : une librairie RPC pour Smalltalk, Introduction à RPC et utilisations de RPC-Talk. Report LAFORIA 94/26, November, 1994;

[20] M. J. Wooldridge et N.R. Jennings. Agent Theories, Architectures, and Languages: A Survey. Knowledge Enginering Review 10 (2), June 1995.

[21] S. Zilberstein and S. Russel. Optimal Composition of Real-Time Systems. AI 82 (1996), pp. 181-213.

Session 1B

Telecommunication Services

Chair

Hendrik Berndt

Co-existence of TMN and CORBA for Service Management

Andreas Dittrich[1], Sonny Rasmussen[2], Declan O'Sullivan[3]

[1] GMD-FOKUS, [2] UH Communication Aps, [3] IONA Technologies Ltd.

Abstract

There are many driving forces which have compelled telecommunication operators and vendors to seek new solutions in telecommunications management. Increased competition has led to an increased focus on how traditional operators can do their business better through the use of integrated service and network management systems. Moreover, there is a need to integrate legacy applications and new applications, all of which could be written in different languages, running perhaps on different platforms, and almost be certainly distributed over a network. CORBA has been recognised as a key technology solution.

Based on the case studies performed for the ACTS project PROSPECT this paper discusses the advantages and disadvantages of TMN and CORBA regarding network and service management and proposes a phased approach for TMN-CORBA co-existence in the telecommunications management arena based on the Management Systems Framework of the Network Management Forum. Moreover, it presents some initial ideas concerning required basic management functionality that serve as a background for the CORBA/TMN gateway functionality to be implemented within the PROSPECT trial network.

1. Introduction

Since the first days of telephony, it has been clear that management of telecommunication services and networks is the key to the profitability of a telecommunications operator. Fundamentally, if you do not make the connection between party A and party B, you do not have a product to sell. If you are not able to track the usage of the network and services, you cannot handle faults or the production of bills.

The functionality required to realise these two tasks has been traditionally implemented in different ways:

- The management involved with call handling, being inherently distributed, has been embedded in the fabric of the network in the switches and terminals. Traditionally, this has been considered as "control" rather than "management" functionality. This software needs to operate in a realtime environment, and is required to have a high degree of performance and fault tolerance.

- The management involved with handling faults, invoicing, billing, customer ordering, for examples, has been traditionally undertaken in a more centralised manner. These management systems have been categorised in terms of business, service, network and network element management systems. Furthermore, these management systems have been typically introduced by different departments in a telecommunications operator organisation, resulting in the proliferation of diverse hardware and software platforms within an organisation.

As well as different departments, there are also different "philosophies" which have been employed. Typically the functionality found at the business and service management levels have come from a traditional "IT" perspective. On the other hand, the management of the network elements and networks themselves have been more influenced by the telecommunications engineering perspective, and also by the introduction in 1985 of the TMN framework for management by the ITU-TS [1]. Initially these management systems were intended to be standalone, but increasingly there is a need to integrate them and to enable interworking between them. These management systems need to meet different performance and fault tolerant requirements than those of the "control" software.

The ACTS PROSPECT project is implementing a trial

network which can demonstrate and validate the management of co-operating and competing services in support of commercial/business end-users.

Experiences from PROSPECT and the RACE II project R2004 PREPARE [2] indicates that for many service management tasks TMN is fairly unwieldy. This seems especially so in the following cases:

- where a customer needs a management interfaces to some service to which they subscribe but the cost and complexity of existing TMN platforms prevents them from adopting a TMN solution just for this;

- in service management applications where OSFs are communicating in both manager and agent roles to perform functionality complex operations. Experience in PREPARE was that designing this using TMN imposed a large complexity overhead over a solution that could be implemented by direct functional calls;

- where management functions need to be developed and modified rapidly;

- where an API rather than a communications interface would be more natural.

TMN solutions seemed to be more suitable in the following cases:

- where standards and agent implementations already existed, e.g. TMN Q3 interfaces;

- where large amounts of information needed to be retrieved on a regular basis;

- where notifications of events need to be handled efficiently;

- where information need to be retrieved selectively (e.g. using scoping and filtering).

In general, TMN would be more suited to the network element and network management level where considerable investment has already been made in TMN based standards, platform and implementations by network operators and deployments have been made. At the service management level however, where the influences of the open service market will be felt most keenly and where there is little in the way of existing standards or platform support for TMN, CORBA technology is being widely used. CORBA's approach [3] allows for the integration of legacy applications, but also the flexible implementation of distributed applications. These two aspects are very impor-

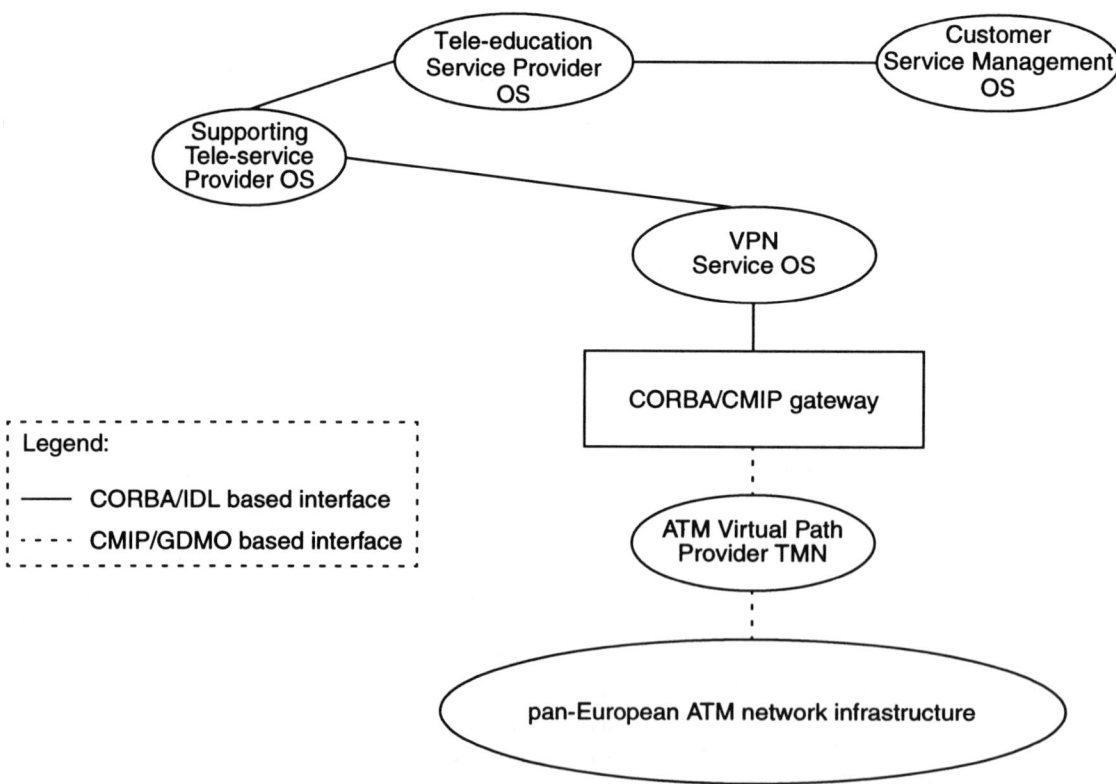

Figure 1: PROSPECT management architecture

36

tant for the implementation of service management solutions.

The management service architecture adopted in PROSPECT reflects the observations listed above as illustrated in Figure 1, which depicts a simplified version of the PROSPECT management architecture. The PROSPECT trial will involve a number of service provider operation systems (OS), which co-operates in order to provide tele-educational services to customers. The tele-educational services provided to the end-users in the trial (i.e., virtual classroom) require access to other tele-services, which are referred to as supporting tele-services. Typical supporting tele-services are multimedia mail, information retrieval, conferencing, etc. These are available to the tele-education service provider at more than one site. A number of value-added services (i.e., VPN, PCS) and ATM bearer services are required to support the operation of the higher-layer tele-services. The operation systems at the service level offers management services via CORBA based management interfaces.

One of the key components in this architecture is the Virtual Private Network Provider OS. The VPN provider will offer VPN Services to customers, which may be end-users or other tele-services providers. Another key component in the PROSPECT trial architecture is the ATM Virtual Path Provider. The ATM Virtual Path service is realised with a TMN based management system, offering Virtual Path services via a TMN X-interface. With this architecture there is consequently a need for a CORBA/CMIP gateway which allow the CORBA based VPN Service Provider OS to use the ATM Virtual Path service via the X-interface offered by the ATM Virtual Path Provider TMN based management system.

Thus, there is a clear need to examine how CORBA and Telecommunications Management approaches will co-exist. The aim must be not to replace TMN with CORBA but to see how best CORBA can be used in conjunction with the existing TMN investment.

2. Adopting the co-existence approach of the NMF

The Network Management Forum (NMF), which is the major forum in the telecommunications management domain, has come out in favour of CORBA's IDL as the interface solution between the elements of its proposed Management System Framework [4]. The *access facilities* and *agents* in the architecture take care of access to management information in GDMO or SNMP MIBs. *Adapter objects* are defined in order to avoid the direct use of the individual application programming interfaces offered by

the access facilities. Therefore, they offer access in a uniform manner, but they still do not hide the differences among resources, i.e. the underlying information models. Other components of the MSF are *common services* that provide functionality that is frequently needed by other framework components as well as management applications. The MSF also describes how NMF envisages the use of CORBA ORB and CORBA services to implement the architecture.

The other major piece of work in this area comes from a joint working group formed by NMF and X/Open. The output from this group [5], first describes the scenarios where co-existence with CORBA will be necessary, and compares the models involved: IDL, SNMP, GDMO. The main body of the work describes "Specification Translation" where syntactic mappings are included: from GDMO to IDL, IDL to GDMO, and SNMP to IDL. Since the mappings have only been published in a stable format in the last year, implementations of GDMO to IDL compilers have not been very common or stable.

Further work is underway which will specify "Interaction Translation" algorithms which together with the "Specification Translation", should enable the implementation of "gateways" which will allow requests/responses in one paradigm to be dynamically translated into the other paradigm.

Taking these activities into consideration, two independent but complementary approaches can be used to enable co-existence of TMN and CORBA in service management:

1. development of access facilities for CORBA objects based on the specification and interaction translation algorithms;

2. specification and implementation of adapter objects that hides the use of specific communication protocols.

The former approach enables access to any kind of TMN management information. However, knowledge about details of the management information model is still required by a developer of a service management application. The latter approach masks not only the details of the underlying communication, but also of the used information model since it allows the definition of a more abstract functional interface.

Combining both methods allows to establish a phased approach for solving the TMN-CORBA co-existence problem. As long as the CORBA access facilities are unavailable, CORBA adapter objects and common services can be defined and implemented that "manually" translate CORBA operations into (possibly a sequence of) operations provided by specific management protocols. After implementations of CORBA access facilities are at hand, the implementation of these adapter objects can be changed to use the "automatic" translation features of these facilities without modifications of the defined CORBA interfaces. Section 3 outlines the approach PROSPECT has chosen for realising the basic CORBA access facilities. Section 4 presents initial ideas about useful adapter objects that support TMN-CORBA co-existence.

3. Access facilities for CORBA objects

A CORBA-to-TMN access facility will be developed in the PROSPECT project to provide CORBA access to TMN Service Management facilities within the PROSPECT TMN System.

The main goals for this service is to provide a seamless CORBA compliant service to any customer application that may have the need of interactive and direct control of telecommunications resources and services offered via TMN compliant X or Q interfaces.

It is important that the access facility is able to operate in either a centralised (one central access facility providing access to a number of TMN agents) or in a distributed fashion (co-location of access facility and TMN agent), in order to support different strategies for issues such as load-distribution, versioning, fault-tolerance, availability etc.

To adopt to the ever increasing market demands for new telecommunication services, the telecommunication providers needs to be able to update and extend the services and therefore also the TMN access facilities very frequently and with a minimum of impact on existing service availability. Another important requirement to the access facility is therefore to allow for easy access to new agents and/or new objects when needed.

An access facility should ensure IDL consistency across different implementations of a given GDMO standard. It should therefore be based on the principle of direct translation of GDMO specifications into equivalent IDL definitions according to the principles dictated by the JIDM working group. The access facility developed within PROSPECT will as directly as possible represent the contents and structure of the underlying TMN agents at the CORBA server interface. The direct translation approach

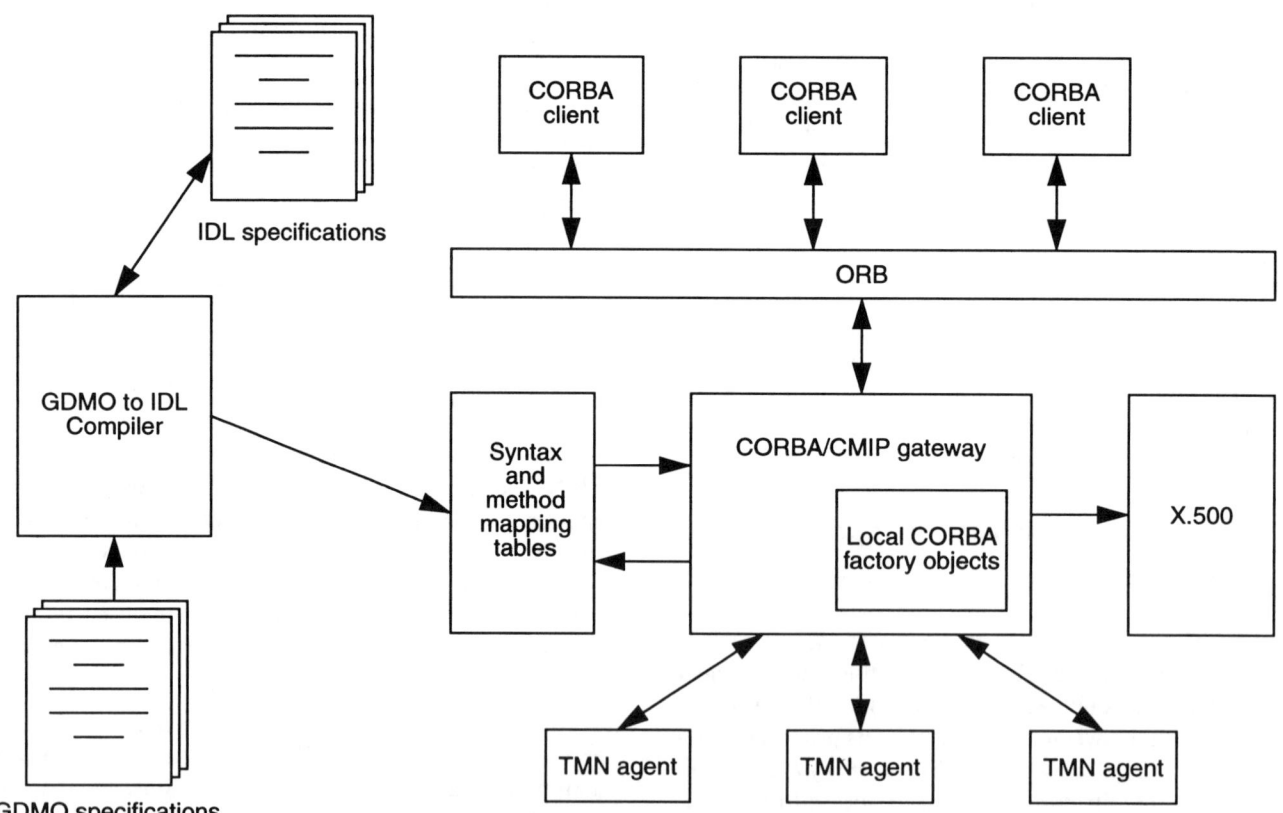

Figure 2: CORBA access facilities

38

facilitates a table driven method which allows for dynamic inclusion of new agents and/or new object classes in the gateway with minimal or no service interruption.

The CORBA-to-TMN access facility developed in PROSPECT will include:

- a GDMO-to-IDL translation utility, capable of translating ASN.1/GDMO specifications into IDL interface definitions;

- a CORBA-to-CMIP gateway, capable of receiving CORBA requests and translating them into appropriate CMIP request.

The access facilities will be developed based on the commercially available software package Q3ADE [6], which is a complete environment for developing TMN/Q3 compliant agents and mediation functions.

The GDMO-to-IDL translation utility is used for two purposes:

1. To generate IDL definitions from a GDMO specification according to the guidelines given by the X/Open and NMF Joint Inter-Domain Management (JIDM) working group [5].

2. To generate syntax and method translation tables for the CORBA-to-CMIP gateway

The CORBA-to-CMIP gateway will act as a CORBA server on behalf of the TMN agents accessible via the gateway and contains the following functions:

1. Data translation

2. CORBA-to-CMIS method translation

3. CORBA Object reference-to-Managed Object Instance Name translation

4. Internal implementations of a number of factory objects representing the ability to create certain managed object classes.

5. Address resolution based on X.500 directory services

6. The CMIP protocol stacks

In order to be able to support CMIS CREATE operations, the gateway needs to have a specific set of so called factory objects, representing the ability to create objects. Such objects are not an integral part of the GDMO specification and therefore will not exist in the managed agents, they consequently need to be managed locally by the gateway. The GDMO-to-IDL translator will therefore in addition to the basic translation of the GDMO specificication generate the IDL definitions for CORBA factory objects on the basis of the CREATE clause of the name bindings contained in the GDMO specifications. The gateway will provide local implementations of the generated CORBA

factory objects definitions. The generated IDL definitions will be used successively in the generation of CORBA client code.

The CORBA-to-CMIP Gateway will register the CORBA object implementations in the ORB implementation repository on the basis of the generated IDL definitions. Restrictions are imposed on the CORBA object naming, which allows the direct translation of the CORBA object names into CMIS globally distinguished name.

The CORBA-to-CMIP gateway uses the generated syntax and method mapping tables to dynamically translate IDL request into CMISE PDUs and CMISE responses PDUs into IDL replies based on the original GDMO specifications. The translation tables can be updated dynamically and allows for the enrollment of new TMN agents in the gateway.

4. Management facilities

This section deals with basic management facilities, i.e. adapter objects, that allow a CORBA-based manager to access specific TMN management information. These management facilities have been modelled as ODP computational objects [7] that may offer several operational interfaces. Afterwards, these computational objects have been implemented as a set of inter-related CORBA objects, where each CORBA object provides one of the interfaces defined for the represented computational object.

Two kinds of specific management information have been identified as being of primary importance for the development of CORBA-based service management applications:

- event reports (e. g. to support proactive service management);

- logs and log records (e. g. for accounting purposes).

Section 4.1 and Section 4.2 presents specific CORBA objects that make that information available in the standard CORBA way.

4.1. Event control

In order to allow CORBA-based managers to be unaware of the source of event information, it is desirable to use standard CORBA services [8] as far as possible. One of the standard CORBA services which is important in this context is the CORBA Event Service. The CORBA Event Service decouples the communication between objects in the supplier role and objects in the consumer role. Suppliers produce event data and consumers process event data. An event channel is an intervening object that allows mul-

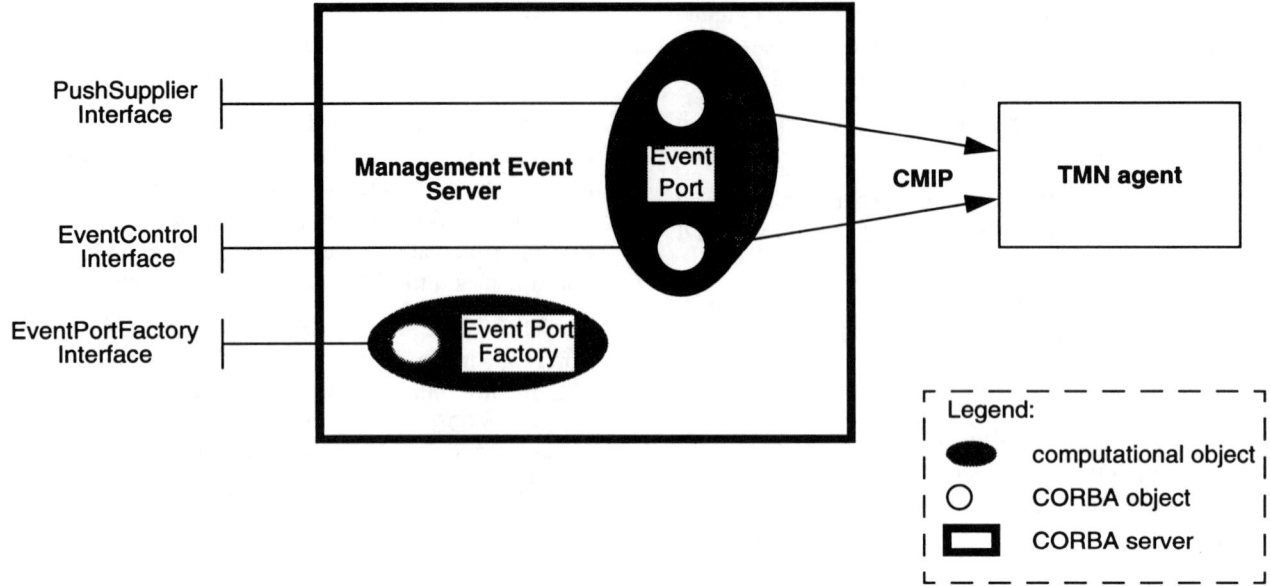

Figure 3: CORBA-based management event control and dissemination

tiple suppliers to communicate with multiple consumers asynchronously. Event channels are standard CORBA objects and communication with an event channel is accomplished using standard CORBA requests.

Two approaches for the integration of management event reports with the standard CORBA Event Service can be identified:

- specification and implementation of a specific supplier object;
- specification and implementation of a specific event channel.

The first approach has the advantage that it also allows the direct communication of events between a management event supplier object and an arbitrary consumer object, i. e. without an intervening event channel. This specific supplier computational object is called *event port* (cf. Figure 3) and provides the standard *PushSupplier* interface as defined by the CORBA Event Service. Subclasses of the event port object may be defined that support typed event communication, e. g. for all notification types defined in the Object Management Function [9], the Alarm Reporting Function [10] or the Security Alarm Reporting Function[11].

Initially, no *PullSupplier* interface will be offered by the event port object. If a pull-style communication of events is needed, the event port can be connected to an event channel that provides the *PushConsumer* interface and that supports the *PullSupplier* interface.

Additionally, the event port object offers an manage-

ment interface called *EventControl* that allows to control which events shall be supplied to consumers. The following capabilities are supported by the *EventControl* interface:

- Status inquiries;
- Suspension of event reporting;
- Resumption of event reporting;
- Termination of event reporting;
- Access and modification of a filter that determines which event reports shall be disseminated.

The last capability supports a finer granularity of event filtering as the standard CORBA Event Service since it does not only allow filtering based on event types like event channels that support typed event communication, but enables to filter according to event parameters, time of the event, etc.

The proposed approach to CORBA-based management event control and dissemination is similar to the TINA Notification Service described in [12]. Both support a finer granularity of event filtering as the standard CORBA Event Service. However, the TINA Notification Service allows *each* recipient object, i.e. an event consumer in CORBA terminology, to specify a filter expression that prescribes the conditions under which it wishes to receive a notification of a given type. The presented approach permits only the definition of a *single* filter per event port by some event control application. Thus, all event consumers attached to the event port will receive an event report that satisfies the specified filter.

40

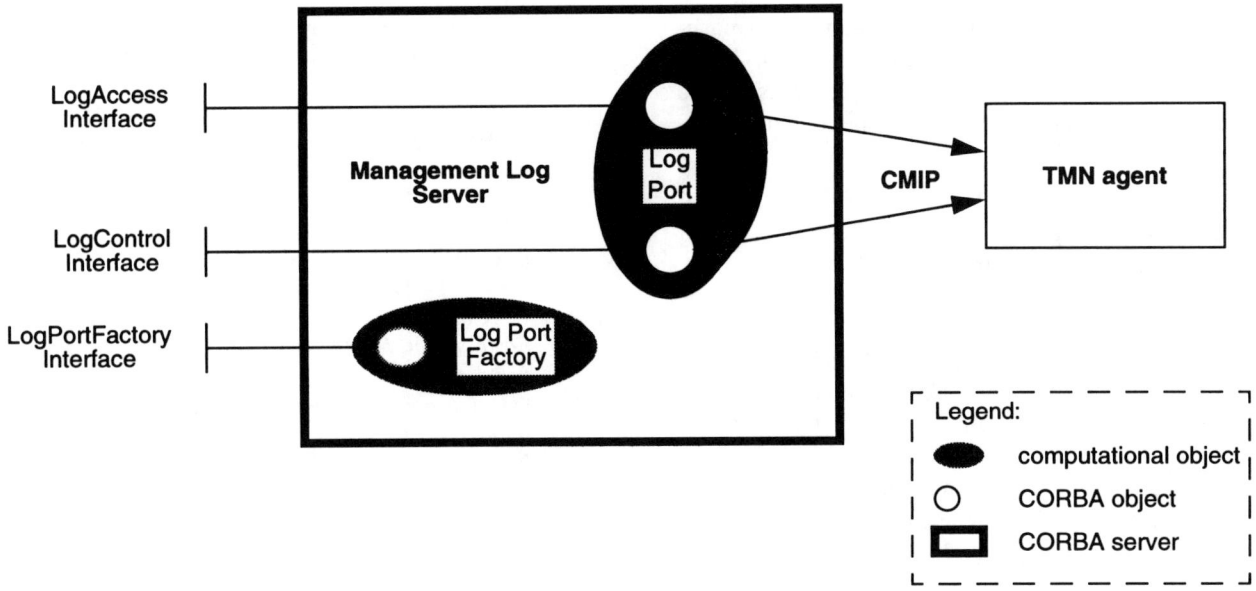

Figure 4: CORBA-based log control and access

4.2. Log control

In order to enable CORBA manager to access and control OSI management logging, a computational object called *log port* is provided (cf. Figure 4). This computational object offers two interfaces: the *LogAccess* and the *LogControl* interface. The *LogAccess* interface provides operations to retrieve log information that fulfils specified criteria like the time of the logging, the type of the event that has been logged, etc. The *LogControl* interface supplies the following capabilities:

- Status inquiries;
- Suspension of logging;
- Resumption of logging;
- Termination of logging
- Access and modification of a filter that determines what shall be logged;
- Access and modification of log configuration information (e.g. allowed maximum size of the log, behaviour if the log is full, etc.).

Operations provided at the *EventControl*, *LogControl*, and *LogAccess* interfaces are mapped onto operations on event forwarding discriminators [13], log and log records [14], respectively. Event port and log port computational objects are created and destroyed by event port factories (cf. Figure 3) and log port factories (cf. Figure 4), respectively).

5. Summary

Clearly CORBA will be a key solution technology for many areas of telecommunication systems, ranging from being embedded in the actual network elements, right through to being the basis for new telecommunication services like multimedia. It is increasingly obvious from research applications that it is desirable and feasible for CORBA to be used as an implementation technology for systems also at the network management and network element management levels. There is however a large amount of TMN standardisation effort put into the network element and network management and a significant number of existing management systems built in accordance with these principles. The PROSPECT trial therefore exemplifies likely future scenarios, where new CORBA based management systems will use management services offered via X-interfaces by TMN based network and network element level management systems.

There is tremendous synergy between CORBA and Telecommunications Management, and it is clear that both will live in perfect harmony. The primary reason for this optimistic outlook, is that CORBA is a very good solution to the significant problems that Telecommunications Management systems developers face: distributed components, diverse programming languages, lack of a single architecture or design paradigm, legacy systems etc. CORBA provides the necessary solution through the marriage of distributed technology and object orientation. The endorsement of CORBA IDL and use of CORBA Services as part of Network Management Forum's architecture for management systems (the Management System Framework) is significant evidence that CORBA has been recognised as a key part of telecommunication management solution implementation.

The appearance of CORBA as part of deployed solutions is likely to be staged. Initially, it will occur as part of integration solutions in the area of service management. Two other steps will then start to emerge in parallel: new proprietary distributed management applications will start to emerge in operator organisations and traditional management platforms built using CORBA will be made available by vendors. Finally, systems which feature CORBA application and TMN application co-existence will become available, together with management platforms which will allow the construction of management applications according to CORBA and TMN principles.

The approach described in this paper recognises the co-existence of the TMN and CORBA world and allows a seamless transition between these two domains. It relies on the work of the X/Open Joint Interdomain Management Group. Through the use of the translators and gateways specified by this group it will be possible to use a generic approach in order to tackle the problems associated with the integration of TMN functionality into a CORBA-based environment. Therefore, the PROSPECT project will be engaged in the development and realisation of CORBA/TMN gateway functionality.

6. References

[1] ITU-T Recommendation M.3010. *Principles for a Telecommunications Management Network*. Draft Revised Version, January 1996.

[2] PREPARE. *Modelling and Implementing TMN-based Multi-Domain Management*. January 1996.

[3] Object Management Group. *The Common Object Request Broker: Architecture and Specification*. Technical Report, Revision 2.0, July 1995.

[4] Network Management Forum. *OMNIPoint Integration Architecture - Delivering a Management System Framework to Enable Service Management Solutions*. Technical Report, Issue 1.0, August 1994.

[5] X/Open. *Inter-domain Management Specifications: Specification Translation*. X/Open Preliminary Specification, Second Sanity Check Draft, June 1996.

[6] UH Communication Aps. *The Q3 Agent Development Environment, The Implementers Guide*. Doc no: A0604.0013

[7] ITU-T Recommendation X.901 | ISO/IEC 10746-1. *Information technology - Open Distributed Processing - Basic Reference Model - Part 1: Overview and Guide to use*.

[8] Object Management Group. *CORBAservices: Common Object Services Specification*. Technical Report, Revision 2.0, March 1995.

[9] ITU-T Recommendation X.730 (1992) | ISO/IEC 10164-1 : 1993. *Information Technology - Open Systems Interconnection - Systems Management: Object Management Function*. International Standard, 1993.

[10] ITU-T Recommendation X.733 (1992) | ISO/IEC 10164-4 : 1992. *Information Technology - Open Systems Interconnection - Systems Management: Alarm Reporting Function*. International Standard, 1992.

[11] ITU-T Recommendation X.736 (1992) | ISO/IEC 10164-7 : 1992. *Information Technology - Open Systems Interconnection - Systems Management: Security Alarm Reporting Function*. International Standard, 1992.

[12] TINA-C. *Engineering modelling concepts (DPE Architecture)*. December 1994.

[13] ITU-T Recommendation X.734 (1992) | ISO/IEC 10164-5 : 1993. *Information Technology - Open Systems Interconnection - Systems Management: Event Report Management Function*. International Standard, 1993.

[14] ITU-T Recommendation X.735 (1992) | ISO/IEC 10164-6 : 1993. *Information Technology - Open Systems Interconnection - Systems Management: Log Control Function*. International Standard, 1993.

TANGRAM: Development of Object-Oriented Frameworks for TINA-C–Based Multimedia Telecommunication Applications

K.-P. Eckert[1], M. Festini[2], P. Schoo[1], G. Schürmann[1]

[1]GMD FOKUS, [2]DeTeBerkom GmbH

Abstract

The paper presents some results of the TANGRAM project, that prototypes and evaluates an environment supporting the object-oriented development of distributed multimedia applications based on the architectures developed by the "Telecommunications Information Networking Architecture Consortium" TINA-C. These architectures suggest how to structure software for information networking. Their strength is the applicability for a wide range of telecommunication services and the independence of system and network technology. To enforce the approach taken in TINA-C, the concepts of abstract object-oriented frameworks are introduced and applied. Such frameworks can be a means to support the design and development of software sharing the advantages of object-orientation like reusability, scalability, and customizability.

1 : Introduction

For the development of telecommunication software, the Telecommunications Information Networking Architecture Consortium (TINA-C) has progressed software architectures for information networking [16] to assist service designers and software developers. The architectures address distributed multimedia telecommunication applications for service logic and switch control, and their operation, administration and management. They aim ensuring adherence to the specific problem domain of information networking, as a part of telecommunications. Software architectures comprise concepts, rules and suggestions for the structuring of processing tasks that application software should follow. The resulting software structures are intended to be amenable to changes when variations are required, and thus support the rapid introduction of new services.

The TINA-C architectures are based on the principles of distributed object-oriented computing technology. They define and specify a set of telecommunications problem domain specific services which are carefully designed to ensure a wide reusability. Based on the Reference Model of Open Distributed Processing (RM-ODP) [6], the TINA-C architectures include primarily information and computational models for a wide range of telecommunication services. Information models formalise problem domain specific concepts and their relationships to specify the semantics commonly used in a variety of information processing tasks. Subsequently, from a computational per-spective so called *components* are identified and specified which fulfil the processing tasks.

The design of components is based on their responsibilities in relation to the processing tasks; or, as result of a pragmatic approach, as computational representations of information objects. The specification of components is typically expressed in terms of computational object (CO) groups. These groups are distributed and deployed in a so called *distributed processing environment* (DPE) superposing the telecommunication network resource infrastructure.

Today the specification of TINA-C computational objects is mostly flat. It neither takes advantage of abstract and concrete properties of other objects, which could be inherited and specialised nor aggregates objects to form the required set of properties. Addressing this topic, this paper presents object-oriented frameworks as a tool that supports the object-oriented design and development of distributed telecommunication applications and their implementation in an ODP-based engineering environment. In fact, two frameworks are introduced: An application framework that comprises TINA-C compliant abstract descriptions, that are meant to be combined and specialised; and an engineering framework concerning the transformation of computational objects into engineering representations. The latter enables the distribution and deployment of manageable groups of engineering objects on DPE nodes.

The paper presents some results of the TANGRAM project [15] which validates the TINA-C suggestions on the Service, Computing and Management Architectures. It is organized as follows: After a brief presentation of the

TINA-C initiative (Section 2) the TANGRAM project is introduced in Section 3. Subsequently Section 4 presents and examines object-oriented frameworks. Section 5 discusses the application of framework concepts in the scope of telecommunications applications and Section 6 concludes the paper.

2 : The TINA-C initiative

The Telecommunications Information Networking Architecture Consortium TINA-C consists of more than 30 telecommunications and information technology companies from all over the world. The consortium started in 1993 and has a duration of five year. The goal of the TINA consortium is to define and validate a software architecture that will enable the efficient introduction and management of new and sophisticated telecommunications services. This activity is based on the consortium members' recognition that the global, information future requires interworking and federation with many other participants.

The TINA-C Overall Architecture [16] encompasses basic design and modelling rules which are applicable to a wide range of services and decomposes the complexity of the problem space. In particular, the TINA-C Service Architecture [17] is based on the enterprise model describing the relationships between service providers, customers and end-users. The latter are not necessarily human being, but may also be of a resource type. This architecture introduces a session concept, which allows to consider the association of users, services they use, and the involved equipment. Further the service architecture elaborates the information models required for services which shall fit into the identified business model, and, finally, suggestions for specifications of computational components are encompassed.

Of particular interest in the scope of this paper is the session model and the related component specifications. The session model enables to distinguish: Access Sessions, the (multiple) association of one user and a service provider; Service Sessions, the association of users to a service, being constraint by the customers requirements, the users preferences and the users actual technical environment; and Communication Sessions, the (multiple) association of end points during a Service Session enabling the transfer of data through the network infrastructure.

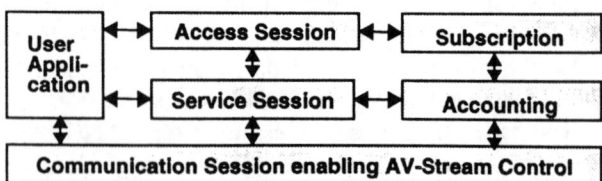

Figure 1: The TINA-C session model:
A foundation for the TANGRAM application framework

3 : The TANGRAM project

The TANGRAM project is part of the R&D programme of DeTeBerkom GmbH, a subsidiary of Deutsche Telekom AG and is carried out at GMD FOKUS. It started in April 1995. The main objectives of TANGRAM are to suggest, prototype, and evaluate an environment supporting the object-oriented development of distributed, multimedia telecommunication applications based on the TINA-C architectures. Secondly TANGRAM exploits the TINA-C suggestions for a DPE and evaluates CORBA based software products for their suitability. Besides the projects main objectives, the TANGRAM DPE is meant to be made available for use by other DeTeBerkom projects, which:

- enhance the access session and their related computational objects to support mobility of users, services and sessions to a further extend [9];
- use TMN service management concepts based on CORBA engineering mechanisms to identify a migration path from CMIP towards GIOP [1];
- integrate security aspects into the TINA-C architectures; and
- exploit video-on-demand services based on DMS-CC communication protocol and MHEG-5 coded content data [4].

3.1 : Service development support in TANGRAM

The first main objective of the TANGRAM project is concerned with the definition, instantiation and evaluation of service independent object-oriented frameworks based on the TINA-C Service and Management Architectures. Avoiding to develop the frameworks from scratch, the TANGRAM project implements on top of the core TANGRAM platform the multimedia communications service T-MMCS [2], which reuses some software developed in an earlier DeTeBerkom project. Adhering to the suggestions made in the TINA-C Service Architecture, the T-MMCS enables the exchange of audio and video data between a group of participants. It typifies the category of communication service which enables users in a group to isochronously exchange data through a network.

The T-MMCS development aims to enable the application to respect the constraints made between the service provider and the customer, the registered users' preferences, and the current technical environment of the users. This information is considered in the service control when requested, i.e. when the service is actually invoked and the service session is initiated. For the T-MMCS, the control of the service is strictly separated from the connection management, since both have to respect different kinds of effectiveness: the control part first of all has to be highly flexible, the communication part should concentrate on performance increase. T-MMCS comprises service management regarding subscription and accounting aspects. Accounting has been implemented according to the relevant ISO | ITU-T released standard [7]. Hence, it has been avoided to add on required service management at a later stage.

The resulting design of the TANGRAM platform is a discrete software structure made out of *components* which can be regarded as customized and implemented design patterns. These components have clearly distinguishable processing tasks. They are units of design, which are reusable in the context of various telecommunication services and can be specialised via application frameworks. Particular effort has been spend to identify a structure of components where each component is amenable to further specialization, for example, to handle policy variations on fault, configuration, accounting, performance and security management addressed by OSI standards or OMG standards.

Each component is realised as a group of interacting objects having multiple interfaces, and any interaction among components happens via the object interfaces. To be supported by the engineering framework and without leaving the object model, the design unit "component" has been mapped to groups of realisation units "object" for the target DPE made out of CORBA platforms.

3.2 : The TANGRAM DPE

The TANGRAM DPE is formed out of interconnected processing platforms. It is an administration and execution environment for groups of computational objects and enables the cooperation of wide area distributed applications. It offers the necessary and required support for applications, that provide services to users and achieve to operate networks within one paradigm. It encompasses hard- and software resources for enabling computing and communication. Regarding such a DPE, the TANGRAM project's aim is to suggest and develop an environment for open distributed processing using existing distributed objects technology platforms, such that

- interoperability between different platform products is achieved, and
- distributed applications providing telecommunication services can be installed in this environment.

Accordingly, the TANGRAM project develops a CORBA-2 based DPE encompassing a variety of hardware and software systems including major UNIX systems, PC (Windows 95), and Apple Macintosh (MacOS). The platform products actually used are: Orbix and OrbixWeb of IONA [5], ParcPlace's [13] Distributed Smalltalk, and VISIGENIC's C++/ Java VisiBrokers [21]. Based on the TINA-C Computing Architecture and an appropriate map-

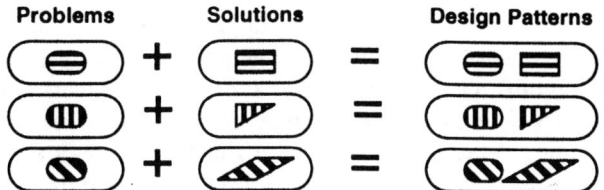

Figure 2: Design patterns

ping of ODP concepts on CORBA concepts, the DPE supports computational distribution transparencies and engineering concepts as defined in the RM-ODP.

4 : Object-oriented frameworks

This section gives an overview on object-oriented frameworks. It introduces the main characteristics of frameworks [12] and shows how they can be designed, implemented, and used. The presented approach is based on frameworks [14] in the problem domain of distributed systems.

4.1 : Design patterns, frameworks, and architectures

An object-oriented *framework* can be described as a construction kit for domain specific applications. It consists of set of prefabricated components, together with some architectural concepts that define the glue to put them together. An important characteristic of the framework approach is the identification of commons subproblems *together* with the presentation of components that provide solutions for these subproblems. The intentional specifications of these components are called object-oriented *design patterns* (Figure 2). They represent abstract solutions for specific problem classes and capture the static and dynamic structure and collaboration of a group of objects, providing a service. Additionally an *overall architecture* specifies how these groups of objects can collaborate to implement the extensional framework as a solution for the whole problem (Figure 3).

While design patterns can be looked at as a horizontal structure over a set of computational objects, frameworks

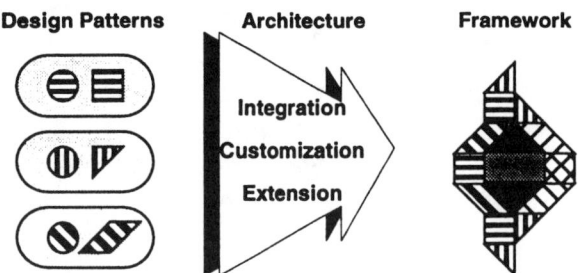

Figure 3: From design patterns to frameworks

can be looked at as vertical, domain specific groups of components. They provide an integrated set of domain specific functionality. The session specific components of the TINA-C Service Architecture are examples for such groups. As explained in Section 1 in TINA-C the description of a component encompasses a distinct (part of an) information model, its relation to other components and a model expression in terms of computational objects. Further specifications could be encompassed by a component description of which one could be a design pattern. It allows to abstractly express component specific properties and to devise different specializations of the component (up to instantiable realisations), while maintaining the inherited properties as given in an architecture. As a consequence, the reuse of components at design time is supported to a further extend since also such specializations are made available for later use.

4.1.1 : A deeper look at frameworks

During the development of frameworks the three stages shown in Figure 4 can be identified. In the beginning the architects of a framework define its essence:
- the overall architecture;
- the reusable design patterns; and
- the requirements on the supporting run-time system.

At the second stage the application designers instantiate the framework. They develop concrete classes that implement the behaviour specified in the more abstract patterns. The application consists of groups of interworking instances of these classes.

In the third stage the framework will be extended by its architects. They introduce new design patterns which can be abstracted from the instantiated application or which do correspond to additional subproblems. Thus a framework can be enhanced and customized due to the requirements occurring in open systems. During the evolution of a framework the stages two and three will occur repeatedly. In practice the first stage is often missed and the advantages of using frameworks are not recognised.

The definition and enhancement of frameworks may follow a step by step "case-study" approach. In a first step the main design patterns and the related core classes are identified and melted by the framework's architect. This allows the instantiation of a first prototype of the framework. This prototype can be tested and evaluated. In the following steps the framework can be extended with respect to the results of the evaluation process and additional case-studies. During these extensions further aspects of the problem space or additional candidates for the definition of new design patterns and abstract classes can be identified and integrated into the framework. It is important to notice, that this description can only be a guideline for the development of frameworks. There is no guarantee that the architects and application designers are able to recognize similar solutions for similar subproblems in different parts of the problem space. There is the need for framework specialists who have an overview on the whole project, who are able to identify potential design patterns, and who are responsible for the extensions of the framework.

4.2 : Benefits and drawbacks

There are several reason why to introduce frameworks. One important reason in the management and realization of large projects like TANGRAM is the

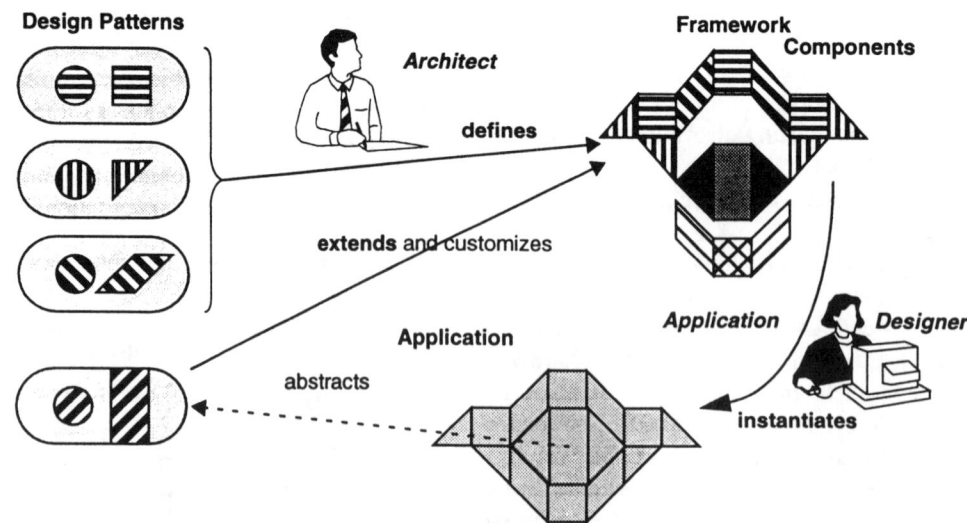

Figure 4: The development, instantiation and extension of frameworks

46

communication of architectural knowledge between the designers and developers. The usage of design patterns and frameworks implies the introduction of a common vocabulary and the definition of common design structures for all persons involved in the project. Hence, designers and developers are made cognizant of essential and decoration details within the solution. Design patterns help to manage the complexity of the problem solving process because they support the procedure of identifying similar problems and similar solutions. It has to be mentioned that these properties of a design process are not new, especially in an object-oriented world. Other reasons for the introduction of design patterns are to ensure the separation of different concerns, the adherence to standards, or to give a related toolset to the, possibly inexperienced, designer. Thus design patterns can be used to introduce object-oriented design principles in the project. Because they are closely related to object-orientation their application is one step towards the benefits of object-orientation like reusability, portability, customizability, and scalability.

But the introduction of design patterns may lead to some drawbacks. Firstly, since design patterns are an abstract concept, their usage may lead to design reusability, design portability, and abstract customizability. However, there is no direct mapping to code reusability, code portability, code customizability or even to code generation and execution. Secondly a framework consists of customized design patterns, implemented in concrete implementation languages and executed in a possibly distributed object run-time environment. The influence of the chosen language and the requirements on the run-time environment may not be underestimated. Although patterns describe only parts of a system and help managing the complexity of the system development process they are only deceptively simple. Last but not least it has to be mentioned that good design patterns, like good inheritance hierachies, can not be invented in an easy way. They have to be chosen and designed very carefully. The introduction of insufficient and wrong chosen patterns may prevent the design and implementation of good frameworks and related applications.

5 : The TANGRAM frameworks

Within the TANGRAM project application and engineering types of frameworks have been distinguished. Application frameworks support the development of distributed applications, that are specific to the telecommunications problem domain. These applications are particular, since, for example in TINA-C, service stakeholders requirements (user, customer and service provider) and connection management have to be adequately respected. Connection management is considered for services that use transfer capabilities of the network resource infrastructure, to establish streams carrying audio or video information between human or non-human Service Session participants. The connection management has to respect available network infrastructure resources plus required qualities.

Engineering frameworks are concerned about the support of ODP distribution transparencies which are of primary interest when transforming a computational description into engineering representations. For example, distribution aspects of object groups not being considered in the computational perspective have to be resolved, such that engineering structures can be distributed on nodes of the DPE.

5.1 : The application framework

With reference to TINA-C's architectures, TANGRAM has identified the following telecommunication specific problem areas and introduces related solutions in the form of design patterns and appropriate components:

- Subscription to a telecommunication service;
- Getting access to a telecommunication service;
- Creation and instantiation of a Service Session;
- Performance of a Service Session;
- Establishment and usage of a stream connection (Communication Session);
- Accounting of a Service Session;
- Extend the Access Session to enable personal mobility.

5.1.1 : The service session design pattern

The application of framework concepts will be illustrated using the example of a Service Session. Related to the Service Session, TINA-C has suggested a group of computational objects which are involved to control a session, i.e. to arrange a controlled association between session participants (humans or resources) and finally establish the required connectivity. A Service Session Manager (SSM) has the global view on a service session, and a User Service Session Manager (USM) has the local view on a service session for each user participating in the session. These components are refined [19] such that session control aspects of USMs and SSM can be distinguished from service specific aspects. The refinement led for a USM to the components User Service Part (USP) and User Service Session Control (USC), where the former is service control and the latter is session control related. Correspondingly the SSM has be refined to the components Global Service Part (GSP) and Global Service Session Control (GSC), distinguishing service and session control aspects. The session control part enables to express connectivity requirements on a high abstraction level. The required connectivity is forwarded to the Communication

Session Manager (CSM), being asked to initiate the appropriate computational streams to enable the isochronous and finally network technology independent exchange of data through a network.

Thus the Service Session is divided into a session specific part consisting of COs managing the connectivity requirements of a session and into a service specific part, regarding the functionality which is particular for the selected service. The former consists of concrete operations which are responsible for the invitation of and the resulting negotiations between new session members. They depend on a set of given policies, which can be defined in an abstract way and which have to be made concrete with respect to the demands of different services. The latter consists mostly of service specific, abstract COs which have to be refined, customized, and configured using service specific supporting objects (SSSO). An example is the introduction of audio-video stream-connections by the CSM. Due to the degree of service dependency it is important to define an abstract pattern that describes the general, reusable parts of a Service Session. Figure 5 shows this division of Service Session specific COs and interfaces into generic, mandatory parts and service specific extensions.

Comparable design patterns have been identified in other areas of the TINA-C session model shown in Figure 1as well. For example, the Subscription Registrar together with the Service Template Handler manages the subscriber specific instance of a Service Description. This description

is divided into two parts. The concrete, service independent part is maintained in the Service Profile and the abstract, service specific part is maintained in the Service Template. The Service Description can be accessed via a set of concrete operations implementing the interface of the Service Profile and a set of abstract operations specifying the interface of the Service Template. The latter have to be customized and implemented individually for each service. The advantage of this approach is, that the CO group of the Access Session can be implemented independent from the subscribed service.

5.1.2 : Advantages of the application framework

The computational specification of CO groups is performed in the TINA-C Object Definition Language (ODL) [20], which is a superset of CORBA IDL [10]. ODL encompasses specifications of operational interfaces, computational objects, and object groups including textual behaviour specifications. The components' specification of the Service Session example is based on an information model (Session Graph) describing the needed information to express connectivity requirements and their particular computational specifications. Service specific aspects have been factored out, since it is assumed that the Session Graph and its computational interpretation are widely usable by services involving audio/video streams for representing computationally the transfer of data through networks.

Considering the traditional development of a service providing distributed application that reuses these specifications, the ODL specifications will probably be taken and code can be compiled for a chosen programming language. This code will only cover basic templates including methods the programming language specific implementation objects must be able to react on. This reaction is in the best case specified in the ODL specification given and must be implemented. For example, an implementation of the GSP will include some functionality, which enables to cooperate correctly with an GSC instance. For each variation of a service this procedure has to be entirely repeated.

When using design patterns the steps taken in the development are different. Based on the same specifications, a framework includes semi-complete implementations of the GSP in a chosen programming language. Interactions between GSP and GSC, to modify the connectivity requirements are implemented *concrete*, while the service specific control of the GSP are specified *abstract*. The abstract methods encompass, for example, the invitation of participants to a session and the session termination. To derive a service specific GSP which can be instantiated, those parts described abstract have to be made concrete. Since the in-

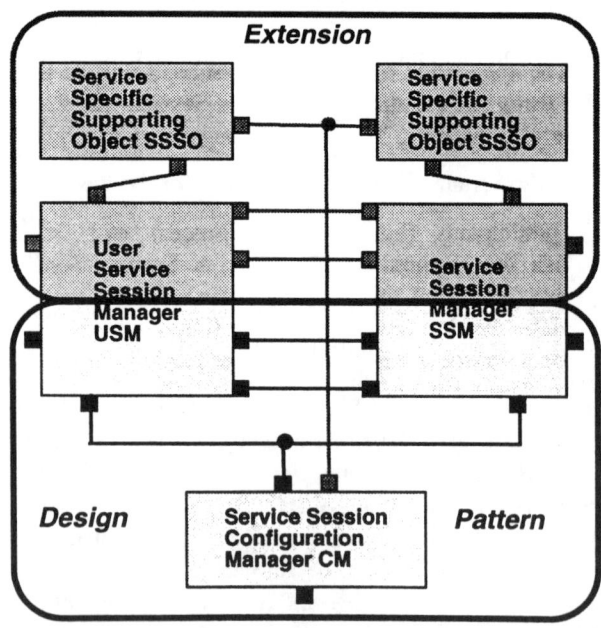

- ■ Generic mandatory interfaces
- ▩ Service specific customized interfaces

Figure 5: Service session design pattern

teractions with GSC are available, the development effort is reduced.

This principle is not specific to the Service Session components, but illustrates the use of frameworks aiming to make the development more efficient and effective for distributed telecommunication applications which are required in many variations. Thus the specification of the TANGRAM components can be looked at as the definition of telecommunication specific design patterns for multimedia applications. Additionally the specification of the relations between these patterns is another important constituent of the TANGRAM application framework. Using so called *use cases* (comparable to the case studies mentioned above) for the specification of allowable sequences of actions in a TINA-C compliant fashion, the TANGRAM framework defines the run-time control and determines which operations and components have to be called in response to the events, triggered by the execution of the different use cases. The addition of new use cases may lead to an extension of the TANGRAM application framework.

Summarizing, the TANGRAM application framework consists of:

- the specification of the concrete and abstract types describing components as groups of COs which together provide solutions for special areas of the TINA-C architectures;
- the definition of the interworking between the COs within each component; and
- the overall architecture defining the glue between the components on the basis of use cases that describe the possible activities in a TINA-C compliant application.

Moreover TANGRAM has developed an engineering framework, that is used to support the implementation of the application framework in a CORBA based engineering environment.

5.2 : The engineering framework

The TANGRAM-DPE provides mechanisms which ensure that the hosted computational objects can be designed in a uniform way, that they can be deployed, located, and that they are able to interwork independent from the underlying heterogeneous software, hardware, and network structures. Therefore the TANGRAM-DPE [3] supports the different distribution transparencies identified in the computational part of the RM-ODP by the provision of so called DPE-functions. These functions are used as enabling mechanisms to achieve distribution transparencies having the desired properties. The functions can be mapped to engineering design patterns. The following subproblems have been identified:

- Support for objects providing multiple interfaces

- Support for the object life cycle incl. object creation, initialization, and deletion
- Support for object deactivation, migration, and reactivation
- Support for object checkpointing, externalization and internalization
- Support for object location

For the sake of brevity, only the main principles are illustrated using a typifying example. It has been mentioned that an application framework consists of the specification of interworking components where every component is a group of computational objects. Thus a related DPE has to support the creation, initialization, execution, and termination of object groups. Therefore TANGRAM defines a Configuration Pattern consisting of three main parts (Figure 6). The Configuration Manager defines the external interface of a group to its environment and is responsible for the creation, initialization, and termination of a group of objects. The accompanying requests are delegated to type specific objects, so called Life Cycle Managers, which are responsible for the creation, initialization, and termination of related computational objects. While a Life Cycle Manager uses type specific factory objects [11] for the creation of computational objects, the initialization and termination of a single object are executed via invocations of a so called Control Interface that every engineering representation of a computational object has to provide. The definition of the Configuration Pattern provides a uniform solution for the introduction of new components in a TINA-C compliant application. In TINA-C this problem has only been addressed to a limited extend. An example is the Service Factory of the Service Session which can be mapped to the Configuration Manager described above. This approach of TANGRAM actually applies the concepts of capsules, clusters, and basic engineering objects (BEO) as introduced in the RM-ODP.

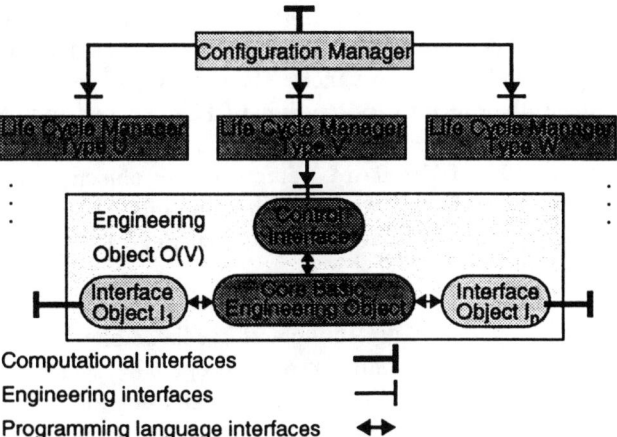

Computational interfaces ⊢
Engineering interfaces ⊢
Programming language interfaces ↔

Figure 6: The TANGRAM engineering framework

As explained in Section 3.2 the TANGRAM-DPE is based on OMG's CORBA2 [10] technology. Because CORBA identifies objects with single interfaces, TANGRAM has defined another elementary engineering design pattern that describes CORBA based multiple interface engineering objects in a programming language independent style, comparable to [8]. The related BEO structure has been extended to include the engineering entities mentioned above. These entities build the basis for the definition of the extended Engineering Object Pattern.

The TANGRAM engineering framework comprises these two design pattern. As a result, each concrete BEO inherits the abstract interfaces defined in these engineering design patterns. The core BEO has to be implemented with respect to the semantics of the object's type, and with respect to the technological constraints resulting from the used CORBA product. Although the realisation of multiple interface objects, object creation, checkpointing of objects, and their activation, deactivation, and reactivation will be implemented in different ways in a C++ and in a Smalltalk environment, the used interfaces, the used mechanisms and the used concepts are the same.

6 : Conclusions

The paper gives a brief overview on the TANGRAM project and introduces the concepts of object-oriented design-patterns and frameworks. Based on the TINA-C's software architectures the concepts have been applied and it has been discussed how frameworks can simplify the development of distributed, multimedia telecommunication applications. Within the TANGRAM project two related frameworks have been developed. The TANGRAM application framework extends the approach of TINA-C in a significant way. It shows a way to integrate and customise different TINA-C compliant components. The TANGRAM engineering framework supports the introduction of these components as well as the distribution transparencies identified in the RM-ODP. It is based on design patterns that describe a micro-structure for basic engineering objects and their management. The framework itself coordinates the execution of the single DPE-functions and defines a homogeneous CORBA2-based environment hosting the TANGRAM computational objects. Thus TANGRAM shows a way to reach the promised benefits of object-orientation like reusability, customizability, and scalability.

To validate these concepts TANGRAM has implemented and demonstrated the multimedia communication service T-MMCS running on various operating systems in a heterogeneous environment. An extension of the T-MMCS supporting mobile participants, the implementation of other services, and the integration of further DPE supporting platform products have shown the advantages of the framework approach taken.

It has to be noticed that TINA-C's new Service Architecture [18] applies some of the ideas presented in this paper. These concepts will be investigated in the near future.

7 : References

[1]   A.Dittrich, M.Höft: *Integration of a TMN-based Management Platform into a CORBA-based Environment*; Network Operations and Management Symposium NOMS'96, Kyoto, Japan, Apr. 1996

[2]   J.Dittrich, K.-P.Eckert, P.Schoo: *Design of a Multimedia Collaboration Service for an Environment of Distributed Processing Platforms*, Proc. of International Workshop on Distributed Object Oriented Computing (DOOC'95), Oct. 1995

[3]   K.-P.Eckert, M.K.Durmosch, K.-D.Engel, P.Schoo: *A CORBA2 based DPE for Telecommunication Applications*, Distributed Object Computing for Telecom Conference; Frankfurt,Germany; Oct. 1996

[4]   K. Hofrichter: *MHEG 5 - Standardized Presentation Objects for the Set Top Unit Environment*, Proc. of European Workshop on Interactive Distributed Multimedia Systems and Services (IDMS'96), Berlin, Germany, Mar. 4-6 1996.

[5]   IONA Technologies: *The Orbix Architecture*, Nov. 1996, http://www.iona.ie/Orbix/arch/index.html

[6]   ISO/IEC JTC1/SC21 IS 10746-3: *Open Distributed Processing - Reference Model, Part 3, Architecture*, 1995

[7]   ISO/IEC DIS 10164-10:1995 | ITU-T Recommendation X.742, *Information Technology - Open Systems Interconnection - Systems Management: Usage Metering Function for Accounting Purposes*, 1995

[8]   B. Kitson: *CORBA and TINA: The Architectural Relationship*; TINA'95, Melbourne, Feb. 1995

[9]   T.Magedanz et al.: *Personal Communications Support in the TINA Service Architecture - A new TINA-C Auxiliary Project*; TINA'96, Heidelberg,Germany, Sep. 1996

[10]   Object Management Group: *The Common Object Request Broker: Architecture and Specification*, Revision 2.0, Jul. 1995

[11]   Object Management Group: *CORBAservices - Common Object Services Specification*, Mar. 1995

[12]   OMG First Class: FOCUS: *Patterns and Frameworks*, Feb./Mar. 1995

[13]   ParcPlace: *Distributed Smalltalk*: http://www.parcplace.com/bod_prod.htm.

[14]   Douglas C. Schmidt: *Using Design Patterns to Develop Reusable Object-Oriented Communication Software*; Communication of the ACM, Vol.38, No.10, Oct. 1995

[15]   TANGRAM: *Project Description*, http://www.fokus.gmd.de/ovma/Tangram/entry.html

[16]   TINA-C, *Overall Architecture*, TB_MDC.018_1.0_94, Feb. 1995

[17]   TINA-C, *Service Architecture*, TINA Baseline, TB_MDC.012_2.0_94, 31 Mar. 1995

[18]   TINA-C, *Service Architecture Version 4.0*, TINA Service Stream, TB_RM.001_4.0_96, 28 Oct. 1996

[19]   TINA-C, *Service Component Specifications*, EN_HK.002_0.1_95, Oct. 1995

[20]   TINA-C, *TINA Object Definition Language*, TR_NM_.002_2.3_96, Jul. 1996

[21]   VISIGENIC: *Technical Reports on VisiBroker*, http://www.visigenic.com/techpubs/

A Data-Driven Implementation of Telecommunication Network Systems

Hiroaki NISHIKAWA

Institute of Information Sciences and Electronics, University of Tsukuba
Tsukuba Science City, Ibaraki 305 JAPAN

Souichi MIYATA, Shinichi YOSHIDA, Tsuyoshi MURAMATSU

IC Development Center, Tenri IC Group, Sharp Corporation
2613-1 Ichinomoto-cho, Tenri-shi, Nara 632 JAPAN

Hiroshi ISHII, Hidetsugu KOBAYASHI, Yuji INOUE

NTT Multimedia Networks Laboratories
3-9-11 Midori-cho, Musashino-shi Tokyo 180 JAPAN

Abstract

What should future information environment be like? At present, it is not so easy to answer the question but an inevitable issue. Undoubtedly, Users' requirements and engineering constraints will give a great influence onto the future information environment.

The authors therefore have started a project named CUE (Coordinating Users' requirements and Engineering constraints). The CUE project took up efficient ultra-multi-processing capability with overload tolerance and maintenability/expandability incorporating run-time reconfiguration as essential requirements to realize distributed networking systems such as TINA(Telecommunications Information Networking Architecture) which is being studied aiming at interoperability, reusability and transparency of application software essentially demanded by users of future infrastructure.

To satisfy these requirements from engineering viewpoints, the CUE project proposes applying to the network systems an autonomous decentralized data-driven processor which has been studied by some of the authors through various VLSI realizations. The data-driven processors adopt autonomous execution control scheme in which packetized data passing through the elastic pipeline including flow-branching/merging mechanisms achieves all data transferring, processing and storing/buffering functions in a completely decentralized manner. The processor can therefore perform high flow-rate multi-media streams with considerable flow-fluctuations and act as a functional module flexibly adding/removing to/from system at run-time without any overheads.

In this paper, essential requirements to realize the distributed networking environment such as TINA are first discussed. The applicability of our data-driven processor is shown through actual performance evaluations by the latest VLSI systems. Autonomous run-time reconfiguration scheme in a multi-processor system is also proposed to demonstrate potential of the data-driven processor which will be introduced to our data-driven processor currently being designed. Finally, the current status and future directions of the CUE project are briefly addressed.

List of Key words

TINA, Distributed Processing Environment, Multimedia Networking, Data-Driven Processor, Autonomous Decentralized Control, Elastic Pipeline Processing

1 Introduction

The authors have started the study project named CUE. From the viewpoint of both telecom network users' requirements and engineering constraints, CUE is clarifying how telecom network systems might best be implemented. This paper introduces the first stage achievements of the CUE project.

As for the telecom users' requirements, we consider recent trend of network environment. The trend is toward realizing an environment which enables users to easily utilize telecommunication network. The internet is one such example. There is another approach, that is TINA, Telecommunications Information Networking Architecture[1][2]. TINA is a Networking Architecture which is aiming at achieving interoperability, reusability and transparency of application based on distributed processing and object-orientation technologies. TINA enables both users and network providers to define and utilize versatile services in the same environment.

TINA itself is a networking software architecture which is defined independently of underlying technologies. However, to achieve real systems, underlying technologies must be taken well into account. At the same time, TINA must clarify requirements to the underlying technologies. For example, messages exchanged between TINA applications are in some cases transferred between physically remote workstations accommodating applications. The transfer media is called kTN (kernel Transport Network) . At this point in time, TINA has not clarified specific requirements to the kTN, which influences total reliability of TINA systems. TINA-DPE (Distributed Processing Environment) has common service functionalities such as naming and trading to which network load may be concentrated. Such traffic related requirements are yet to be well defined. In the distributed systems, fault in some specific application may influence several applications or services and it is generally difficult to identify the fault and influenced area without any centralized surveillance. Hence some effective way of autonomous fault management at run-time is essential. Substantial problem in current systems is that the execution of complex management related processing, e.g., traffic control, load balancing, testing, reporting, status checking in addition to ordinary process requires more multi-processing capability, which causes frequent thread or context switching resulting in much overhead in processors.

On the other hand, from the engineering constraints, the authors have been studying autonomous decentral-

51

ized data-driven processor for more than ten years[3]. Multi-processing capability without any overheads which is native characteristics of data-driven processing scheme has much possibility in providing communication system with superior processing capability. The data-driven processor seems very suitable for such systems with unpredictable nature rather as the telecommunication systems which have dynamic behavior at run-time. The TINA network system is based on distributed processing in which the unpredictable nature appears much more than ordinary telecom systems.

Our data-driven processors adopted the autonomous execution control scheme that gave rise to unique "flow-thru processing" scheme in which all processing, data-transfer, and storage functions are carried out in a highly parallel manner by a packetized data-flow through the elastic pipeline processing stages with completely decentralized controls[4]-[7]. The latest VLSI realization demonstrates more than 600MOPS(Mega operations per second) in an actual video signal processing application[8]. The processor is realized by 4 PE super-integration technology which seems highly applicable to realize robust distributed network environment. The data-driven principle states that functions are not fired until necessary data are ready. Improvement in network reliability is possible by taking advantage of this characteristic by which faulty functions based on the data-driven scheme is easily isolated, and when no data is inputted to it, will not influence other external functions.

In this paper, essential requirements to realize distributed networking environment such as TINA are first discussed. The applicability of our data-driven processor is then shown through actual performance evaluations by the latest VLSI systems. Autonomous run-time reconfiguration scheme is also proposed to demonstrate potentiality of our implementation of TINA. The current status and future directions are briefly addressed.

2 Requirements for Telecommunication Networks

2.1 Reliability and Maintenability

TINA defines the concepts and principles used to structure the software in telecommunications systems. It defines the constraints that should be applied during the specification, implementation, execution, and operation of software. TINA thus defines the basic structuring and operation of software. In TINA, telecommunications network is considered to be a large distributed software system to which distributed computing and object oriented software techniques can be applied.

Fig.1 shows schematically the network based on the TINA concept. TINA is a networking architecture that has been studied since 1993 in TINA-C. TINA is aiming at achieving interoperability, reusability and transparency of application objects based on distributed processing and object orientation technologies.

A TINA architecture consists of: 1) TINA application that provides versatile telecom service and controls/manages transport network, 2) DPE, Distributed Processing Environment and 3) NCCE, Native Computing and Communication Environment.

Objects can communicate with each other via DPE without considering physical locations. The kernel

Transport Network (kTN) interconnects each physically remote node accommodating DPE and applications over it. The kTN transports all messages and data (called operational data) that are exchanged among applications over DPE. The messages and data will cover, for example, connection control message from TINA control node to switching node, management messages issued from TINA management node to the transport network or exchanged between TINA nodes, and service session control message between user and transport. Since the kTN and TINA nodes play very important role to control and manage transport network, it is indispensable to achieve high reliability.

A service is realized by several objects residing in geographically distributed TINA nodes, such as workstations. For example, a fault is detected by a service user, there are several possibilities of fault occurrence: a fault in the hardware, kTN node, kTN itself, communication software, DPE, and application object itself. Then, localization of fault is needed. Isolation or deactivation of the faulty item follows. Substitution of faulty item to sane one (run-time reconfiguration) will be needed to restore faulty condition. After a service becomes available, restoration of faulty item would follow. Among these fault management, run-time reconfiguration with fault isolation is a key requirement to achieve reliable and maintainable TINA systems as the future infrastructure.

2.2 Ultra-Multi-Processing Capability

Telecommunication nodes must in nature process multiple calls, connections and messages simultaneously. In addition to call and protocol processing, management processing such as diagnosis and failure restoration must be executed simultaneously. Same kind of multiple contexts and multiple kinds of contexts must be handled simultaneously in the telecommunication node.

The TINA nodes, especially, TINA-DPE servers such as trader and name server will have many simultaneous contexts, because many clients will access those servers. Increasing the processing capability will not solve the problem but increases inefficiency. It is because the bigger the number of contexts grows, the more the overhead of context switches becomes. Fig.2 shows an experimen-

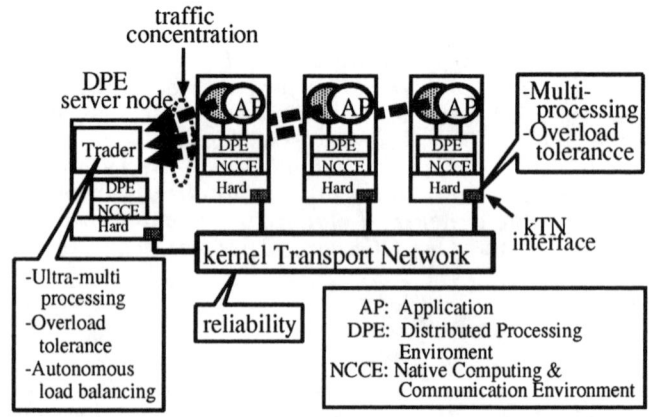

Fig.1 TINA Based Network

tal result of CORBA (Common Object Request Broker Architecture)-based simple client-server job processing time on the ordinary workstations (Von Neumann processors). Processing time per thread is increasing in proportion to the increase in number of threads. The increase in processing time per thread is apparently caused by overhead of thread switching. Hence, efficient processing avoiding overhead is required.

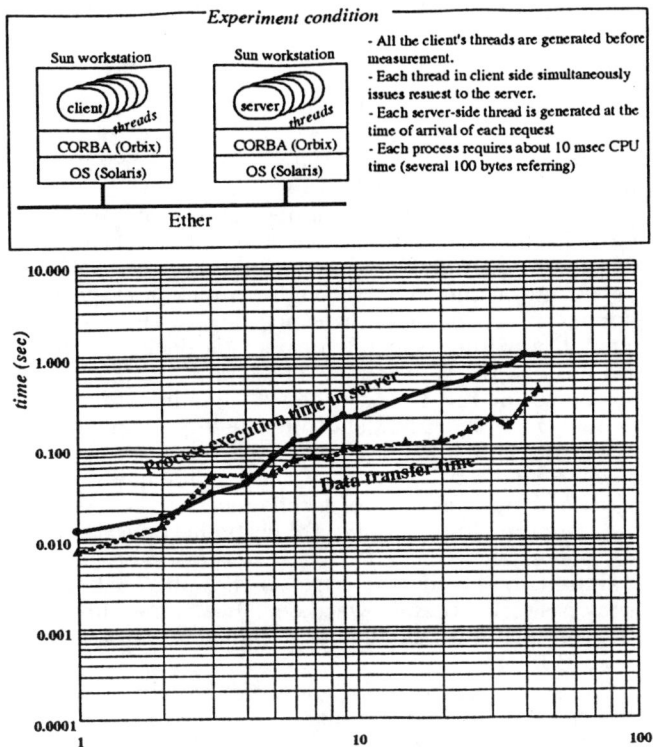

Fig.2 Process execution in Von Neumann processors

2.3 Expandability

System expandability (software and hardware renewal, addition and removal) must be taken into account, as the size of the network grows. Unlike stand-alone equipment, network equipment requires expandability which gives as little influence as possible to other equipment. It is needed to expand the system without suspending service provision. It is indispensable to achieve expandability or flexible reconfiguration.

2.4 Expectation of Data-Driven processor

The three requirements for telecommunication networks discussed in the previous sections, i.e., reliability/maintainability, ultra multi-processing capability, and expandability are essential and universal to any telecom systems. These requirements have been studied and in a sense achieved in "public carrier network". However, important points for ordinary users, the number of whom is much growing due to recent expansion of network including internet, are to be highly cost-effective and easy-to-install. From other viewpoints, the

autonomous and decentralized manner seems much welcome by them to avoid complexity and troublesome interface with providers. Considering the above, this paper proposes to adopt VLSI oriented data-driven processor to the networking systems.

3 Data-Driven Implementation of Autonomous Decentralized System

Data-driven scheme[9]-[11] is a simple concurrent/distributed processing principle which states that execution occurs only when input tokens necessary for execution become available. Our data-driven processor is based on the so-called dynamic data-driven principle to enhance stream processing capability essentially demanded by multi-media networking environment. Also in terms of VLSI realization, the processor controls tagged token or packetized data-stream through autonomous routing scheme by referencing tags and using hand-shake controls between packet-data latches. The suitability of this scheme to VLSI realization has been already shown with data-driven processors based on autonomous decentralized pipeline scheme. We have already demonstrated through implementation of 32-bit, 20MOPS (Mega Operation Per Second)@1.2μm rule[5]; 12-bit 40MOPS@0.8μm rule[6]; and 12-bit 600MOPS@0.6μm rule[8]).

Currently the authors have super-integrated several processing elements into a single chip. This not only simply improves processing performance, but also is effective for fault tolerance in the chip level as well as among chips if the same functional element is redundantly maintained.

This chapter first explains elemental architecture of autonomous decentralized data-driven processor to show effectiveness of a super-integration as well as applicability of the processor to multi-media networking environment. Characteristics of the autonomous decentralize-controlled pipeline are then shown through performance evaluations to demonstrate ultra-multi-processing capability of the processor. Finally, autonomous run-time reconfiguration scheme, which will be introduced to our data-driven processor currently being designed, is proposed to show reliability, maintainability and expandability of the data-driven TINA environment.

3.1 Super-Integrated Data-Driven Processor

Fig.3 shows the architecture of the data-driven processor. The data-driven processor is comprised from multiple unit processors called nano Processing Element(nano PE) and I/O Control which interconnects with them. The chip configuration shown in Fig.3 consists of nano PE 0 through 3 and two parallel input ports IA and IB, as well as two parallel output ports OA and OB.

Fig.3(b) shows the block diagram for the I/O Control. I/O Control receives and sends data packets off the chip via the IA/IB and OA/OB, respectively. The I/O Control is also responsible for routing of packets among the nano PE within the chip. RT1 receives incoming packets and determines whether the packet is destined for it. If it is determined that the packet was not destined for it, the packet is sent to RT4 where it is sent outside the chip through either OA or OB. If RT1 determines that the packet was destined for it, the packet is sent to the input processing part IN where it fetches the next instruction and is routed to the proper nano PE via branch B2. The identifier for RT1 is initially set, and this is compared with the destination PE number of the packet. Likewise, RT4's packet output condition has already been set, and determines which port to output by comparing certain bits in the input packet with output conditions, for example. Similarly, branch B2 outputs to nano PE within the processor by referencing the packet's instruction code. Because instructions in each nano PE is local, a portion of the instruction code can be used for branch condition. Although various routing mechanisms are possible, fundamentally, branching is determined by comparison result between the conditional parameter maintained by the routing mechanism and input packet's field.

Fig.3(c) shows the block diagram of the nano PE. Each nano PE is comprised from Firing Control (FC), Functional Processor (FP), and Program Storage (PS) which are key components in a data-driven processor. Each of these components are connected in a circular manner in conjunction with Junction (Jx) and Branch (Bx), where processing continues by packets circulating the pipeline. The data transfer paths between nano PEs within the processor, the data transfer paths between components in the nano PE, and the components themselves are all based upon the self-timed data transfer mechanism. By exploiting the elasticity of the pipeline, a processor that is able to withstand fluctuations in data stream is realized.

By allocating unique operations to the Functional Processor of each of the nano PE, a reduction in floor space is possible, as compared to allocating the same opera-

tions to each of the nano PE. On the other hand, an advantage of allocating the same operations to each of the nano PE is improved fault tolerance among the nano PE, that is, improved reliability on the chip level. Future development plans will super-integrate combination of six to eight heterogeneous/homogeneous nano PEs to improve processing performance while at the same time, achieve highly reliable super-integrated processors.

3.2 Characteristics of the Autonomous Decentralize-Controlled Pipeline

Fig.4 shows measured result of the relation between the number of input data packets per nano PE and the processing performance on the data-driven processor implemented through the half-micron-level wafer process. The horizontal axis is the number of data packets circulating in the circular ring of the nano PE and the vertical axis shows the number of data (2 data per packet) passing during a unit time through a pipeline stage of the pipeline ring, which is corresponding to the processing capability of the nano PE for a single packet to process two data at the same time. The circular ring of a nano PE shown in Fig.4 consists of 26 pipeline stages. The inclination of the graph showing the improvement of the processing capability is determined by the ratio of forward to backward latency of the hand-shake signals of self-timed transfer control mechanism. If the two latencies were the same, the shape of graph becomes that of an isosceles triangle. The reason that the top of the graph is shifted to the right in Fig.4 is that the backward latency is shorter than forward one. In the measured nano PE, the backward latency is designed to be 1/5 of the forward latency.

There is a characteristics of the data-driven processor that the processing performance is improved as the number of input data packet or number of threads processed simultaneously increases under the condition the filling rate is below a constant value. The paper concludes by discussing a data stream path's diagnostic algorithm which makes possible self-diagnosis during execution and dynamic reconfiguration.

Fig.4 shows the performance is linearly improved until 70% pipeline filling rate (the case where the number of

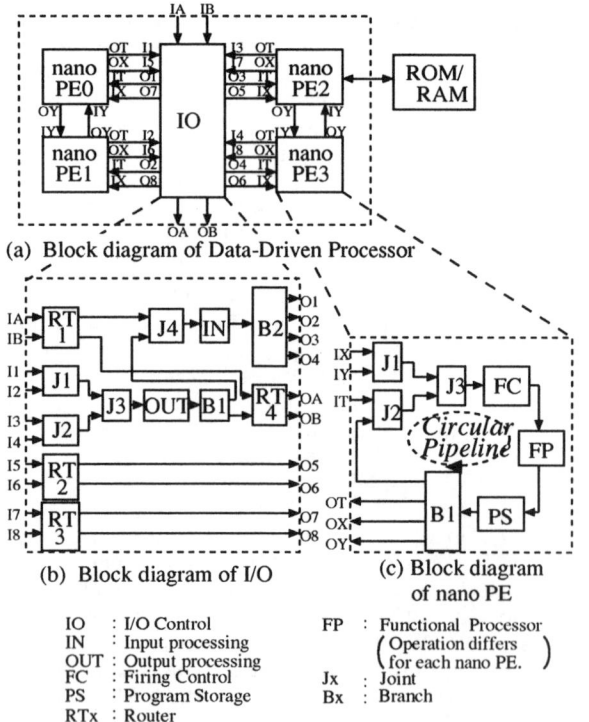

(a) Block diagram of Data-Driven Processor

(b) Block diagram of I/O

(c) Block diagram of nano PE

IO	: I/O Control	FP	: Functional Processor (Operation differs for each nano PE.)
IN	: Input processing		
OUT	: Output processing		
FC	: Firing Control	Jx	: Joint
PS	: Program Storage	Bx	: Branch
RTx	: Router		

Fig.3 Data-Driven processor architecture

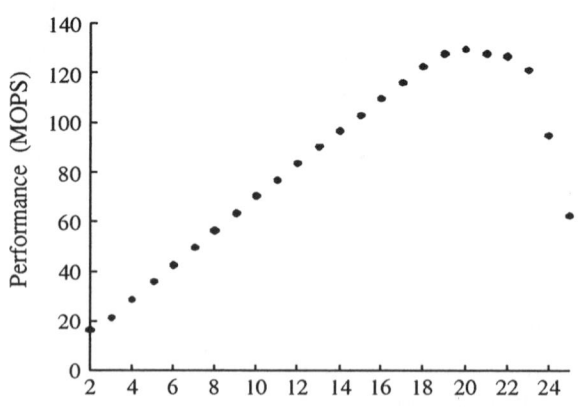

Number of Standing Packets in Circular Pipeline
(Total number of Circular Pipeline stages : 26)

Fig.4 Measured performance of a nano PE

input data packets is 18). In the range where the number of input data packets is 19 to 22 (filling rate is 85%), the maximum performance (130MOPS) is kept. The reason the shape of graph becomes not isosceles triangular but a trapezoidal is caused by variance of delay characteristics of the transfer control elements that construct the circular ring. Fig.4 also shows that the circular pipeline is tolerant of overload where the number of data packets momentarily increases beyond 20 which is caused by variance in the number of threads executing simultaneously.

Next, we will demonstrate that threads executing in parallel will not affect each other. Given the 3 threads described below that will be executed in parallel, P10: a program that executes 10 consecutive instructions from the input to output (execution time on its own:3.0μs) P15: a program that executes 15 consecutive instructions from the input to output (execution time on its own:4.4μs) P20: a program that executes 20 consecutive instructions from the input to output (execution time on its own:5.9μs)

These programs are invoked at fixed intervals(i.e launching of a new thread), where Fig.5 shows the observed execution times of each threads. Horizontal axis represents the identifier of each thread which is invoked at fixed intervals (every 200ns for this experiment). The execution sequence is in order of the identifier, that is, in the order of their launch. The vertical axis represents the time required for execution of each thread.

For example, in the figure, P10_30 indicates 30 consecutive execution of program P10. Similar label is used for others. In the case for P10_30, because execution time for each thread is 3.0us and a thread is launched every 200ns, 15 threads are continuously being parallel processed. Concurrent execution of 15 threads corresponds to 15 standing packets in the circular pipeline (see Fig.4). Note that in Fig.5, the execution times for each of the threads is 3.0us which is the same time as for a single thread's execution. This shows that the execution time is not affected by the number of parallel threads in execution as long as there is room in the circular pipeline.

Similar observation can be made for P15_20 and P20_20. In this case, because the number of launched thread is 20, 20 threads are continuously being parallel processed. Similar to the case for P10_30, because there is room in the circular pipeline, execution time for each of the threads is unaffected even when the number of parallel threads in execution is 20.

However, in P15_30, because the execution time for a single thread is 4.4us and a new thread is launched every 200ns, 23 threads are continuously being parallel processed. According to Fig.4, when the number of standing packets reaches 23, the number of unoccupied pipeline decreases slightly decreasing the processing performance. As a result of this, the processing time for each thread slightly increases to a maximum of 4.9us. The reason for the decrease in the execution times for thread identifier nearing 30 is due to the fact that the threads launched earlier completed their execution and leave the pipeline, thus increasing the number of unoccupied pipeline stages and restoring the processing performance. Similar observation can be made for P20_22.

With respect to P20_24 in which its pipeline stages are even more congested, maximum of 24 threads are continuously being parallel processed. Referring to Fig.4, the number of unoccupied pipeline stages decreases resulting in increase in execution time with a maximum of 8.0us. Although execution continues for up to 24 parallel threads in execution, when the 25th thread is launched, an overflow in the pipeline occurs and all activities in the pipeline cease. Note that the reason for launching 24 - 25 threads was to experimentally observe their effect on the pipeline. By employing an algorithm which avoids congestion of data stream, it should be possible to maintain appropriate number of parallel executable threads in the actual implementation of the data-driven processors to a network system.

Summarizing the above observations and referring to Fig.4, it can be said that as long as the number of packets in the circular pipeline, or the number of parallel threads in execution, is 20 or less, threads executing in parallel will not affect each other's execution time.

This feature is unseen for von Neumann processor's behavior shown in Fig.2, where there was a linear increase in execution time with the increase in the number of executed threads. If, in Fig.4, the number of parallel thread in execution exceed 20, the elasticity of the circular pipeline can withstand momentary increase up to 24 threads. The possibility of threads with short execution time causing an overflow and stopping the circular pipeline is small since the time that the packets stay in the pipeline is short. Thus, if the processing load of a thread is relatively small for such threads as self-diagnosis of data path in a data-driven processor, it will not greatly affect the processing as a whole.

3.3 Autonomous Run-Time Reconfiguration

The responsibility of the lower layers of the network is to ensure that the transmitted stream from the input side reaches its destination via a reasonable route.

In order to satisfy this demand, a scheme will be proposed in this section in which each node units in the network executes accordingly without employing a centralize-control that acts as a supervisor to enable the entire network to operate dynamically. Sub section(a) will describe the physical topology of the proposed network. Sub section(b) will describe how the input stream in the physical network is autonomously transmitted to

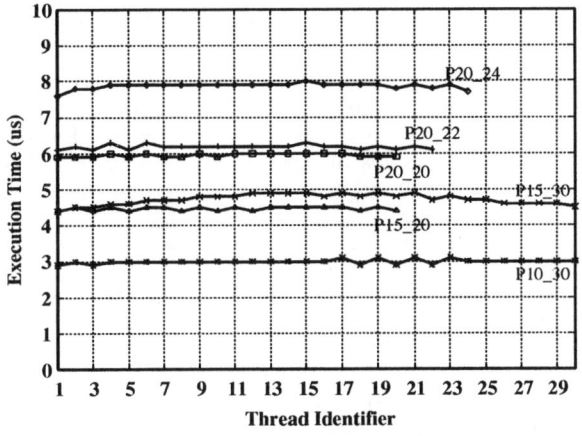

Fig.5 Process execution in Data-Driven processor

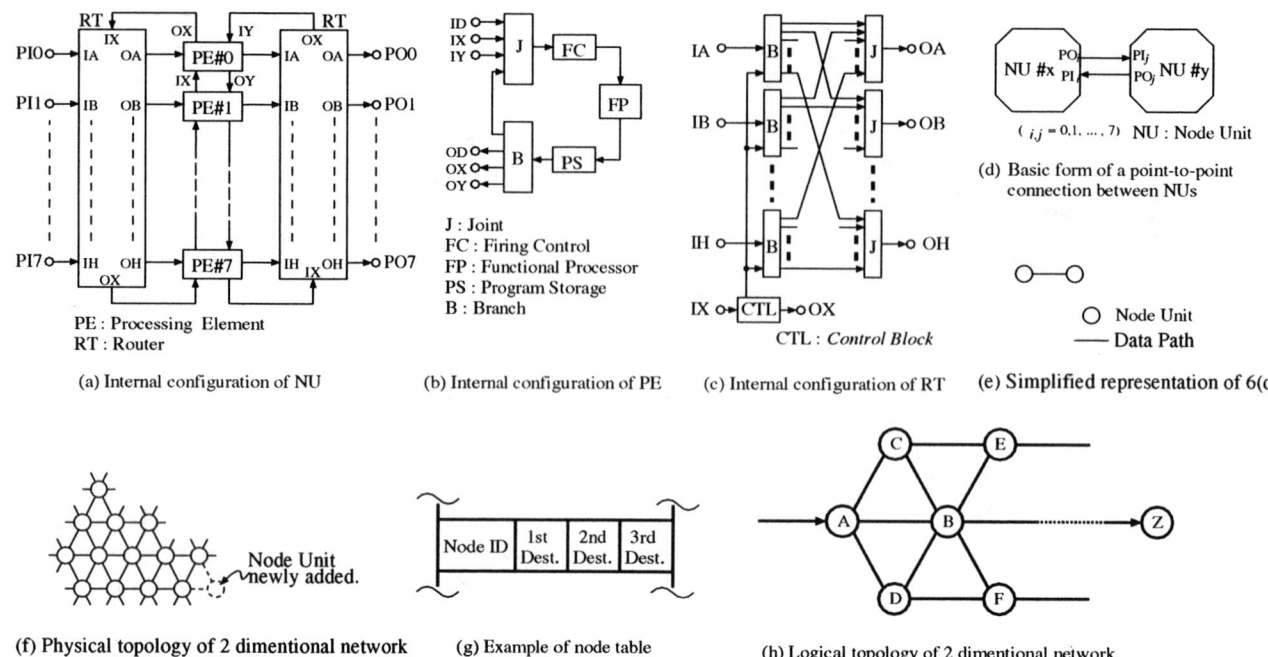

PE : Processing Element
RT : Router

(a) Internal configuration of NU

J : Joint
FC : Firing Control
FP : Functional Processor
PS : Program Storage
B : Branch

(b) Internal configuration of PE

CTL : *Control Block*

(c) Internal configuration of RT

($i,j = 0,1, \dots , 7$)   NU : Node Unit

(d) Basic form of a point-to-point
connection between NUs

○ Node Unit
—— Data Path

(e) Simplified representation of 6(d)

Node Unit
newly added.

(f) Physical topology of 2 dimentional network

| Node ID | 1st Dest. | 2nd Dest. | 3rd Dest. |

(g) Example of node table

(h) Logical topology of 2 dimentional network

Fig.6 Node unit and 2 dimensional network configuration

its destination using appropriate routes. Sub section(c) will describe how a fault in the network is detected as well as its process in the event of a detected fault. Lastly, in sub section(d), a discussion will be on how the function of the network is autonomously maintained with an existence of a fault.

(a) Physical topology of 2 dimensional network

Fig.6(a) shows the internal configuration of a node unit which is the fundamental element in configuring a 2 dimensional network. PE#n represents a data-driven processor core in which 8 are used for each unit. Also RT is a physical data switch with 8 inputs and 8 outputs. In configuring a network in which its connectivity is triangular, each node unit would require 6 input/output ports. In addition to the 6 input/output ports, 2 ports are reserved as spares.

Fig.6(b) shows the internal configuration of a PE. Data inputted from ID/IX/IY input ports are processed as it travels through the circular pipeline and is outputted via OD/OX/OY output ports. With respect to Fig.6(a), execution stream flows from RT to PE to RT. Also, to transmit control data within a node unit, a 2 circular paths connecting PEs to each other are used. In that case ID/OD port, IX/OX/IY/OY ports shown in Fig.6(b) are used. Fig.6(c) describes the specifics of the RT. Data inputted from IA - IH are routed according to branching conditions previously set in circuit block B and are outputted to one of the output ports OA - OH. There is no priority of input/output with respect to A - H, all are treated equally. IX/OX port is used for transmitting control data for RT.

Streams inputted from PIn basically passes through the RT on the input side and is sent to PE#n. PE determines the destination and transmits the stream along with the branch specification to the RT on the output side. The RT on the output side transmits the stream to appropriate POi according to the branch specification given to it using its data switch.

Fig.6(d) shows the basic form of a point-to-point connection between node units. The connection is basically a full duplex connection with an input port and an output port. Since a node unit has 8 ports, it is possible to connect with maximum of 8 neighbor node units.

With respect to a 2 dimensional network as a whole, Fig.6(e) is a simplified representation of Fig.6(d) with the circle symbolizing a node unit and the line connecting the circle symbolizing the full duplex stream path. Employing this symbol, Fig.6(f) shows a 2 dimensional network. By providing 6 neighbor nodes for each node unit, it can be seen that a highly flexible and uniform connection is realized. Also with the node unit referred to in this paper, the 2 reserve ports, for example, can be used to configure multiple 2 dimensional network in the Z axis, that is a 3 dimensional network becomes possible.

(b) Logical topology of 2 dimensional network and "Hello Packet"

This subsection will discuss how the input stream in the physical network is transmitted to its destination. This basically is how transmission paths of streams between arbitrary node units are defined, how the network's logical topology is described so that each node unit is able to recognize each other. In describing the network's logical topology, the authors have proposed a node table where each node posses one to recognize each other. Furthermore, in order to autonomously configure/reconfigure each node unit's node table, the authors will propose a control stream to configure the logical network, "Hello Packet", and describe its processing algorithm.

In Fig.6(f), upon reception of a stream from one of its input port, each node unit in the network only needs to know to which of the 5 remaining ports it should output to in order for the stream to reach its destination. As an example of how a network is configured, suppose that a new node unit or a new line has been added to the network shown in Fig.6(f). After the power is turned on, the added node unit broadcasts a "Hello Packet" which basically is its node ID to all its physically connected neighbor nodes(2 nodes in this example) in order to notify the network of its presence. The existing node units configure a logical network following the algorithm described below.

(1) With respect to the first arrival of a "Hello Packet", the node unit refers to its node table and if (a) there exists no such node ID as that of the "Hello Packet", or (b)such a node ID exists but is old. (determined by such information as time stamps), the node number of the "Hello Packet" is used to update the node table. Also the input port that the "Hello Packet" came from is used to update the 1st Dest. of the node table. After updating, the "Hello Packet" is copied and sent out to all ports except the port that it came from.

(2) With respect to "Hello Packet" received after the first one, the port that it came from is used to update the 2nd Dest. of the node table and the "Hello Packet" is deleted.

(3) Likewise, the 3rd "Hello Packet" is used to update the 3rd Dest. and the "Hello Packet" is deleted.

(4) For all ensuing "Hello Packet" reaching a node unit, no updating is done and "Hello Packet" is absorbed.

The resulting node table of the example discussed above is shown in Fig.6(g). 1st - 3rd Dest. field contains corresponding port ID for PO0 - PO7. The node ID of neighboring nodes need not be listed in the node table. 1st Dest. serves as the stream destination under normal conditions where 2nd and 3rd Dest. serves as bypass stream destinations.

With the transmission of "Hello Packet" by the newly added node, and with the appropriate processing of that and following packets by all the existing nodes, all existing nodes are able to recognize the newly added node.

For a consistent and complete logical network configuration 2 processings remain.

One is when a node is newly added to a network such as one described above. The newly added node must also posses its own node table. In order for the new node to create a node table, each existing nodes in the network must transmit a "Hello Packet" as described above, beginning with step (1). Although this will merely serve as a confirmation with respect to existing nodes sending "Hello Packet" to other existing nodes, this will enable the newly added node will create a node table.

The other point is not when a new node is added, but when a new line is added, creating a possible shortcut path from one node unit to another. In such a case, it can be said that the logical network configuration has been altered.However, as it is, unless a new node is added, no existing node unit will transmit a "Hello Packet" thus the logical network understood by all the node unit remains unchanged. In this case, one of the node unit should be made manually to transmit a "Hello Packet". By following this step, reconfiguration of the logical network is made possible.

By utilizing the algorithms, addition of a new node to a logical network is dynamically and autonomously configured. Not only that, reconfiguration of a logi-cal network resulting from addition of a new connection which improves the stream transmission path is likewise dynamically and autonomously realized by according to this algorithm. Fig.6(h) is an example to describe the transmission mechanism of the execution stream. In this example, a stream with destination node Z is inputted to node A. According to the algorithm discussed earlier, there exists a node table in node A where Node ID=Z's 1st Dest. is node B. To be more precise, the 1st Dest. would be listed with "go right". Node A would, in accordance with its node table, transmit the stream to its right. Similarly in node B's node table, there is a Node ID=Z where it indicates "go right", and so on to enable the input stream to autonomously traverse the near shortest path to node Z.

(c) Self diagnosis and logical disconnection/restoration

A discussion on various fault modes such as fault detection(self-diagnosis), logical disconnection, substitution method, and restoration method for each element in the network will be touched upon.

Firstly, the discussion will be on method of self-diagnosis within the node unit. The 8 PEs in Fig.6(a) exchange self-diagnostic data among themselves via the circular path to perform a surveillance diagnosis on each other. If, for example, a fault was detected in PE#3, the RT on the input side of the node unit will not transmit stream out of port OD but will instead have its branching conditions altered to have a substitute port from one of the other ports OA - OC or OE - OH. By doing this, the stream input from the PI3 will not be transmitted to PE#3, thus a logical disconnection has been enabled. The stream will be transmitted and processed at one of the other 7 PEs. PE#3 on the other hand, since it has been logically disconnected from the network and receives no input stream, is autonomously de-activated. This conforms to the nature of the data-driven processing principle in which no processing is done until data is supplied to the processor.

Although a diagnosis to a certain degree is possible for the 2 RTs by utilizing the circular paths within the node unit, since diagnosis between unit nodes is in effect a diagnosis of the RT, full diagnosis is left to diagnosis among node units.

A point-to-point self-diagnostic method between 2 node units which is the basic form of connection will next be described. The discussion will be on the inter-connection between PE#i in NU#x and PE#j in NU#y shown in Fig.6(d). By way of periodically exchanging diagnostic data which indicates the status of the line or the node unit on the other side, each node unit is able to detect faults. From the viewpoint of NU#x, if a fault is detected in the connecting destination of PIi/POi whether the fault be in the line or the destination node, transmission of stream via PIi/POi is no longer allowed which logically disconnects the connection with NU#y. The decision of whether NU#x will logically disconnect and its appropriate processing is carried out by PE#0 - 7 within NU#x. Ensuring a substitute path following a logical disconnection will utilize bypass paths. Description of how a bypass is ensured is discussed later.

With regard to restoration of lines between node units, the periodic exchange of diagnostic data continues even while logical disconnection is taking place. When the fault in the line or the opposing node unit continues, the symptom is maintained. If, for example, the faulty line is physically replaced or NU#y is replaced with a working node unit, then exchange of diagnostic data is restored. Upon detection of this, NU#x and NU#y reconnects the logical lines to complete the restoration mode.

(d) Maintaining functional reliability of network

A description of how a bypass path is ensured is discussed. Assuming that the line between node A and node B in Fig.6(h) has been logically disconnected as a result of self-diagnostic method described earlier. In this case, node A refers to either 2nd Dest. or 3rd Dest. field in the node table shown in Fig.6(g) which in this case would indicate "upper right" or "lower right", and transmit the stream accordingly. In this case, when the stream is inputted from 2nd Dest., it must be transmitted to the port pointed to by 3rd Dest. and vice versa. For input stream from other ports, it is desirable to randomly select the port designated by 2nd or 3rd Dest.

By employing this mechanism, when a stream encounters a faulty line or a faulty node unit as it traverses the network, it should be possible to autonomously select and traverse the bypass route.

In order to maintain stable operation of a network, there must be countermeasures for the following 2 phenomena. One is that there must be a way to avoid streams getting lost in a network when a logical disconnection takes place. The other is that streams should be able to avoid congested node units or lines. Although specific measures for both have been thought out, the authors would like to leave that topic for future discussions.

4   Conclusion

This paper reported the state-of-the-art results of CUE project. The paper addressed the requirements, reliability/maintenability, ultra-multi-processing capability and expandability to the network environments that cover not only public network providers but users' networks and systems. To satisfy these requirements, it discussed the technology achievements of data-driven processor with autonomous decentralize-controlled pipeline mechanism. This paper showed the high applicability to the required network environment of data-driven processor with cost-effective and size effective nature.

The characteristics of autonomous decentralized data-driven processor have great potentiality to satisfy the requirements identified in section 2. The reliability and maintenability will be achieved by overload tolerance, fault tolerance, self diagnosis and autonomous rerouting by use of redundancy. The ultra-multi-processing capability will be naturally achieved by the processor with characteristics of linear increase of processing performance to the increase of simultaneous threads. The reconfiguration capability will be realized by expandability. The chip configuration of the processor may solve cost and size problems which are issues for ordinary users.

In the CUE project, we will realize a super-integrated data-driven processor performing around 5 Giga Operations Per Second within a year and will apply the data-driven processor to the protocol processing (lower layer) for the reliable and reconfigurable network systems.

Acknowledgements

Although it is impossible to give credit individually to all those who organized and supported the CUE project, the authors would like to express their sincere appreciation to all the colleagues in the project. The authors would also like to give thanks to R.T.Shichiku for his efficient proof-reading in preparation of this paper. One of authors, H.Nishikawa are very grateful to Prof. Arvind and Prof. Jack B. Dennis of Laboratory for Computer Science, MIT for their fruitful discussions on data-driven architectures.

This research is partially supported by the Scientific Grant of the Japanese Ministry of Education, the Telecommunication Advancement Foundation and International Communications Foundation.

References

[1] G. Nilsson, F. Dupuy, M. Chapman, "An Overview of the Telecommunications Information Networking Architecture", TINA95, Melbourne (Feb. 1995).

[2] E. Kelly, N. Mercouroff, P. Graubmann, "TINA-C DPE Architecture and Tools", TINA95, Melbourne (Feb. 1995).

[3] H. Nishikawa, K. Asada and H. Terada, "A decentralized controlled multi-processor system based on the data-driven scheme" Proc. of the 3rd Int'l Conf. on Distributed Computing System, pp.639–644 (Oct. 1982).

[4] H. Nishikawa, H. Terada, K. Komatsu, S. Yoshida, T. Okamoto, Y. Tsuji, S. Takakura, T. Tokura, Y. Nishikawa, S. Hara, and M. Meichi, "Architecture of a one-chip data-driven processor: Q-p," Proc. of the 16th ICPP, pp.319–326 (Aug. 1987).

[5] H. Nishikawa, H. Terada, S. Komori, K. Shima, T.Okamoto, and S.Miyata, "Architecture of a VLSI-oriented Data-Driven processor: The Q-v1", "Advanced topics in data-flow computing", Chapter 9, pp.247-264 Printice-Hall, Inc. (Jan. 1991).

[6] H. Kanekura, S. Miyata, "An Evaluation of Parallel-Processing in the Dynamic Data-Driven Processor", Tech. report on "Micro Computer Architecture", IPSJ, pp.9–18 (Nov. 1991) [in Japanese]

[7] H. Nishikawa, H. Ishii, and Y. Inoue, "A Stream-Oriented Data-Driven Processor Realizing Hyper-Distributed Systems", Proc. of the 8th IASTED Int'l Conf., pp.47–51 (Oct. 1996).

[8] S. Yoshida, R. T. Shichiku, Y. Matsuura, T. Muramatsu, T. Okamoto, S. Miyata, "Video Signal Processing Oriented Data-Driven Processor", Tech. Report of IEICE, ICD95-158, pp.39–46 (Nov. 1995) [In Japanese].

[9] J. B. Dennis, "Dataflow Schemes", Project MAC, pp.187-216, M.I.T. (July 1972).

[10] Arvind, R. S. Nikhil, "Executing a Program on the MIT Tagged-Token Dataflow Architecture", IEEE, Trans. on Computers, Vol.39, No.3, pp.300-318 (Mar. 1990).

[11] J. R. Gurd, C. Kirkham and I. Watson, "The Manchester Prototype Dataflow Computer", Commun.ACM, Vol.28, No.1, pp.34-52 (Jan. 1985).

Session 1C

Distributed Algorithms

Chair

Paulo Verissimo

Heuristic Token Selection for Total Order Reliable Multicast Communication

Weijia Jia   *Jiannong Cao*   *Xiaohua Jia*

Department of Computer Science
City University of Hong Kong, 83 Tat Chee Avenue, Hong Kong
E-mail: wjia@cs.cityu.edu.hk

Abstract: An efficient multicast protocol is presented by using a heuristic function for passing a virtual token to decide the message total ordering. Unlike existing token-passing based algorithms, there is no physical token passing message in the protocol, instead, the token holder piggybacks token passing onto normal multicast messages. Executing the heuristic function does not incur any communication overhead and only relies on local information of a token holder. For a group of n processes and k multicast messages, the protocol is able to achieve atomicity of one message by average (n–2)/k point-to-point control messages. Dynamic membership is non-blocking, namely, each individual process in the group can take unilateral decisions at each step of the membership algorithms. System-wide consistent group configuration can be obtained in consistent order of normal multicast messages. Performance of the protocol is shown by implementation figures.

1. Introduction

Multicast message ordering and fault-tolerance are major concerns to the design of multicast (group) communication. Many applications, such as distributed transactions of databases, do require these properties. The main requirement for designing such a protocol is that the processes in the group should have a consistent view. That is, they all receive the same set of messages (atomicity) in the same specific order (causal or total ordering). Reaching a consistent view is very difficult, especially in an asynchronous environment. For this purpose, various reliable multicast protocols (e.g., the ISIS protocols developed by Birman et al. [2]) have been proposed in the literature [1, 3, 5, 8-14].

Among those existing approaches, logical token passing ring is a simple and efficient way to design the single group protocols [1-3, 6-8, 13]. We have designed and implemented a logical token ring based group multicast protocol (RMP) [6, 7]. The protocol ensures reliable multicast messages of total ordering. However, there are some drawbacks related with RMP and existing token-passing approaches: The token must be transferred sequentially around the ring in order to guarantee that a message is received by every member. This causes some unnecessary synchronization and delay of messages. Another problem is that the protocol does not scale well when the ring becomes larger.

To overcome the problems, this paper describes a novel totally-ordered, atomic multicast protocol (called HRMP) for single group based on logical token passing-approach. Given a process group $G = \{ p_0, p_1, ..., p_{n-1} \}$, its membership list is known by every process in the group. In contrast to other token-passing protocols, no real ("physical") token-passing message is sent in G, instead, the *token passing* information is carried out onto the normal multicast messages. The token holder (called *token process*) is allowed to multicast totally-ordered messages and decides who will be the next token process by applying a *heuristic function* on its local information (called *weight* array, see next section). Upon reception of an ordered message, receivers are informed the position of the token process. Therefore, all the receivers in the group are able to achieve a consensus to monitor the activity of the token process and expect some totally-ordered messages from it. Any member wishing to multicast messages can send a *request* message to the current token holder, requesting to hold the token. Token information will never get lost as long as the normal multicast messages are received at the receivers. Lost messages can be detected effectively by the gap between an incoming message order and the order expected by a receiver. With this approach, token repositioning is flexible and takes no overhead. Failed stop of the token process can be easily detected by the individual members. In summary, HRMP has the following characteristics:

- Achieving total ordering and atomicity of multicast message efficiently: token process can multicast multiple totally-ordered messages. It entails at most $n-2$ point-to-point control messages to establish delivery atomicity for k multicast messages.
- Enable Scalability: Token transfer to obtain the total order of multicast messages is not affected by the size of the group.

- Efficient fault detection: The location of a failed process can be identified by repositioning the token in the membership list at most once.
- Dynamic group: The cost of recovering from process failures or network partitioning is rather low. *Virtual synchrony* [2] is obtained by arranging the membership change in the same relative order as normal multicast messages.

2. System Environment and Assumptions

Communication models and fault-tolerance: The network (e.g., a local area network) does not guarantee that a site will receive every message. Message loss may be due to site fail, buffer overflow, or unreliable communication media. The following assumptions are made in designing this protocol. The network consists of a set of sites that communicate through channels. Processes residing on the sites can transmit messages in both point-to-point and broadcast (multicast) modes. Communication is asynchronous. That is, a message (either point-to-point or multicast) requires a variable amount of transmission time for completion. A message may be lost but will not be modified during transmission. A process may fail by halting prematurely. Until it halts, it behaves correctly. A process failure does not have malicious actions. Failed process can restart as a completely new process. A network may partition into segments and the processes in one segment may appear to have failed from the viewpoint of the processes in the other segments. *It is assumed that the communication paths between the partitions will be reconnected eventually.*

Process states: Let the membership list be $L=(p_0,...,p_{n-1})$. L can be dynamically reconfigured, resulting in a new *version* and L is said to be operational if its structure contains a majority of the operational processes of G (i.e., $|L| > n/2$). Every process p_i maintains L and following information:
— V_i: The version of L, initially 0, and incremented by 1 after each reform of L.
— Q_i: A buffer for storing the totally-ordered multicast messages received by p_i but not yet delivered to the applications.
— B_i: A buffer for storing the out-of-order multicast messages received at p_i.
— S_i: A global sequence number of messages received in total order at p_i, initial 0.
— $count_i$: An integer used to count the number of messages multicast by p_i each time p_i holds the token;

— *Current_token* records id (address) of the current token process.
— $W_i(0, ..., n-1)$: The weight array, initially all 0. p_i uses W_i in a heuristic function to locate the next token position. $W_i(j)$ (>0) records the time of process p_j's request for holding the token and $W_i(j)=0$ indicates that p_j has no request.
— $A_i(0, ..., n-1)$: The acknowledgment array of p_i, initially all 0. $A_i(j)$ records the message order number received at process p_j's to the knowledge of p_i.
— τ. A time-out setting for preventing situations in which a process has waited too long for a message, due to message lost or failure stop of a sender. A message taking longer than τ time units for transmission between two processes is considered as having been lost (logically) at the receivers.

Message format: Suppose p_i is the sender of message m, m consists of the components in tuple (*version, sender, type, info, order, next, time, atom*) where
— *version* is $V_i(L)$ at the time when m is sent.
— *sender* = p_i is the sender id.
— *info* is the data value of m.
— *order* is sequence number of m and $m.order = S_i$. If $m.type$ = MCAST, then *order* is the total order of m, otherwise, *order* is the maximum order of messages received by p_i and can be taken as p_i's acknowledgment.
— *next* records the next token process id (or IP-address).
— *atom* is the minimum value of A_i, recording the order of messages received by all processes on L in the viewpoint of *sender* p_i, i.e., it is "all-received-up-to" parameter.
— *time* is an integer used to timestamp the *request* packet and $request.time := S_i$ when a *request* is initiated and *request.time* will not be changed as along as it exists.
— *type* is a message type from the set: {*MCAST, reques, ping, ack, REFORM, nack, request_join, NML*} where *MCAST* denotes multicast message with ordering; *request* requests to hold the token; *ping* is a probe message to ping a process when the *sender* suspects any failure; *ack* is an acknowledgment message; *REFORM* indicates a multicast reformation intention; *nack* is a *negative ack*; *request_join* is used by a new member for joining G and *NML* is multicast message containing new membership list. Note that *MCAST, REFORM* and *NML* are multicast typed messages and have components of (version, sender, type, order, next,

atom, info) and others are point-to-point messages and with form of (version, sender, type, order, time).

3. Algorithms of Ordered and Atomic Multicast

Following rules apply to process p_i while $i \in \{0,..., n-1\}$ during normal operation. Assume that process p_j holds the token and multicasts message m:

- *Receive Multicast Message*: p_i, upon reception of m ($m.type = MCAST$) from p_j, when $m.order = S_i + 1$, it indicates that m is received in total-order at p_i, thus m is buffered in Q_i. p_i increases S_i by 1 and assigns $W_i(j):= 0$ (to indicate that p_j multicast its messages). Whereas $m.order > S_i+1$ reflects that there exist some messages m' such that $m'.order = S_i+1$ that has not been received at p_i. Therefore, m is "out-of-order" and buffered in B_i temporarily. At the same time, p_i requests one retransmission of m'. For $m.next = p_k$, consider three cases: (1) $k = i$: p_i is appointed as the next token holder by p_j, and it is going to multicast its messages (see rule SMM below). (2) $k \neq j \neq i$: p_i expects the next totally-ordered multicast message from p_k. (3) $k = j$: p_j continues to hold the token and multicasts its messages. In case (3), if p_i wants to hold the token, it sends request message as described before.

- *Send Request*: When p_i wishes to multicast, it assigns $W_i(i) := 1$. If p_i did not sent any request before, i.e., $request=\emptyset$, it constructs the *request* packet by $request.time:= request.order := S_i$ and sends it to p_j. If the *request* has been sent before (i.e., $request \neq \emptyset$), p_i adjusts $request.order:= S_i$ and sends it to p_j, keeping $request.time$ unchanged.

- *Receive Request Token*: Once receiving the *request* message from p_i, p_j assigns $W_j(i):= request.time$.

- *Send Multicast Message*: If p_i is appointed to hold the token via reception of a message, i.e., $p_i=m.next$, it assigns $S_i:=S_i+1$, $request:=\emptyset$ and initiates a message m' such that $m.type=MCAST$ and $m.order:=S_i$. If p_i has message from its own application then $m.info:= application\_data$, otherwise, $m.info:= NULL$. It is p_i's responsibility to decide the next token holder id and attaches it as $m.next$. A heuristic function *decideNext* is applied by each of the token holder with weight array W as

parameter. The function returns id of the next token process:

decideNext(W_i) -- where p_i holds the current token;
begin
1: **if** ($W_i(i)>0$) **then** return(p_i);
2: **else if** ($\exists j: j \in \{0,..., n-1\}: W_i(j) >0) \wedge (i \neq j)$ **then**
   return (p_k) such that
   ($W_i(k) = min \{W_i(j)\}) \wedge (W_i(j) > 0)$
   $\wedge (k = min(j - 1) \, mod \, n)$);
3: **else** return(p_{i+1});
end.

The function *decideNext* indicates that (1) p_i has the priority to hold the token continually and multicast its own messages. (2) p_i releases the token to some process that requests to hold the token. To p_i's knowledge, it will choose process p_k which has requested the token "*earliest*" (i.e., $W_i(k)>0$ and $W_i(k) = min\{W_i(j)\}$) and is "*most close*" to p_i (i.e. $k = min(j - i) \, mod \, n$ where $n>j \geq 0$ and $k \neq i$). (3) if there is no process ever sends their request (to p_i's knowledge), it passes this token to its neighbor p_{i+1} in L. Figure 1 illustrates an example:

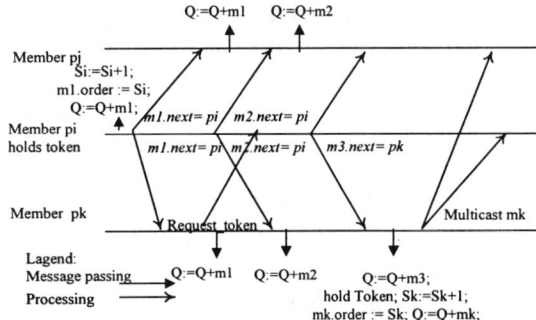

Fig. 1 Regular operation of the protocol

Achieving atomicity. Atomicity of multicast message m is defined as when a process in G is going to deliver m to its applications, it knows that all other processes in G have received the same m and will deliver it. A general approach to achieving atomicity of m may require that it be buffered until every process in G has acknowledged it. In fact, algorithm of Ordered Multicast/Receive is able to achieve atomicity of m eventually by buffering m in Q_i until every process holding the token and multicasting their messages (which implicitly acknowledged m). If some processes have many messages to multicast without releasing the token, acknowledgment of m cannot be achieved and m may have to be buffered indefinitely.

As stated before that every multicast message is attached with the next token holder indicator, (i.e., $m.next$). Upon reception of a totally-ordered multicast message m from the previous token process p_i, all receivers know that the next multicast message will come from $p_j=m.next$. In order to fast achieve atomicity of m, a scheme called AA (Advance-Acknowledgment) is devised here that uses $atom$ parameter. AA works as follows: Except sender of m (here p_i) and p_j, all receivers send their acks about the reception of m to p_j by attaching $ack.order=m.order$, so called "advance ack" because p_j will transmit the next totally-ordered message m'. Before m' is multicast, p_j collects the acks and records them in ack array A_j. Note that the minimum value of $A_j[0, ..., n-1]$ represents "all-received-up-to" message orders, i.e., the lowest-water mark of the message order received by all the processes in G to the knowledge of p_j. Therefore, p_j chooses $min\{A_j[0, ..., n-1]\}$ as $m'.atom$ and multicasts it. Upon reception of m', the receivers picking up $m'.atom$ are able to detect that the messages in their queues Q with orders less than or equal to $atom$ have been received by all the processes. Therefore, the messages can be delivered to the application processes and taken off Q. In general, for any two successive multicast messages m and m', if $m.atom < m'.atom$, any uncommitted message msg in Q such that $m.atom \leq msg.order < m'.atom$ can be committed atomically. One of the important properties of the $atom$ parameter is that every process, on receiving a multicast message, can capture, at least, the current global view of G without additional communication overhead.

The AA approach described above requires a process to send ack for each of the multicast messages. This may introduce many ack packets in the network. To save the ack transmission, instead of sending ack every time to the token holder, receivers send ack only once if token process changes. For example, assume that p_i receives m from p_j. p_i sends ack to p_k only when p_i learns that $m.next \neq p_j$. Thus, atomicity of k messages multicast by p_j can be achieved by only $n - 2$ ack packets (control messages). Assume that p_j holds the token, the following rules are applied to process p_i where $i = \{0, ..., n - 1\}$:

Atom Receive: Upon reception of a totally-ordered message m from p_j, p_i records $m.order$ by $A_i[j]:=max\{m.order, A_i[j]\}$. If $m.info \neq NULL$ and $Current\_token \neq m.next$ (token has been transferred), p_i sends ack with $ack.s = S_j = m.order$ to p_k where $p_k = m.next$. If p_i already sent $request$ to p_k, it does not send ack because $request$ packet can be taken as the ack message.

Atom Multicast: Once p_i holds the token, it first makes sure that it has received all ordered messages and assigns $A_i[i]:=m.order$. Before multicasting its message m, p_i collects ack messages from any of the process p_k. Once such a message is received at p_i, it assigns $A_i[k]:= max\{ack.order, A_i[k]\}$, then m is multicast with $m.atom := min\{A_i[0], ..., A_i[n-1]\}$.

Note that although p_i is required to collect the ack messages on holding the token initially, AA algorithm will not block itself indefinitely for waiting for acks without multicasting its messages. Once any ack cannot be received at some specific time interval, p_i still multicasts ordered message. Collection the delayed acks for the $atom$ parameter can be achieved subsequently. If an ack never received at p_i, fault detection mechanism come into act.

Solving problems of fairness and bounded buffer: Two problems associate with the algorithms described above as *fairness* and *limited buffers*. By *fairness*, we mean that if any process in the group wishes to multicast, it should be able to do so eventually. In the algorithms described above, such a process may be blocked indefinitely because some processes may multicast many messages without releasing the token. By *limited buffer*, we mean that buffers Q and B may overflow if many messages were buffered. For example, some processes transmit many bulky messages. In order to solve these problems, we impose a bound π to restrict the number of messages that a process can multicast on holding the token. The process keeps counting the messages it sent and releases the token when the bound is reached. This strategy is reflected in function *decideNext*:

decideNext(W_i) -- where p_i holds the current token;
begin
1:**if** $(W_i(i)>0) \wedge(count_i < \pi)$ **then**
$count_i : = count_i +1$; return(p_i); -- if p_i wishes to multicast continually.
2: **else** $count_i := 0$;
        if $(\exists j: j\in\{0,..., n-1\}: W_i(j) >0) \wedge(i \neq j)$
        then return (p_k) such that $(W_i(k) = min\{W_i(j)\}) \wedge (W_i(j) > 0) \wedge (k = min(j-i) \bmod n)$;
3:        **else** return(p_{i+1});
end.

where $count_i$ is maintained by function and becomes 0 whenever p_i release the token to another process.

4. Algorithms Supporting Fault Tolerance and Dynamic Group

We discuss the fault tolerance in the context of a majority of operational processes. In HRMP, timeout is used by all the members to monitor the token process and detect its possible failure. Assume that token process p_i crashes. The failure can be detected by the operational process p_j since it is expecting ordered multicast message(s) from p_i. As defined in Section 2, timer τ is used to monitor a message transmission from p_i. Considering any message retransmission, additional time interval $(l-1)\tau$ is added to the timeout setting. Therefore, $T=l\tau$ is employed by each of the processes to monitor p_i. On expiring the timer, a (logical) failure of p_i is suspected by p_j and it sends a *ping* message to p_i, setting another timer 2τ to monitor p_i. If p_i is still alive, it responds immediately to p_j. Otherwise, a failure of p_i is detected and p_j enters into a reformation algorithm (see below) to reconfigure G.

Detecting failure of a non-token process can be transformed into the detection of token process. Recall that in order to achieve message atomicity, token process p_i will collect *acks/requests* from all receivers. Suppose *ack* (or *request*) expected from p_j is never received, p_i can realize this by checking if $A_i[j]$ is the minimum among all $A_i[0, ..., n-1]$ and if $(S_i - A_i[j]) > u$ where u ($u \leq \pi$) is the number of multicast messages sent by a token process prior to p_i. The conditions indicate that p_i never receives any *ack* from p_j since the last u ordered messages were sent. The next step is that p_i dispatches p_j as the next token process and brings all processes to moinitor p_j unanimously. This is done by the updated heuristic function *decideNext* below:

decideNext(W_i) -- where p_i holds the current token;
begin
1: **if** $(W_i(i)>0) \wedge (count_i < \pi)$ **then**
   $count_i := count_i +1$;   return(p_i); -- if p_i wishes to multicast continually.
   else $count_i := 0$;
2: **if** $(A_i[j]=min\{A_i[0, ..., n-1]\}) \wedge (S_i - A_i[j] > u)$ **then** return(p_j);
3: **if** $(\exists j: j \in \{0,..., n-1\}): W_i(j) >0) \wedge (i \neq j)$
   then return (p_k) such that $(W_i(k) = min\{W_i(j)\}) \wedge (W_i(j)> 0) \wedge (k=min(j-i) \bmod n)$;
4: **else** return(p_{i+1});
end.

In case of several $A_i[j]$, $A_i[k]$, $A_i[l]$,..., are minimal, p_i may choose one of p_j, p_k, p_l ,..., randomly as the next token process.

4.1 Coordinator-Subordinator algorithm

On detecting a failure of p_j, the operational processes will enter a reformation algorithm and eliminate p_j out of L. A group coordinator must be chosen to carry out the following tasks:

- Acquire an agreement from all the operational processes to elect it as the group coordinator.
- Notify all the operational processes in the group to execute a reformation algorithm.
- Retransmit any lost messages and ensure that all the operational processes receive the same set of messages.
- Command the operational processes to deliver the same set of messages in total order.
- Construct a new membership list L', containing all the operational processes and guaranteeing L' to be received, installed and delivered at the operational processes in the same relative order with other totally-ordered messages.

Recall that in the previous subsection, $l\tau$ is set as the timeout for receivers to expect a message from the token process. On expiring the timer, a process p_i detecting a failure of token process p_j enters procedure *Coordinator*, trying to take the coordinator role by multicasting a *reform* message. The other operational processes, upon reception of the *reform* message, if admit p_j as the coordinator will enter a procedure *Subordinator*. *Coordinator-Subordinator* runs in two phases: In the first phase, the coordinator multicasts a *reform* message and invites the other operational processes to join a new membership list L'. On receipt of such a *reform* message, a process that wishes to join L' responds to the coordinator. Lost messages can be retransmitted during that phase. When the coordinator has finally invited a majority of the operational processes within or by the timer T, it increments $V(L)$ as version of L', multicasts and installs L' in the second phase. Note that in *Coordinator*, timer $D\tau$ (D is a constant), is used as the maximum time interval for the coordinator to collect all *acks* from the operational processes and to construct and disseminate L' over G. Similarly, $T'=E\tau$ where $E = D+1$ in *Subordinator* is also taken as the timer for the rest processes to receive L' from the coordinator. New membership change is buffered in Q and will be delivered atomically by the members in L' in normal operation. Race case may occur when several processes timeout and multicast *reform* messages simultaneously. Methods of selecting a unique coordinator has been imposed as Criteria (Cr1-

Cr4) below are applied in a election function $Elect(p_i, p_j)$ which elects either p_i or p_j as the coordinator candidate. Denoting "v !" as "vote in favor", the criteria has been denoted as

$(Cr1)$ $V(L_i) < V(L_j) \Rightarrow p_i \, v! \, p_j;$
$(Cr2)(V(L_i) = V(L_j)) \wedge (S_i < S_j) \Rightarrow p_i \, v! p_j$;
$(Cr3)$ $(V(L_i) = V(L_j)) \wedge (S_i = S_j) \wedge (id(p_i) < id(p_j))$
$\Rightarrow p_i \, v! p_j;$
$(Cr4)$ $(p_i \, v! \, p_k) \wedge (p_k \, v! \, p_j) \Rightarrow p_i \, v! \, p_j.$

For details discussion of the criteria, interested reader is referred to Reference [7].

4.2 Dynamic membership

We consider broadcast and point-to-point networks. In the broadcast network, a multicast packet is broadcast over the entire network and only the processes in a group executing the protocol are able to receive the multicast messages. Assume process p is a new process that joins G. By procedures *Coordinator-Subordinator*, L can be reformed by steps:

1. A new process p executing the protocol opens a port to detect any multicast messages if it wishes to participate G.
2. Once p receives m from G, it sends *request_join* packet to $p_i = m.next$.
3. Upon reception of the request, p_i enters *Coordinator*, multicasts *REFORM* message to G.
4. All process, including p, upon receipt of *REFORM* from p_i enter *Subordinator*.

The algorithms *Coordinator-Subordinator* have to go under several round of message exchange in two phases. In order to quickly incorporate p into the group. We propose a new algorithm called *non-blocking membership algorithm*. By *non-blocking*, we mean that the processes in G make decision individually to change their L to include p in G. Agreement (consensus) of new L is gradually recognized by all operational members. Assume that p_i holds the token and receives *request_join* from p, the protocol can be described below:

- p_i assigns p as the $(n+1)$th member in L. L becomes L' and $V(L'):= V(L)+1$. At the same time, p_i constructs m with total order $S_i := S_i +1$ such that $m.order:= S_i$; $m.type:=NML$ and $m.info:=L'$. Then p_i assigns p as the next token holder by $m.next:= p$ and multicasts m to all members in L'

- Upon reception m, any p_i, $p_i \in L'$ and $p_j \neq p$, buffers m in Q_j in the total order with normal ordered multicast messages and sends ack to p because $p=m.next$.
- On the part of p, it does the following steps: 1. Upon reception m while $m.type = NML$ and $m.next=id(p)$, it records L' in Q, initiates its sequence number by $S_p:= m.order$ and expects *ack* from any p_j. 2. Assume that no message is lost, reception of the *ack* message from all the members shows an agreement of accepting p as new member by all processes in L. p multicasts a message m', attaching $m'.atom$ to achieve the atomicity of L'.
- On reception m', all members in L' will install L' as L and deliver L' to their applications.

This protocol only uses one phase of message transmission to establish new L. In case of point-to-point network, a member is required to service as the group interface and acts as an entrance to the group. Any new process p intending to join the group sends its application *request_join* to that process. The process then delegates the applicant and forwards such "join" *request* to the current token holder. The rest steps are the same as described above. An example of the protocol is shown in Figure 2.

Dealing with failures in the membership change: During the execution of dynamic membership algorithm, message may get lost, processes may fail and network may partition. We consider the cases in turn:

- *Lost messages.* All messages (except the *request_join* message of p), if lost, will be detected by the receivers. Retransmission of the messages are carried out by either of the processes. If *request_join* is lost, p will resend it to the current token holder.
- *Process fail-stop.* Assume p is the applicant that wants to join G, consider that (1) Token process fails before it transmits any message. (2) Member p_j fails before sending *ack* to p. p holds the token without receiving *ack* from p_j. By function *decideNext*, p will pass the next token to p_j to detect its failure.(3) On holding the token, p fails before its multicast, all old members will monitor its activity by setting timer. All the cases above will be handled by *Coordinator-Subordinator* algorithm described before.
- *Handling Network Partitioning*: Without losing generality, assume G is partitioned into G_1 and G_2. Consider two cases: (A) G_1 contains the majority of the operational processes. Thus, $|G_1|>n/2$ and $|G_2| <$

$n/2$. To tolerant the failure, a coordinator $p_i \in |G_1|$ will be elected eventually. Recall that a majority of the processes are operational and p_i will reconfigure L as L_1 that proceeds normal operations in G_1 eventually. Since G_2 does not contain the majority of the processes, operational processes in G_2 will try repeatedly to elect a coordinator and abort every time because no majority can be formed. Consequently, no two groups can run in parallel and nor inconsistency of multicast messages can exist concurrently. Since L_1 is operational, for any $p_i \in G_1$ and $p_j \in G_2$, there exists $V(L_i) > V(L_j)$. From the fault models described in Section 2, a communication path will be reestablished between G_1 and G_2 eventually. Operational processes in G_2 start as new processes in terms of G_1 and join G_1. A consistent new membership list L' will be achieved eventually from L_1. (B) There is no subgroup with the majority of the processes; all the processes in the subgroups are blocked. Survived processes in G_1 and G_2 will periodically try to elect a coordinator by multicast *reform* messages. Since the communication route will be established eventually between G_1 and G_2, as illustrated before, a new operational membership list L' will be established eventually.

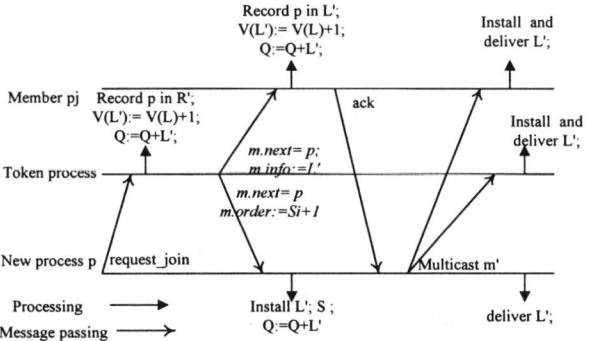

Fig. 2. Example of dynamic membership protocol

5. *Implementation and Performance*

HRMP has been implemented in an 10 Mbit/s Ethernet over a group of workstations. The most recent measurements were taken with message sizes from null packet to 4 Kbytes (excluding the message header of 64 bytes). The experiments have shown that the protocol has achieved excellent performance (throughput about of 1,500 total ordering message/sec for Null packets in a group of 5 processes/sites) which is faster than any other total-order protocols existing in the literature. The experiments have been conducted on a cluster

workstations of Sun 4C/Sun 4M and DEC stations (MC68020-30s, 16.67-20 MHz) connected to a 10 Mbit/s Ethernet by AMD (Advanced Micro Devices) Lance chip interface. The sites used in the experiments were able to buffer 32 Ethernet packets before the Lance overflowed and dropped packets. The most measurements were taken with message sizes from null packet to 4 Kbytes (exclusive the protocol message header 64 bytes). The experiments showed that HRMP has achieved excellence performance. The first experiment as depicted in Figure 3 in which various senders continually multicast Null, 1 Kbytes and 4 Kbytes packet messages. We have compared our experiment results with Ameoba [8] protocols and our previous RMP protocol [7]. The figure indicates the committed message number received by application programs with total order, showing the maximum throughput of the protocol. For a group of five processes, it can achieve the peak throughput more than 1,400 ordered messages which is higher than any prior protocols.

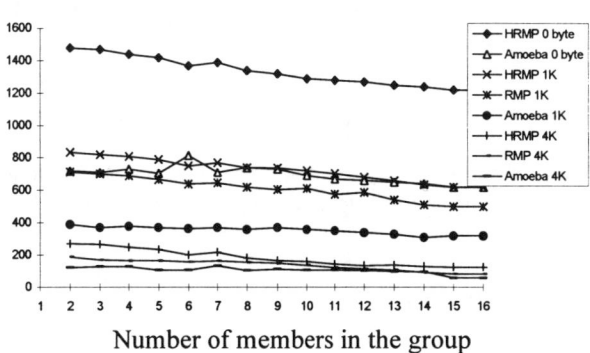

Number of members in the group

Fig. 3. Throughput of total ordered multicast messages per second. HRMP is compared with RMP and Amoeba. The sender number is the same as the group size.

In HRMP, a process holding the token can continually multicast many messages. This is unlike RMP that one process can send only one message on holding the token. Another difference from existing token solutions is that the large group size will cause no much delay on token transfer and the throughput of HRMP is constant. When multiple senders want to multicast their messages, some senders may have to wait until they hold the token. This may block the individual senders but will not affect the overall performance of the protocol. Total ordering of multicast messages can be achieved quickly among the group. In our experiments, the senders each sends 1,000 messages, for a group of three processes, the measured delay for the NULL message is about 0.7 msecs. Compared with Amoeba protocol [8] on the same architecture, their delay is 2.7 msec. For a group of 16 members, the delay is about 2 msecs. This delay includes that a sender requests to hold the token. When the message size increases to 1 Kbytes, the average delays are 1.9 msecs for a group of member 3 and 3.1 msecs for a group of members 16 (we have run out our testing workstations). Assume that there is no message lost, the multicast message carrying the token pass to the next token holder causes no additional transmission cost. To achieve atomicity of k messages, $n-2$ control messages are required which cause no much delay because the receivers can transmit such *ack* messages concurrently with the token process multicast. HRMP has achieved a good balance between the extra packets and the delay as shown in Figure 4.

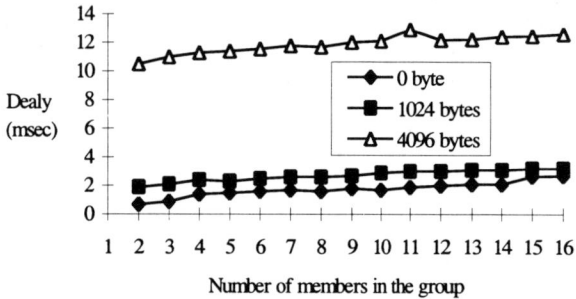

Fig. 4. Message Delay. The senders number is the same as the size of the group and each sender sends 1,000 messages.

6. Conclusions

We have proposed a novel, heuristic logical token passing multicast protocol, which is designed for a single process group in the asynchronous systems. By using the virtual token and heuristic method, the characteristics of the protocol are its efficiency of achieving message multicast, atomicity and member mointoring. The protocol is "smart" because the current token process is able to decide the next token process by piggyingback the token passing information to its normal multicast message. Without token loss problem as associated with other token-passing algorithms, it can detect and recover from communication faults, process crashes and network partitioning. With respect to message atomicity, it has minimized the communication costs while incurring a relative short delay. Up to $n-2$ point-to-point control (*ack*) messages are required to establish atomicity of k totally-ordered messages.

References

[1] Y. Amir, L.E. Moser, P.M. Melliar-Smith, V. Agrawala and P. Ciarfella. The Totem single-ring ordering and membership protocol. *ACM Trans. on Computer Systems*, 13(4), November 1995, pp.311-342.

[2] K.P. Birman and R. van Renesse, "Reliable distributed computing with the Isis toolkit", *IEEE Computer Society Press*, 1994.

[3] J.M. Chang and N.F. Maxemchuk, "Reliable broadcast protocols", *ACM Trans. on Computer Systems*, 2(3), Aug. 1984, pp. 251-273.

[4] M. J. Fischer, N. A. Lynch and M. S. Paterson, "Impossibility of distributed consensus with one faulty process", *J. ACM*, vol. 32, no. 2, April 1985, pp. 374-382.

[5] H. Garcia-Molina and A. Spauster, "Ordered and reliable multicast communication", *ACM Trans. on Computer Systems*, 9(3), Aug. 1991, pp.242-271.

[6] W. Jia, J. Kaiser and E. Nett, "An efficient and reliable group multicast protocol", *Proc. 2nd IEEE Symp. on Autonomous Decentralized Systems*, April 25-27, 1995, Phoenix, USA, pp. 127-133.

[7] W. Jia, J. Kaiser and E. Nett, "A fault-tolerant efficient group multicast communication", *IEEE Micro*, April 1996, pp. 59-67.

[8] M. Kaasshoek and A.S. Tanenbaum, "An evalution of the AMOEBA Group communication system", *Proc. 16th Int. Conf. on Distrib. Syst.*, Hong Kong, May, 1996, 436-447.

[9] S. Luan and V.D. Gligor, "A fault-tolerant protocol for atomic broadca", *IEEE Trans. on Parallel and Distributed Syst.*, vol. 1, no. 3, July, 1990, pp. 271-285.

[10] P.M. Melliar-Smith, L.E. Moser and V. Agrawala, "Broadcast protocol for distributed systems", *IEEE Trans. on Parallel and Distributed Syst.*, vol.1, no.1, Jan. 1990, pp. 17-25.

[11] S. Navaratnam, S. Chanson and G. Neufeld, "Reliable group communication in distributed systems", *Proc. 8th Int. Conf. on Distributed Systems*, San Jose, CA, June 1988, pp.439-446.

[12] L.L. Peterson, N. Buchholz and R. Schlichting, "Preserving and using context information in interprocess communication", *ACM Trans. on Comput. Systems*, vol. 7, no. 3, Aug. 1989, pp. 217-246.

[13] B.Rajagopalan and P.K.McKinley, "A token-based protocol for reliable, ordered multicast communication", *Proc. 8th IEEE Symp. on Reliable Distributed Systems*, Seattle, Oct.,1989, pp. 84-93.

[14] P. Versimo, L. Rodrigues and M. Baptista, AMp: A highly parallel atomic multicast protocol, *ACM SIGCOMM Symposium* , 1989, pp.83-93.

An Autonomous Decentralized Scheduling Algorithm for a Job Shop Process with a Multi-Function Machine in Parallel

Hitoshi Iima, Ryoichi Kudo, Nobuo Sannomiya   and   Yasunori Kobayashi

Kyoto Institute of Technology
Matsugasaki, Sakyo-ku, Kyoto 606, Japan
{iima, kudou2r, sanmiya}@si.dj.kit.ac.jp

Fujifacom Corporation
Fuji-machi, Hino, Tokyo 191, Japan
hhg03471@niftyserve.or.jp

Abstract

This paper deals with a job shop scheduling problem with a multi-function machine. In this problem, the multi-function machine can process all operations, and each operation is processed by either a single-function machine or the multi-function machine. This problem has two objective functions. One is to minimize the sum of the tardiness of each job, and the other is to maximize the working time of the multi-function machine because of the operating cost of machines. An autonomous decentralized scheduling algorithm is proposed to obtain a compromise solution of this problem. In this algorithm, a number of decision makers are called subsystems which cooperate with one another in order to attain the goal of the overall system. In our algorithm, all jobs and the multi-function machine are defined as the subsystems because their objective functions are competitive. They determine the scheduling plan on the basis of their cooperation and their own objective functions. The effectiveness of the algorithm is investigated by examining numerical results.

1 Introduction

For controlling a system, a hierarchical structure is needed to supervise and regulate all parts or all subsystems of the system. All information must be concentrated and analyzed in order to produce commands needed for the subsystems working under the supervision of the entire system. In the case of a large-scale system, centralization and concentrated management lead to many serious problems such as deterioration of system reliability and flexibility. Therefore, the idea of autonomous decentralized systems is offered as an alternative [1]. The autonomous decentralized system is a system whose functional order is generated only by cooperative interactions among its subsystems. In other words, the system does not have any supervisor for the entire system. Instead, each subsystem has the autonomy to control each part of the system.

This paper deals with a job shop scheduling problem with a multi-function machine [2]. In this problem, the multi-function machine can process all operations. The job shop process consists of several single-function machines, each of which can process only its own operation. Then each operation of any job is processed by either the single-function machine or the multi-function machine. This system has two objective functions. One is to minimize the sum of the tardiness of each job. The other is to maximize the working time of the multi-function machine because the operating cost for the multi-function machine is less than that for the single-function machine.

An autonomous decentralized scheduling algorithm is proposed to solve the multiobjective problem. The algorithm is applied to solving a scheduling problem for a metal mold assembly line. The effectiveness of the algorithm is investigated by examining numerical results.

2 Basic Assumption

In order to construct an autonomous decentralized scheduling algorithm (ADSA), we assume the following basic conditions for our system :

(A) There is no decision maker who can construct the global optimal scheduling plan of the overall system. However, there are several decision makers who can propose a locally optimal plan individually. Such a decision maker is called a subsystem.

(B) Each subsystem has its own objective function.

(C) All subsystems intend to cooperate with one another in order to attain the goal of the overall system. For this purpose, the information exchanges are performed among subsystems. The blackboard is used as the medium of exchange of information.

(D) The decision of each subsystem is made either individually by an exchange of opinions or unanimously with the participation of all subsystems.

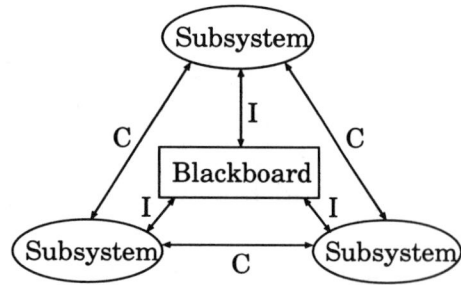

Fig.1 Autonomous decentralized system
C : Cooperation
I : Information exchange

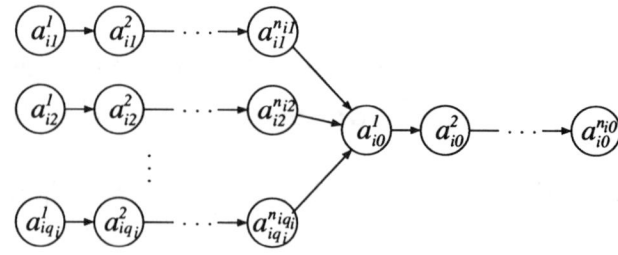

Fig.2 The precedence relation among the operations.

From the above conditions, we construct an autonomous decentralized system which consists of several subsystems and a blackboard, as shown in Fig.1. In this procedure, the problem is to define the subsystems and then to design ADSA in such a way that the conditions (C) and (D) are realized. This algorithm has to be constructed individually for each case study. In our earlier paper [3], we constructed an ADSA for a parallel machine problem. In this paper, we propose a different kind of ADSA for a job shop problem.

3   Problem Statement

Our metal mold assembly process is of job shop type and consists of K single-function machines m_k ($k = 1, 2, \cdots, K$) for processing jobs. In parallel with this process, a new multi-function machine (called m_0) is introduced for improving the production. The machine m_0 can deal with any job and then may be substituted for any single-function machine. In what follows, a job means producing a product.

In this assembly process, I products a_{i0} ($i = 1, 2, \cdots, I$) are produced. The product a_{i0} is made by assembling q_i parts a_{ij} ($j = 1, 2, \cdots, q_i$). Each a_{ij} ($j = 0, 1, \cdots, q_i$) is produced by applying the operations $a_{ij}^1, a_{ij}^2, \cdots, a_{ij}^{n_{ij}}$ in this order. Namely, each a_{ij}^ℓ ($\ell = 1, 2, \cdots, n_{ij} - 1$) must be completed before the start of $a_{ij}^{\ell+1}$. Moreover, $a_{ij}^{n_{ij}}$ for all $j \in \{1, 2, \cdots, q_i\}$ must be completed before a_{i0}^1. Fig.2 shows the precedence relation among the operations.

The operation a_{ij}^ℓ is processed by the machine M_{ij}^ℓ which is either the single-function machine $m_{Z_{ij}^\ell}$ ($Z_{ij}^\ell \in \{1, 2, \cdots, K\}$) or the multi-function machine m_0. The processing time for a_{ij}^ℓ is given by p_{ij}^ℓ, which is common to $m_{Z_{ij}^\ell}$ and m_0. The release date, on which a_{ij} becomes available for processing, is defined as r_{ij}. The due date for a_i is defined as d_i.

The problem is to assign each operation a_{ij}^ℓ to the machine M_{ij}^ℓ and to decide the processing order X_{ij}^ℓ and the start time S_{ij}^ℓ of the operation a_{ij}^ℓ in such a way that the sum T of the tardiness of each product should be minimized and also the processing time P for the multi-function machine should be maximized. We define the total number of parts and the total number of operations respectively as

$$J = \sum_{i=1}^{I} q_i, \quad L = \sum_{i=1}^{I} \sum_{j=0}^{q_i} n_{ij} \qquad (1)$$

Then the order of the operation a_{ij}^ℓ is given by $X_{ij}^\ell \in \{1, 2, \cdots, L\}$.

The finish time E_{ij}^ℓ of the operation a_{ij}^ℓ is

$$E_{ij}^\ell = S_{ij}^\ell + p_{ij}^\ell \qquad (2)$$

Then, the tardiness T_i of the product a_{i0} is given by

$$T_i = \max\{0, \ E_{i0}^{n_{i0}} - d_i\} \qquad (3)$$

The objective functions of the problem are defined as

$$f_1 = \sum_{i=1}^{I} T_i \qquad (4)$$

$$f_2 = \sum_{a_{ij}^\ell \in A} p_{ij}^\ell \qquad (5)$$

where A is the set of operations processed by m_0. The decision variables are $M_{ij}^\ell \in \{Z_{ij}^\ell, 0\}$, $X_{ij}^\ell \in \{1, 2, \cdots, L\}$ and S_{ij}^ℓ for $i = 1, 2, \cdots, I$, $j = 0, 1, \cdots, q_i$ and $\ell = 1, 2, \cdots, n_{ij}$.

4   Algorithm

We have a multiobjective optimization problem. Since the objective functions (4) and (5) are mutually competitive, we have to obtain a compromise solution for the overall system. In such a situation an autonomous decentralized system is considered to be a good and practical structure of decision making. The subsystems should be defined by decision making units which have competitive objectives.

4.1 Definition of subsystem

The subsystems in this autonomous decentralized system are defined by the multi-function machine m_0, I products a_{i0} and J parts a_{ij} ($i = 1, 2, \cdots, I$, $j = 1, 2, \cdots, q_i$). According to the condition (B) in Section 2, the objective function of each subsystem is given by Subsystem m_0 :

$$\max \quad f_0 = \sum_{a_{ij}^\ell \in A} p_{ij}^\ell \qquad (6)$$

Subsystem a_{i0} :

$$\min \quad f_{i0} = \max\{0, \ E_{i0}^{n_{i0}} - d_i\} \qquad (7)$$

Subsystem a_{ij} :

$$\min \quad f_{ij} = \max\{0, \ E_{ij}^{n_{ij}} - d_i - \sum_{\ell=1}^{n_{i0}} p_{i0}^\ell\} \qquad (8)$$

$$\text{for} \quad i = 1, 2, \cdots, I$$
$$j = 1, 2, \cdots, q_i$$

We have $(I+J+1)$ subsystems in this problem. The subsystem m_0 has information such as the set A and the objective value f_0. On the other hand, the subsystem a_{ij} has information such as M_{ij}^ℓ, X_{ij}^ℓ, S_{ij}^ℓ ($\ell = 1, 2, \cdots, n_{ij}$) and the objective value f_{ij}.

4.2 Function of blackboard

The blackboard shows the result of the decision of each subsystem for reference by all subsystems. In this problem, the following information is shown on the blackboard.

$$X_{ij}^\ell = b, \ M_{ij}^\ell = k \qquad (9)$$

$$\text{for} \quad i = 1, 2, \cdots, I$$
$$j = 0, 1, \cdots, q_i$$
$$\ell = 1, 2, \cdots, n_{ij}$$
$$b = 1, 2, \cdots, L$$
$$k = 0, 1, \cdots, K$$

In order for the subsystem a_{ij} to determine S_{ij}^ℓ ($\ell = 1, 2, \cdots, n_{ij}$), the blackboard gives the following instructions to the subsystem.

Step 1. Set $b \leftarrow 1$.

Step 2. The blackboard informs the subsystem a_{ij} of the operation a_{ij}^ℓ such that $X_{ij}^\ell = b$.

Step 3. From the blackboard, the subsystem a_{ij} finds the machine number k such that $M_{ij}^\ell = k$.

Step 4. If a_{ij}^ℓ is the earliest operation assigned to m_k, set $C \leftarrow 0$. If not, let C be the completion time of the latest operation assigned to m_k. In this case, the subsystem a_{ij} knows the value of C from the subsystem which the latest operation belongs to.

Step 5. The subsystem a_{ij} determines S_{ij}^ℓ as follows.

$$S_{ij}^\ell = \max\{C, E_{ij}^{\ell-1}, r_{ij}\} \qquad (10)$$

$$(\text{for} \quad \ell = 1, 2, \cdots, n_{ij})$$

where

$$E_{ij}^0 = \begin{cases} \max \ (E_{i1}^{n_{i1}}, E_{i2}^{n_{i2}}, \cdots, E_{iq_i}^{n_{iq_i}}) \\ \hspace{3em} (\text{for} \quad j = 0) \\ 0 \hspace{2.5em} (\text{for} \quad j > 0) \end{cases} \qquad (11)$$

Step 6. If $b = L$, terminate the calculation. If not, set $b \leftarrow b + 1$ and return to Step 2.

It is noted that the function of the blackboard is not to make a decision but only an indicator.

4.3 Algorithm for the overall system

The algorithm for the overall system is as follows.

Step 1. Set $t \leftarrow 0$. Under unanimity of all subsystems the initial values of X_{ij}^ℓ and M_{ij}^ℓ are determined such as the constraints for the precedence relation are satisfied. These values are shown on the blackboard.

Step 2. The subsystem a_{ij} determines S_{ij}^ℓ and then the objective value f_{ij}. The subsystem m_0 calculates f_0.

Step 3. A subsystem is selected at random with a uniform probability.

Step 4. If the selected subsystem is m_0, the subsystem proposes a revised version of A, which is the set of operations processed by m_0. If the selected subsystem is a_{ij}, the subsystem proposes a revised version of X_{ij}^ℓ. The proposition is shown on the blackboard.

Step 5. If all subsystems agree with the proposition, it is accepted by setting $\nu=2$. If not, set $\nu=3$.

Step 6. If $t = t^*$, terminate the computation. If not, set $t \leftarrow t + 1$ and return to Step ν.

In Step 6 t^* is the final iteration number, which is given in advance a sufficient large number so as to attain the agreement from all subsystems. It is noted that the procedure of Step 5 is based on the conditions (C) and (D) in Section 2.

Table 1 Numerical results

Case	(I, J, L, K)	t^*	f_1	f_2
1	(2,8,50,5)	3000	0	56
2	(4,6,50,5)	3000	56	62
3	(5,5,50,5)	3000	88	53
4	(14,120,702,19)	70000	4776	18846

4.4 Procedure for proposing a revised version of A

The subsystem m_0 proposes a revised version of A in Step 4 in the overall algorithm. The procedure is done by the following three steps.

To begin with, m_0 selects at random an operation $a_{i_1 j_1}^{\ell_1} \in A$ and an operation $a_{i_2 j_2}^{\ell_2} \notin A$. Then, the first step of proposition is that the multi-function machine processes both $a_{i_1 j_1}^{\ell_1}$ and $a_{i_2 j_2}^{\ell_2}$. If the proposition is rejected, the second step is that the multi-function machine processes $a_{i_2 j_2}^{\ell_2}$ and does not process $a_{i_1 j_1}^{\ell_1}$. If the proposition is also rejected, the final step is that the multi-function machine processes neither $a_{i_1 j_1}^{\ell_1}$ nor $a_{i_2 j_2}^{\ell_2}$.

4.5 Procedure for proposing a revised version of X_{ij}^{ℓ}

The subsystem a_{ij} proposes a revised version of X_{ij}^{ℓ}. The procedure is as follows.

Step 1. Set $\ell \leftarrow 1$.

Step 2. Find an operation $a_{i_1 j_1}^{\ell_1}$ such that $X_{i_1 j_1}^{\ell_1}$ is the maximum number in the set $\{X_{i \bullet j \bullet}^{\ell \bullet}\}$ satisfying the following equations:

$$X_{i \bullet j \bullet}^{\ell \bullet} < X_{ij}^{\ell}, \quad M_{i \bullet j \bullet}^{\ell \bullet} = M_{ij}^{\ell} \qquad (12)$$

Step 3. Propose setting $X_{ij}^{\ell} \leftarrow X_{i_1 j_1}^{\ell_1}$ and $X_{i_2 j_2}^{\ell_2} \leftarrow X_{i_2 j_2}^{\ell_2} + 1$ ($X_{i_1 j_1}^{\ell_1} \leq X_{i_2 j_2}^{\ell_2} < X_{ij}^{\ell}$).

Step 4. When the proposition does not satisfy the constraints for the precedence relation, it is rejected.

Step 5. If $\ell = n_{ij}$, terminate the procedure. If not, set $\ell \leftarrow \ell + 1$ and return to Step 2.

4.6 Condition for agreement

In order to carry out Step 5 in the overall algorithm, the condition for agreement to the proposition is given as follows :

Step 1. Each subsystem calculates its own objective value for the revised version of the scheduling plan which was given by Step 4 in the overall algorithm.

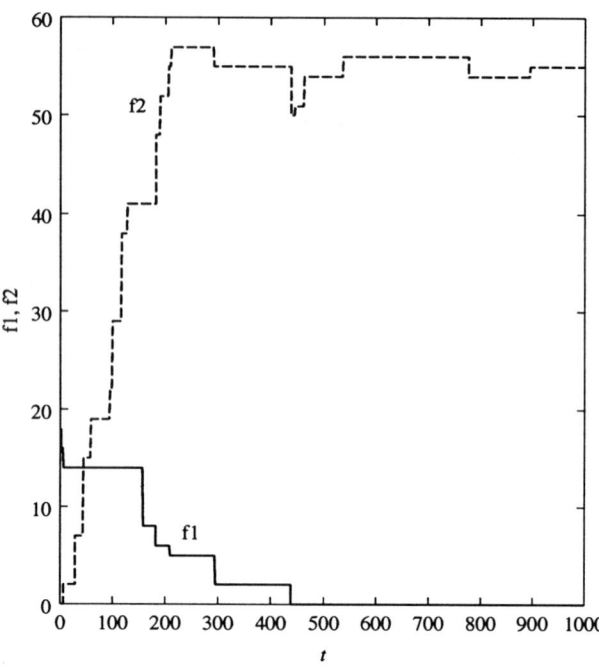

Fig.3 Convergence process of the algorithm (Case 1)

Step 2. If the objective value is better or is not worse within e percent deterioration than before, the subsystem expresses the approval of the revised plan.

We set $e=10\%$ in the following numerical experiment. If we set $e=0\%$, we had a bad local optimal solution. Then the subsystem accepts a little deterioration of its own objective value in order to attain better goal of the overall system.

5 Numerical Results

The proposed algorithm is applied to four case studies as shown in Table 1. Cases 1, 2 and 3 are test problems of small scale, which are constructed artificially. On the other hand Case 4 is obtained from the real metal mold assembly line. The details of the numerical data for problem parameters are omitted here.

Fig.3 shows the convergence process of the algorithm for Case 1. It is observed from the figure that both objective values f_1 and f_2 are improved as the calculation step goes. Fig.4 shows the convergence process for Case 4. Similarly, both objective values are improved in this case. However, the convergence speed for Case 4 is slower than that for Case 1.

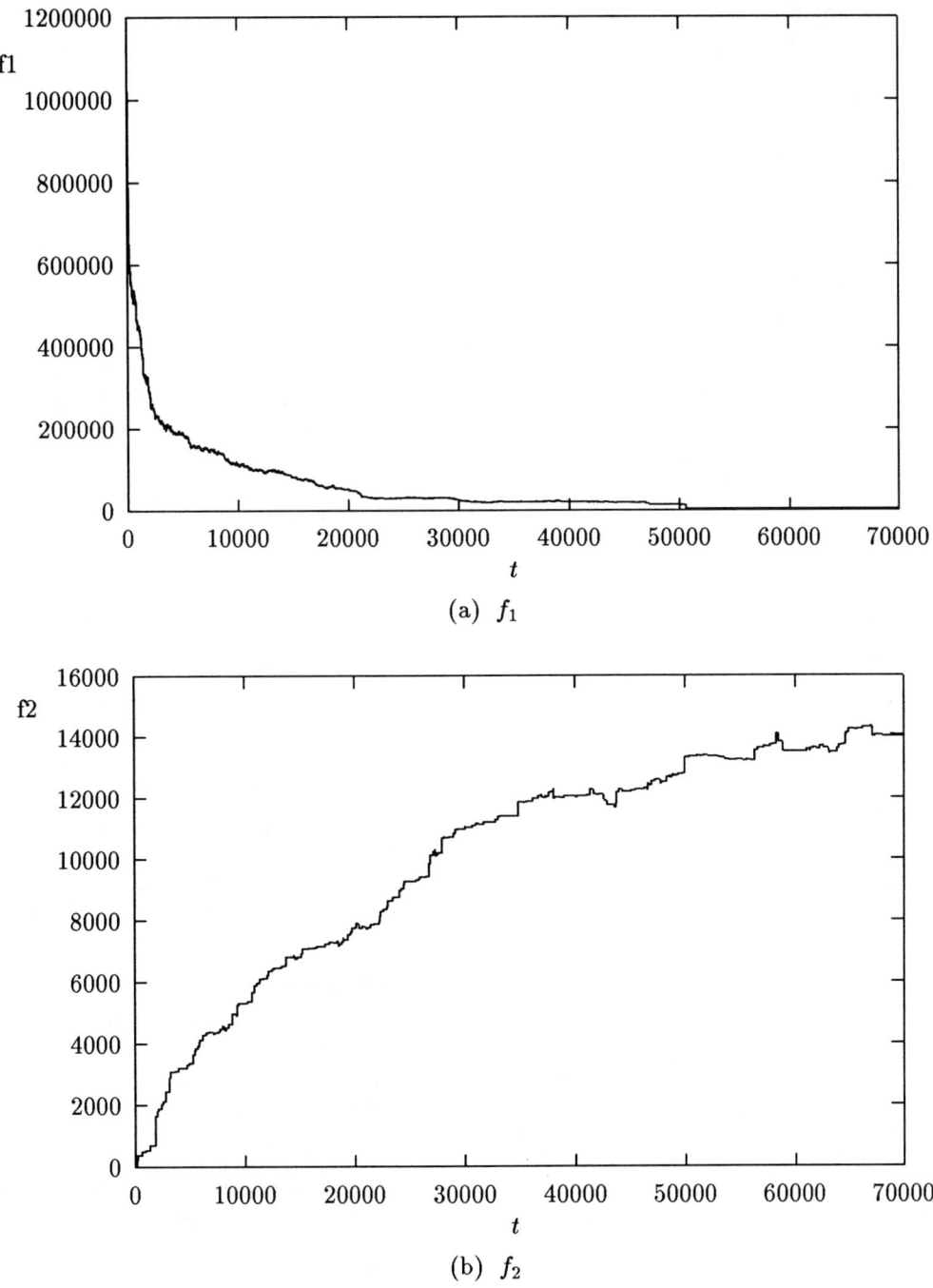

(a) f_1

(b) f_2

Fig.4 Convergence process of the algorithm (Case 4)

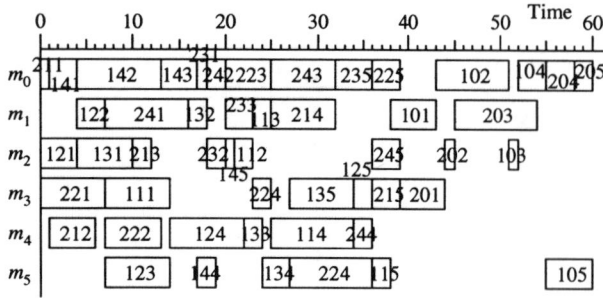

Fig.5 Gantt chart for the solution obtained (Case 1)

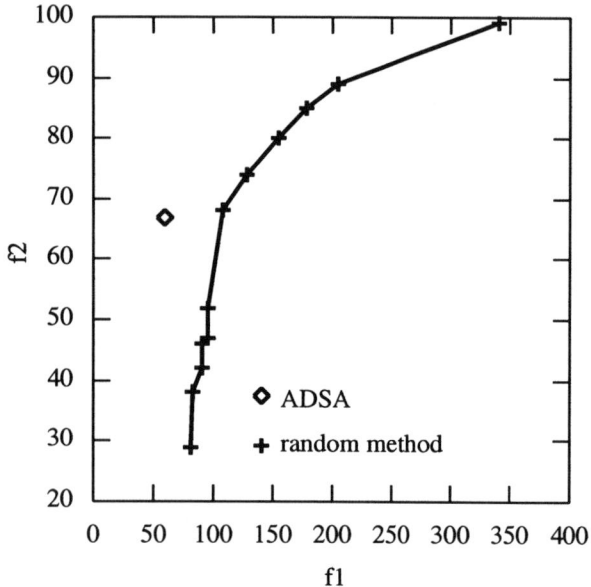

Fig.6 Comparison the objective values between ADSA and the random method (Case 2)

Fig.5 shows the Gantt chart obtained at $t=t^*$ as the result of Fig.3. A number in the rectangular box indicates $100i+10j+\ell$, which means the operation a_{ij}^ℓ. As shown in the figure, the multi-function machine is used almost in full time. Because of the constraint for the precedence relation between a product and the parts, the operations of a_{10} and a_{20} is processed later than those of the parts.

We compare the objective values between ADSA and the random method in order to examine the effectiveness of ADSA. In the random method, 100000 solutions are searched. Fig.6 shows the non-dominated solutions obtained by the random method and the so-

lution obtained by ADSA. It is observed from the figure that the solution obtained by ADSA gives better value of f_1 and a moderate value of f_2 as compared with the solution obtained by the random method. Consequently the proposed algorithm is considered to be effective for obtaining a compromise solution.

6 Conclusion

An autonomous decentralized scheduling algorithm (ADSA) has been proposed for a job shop line with a multi-function machine in a metal mold assembly process. The problem is multiobjective; one is to minimize the sum of the tardiness of each product and the other is to maximize the working time of the multi-function machine. A compromise solution of the problem has been obtained for four examples, i.e. three artificial problems of small scale and one problem constructed from a real data. The effectiveness of the algorithm has been examined by comparing the numerical results between ADSA and the random method.

Generally a new machine with high performance is introduced for development of equipments in a process. In such cases a scheduling plan of the existing process must be updated so as to use both a new machine and old machines efficiently. In this case, the goal of the scheduling system becomes complicated. A new concept of optimality is needed although a good part of the existing scheduling plan should be kept. The global scheduling plan of the overall system is difficult to obtain because the optimality of the solution can not be defined definitely. In such situations ADSA obtains a compromise solution easily by information exchange and cooperation between subsystems.

This work is partly supported by NGMS IMS Program.

References

[1] M. Ito, "Autonomous decentralized systems", *Proc. of 35th Annual Conf. of the Institute of Systems, Control and Information Engineers*, pp. 9-16 (1991, in Japanese).

[2] A. Aoki, H. Iima, N. Sannomiya and Y. Kobayashi, "Application of genetic algorithm to a job shop problem with a universal machine", *Proc. of 23rd SICE Intelligent System Symp.*, pp. 91-96 (1996, in Japanese)

[3] H. Iima, A. Fukui and N. Sannomiya : "A method for constructing an autonomous decentralized scheduling system in a parallel machine problem", *Trans. of the Institute of Systems, Control and Information Engineers*, Vol. 9, No. 2, pp. 97-99 (1996, in Japanese)

An Autonomous Decentralized Recognition System Having a Dispersive Wave Property

Hideo YUASA
School of Engineering
Nagoya University
Nagoya, 464-01, Japan
yuasa@nuem.nagoya-u.ac.jp

Satoshi ITO
Bio-Mimetic Control Research
Center, RIKEN
Nagoya, 456, Japan
satoshi@nagoya.riken.go.jp

Hideo MARUYAMA
NTT Corporation
Musashino, Tokyo, 180 Japan
maruyama@tnlab.ntt.jp

Masami ITO
Bio-Mimetic Control Research
Center, RIKEN
Nagoya, 456, Japan
itom@nagoya.riken.go.jp

Abstract

The concept of an autonomous decentralized system can be applied to recognizing various complicated and skillful actions performed by humans. This concept is the basis for a proposed pattern recognition system, which can be regarded as an associative memory system modeled by a reaction-diffusion equation system.

This paper will improve upon this pattern recognition system using the property of a dispersive wave instead of diffusion. This improvement has three advantages: acceleration of recognition, spread of similar regions, and human-like hysteresis in recognition.

Computer simulation has shown that propagation of a dispersive wave is faster than that of diffusion, and that the dispersive wave system exhibits the above advantageous properties of recognition.

Key words:
dispersive wave, associative memory, autonomous decentralized system, hysteresis

1 Introduction

Although parallelism has gained much popularity recently in system design, the explosive increase in connections has become a serious problem in the construction of parallel systems. One method to avoid this problem is to restrict these connections locally. It is not yet clear, however, how to achieve a global task with only local processing. This is one of the most important problems to be solved in autonomous decentralized system theory.

Such local processing has been studied in early vision systems. Poggio et al. have formulated lower-level image processing for such tasks as edge detection, fine-line formation, and stereo vision, as regularizing various ill-posed functions suitable for these problems[1]. The regularization theory, however, has seldom discussed dynamical or transient processes in spite of their importance. On the other hand, higher level processing, such as pattern recognition, requires global information covering the whole image, which means that all processing units must be connected to each other[2][3].

The brain would seem to use only local operation to accomplish the cognitive process that requires global visual information, since the number of connections or synapses in one neuron is much less than the total number of neurons in the brain. At the same time, the presence of parallelism in the brain is borne out by computing speed. If the speed of human image recognition is contrasted with the delay in neural processing, it seems highly likely that most image

75

recognition is performed by local and parallel processing and not by global and sequential processing[4][5][6].

Various image recognition methods based on parallel processing have been proposed, such as associative memory[7][8], neural networks[9][10][11] [12], Neocognitron[13][14] and synergetic computation [15][16]. In Neocognitron with a multilayered neural network, the receptive region up to a layer can be made large so that the range of local operation expands and the cells that selectively respond to specific stimulus are self-organized. The other methods, however, neglect the spatial connection in images. In fact, natural images have the property that the correlation between pixels becomes greater the closer pixels are to each other, and smaller the farther they are from each other. However, many methods quantize the images and treat them as vectors (one-dimensional data set), losing the two-dimensional connection in the image. In addition, vector operations, like inner products and correlations, must link all vector elements to one another, implying that all pixel values for both distant and close pixels are treated equivalently and simultaneously in spite of the difference in distance within the image. This is because integration (or summation) calculation is essentially a global task covering all function (vector) values. When considering neural network hardware, such vector operations lead to an explosive growth in the number of connective lines between pixels.

Here, we consider that topographical correlation, or spatial continuity in an image, inhibits explosion of connections, i.e., restricts connection to local areas, while maintaining performance. In fact, it is demonstrated that neural circuits in the visual cortex maintain the topography of the retinal image[17]. In study of visual computation, the stereo vision model by Marr[18] utilizes the spatial continuity for depth perception. In addition, standard regularization theory for early vision problems[1] used many types of spatial differential operators reflecting various types of spatial smoothness.

From this point of view, we have already proposed a pattern recognition system based on evolution equations[19] as an associative memory system modeled by reaction-diffusion equations. The diffusive process is introduced as local parallel operations that maintain a neighboring relation. In this paper, we will improve upon this system using the property of a dispersive wave instead of the diffusive process. This improvement has three advantages:

acceleration of recognition, spread of similar regions, and human-like hysteresis in recognition.

The following section explains the basic idea behind combining an associative memory with an evolution equation system. The third section compares the spreading motion of waves and dispersive waves with that of diffusion, the fourth section analyzes the dynamic property of dispersive waves, the fifth section constructs a recognition system using the dispersive wave property, and the sixth section demonstrates the powerful associative ability of this recognition system using several computer simulations.

2 Autonomous decentralized image recognition system

Autonomous decentralized system theory aims at global pattern formation only with locally interacting active units[20]. As mentioned in the previous section, the spatial relation of each pixel in the image is lost when regarding an image as a vector. One way to avoid this is to regard an image as a function of position[19].

Some dynamic associative memory systems are modeled by a gradient system[15][16][21]. A gradient system in function space means a reaction-diffusion equation system using a suitable potential functional. This diffusive process is introduced as local parallel operations that maintain the neighboring relation. This system is able to satisfactorily recall a stored image only with local parallel operations, though it requires some global judgment.

An associative memory based on the correlation matrix requires that stored vectors be orthogonal to one another in order to recall them correctly. On the other hand, a memory based on orthogonal projection does not require such a condition, but it cannot remove the noise present in the space spanned by stored vectors[7]. Such noise can be removed by adding nonlinear dynamics to such associative memory in a process called synergetic computation[16]. It has been shown that the reaction-diffusion recognition system with only local parallel operations achieves the same performance as recognition by synergetic computation[19].

3 Comparison of diffusion and waves

A diffusion system can be described as a gradient system in function space. Although it is one of the simplest evolution equations consisting of the first derivative with respect to time and the second derivative with respect to spatial variables, it does not converge well to an equilibrium point, indicating a low propagating velocity. One way to improve convergence is to add an accelerating term, that is, to use the second derivative with respect to time. In function space, this corresponds to using a wave equation. It is well known that a wave spreads spatially faster than diffusion.

Here, we use the dispersive wave equation,

$$\frac{\partial x}{\partial t} = -(x^3 - x) + y \qquad (1\text{-}1)$$

$$\frac{\partial y}{\partial t} = \alpha \Delta y + \beta \Delta x + F \qquad (1\text{-}2)$$

where α, β (>0) are parameters which specify the amount of dispersion and wave motion respectively, F is an external input, and Δ is the Laplacian operator. The first term of the right hand side in (1-1) can be regarded as a reaction term. Note that $x = x(\xi, \eta)$ and $y = y(\xi, \eta)$ are state functions at the position (ξ, η).

If the reaction term, dispersive term and external input are removed, (1) becomes a typical wave equation as follows.

diffusion   wave   dispersive wave

Figure 1   Comparision of diffusion, wave, and dispersive wave

$$\frac{\partial^2 x}{\partial t^2} = \beta \Delta x . \qquad (2)$$

Figure 1 shows time evolution in the integral space using the diffusion equation, the wave equation, and the dispersive wave equation, respectively. The diffusion equation used here has the form

$$\frac{\partial x}{\partial t} = \beta \Delta x . \qquad (3)$$

The wave spreads the black area faster than diffusion. This wave, however, continues to oscillate and does not damp. On the other hand, the dispersive wave spreads as fast as this wave and its oscillation dampens.

4   Dynamic property of dispersive waves

The following discusses the dynamic property of dispersive wave equation (1). First, it is assumed that each element does not interact spatially, that is, $\alpha = \beta = F = 0$. This assumption implies that y is not changed from its initial value. If y is in the range

$$|y| < \frac{2\sqrt{3}}{9} , \qquad (4)$$

then x has three equilibrium points; one is unstable and the other two are stable near ± 1. However, if y is out of that range, x has only one stable equilibrium point. The gradient system (5) determines the dynamics of x.

$$\frac{\partial x}{\partial t} = -\frac{\partial U}{\partial x} \qquad (5\text{-}1)$$

$$U = \frac{x^4}{4} - \frac{x^2}{2} - yx \qquad (5\text{-}2)$$

Figure 2 shows the shape of the potential function U that changes according to y, where y is regarded as a parameter. The x value varies towards one local minimum point of U according to the gradient of U. Here, we define the state in the integral region by the value of x. If $x > 0.9$, then this area is firing. On the other hand, if $x < -0.9$, then this area is silent; otherwise, we consider that the area is still in a transient state.

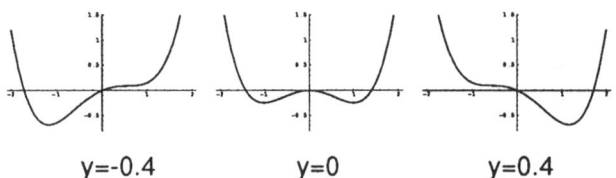

y=-0.4   y=0   y=0.4

Figure 2   Bifurcation of the potential function

77

Now, we add positive external input $F(>0)$. In such a situation, y increases according to the equation of (1-2) implying that the potential function changes to the right function in figure 2. Then, all elements tend to be firing. On the contrary, if we add negative external input $F(<0)$, all elements tend to be silent.

Next, we consider the situation where the spatial interactions work ($\alpha{\neq}0$, $\beta{\neq}0$) instead of external input ($F=0$). Obviously, β influences the wave motion, as does α on the diffusion of y. This diffusive process makes the value of y fixed. Because y determines which value (1 or −1) x will converge to, x will settle with y becoming averaged.

5 Construction of a recognition system

Using this dispersive wave property, we have constructed an image recognition system.

Figure 3 shows the structure of this system. This is constructed by N element planes, where N is equal to the number of stored images. Each element plane is constructed by $S{\times}S$ dynamic elements which obey eq.(1).

As mentioned in section 4, each element enters either the "firing" or "silent" state according to the reaction term, and enters the same state as neighboring elements under the influence of the dispersive wave term. Here, we introduce "winner-take-all" interaction among the planes in order to recall only one image. We define it as follows.

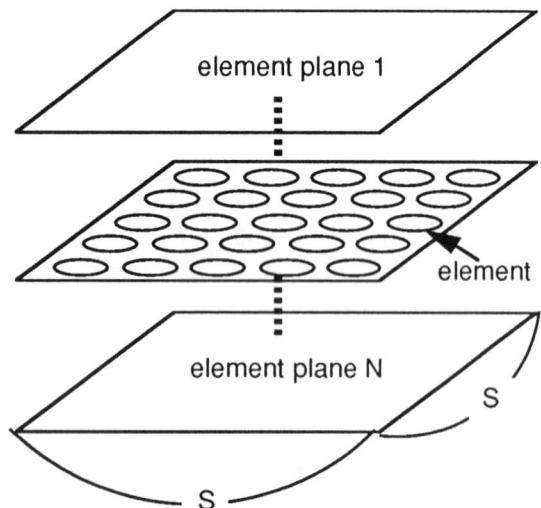

Figure 3   Structure of the autonomous decentralized image recognition system

$$F_k = \sum_{j=1}^{N} F_{jk} \qquad (6)$$

where

$$F_{jk} = \begin{cases} F^- : 0.9 < x_j < 1.1 \text{ and } |\Delta x_j| < \varepsilon \\ 0 \ : \text{Otherwise} \end{cases} (j{\neq}k)$$

$$F_{kk} = \begin{cases} F^+ : 0.9 < x_k < 1.1 \text{ and } |\Delta x_k| < \varepsilon \\ 0 \ : \text{Otherwise} \end{cases},$$

x_j is the variable in the j-th element plane at the same position, F^+, F^- and ε are positive parameters, as shown in figure 4. This means that "firing" elements in the "firing" area are easily fired, and all elements at the same position in other element planes are inhibited from firing. Obviously, this external input plays a competitive role among element planes. Note that to avoid an infinite increase of x, external input has an upper limit.

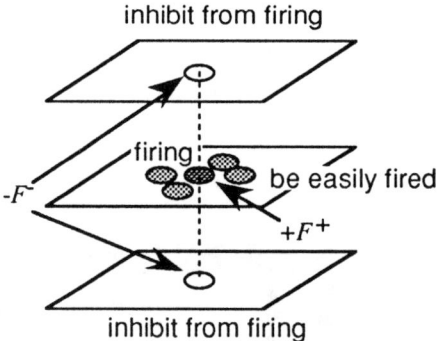

Figure 4   External input which force to compete among element planes

This external input has the following properties.
1. It accelerates the speed at which the state goes to equilibrium, that is, for the image to be recalled.
2. It spreads the effects of a firing region throughout the element plane.
3. It works so that the winner takes all among element planes.
4. It enables the elements to fire with hysteresis.

Properties 1, 2 and 3 should be easy to understand from eq.(1) and eq.(6) or figure 2 and 4. Property 4 is investigated here.

If the boundary is free end, the following condition is satisfied.

$$\iint_\Omega \frac{\partial y}{\partial t} d\xi d\eta = \iint_\Omega F d\xi d\eta \tag{7}$$

where Ω is a integral region, that is, a whole image plane, and (ξ, η) is the coordinates of a point on the image plane.

If there is no external input, that is, $F=0$, then condition (7) changes into

$$\iint_\Omega y \, d\xi d\eta = \text{Const.} \tag{8}$$

If initial values of y are equal to 0 in all elements, the right hand side of eq.(8) remains 0 throughout the evolution of eq.(1). Moreover, owing to the diffusive effect of y, the values of y are close to 0 in almost all elements after some transient. This means that almost all elements are bi-stable and have an equal probability whether they are firing or silent, since the potential function is symmetric about $x=0$ (see figure 5).

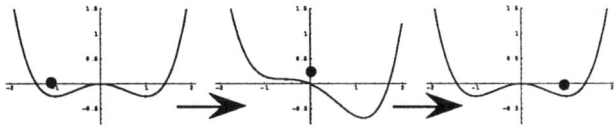

initial state   transient state   final state

Figure 5   Change of the local potential function ($y^0=0$)

On the other hand, if some external inputs are added, each plane either gives F^- or takes F^+ with respect to one another. The value of the right hand side of eq.(7) then becomes different among these planes. The plane containing many firing elements gets much F^+ so that the value of the right hand side of eq.(7) becomes large. Conversely, the plane containing many silent elements gets much F^- so that the value becomes negatively large. The firing plane then tends to fire, and vice versa for the silent plane. This means that the system has the tendency to easily recall the image that was just recalled before, i.e., it has hysteresis. It is assumed that the value of y decreases as follows,

$$y = y^\infty exp(-T_0 \tau) \tag{9}$$

where y^∞ is the value at finished time, T_0 is a time constant of oblivion, and τ is the interval between the end of the first recognition and the beginning of the second one.

Lastly, we have to decide the initial value of x. This value must reflect how close the input image is to the stored one. It is decided that firing elements indicate that their image resembles the input image, and silent ones do not. Here, the initial value of x is decided as follows.

$$x_k = 2exp\{-\lambda(I - M_k)^2\} - 1 \tag{10}$$

Here, I and M_k denote the intensity of input and k-th stored images, respectively, and λ is a parameter of resemblance (see figure 6).

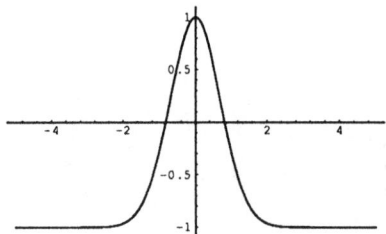

Figure 6   A function which decides the level of similarity.

The output image is expressed as follows.

$$O = \frac{\sum_{k=1}^{N} u(x_k - 0.9)M_k}{\sum_{k=1}^{N} u(x_k - 0.9)} \tag{11}$$

Here, O is the intensity of the output image, and $u(\cdot)$ is a unit step function.

$$u(s) = \begin{cases} 1 : s \geq 0 \\ 0 : s < 0 \end{cases} \tag{12}$$

If there are no firing elements, that is,

$$\sum_{k=1}^{N} u(x_k - 0.9) = 0 \quad ,$$

then the output at this position vanishes, that is, $O=0$. For these settings, if the k-th plane is only firing and all other planes are silent, the output image corresponds to the k-th stored image completely.

6 Simulation Results

The five images shown in figure 7 are stored; they have 100×100 pixel elements, and each element has 256-level gray-scale intensity. The following parameter values are used:

λ=0.002, α=0.8, β=5.0, ε=0.01, F^+=F^-=0.05. Note that this system is robust with respect to these parameters, i.e., some changes in these values have little effect on recognition performance.

6.1 Input having a mixture of two stored images

The input image features 70% intensity of stored image 1 and 30% intensity of stored image 5. This simulation result is shown in figure 8. The last output image is equal to stored image 1. This shows that the proposed system recognizes mixed input properly.

6.2 Input having a part of one stored image

The input image contains 36% of stored image 4 from the center. The simulation result is shown in figure 9. The last output image is equal to stored image 4 although 64% of the input image is different. Note that if external input F is not used, the output image becomes all black. This shows that external input F spreads some firing regions over the plane.

Stored image 1   Stored image 2   Stored image 3   Stored image 4   Stored image 5

Figure 7   Stored images used in simulations.

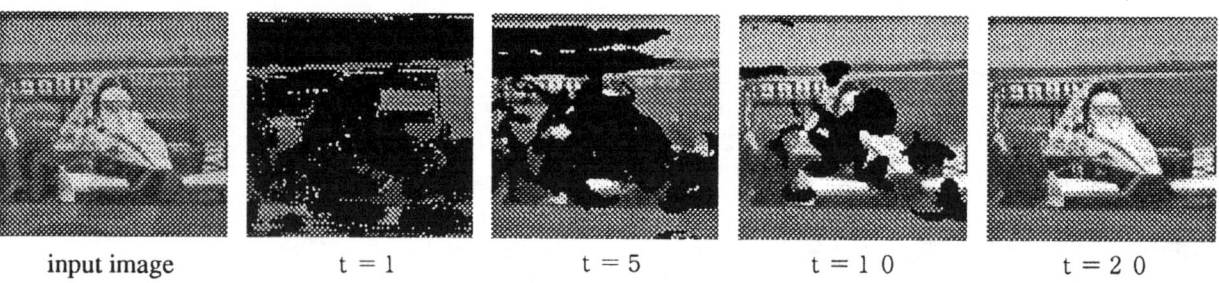

input image   t = 1   t = 5   t = 1 0   t = 2 0

Figure 8   Recognition result of mixed image

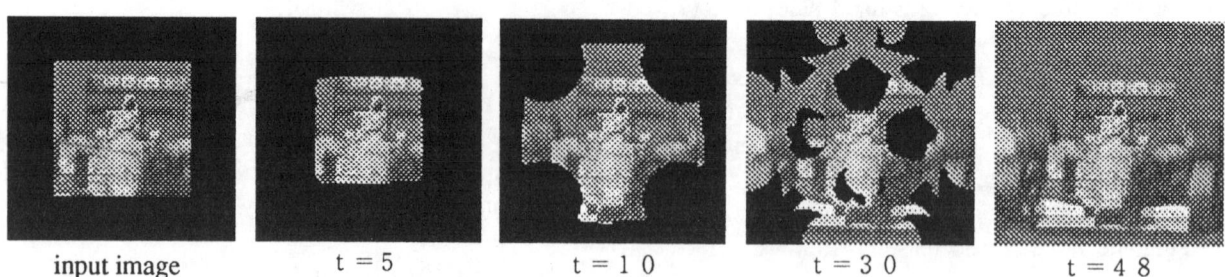

input image   t = 5   t = 1 0   t = 3 0   t = 4 8

Figure 9   Recognition result of a part of stored image

6.3 Experiment on hysteresis

Figure 10(a) and (b) are stained images of stored images 1 and 2 with uniformly distributed noise whose maximum amplitude is 32. Figure 10(c) was constructed by taking every other line from stored images 1 and 2 alternately, which means that in general, figure 10(c) cannot be recognized as either image 1 or 2. However, it can be said that an individual who has just saw stored image 1 would recognize figure 10(c) as image 1, and likewise for stored image 2.

After recognizing figures 10(a) and (b), the distributions of y are shown in tables 1 and 2, respectively. The value of 0.38 $\left(=2\sqrt{3}/9\right)$ is the mono-stable and bi-stable bifurcation value. For external input F, distribution of y is biased to plane 1 (after figure 10(a) is recognized) or to plane 2 (after figure 10(b) is recognized).

The initial values for input image 10(c) are distributed nearly equally over images 1 and 2 (see table 3), so that it cannot be recognized as either image 1 or

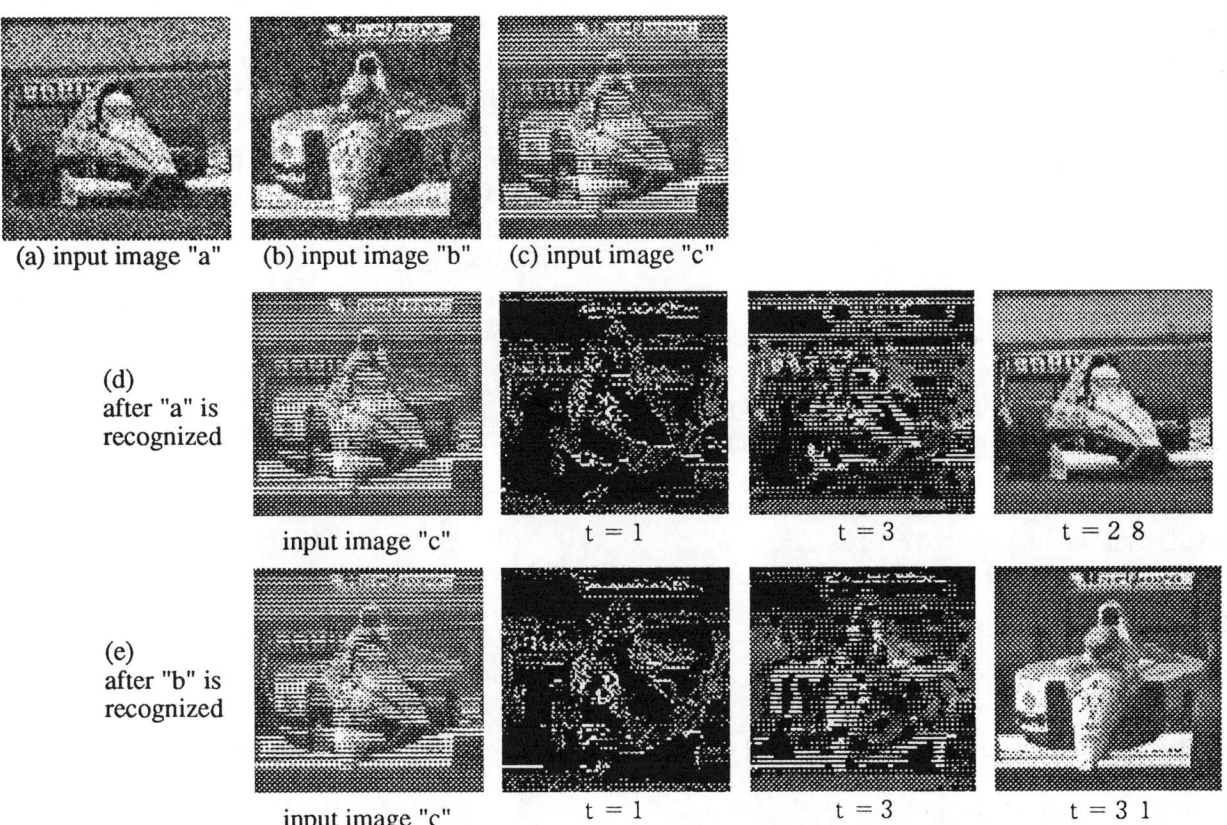

(a) input image "a"   (b) input image "b"   (c) input image "c"

(d) after "a" is recognized

input image "c"   t = 1   t = 3   t = 2 8

(e) after "b" is recognized

input image "c"   t = 1   t = 3   t = 3 1

Figure 10   Recognition result with desirable hysteresis

image	~0.38	0.38~0	0~-0.38	-0.38~
1	0	10000	0	0
2	0	0	9700	300
3	0	0	9475	525
4	0	0	9501	499
5	0	0	9712	288

Table 1 Distribution of y after recognition of "a"

image	~0.38	0.38~0	0~-0.38	-0.38~
1	0	0	9877	123
2	0	10000	0	0
3	0	0	9655	345
4	0	0	9627	373
5	0	31	9409	560

Table 2 Distribution of y after recognition of "b"

2. But as shown in figure 10, 10(c) is recognized as image 1 after recognizing figure 10(a), and as image 2 after recognizing figure 10(b). Here, $exp(-T_0\tau)=0.3$ is used.

image	1~0.5	0.5~0	0~-0.5	-0.5~-1
1	5552	357	414	3677
2	5526	335	479	3660
3	1618	929	1083	6370
4	1659	888	971	6482
5	1513	1091	1040	6356

Table 3 Initial distribution of x for "c"

This result shows that this system recognizes images using human-like hysteresis.

7 Conclusion

This paper has introduced an image recognition system using a dispersive wave property. Each element changes its state by its own bi-stable properties and local information. This dispersive wave property results in rapid spread of local information. External input F plays a cooperative role within planes and a competitive role between planes. The system also features rapid spread of spatial information and exhibits hysteresis.

Computer simulations show that this system has good recognition ability.

References

[1]Poggio T, Torre V and Koch C (1985) Computational Vision and Regularization Theory, Nature, 317:26 314-319

[2]Gelsema ES and Kanal LN(eds.) (1988) Pattern Recognition and Artificial Intelligence, Elsevier Science Publishers B.V.

[3]Oshersen DN et al.(eds.) (1990) Visual Cognition and Action, The MIT Press

[4]Aertsen A(eds.) (1993) Brain Theory, Elsevier

[5]Kanizsa G (1979) Organization in Vision : Essays on Gestalt Perception, Praeger Publishers(1979)

[6]Churchland PS and Sejnowski TJ (1992) The Computational Brain, The MIT Press(1992)

[7]Kohonen T(1977) Associative Memory A System-Theoretical Approach, Springer-Verlag, Berlin Heidelberg

[8]Kohonen T (1989) Self-Organization and Associative Memory (third edition). Springer-Verlag

[9]Arbib MA, Hanson AR(eds.) (1987) Vision, Brain, and Cooperative Computation, The MIT Press

[10]Anderson JA, Pellionisz A, Rosenfeld E(ed.) (1990) Neurocomputing 2, The MIT Press

[11]Matsuoka K (1989) An Associative Network with Cross Inhibitory Connections, Biol Cybern, 61:393-399

[12]Matsuoka K (1990) On Various Structures of Orthogonal Projection Type of Associative Network IEICE Trans J73-D-II:641-647

[13]Fukushima K (1980) Neocognitron: a self organizing neural network for a mechanism of pattern recognition unaffected by shift in position. Biol Cybern 36:193-202

[14]Fukushima K (1988) Neocognitron: A hierarchical neural network capable of visual pattern recognition, Neural Networks, 1[2]:119-130

[15]Haken H (1988) Neural and Synergetic Computers. Springer-Verlag

[16]Haken H (1991) Synergetic Computers and Cognition A Top-Down Approach to Neural Nets, Springer-Verlag

[17]Nicholls JG et al.(1992) From Neuron to Brain, 3rd, Sinauer Associates, Inc.

[18]Marr D (1982) Vision A Computational Investigation into the Human Representation and Processing of Visual Information, W. H. Freeman and Company

[19]Ito S, Yuasa H, Ito K and Ito M (1995) A Pictorial Pattern Recognition Based on an Associative Memory by Use of the Reaction-Diffusion Equation, Proc ISADS95 USA :249-256

[20]Ito M (eds.) (1994) Reports for Grant-in-Aid for Scientific Research on Autonomous Distributed Systems

[21]Hopfeild JJ (1982) Neural network and physical system with emergent collective computational abilities Proc Natl Sci USA 79:2554-2558

Session 2A

Multiagent Systems II

Chair

Kurt Rothermel

The Immune System as a Prototype of Autonomous Decentralized Systems: An Overview

Yoshiteru Ishida

Nara Institute of Science and Technology, Ikoma, Nara, 630-01 Japan,
Email: ishida@is.aist-nara.ac.jp, URL: http://genesis.aist-nara.ac.jp/~ishida

Abstract

In this paper, we discuss the features of the immune system; its system aspect (compatible with Jerne's network view), its process aspect, and its design aspect (compatible with Metchinikoff's self-defining view). Our applications of these three aspects are also presented briefly. Since these features of the immune system agree with the concept of autonomous decentralized systems, we suggest that the immune system can be a typical model for autonomous decentralized systems.

In the self-defining process, agents can refer to the self-information, and can interact with the environment. We propose that this self-defining process extracted from the immune system can be a candidate for design paradigm for autonomous decentralized systems where full specification of the total system is not only available but also inadequate. We also suggest that the "INTERNET" would be a typical example designed by this paradigm.

1 Introduction

Recently, computational method based on the analogy of biological information processing has been paid much attention. *Genetic Algorithm*, which mimic the adaptation of biological systems in genetic level, has been collecting wide attention as a new search method in many fields. *Neural Networks*, which hinted from neural systems in brain, has been used in the patten recognition. Several parallel and distributed models based on the neural net have been proposed and their learning algorithms studied, with the hope of giving a breakthrough for pattern recognition problems such as voice recognition and image recognition. Further, multiagent systems have been studied for investigating complex systems. Researches have been made in interdisciplinary fields including biological systems such as predator-prey models and models of the insect society. Simulation techniques for multiagent system combined with computer technology made several analysis possible which had been difficult due to modeling difficulty for unstructured problems. *Artificial World* or *Artificial Society* can visualize, for example, the behavior of social systems in *Cyber Space* where agents interact with each other.

Yet another important information processing can be found in the immune system, although we cannot be conscious of the activity. Models for the immunity-based systems have been proposed for information processing (e.g. [1, 2, 3, 4, 5, 6, 7, 8, 9, 10] to mention only few). Immunity-based systems consisting of agents (immune-related cells) may have adaptation and learning capability, similar to *Neural Networks*, but it is based on dynamic cooperation of agents. Immunity-based systems may have evolutionary mechanism similar to *Genetic Algorithm*, but it has a sophisticated control of diversity and specificity of populations.

Autonomous decentralized system (ADS) [11, 12] is a concept extracted not only from the self-organizing biological systems but also from the self-organizing economic systems of free market, organization of nations, development of enterprises, etc. The essence of ADS (as opposed to central processing systems) is that they consist of autonomous agents, to whom most of decisions are left to their direction. These autonomous agents form a kind of field (called autonomous field) that implicitly restrict their behavior. Another significant feature of ADS is that its structure is dynamic and hence cannot be designed or specified beforehand. In this sense, the design specification may be embedded in autonomous agents and autonomous field formed by them. It must be left to the dynamic developmental process of the autonomous field. This essence of ADS agrees with the character of the immune system (that will be discussed in section 2).

We have suggested that typical example of ADS taken from biological systems is the immune system, where agents (recognizing units consisting of B-cells and T-cells of specific type) communicate dynamically as opposed to communicate with hard-wired link in the neural networks. More importantly, agents of the immune system self-define and self-maintain by referring to the self-information and the environmental information. This character of the immune system would provide a design paradigm suited for ADS (as will be discussed in section 5).

The aim of this paper is to elaborate the immune system as a model of ADS. System aspect, process aspect and design aspect of the immune system will be discussed in section 3, 4, and 5 respectively. System aspect of the immune system, motivated by the concept of immune network focuses on self-reference where self-information is specified relative consistency among data, control of identification process by activating and inactivating agents, and dynamic interaction among agents. Application to process diagnosis will be briefly reviewed.

85

Process aspect of the immune system focuses on how the immune system adapts to the environment by changing state as well as population of agents. Immune algorithm, proposed elsewhere [10], will be briefly introduced. Application to the active noise control will be briefly reviewed.

Design aspect of immune system focuses on how the immune system constructs and maintains its identity by referring to both the self-information and the environmental information. We discuss the "INTERNET" as a typical example considered to be designed in this way. The "INTERNET" develops with the free hand of each site but restricted so that each site at least must follow a certain protocol to be connected with the rest of the "INTERNET". We try to formalize the immunological design paradigm that may be applicable to design ADS.

Before discussing these three aspects, we briefly review informational features of the immune system in the next section.

2 Features of the Immune System

As a phenomenon, the information processing done by the immune system may be characterized by: Memory (Huge amount of patterns are memorized and distinguished.); Recognition (System level recognition may be attained by recognition chain among agents.); Learning (Pattern recognition capability is acquired by learning not by copying through generations but acquired in one generation.); and Diversity (Diversity [1] needed for the discrimination of huge amount of patterns is generated by genetic recombination [13].)

2.1 The immune system as a self-defining process

The immune system is considered the self-identifying/defining process that continuously monitor the self, discriminate self/nonself, and maintain identity materially. It is not the process targeting for the nonself, but rather targeting for the self. Essence of the immune system as an self-defining process may be traced back at least to Metchinikoff:

> Briefly, Metchinikoff's argument is that immunity resides in the fundamental concept of organism as an intrinsically disharmonious entity striving for harmony. .... But for Metchinikoff, immunological process were primarily those activities that established the organism, and only as a result of secondary phenomena, do they protect. [A.I. Tauber, Immunology Today, 1994]

Thus, eliminating nonself is not the main job of the immune system; it is a subordinate job included in the main job: identifying; defining; and maintaining the self. We used a set of agents as a primitive for our immunity-based approach where agents are assumed

to be; diverse in its specificity in interacting with environment; capable of referring to the self-defining information; and specialized to a function by interacting with environment.

Agents, in general, can communicate with each other and move around allowing the communication network unspecified beforehand. Most importantly, self-defining process is not fully determined by the self-information but somehow open for the environment.

2.2 The Concept of Immune Network and Its Implications

Theory of the immune network proposed by Jerne [14] provides the network view that lymphocytes are mutually and dynamically connected by antigen antibody interaction. Not only antigen but also antibody generated by lymphocytes will act as an antigen against the other lymphocytes, thus presupposing *internal image* of the antigen. Both antigen and its internal image activate the same specific type of lymphocytes. The character of the network theory is that it places an importance on (1) the homogeneous network connected by antigen antibody interaction, and (2) the existence of internal images for antigens.

2.3 Immune network approach to measurement

The network view of the immune system can be understood with the metaphor of weighing by the balance; regarding many types of balance weights as recognizing agent (immune related cells such as B-cells and T-cells that react only with specific antigen) and action of weighing by balance as recognition by paratope and epitope with spatial and chemical affinity. The result of recognition is used to activate other recognizing agent, similarly to the fact that the result of balance is used to determine more appropriate balance weight against the target object. This object-against-object weighing is robust, since the weighing mechanism is a simple comparison and that information is distributed into many balance weights. It may be more appropriate to consider the immune network more sophisticated than the simple balance weighing system in the following points:

1. Each agent has not only information but a recognizing mechanism itself. (Each recognizing agent is comparable with a balance weight equipped with balance rather than only balance weight.)

2. Recognizing agent activated by an encounter with the antigen will reproduce its clone to enhance the ability of elimination of the antigen. (Balance weights, when used, can self-reproduce the balance weights of the same type.)

3. The reproduction above will be performed with mutation to increase the affinity with the antigen. (Balance weight, when used, can reproduce not only the same type, but slightly different types, hence enhancing precision in weighing.)

We use this object-against-object measurement, and applied it to online diagnosis by sensor network [15, 16, 17, 19] in section 3 . Further, we will pursue

---

[1] Acquiring diversity by combinatorial generation may potentially give insight to building new information models, since most of the current models depend upon copying for storage and transfer of information.

86

the adaptive change of the population of specific agent mentioned 2 and 3 above. We have implemented an immune algorithm [10] which uses 2, 3 and diversity generation by genetic recombination as explained in section 4.

3 The Immune System as a Dynamic Recognition System

The concept of immune network implies that agents (corresponding to a set of immune related cells, such as B-cell and T-cell) mutually and dynamically interacts with each other. The continuous and dynamical interactions will be used to; (a) maintain consistency among self and nonself; (b) memorize the encounter with nonself; and (c) regulate the self-identification process by activating/suppressing the activity of related agents.

The features of immune systems we have tried to use may be summarized as follows;

(i) Recognition of nonself is done by distributed units which dynamically interacts with each other in parallel. (ii) Distributed units carry redundant information. (iii) Each unit reacts based only on its own knowledge, but this local information processing connected by evaluation chain leads to emergent behavior: identification of the identity. (iv) Memory is realized as stable equilibrium points of the dynamical network. Recognition of the network is done by changing the state of the network from one stable equilibrium to another one by disturbances on the network.

3.1 Dynamic Interaction Among Agents

Figure 1 shows the evaluation chain of five agents. Each agent tries to identify the identity (self/nonself for immune network and normal/faulty for diagnosis) other agents. State variable (r_i and its normalization R_i) indicating the identity of agents are assigned to each agent. We call the value R_i calculated by the diagnostic models the reliability measure to distinguish it from probabilistic concept of reliability. The continuous value R_i between 0 and 1, indicating inactive and active (faulty and normal) of the agent respectively, is assigned to each agent. Only active agents can affect the states of the other agents. We consider that active agents are accepted as self by the network and that inactive agents are rejected as nonself.

When the agents 4 and 5 are faulty, we have the test result as shown in Figure 1. Simple voting at each agent does not work, since three agents 2, 3 and 5 are all evaluated as unreliable by two other agents and hence cannot be ranked in terms of reliability. Since unreliable agent may give unreliable results, these votes should be weighted. Next, let us introduce the binary weight for each agent: 0 (inactive or unreliable) when the sum of votes for the agent is negative, and 1 (active or reliable) when the sum of votes for the agent is zero or positive. Starting from all the agents active, and evaluate the weight synchronously would result in the following sequence of state vector ($R_1 R_2 R_3 R_4 R_5$).

(1 1 1 1 1) => (1 0 0 0 0) => (1 1 1 0 0)

Thus, weighting vote and propagate the information identifies the unreliable agents correctly. Let us map the above discrete model to the continuous dynamical model. Continuous dynamic network is constructed by associating the time derivative of the state variable with state variables of other agents connected by the evaluation chain. After elaboration for engineering use, we developed several dynamical models for the mutual evaluation [15, 16]. A possible association corresponding to the evaluation chain of Fig. 1 would be the following continuous dynamic network.

$$
\begin{aligned}
dr_1(t)/dt &= -4R_4 - r_1(t) \\
dr_2(t)/dt &= -4R_4 - 4R_5 - r_2(t) \\
dr_3(t)/dt &= -4R_4 - 4R_5 - r_3(t) \\
dr_4(t)/dt &= -4R_1 - 4R_2 - 4R_3 - r_4(t) \\
dr_5(t)/dt &= -4R_2 - 4R_3 - r_5(t)
\end{aligned}
$$

The evaluation of the agent 5, for example, depends upon two variables: R_2 and R_3. As seen in the last equation, since the agent 5 is evaluated negatively by agents 2 and 3, the evaluation of agent 5 will be low if the agent 2 is actually evaluated high.

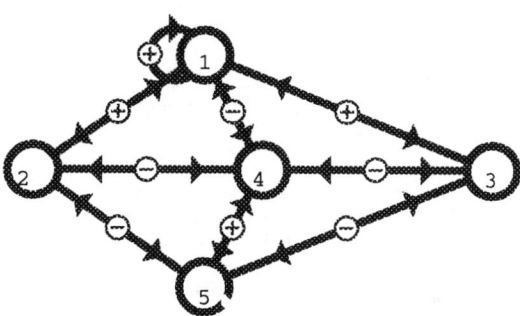

Figure 1: An example of evaluation chain

The significant feature of our model is that it can correctly diagnose the above case (when the agents 4 and 5 are faulty) by dynamically and continuously propagating evaluation through agents. The bottom line is that propagating and sharing information really counts, although instantaneous and local voting does not work.

3.2 Agents on the Sensor Network

In this section, we relate the agents (that interact based on the dynamics presented in the previous section) to the sensor network for the application to process diagnosis.

In process diagnosis, it is often the case that measurements such as temperature, pressure, flows, which are independently measured have relation with each other. In other words, some of these measurements are redundant. Using the dependencies, many constraints among sensors can be constructed. In the sensor network proposed in this section, each agent (corresponding to sensor) evaluates other agents using these constraints, rather than just measuring the process values [16].

In our approach, the agents of the network naturally correspond to the sensors (processes originated at sensors), and the constraints are obtained from simple process knowledge. These constraints between the values of the sensors may be expressed by equations or inequalities. From such a constraint, we can build a link between the sensor monitoring the value A and that monitoring the value B, and say that agents of the previous sensors can evaluate with each other.

When sensor values do not satisfy the constraint among these values, then it would imply that sensors or process corresponding to the constraint may be faulty. Therefore, one natural way of detecting process fault by sensor net is to introduce agent and reliability measure for constraint. Let $R_{T_{ji}}$ denote the reliability measure of the test T_{ji}. Then the dynamical model becomes as follows:

$$dr_i(t)/dt = \sum_j T_{ji}^+ R_j(t) R_{T_{ji}} - r_i(t) \qquad (1)$$

$$dr_{T_{ji}}(t)/dt = T_{ji}^+ R_j(t) R_i(t) - r_{T_{ji}}(t) \qquad (2)$$

$$R_i(t) = \frac{1}{1 + \exp(-r_i(t))}$$

$$R_{T_{ji}}(t) = \frac{1}{1 + \exp(-r_{T_{ji}}(t))}$$

where

$$T_{ij}^+ = \begin{cases} T_{ij} + T_{ji} - 2 & \text{tests i to j and j to i exist.} \\ T_{ij} + T_{ji} - 1 & \text{one of the tests i to j or j to i exists.} \\ 0 & \text{neither test i to j nor j to i exists.} \end{cases}$$

and T_{ij} in this model is as follows:

$$T_{ij} = \begin{cases} 1 & \text{i and j are normal.} \\ -1/1 & \text{i or j is faulty .} \\ 0 & \text{test from i to j does not exist.} \end{cases}$$

The equation 1 is a modification naturally resulted by considering the effect of the reliability measure of the test T_{ji}. The change rate of the agent i; $dr_i(t)/dt$ should reflect all the opinions of other agents weighted not only with the reliability of these other agents but with those of their evaluations. The equation 2 comes from the fact that the evaluating constraint is considered unreliable only when $T_{ji}, R_j(t)$ and $R_i(t)$ are contradictory; $T_{ji} = -1, R_j(t) = 1$ and $R_i(t) = 1$.

We have further extended for the case when many variables appear in a constraint [17, 18]. As an illustrative example, consider the process of keeping the level and temperature in a tank as shown in the left of Figure 2. The right of Figure 2 shows the sensor network of this process consisting of eight sensors and eight constraints among the sensors. The model of this sensor network follows:

$$dr_{F_O}/dt = R_L R_{T_0} + R_{F_I} R_{T_7} + R_L R_{F_I} R_{T_1} - r_{F_O}$$

$$dr_{F_I}/dt = R_{F_O} R_{T_7} + R_L R_{F_O} R_{T_1} + R_{F_H} R_{F_C} R_{T_6} - r_{F_I}$$

$$dr_L/dt = R_{F_O} R_{T_0} + R_{F_I} R_{F_O} R_{T_1} + R_{V_H} R_T R_{T_2} + R_{V_C} R_T R_{T_3} - r_L$$

$$dr_{F_H}/dt = R_{V_H} R_{T_4} + R_{F_I} R_{F_C} R_{T_6} - r_{F_H}$$

$$dr_{F_C}/dt = R_{V_C} R_{T_5} + R_{F_I} R_{F_H} R_{T_6} - r_{F_C}$$

$$dr_{V_H}/dt = R_{F_H} R_{T_4} + R_L R_T R_{T_2} - r_{V_H}$$

$$dr_T/dt = R_{V_H} R_L R_{T_2} + R_{V_C} R_L R_{T_3} - r_T$$

$$dr_{V_C}/dt = R_{F_C} R_{T_5} + R_T R_L R_{T_3} - r_{V_C}$$

$$dr_{T_0}/dt = \frac{1}{2}(T_0 - 1) R_{F_O} R_L - r_{T_0}$$

$$dr_{T_1}/dt = \frac{1}{2}(T_1 - 1) R_{F_O} R_{F_I} R_L - r_{T_1}$$

$$dr_{T_2}/dt = \frac{1}{2}(T_2 - 1) R_T R_{V_H} R_L - r_{T_2}$$

$$dr_{T_3}/dt = \frac{1}{2}(T_3 - 1) R_T R_{V_C} R_L - r_{T_3}$$

$$dr_{T_4}/dt = \frac{1}{2}(T_4 - 1) R_{F_H} R_{V_H} - r_{T_4}$$

$$dr_{T_5}/dt = \frac{1}{2}(T_5 - 1) R_{F_C} R_{V_C} - r_{T_5}$$

$$dr_{T_6}/dt = \frac{1}{2}(T_6 - 1) R_{F_H} R_{F_C} R_{F_I} - r_{T_6}$$

In the sensor network approach, the self is defined mainly by the relative relation among data (it could be defined by the absolute reference, though). When the model is implemented in a distributed processing environment, the evaluation of the reliability measure can be done in a fully distributed and autonomous manner in the sensor network. We have developed several variants and extensions [18, 19] of the model that could not be stated in this section.

4 The Immune System as an Adaptive System

When viewed as a process, the immune system uses genetic control in adaptation. In this section, we review immune algorithm extracted from the adaptation process in the immune system. An application to adaptive disturbance neutralizer is also briefly mentioned.

4.1 An immune algorithm

First, the difference from *Genetic Algorithm* should be mentioned. For the immune system, the environment with which it must interact is not only nonself from outer world but also the self. Another important difference from *Genetic Algorithm* is that each agent may interact (e.g. stimulation, inhibition) with each other while in *Genetic Algorithm* genes are selected by environment. Finally, in *Genetic Algorithm* crossing-over is used to mix genes, however in the immune system, only genetic recombination is used for attaining diversity. In the battle between host and parasite, host has two defense systems: systematic

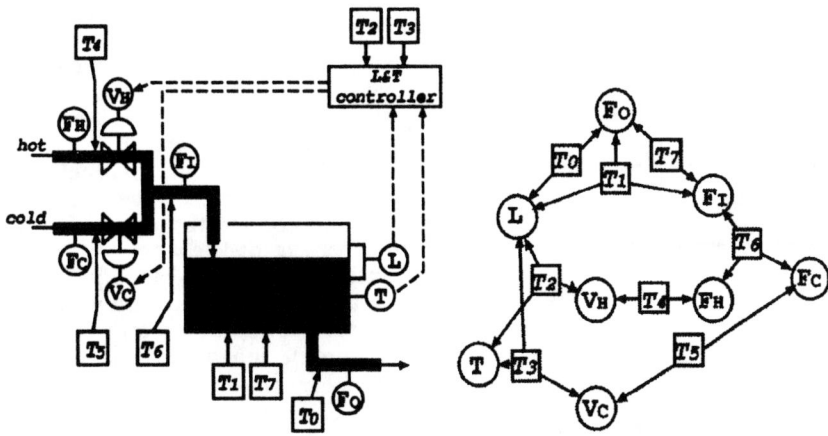

Figure 2: Tank with level and temperature controlled(left) and its sensor net(right)

immune defense and cellular resistance. In the immune defense, the host tries to catch up by evolving molecular that bind, and the parasite tries to escape by evolving proteins that are not bound. In cellular resistance, the roles are reversed; the parasite tries to catch up. Thus, it is hypothesized that the host uses sexual combination for escape [20]. We have proposed

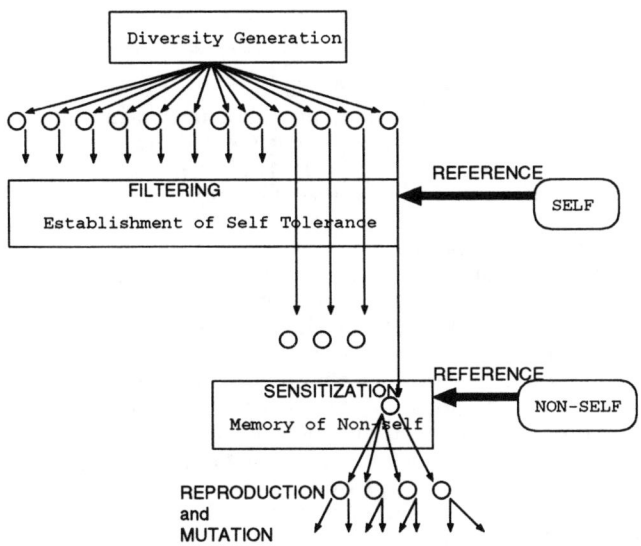

Figure 3: Immune Algorithm for An Agent-based Architecture

an immune algorithm [10] based on the significance of the immune system using genetic recombination for diversity, negative selection for self-tolerance, and mutation for affinity. The most naive immune algorithm has the following three steps.

1. *Generation of Diversity*: Diversity of agents in its specificity is generated.

2. *Establishment of Self-Tolerance*: Agents are adjusted to be insensitive to *known pattern*(self) during developmental phase.

3. *Memory of Nonself*: Agents are adjusted to be more sensitive to *unknown pattern* (nonself) during working phase.

In the step of *Generation of Diversity*, genetic coding for agent could be used in a similar manner to that of *Genetic Algorithm*. In that case, the genetic code of agents will be recombined to guarantee the diversity of their specificity. Then, agents are developed from their coding in this step. In the step of *Memory of Nonself*, activated agents would have any of the following properties resulting in higher affinity with the encountered nonself (for local and cell-based memory as opposed to network memory).
(1) Elongation of life span or lower death rate to attain immune memory,
(2) Reproduction of clones (i.e. agents of same type),
(3) Higher rate of mutation.

In the metaphor of weighing by balance, resolution of weighing (corresponding to affinity) will be increased because a balance weight when used can reproduce not only the same type but slightly different types of balance.

In the step of *Memory of Nonself*, the agents which actually recognized *the unknown pattern* during working phase will be activated. In the step of *Establishment of Self-Tolerance*, the agents which are specific to *self* pattern are removed. Figure 3 shows an schematic diagram of the immune algorithm for the agent-based architecture.

The next section briefly presents the adaptive disturbance neutralizer based on the *agent-based* architecture discussed in the this section.

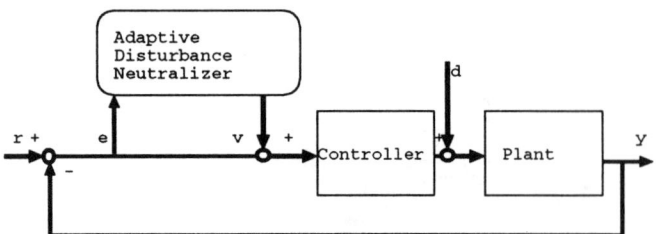

Figure 4: Block Diagram of Adaptive Disturbance Neutralizer

4.2 Disturbance Rejection by Immune Algorithm

Each agent consists of a pair of informations. One of its pair, $x_j(n)$ represents a pattern of input signal to the plant. Another element $y_j(n)$ of the pair represents the output of the plant for the input $x_j(n)$. The pairs of signals We call $x_j(n)$ genes. This pair is actually acquired during long evolutionary process in real biological system. Fig. 5 shows detailed diagram of the disturbance rejection system shown in Fig. 4.

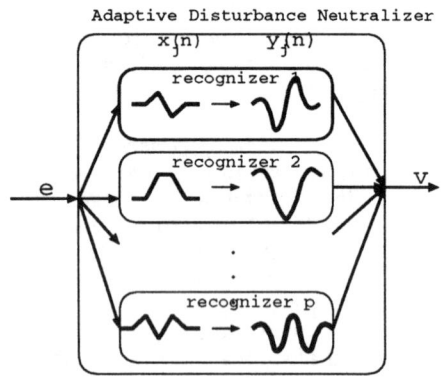

Figure 5: Adaptive Disturbance Neutralizer Consisting of Agents Working in Parallel

In the simulation we conducted elsewhere [10], ten different genes are used. Since genes will evolve in the immune algorithm, initial set of gene may be arbitrary. However, diversity is required so that they can compose many types of disturbance signals.

Noise is assumed to be a periodic disturbance. The periodic disturbance is successfully rejected by the system. This means that there is an agent that corresponds to the imposed disturbance. The disturbance is also known to be rejected gracefully from a period to period as step increases. This comes from the adaptation of agents, which is the result from adaptation of activated agents; higher affinity by higher rate of mutation in this case (as in the step of *Memory of Nonself* in the immune algorithm).

In another simulation, the responses to the initial (at 0 step) and second encounter are compared. The disturbance at secondary encounter is more effectively rejected than that at the first encounter. This again comes from the adaptation of the agent; elongation of the life time in this case.

In the simulation conducted elsewhere [], agents are only selected. However, agents could communicate and cooperate in elimination of disturbance and memorizing disturbance pattern. For example, if the neutralizing signal can affect the other agents. For the actual biological immune system, different types of cells communicate and cooperate extensively during elimination process as known under the name of *interleukin*[2].

In conventional control system, disturbance agent and controller are physically and conceptually separated. However, in this adaptive disturbance neutralizer, they are identified as agents. Agents are rather divided by the disturbance patterns upon which agents act. To each disturbance pattern of signals, different agent which is capable of producing neutralizing signal is supposed to be activated. That is, each agent has the capability of recognizing one specific disturbance signal and that of neutralizing the disturbance.

The immune algorithm proposed may apply to several fields, and seems to be promising for the system where the models of the environment are not available. The adaptive disturbance neutralizer is formalized by applying the *agent-based* architecture to a certain control system.

5 The Immune System as a Self-Defining Process

Since biological systems are information-intensive compared with non-biological systems, they are expected to give insights to artificial systems as they become information-intensive. The immune system is a super adaptive system in the sense that it changes its configurations and parameters to the self as well as to the nonself. We have suggested its significance for providing implications for ADS. The importance of the concept of ADS may be that it may provide a new design paradigm for large-scale systems, to which the conventional design that requires a full specification beforehand, cannot apply. In other literature [12], we have pointed out the importance of the concept of field in ADS, especially biological field (as opposed to physical field) that is formed by the active and intelligent agents. In this section, we elaborate the biological design paradigms and suggest that immunological design paradigm would be a candidate for designing ADS.

One of the most remarkable character of biological systems can be found in their designing strategies.

---

[2]Interleukin is a general term for molecules to communicate among leukocyte. Unlike the *network hypothesis*, they do not correspond to a specific antigen. Discovery and study of interleukin leads to dismissing the naive *idiotypic network* theory [14] that heavily depends on the specificity may be superficial. Later, they are found to react not only with leukocyte but with many other types of cells.

Several new design paradigms can be identified in biological systems such as; ontogenetical one, evolutional one, and immunological one. Ontogenetical design is motivated by "Ontogeny recapitulates phylogeny". As pointed out by Simon [21], since it is a design by repeatedly adding a new feature on the top of the current design, the design process roughly repeats the evolution process.

Evolutional design paradigm is motivated by the concept of *Selfish Genes* [22]. It prepares one self-reproducing agent that allows some noise in self-reproduction. Complex system would emerge by evolution that will be driven by the difference of fitness of the program implemented in self-reproducing agents and the diversity by the random noise in the program (mutation). However, since the system would evolve by the selfish genes, they would hardly evolve for engineering use or for human benefit. Computer viruses or internet worms could be designed by this paradigm.

The immunological design paradigm may be considered a special type of this evolutional one; evolution within one generation and one individual aimed mainly for maintaining the identity. Although the system evolved by the evolutional paradigm mostly reflect the environmental information, those by the immunological one would be specified by the self-information as well as the environmental information.

In the immunological design, each agent can refer to self-information (MHC in real immune systems) when deciding to accept or reject information from the environment. Agents that are initially equivalent can be different component by the flick of a switch trigered by the environment. "INTERNET" would be one example that can be considered to be formed by the immunological design paradigm. The most remarkable feature for these biological designs (and construction) is using the blueprint or program that can operate itself. The self-operating programs have many switches that can be trigerred by the environmental conditions; hence drives specification of each agents.

The immunological design paradigm may be stated as: Without a blueprint for the total system, the system self-design and self-define itself by adaptive change of agents specified by the self information and by the interaction with the environment. Thus, the following elements are critical for the immunological design.

- Self-information that can be referred by each agent.

- Environmental information that interacts with the system all the time.

- Field composed of agents and the environment.

For the case of the "INTERNET", agent corresponds to site; and self-information corresponds to software and hardware restriction for a site to be connected to the "INTERNET". For a site to be connected, there is a minimum requirement that must be met by the site, such as allowable protocols and hardware/software that enables the site to follow the protocols.

One feature of the current "INTERNET" is that the restriction by the self-information is not severe, thus allowing many types of hardwares (computers) to be connected to the "INTERNET". If the self-information is more restrictive, rejecting many types of hardwares then the "INTERNET" would not become such a large-scale network.

Field composed by the "INTERNET" benefits both sites to be connected and the rest of "INTERNET", thus driving the "INTERNET" to grows up. Field for the "INTERNET" would promote specialization process for each site depending upon asymmetry due to the regional difference(i.e. the environmental condition). In the current stage of the "INTERNET", this specialization has not become remarkable. However, as expected from the immune system, some sites would be specialized to; e.g. creating information, processing information, distributing information, and consuming information. A sign of this trend has already been seen in the current "INTERNET".

The immunological design may be characterized by the following features resulted from the self-defining and self-referring feature.

- Designer and the object to be designed are not separated.

- Phases of design, construction, and working are not temporally separated.

- Design and construction process is irreversible.

It may be appropriate for designing and constructing a large-scale system only once, as seen for the "INTERNET", rather than for mass-production of the same goods. As discussed in section 2, since the immune system is essentially ADS (as opposed to hierarchical systems such as the neural systems), the immunological design paradigm may be applied to design ADS.

As another example that may be considered by immunological design paradigm above is "consciousness". Immune system is self-defining, identifying, maintenance process where the self is the material identity. Consciousness may be constructed by the similar process where the self is the mental identity in this case. Festinger proposed a theory of cognitive dissonance [23] that claims human tries to avoid internal inconsistency (cognitive dissonance). In our immunological design paradigm, consciousness is the process to select the mental constructs, and only when the mental construct can be harmonious with so far constructed self, it will be assimilated into the self. Thus, what is important for the mental identity is that the matter can be accepted referring to the mental self rather than the matter is true or false. When accepting a new matter, reconstruction and adjustment in the mental self occurs to some extent to harmonize the new matter with the self.

In sum, ADS may be designed and constructed in the following manner:

- Agents should be designed so that it can be specified by the self-information (program) as well

91

as environmental condition. Rather than incorporating all the functions needed, the program trigerred by environmental conditions work to specify the agent to fit the condition.

- Field should be designed to drive addition or self-reproduction of agents and the specification process of each agents. It should also restrict the behavior of agents so that some agents does not disturb the harmony of agents.

6 Conclusion

We discussed the immune system focusing on the informational features that can be prototype for the autonomous decentralized systems. By exploring the system aspect of the immune system outlined by the network theory, dynamic recognition model where agents mutually control active/inactive states of the other agents by dynamic interaction is presented. By exploring the process aspect, adaptive model where agents generate diversity by genetic recombination and control affinity with nonself by mutation is also presented. Finally, by exploring the design aspect, design model where agents self-define, identify and maintain by referring to the self-information and the environmental information is also presented. Throughout these three aspects, agents (that can be diverse in its specificity; can refer to the self-information; and can be specialized to a function by interacting with the environment) as well as field formed by the agents and environment are primitives extracted from the immune systems for the autonomous decentralized systems.

Acknowledgments

This work has been supported in part by the SCAT (Support Center for Advanced Telecommunications Technology Research) Foundation.

References

[1] J. D. Farmer, N. H. Packard, and A. S. Perelson, "The Immune Systems, Adaptation, and Machine Learning" *Physica,* 22D, 187, 1986.

[2] G. W. Hoffman, "A Neural Network Model Based on the Analogy with Immune System," *J. Theor. Biol.,* 122, pp.33-67, 1986.

[3] H. Atlan and I. R. Cohen, Eds, *Theories of Immune Networks*, Springer-Verlag, 1989.

[4] Y. Ishida, "Fully Distributed Diagnosis by PDP Learning Algorithm: Towards Immune Network PDP Model," *Proc. of IJCNN 90*, San Diego, 1990.

[5] H. Bersisni and F.J. Varela, "The Immune Recruitment Mechanism: A Selective Evolutionary Strategy" *Proc. ICGA 91*, 1991.

[6] F.T. Vertosick and R. H. Kelly, " The Immlune System as a Neural Network: A Multi-epitope Approach," *J. theor. Biol.,* 150. pp. 225-237, 1991.

[7] S. A. Kauffman, *The Origin of Order*, Oxford University Press, 1993.

[8] S. Forrest, A.S. Perelson, L. Allen, and R. Cherukuri, "Self-Nonself Discrimination in a Computer," in *Proceedings of 1994 IEEE Symposium on Research in Security and Privacy*, 1994.

[9] J. O. Kephart, "A Biologicaly Inspired Immune Systems for Compouters," in *Artificial Llife IV*, MIT Press, 1994.

[10] Y. Ishida and N. Adachi, "Active Noise Control by an Immune Algorithm" *Proc. International Conference on Evolutionary Computation 96*, pp. 150-153, 1996.

[11] K. Mori, "Autonomous decentralized systems: Concept, data field architecture and future trend," *Proc. of International Symposium on Autonomous Decentralized Systems*, (1993) 28-34.

[12] Y. Ishida, "An Information Character of Autonomous Decentralized Systems ," *The Journal of The Society of Instrument and Control Engineers,* **10**(1993) 830-836, (in Japanese).

[13] S. Tonegawa, "Somatic Generation of Antibody Diversity," *Nature*, 302, pp. 575-581, 1983.

[14] N. K. Jerne, "The Immune System," *Sci. Am.,* Vol. 229, No.1, pp. 52-60, 1973.

[15] Y. Ishida and F. Mizessyn, "Learning Algorithms on Immune Network Model: Application to Sensor Diagnosis," *Proc. IJCNN 92*, Bei Jing, pp. 33-38, 1992.

[16] F. Mizessyn and Y. Ishida, "Immune Networks for Cement Plants," *Proc. of Interational Symposium on Autonomous Decentralized Systems*, pp.282-288 , 1993.

[17] Y. Ishida, An Immune Network Model and its applications to process diagnosis, Systems and Computers in Japan, vol. 24,no. 6, pp38-46, 1993.

[18] Y. Ishida and Y. Tokimasa, "Diagnosis by a Dynamic Network inspired by Immune Network," *Proc. WCNN 96*, 1996.

[19] Y. Ishida and Y. Tokimasa, "Data reconcilliation combined with data evaluation by sensor networks with immune net metaphor," *Proc. CDC 96*, 1996.

[20] H. J. Bremermann, "The Adaptive Significance of Sexuality," in S.C. Stearns Eds, *The evolution of Sex and its consequences*, Birkhauser, Basel, pp. 135-61. 1987.

[21] H. A. Simon, *The Sciences of the Artificial*, The MIT Press, 1989.

[22] R. Dawkins, *The Selfish Genes*, Oxford University Press, 1989.

[23] L. Festinger, *A Theory of Cognitive Dissonance*, Row Peterson and Company, 1957.

MAGNA - A DPE-based Platform for
Mobile Agents in Electronic Service Markets

Sven Krause[1], Flávio Morais de Assis Silva[1,†], Thomas Magedanz[1],

Radu Popescu-Zeletin[1], Orandi Mina Falsarella[2], Carlos Raul Arias Méndez[3]

[1]GMD FOKUS
Hardenbergplatz 2
10623 Berlin, Germany
Phone: +49-30-25499200, Fax: +49-30-25499202
Email: (krause,flavio,magedanz, zeletin)@fokus.gmd.de

[2]PUCCAMP - Instituto de Informática
Caixa Postal 317, 13020-904
Campinas, S.P. Brazil
Email: orandi@dca.fee.unicamp.br

[3]Universidad de Magallanes
Av.Bulnes 01855
Punta Arenas, Chile
Email: carias@dca.fee.unicamp.br

Abstract

Driven by technological advancements, market growth and deregulation, the global telecommunications environment is rapidly changing its face towards a highly dynamic open environment, where a multitude of competing and/or cooperating service providers offer a steadily growing number of increasingly powerful communication and information services. Due to the number of relevant stake holders, this environment is characterized by great technological and organizational diversity. The need for integration, ubiquitous service access and rapid service provisioning requires an open, universal service platform which supports both, telecommunications and management applications. The development of such a platform based on the concept of distributed processing environments (DPEs) has been a major research area for some time now, and has also yielded important results (e.g. CORBA, TINA). However, since DPEs are based on the centralized client/server paradigm, they are relatively inflexible, and do not adapt well to rapidly changing, or customer-specific requirements. An entirely different platform approach is based on the application of agent technology. Of particular interest are mobile agents, autonomous and thus intelligent software entities, which may roam the network in the course of performing their designated tasks. Due to their inherent autonomy and mobility they offer new opportunities for the provisioning of services in the emerging electronic service markets. The prerequisite for the execution of mobile intelligent agents in distributed service environments is the provision of a distributed agent environment platform. Since both, distributed processing-based and agent-based service provisioning have their specific advantages, such a platform should support both paradigms.

This paper provides an overview of the mobile agent architecture MAGNA, and the accompanying agent platform. Both are currently under development by GMD FOKUS in the context of the MAGNA project[1]. The distributed agent environment provided by the MAGNA platform is based on current DPE concepts and technologies. It therefore enables coexistence and integration of distributed and agent-based implementations of telecommunications and management applications.

1. Introduction

The term „agent" has become a real buzzword in the field of software design over the last years, with numerous, quite different applications claiming agent status for themselves. As with many other innovative concepts, the number of applications claiming agent support is only surpassed by the number of attempted definitions of „agenthood". Still, a commonly agreed definition of what an agent is, i.e. which characteristics it has, and which behavior it exhibits, has not been found yet. And overlooking the apparent „agent hype" it is also hard to come across agent applications which could not have equally well been realized using traditional software technology. Nonetheless, the prospective advantages of agent technology justify a closer look which will in any case demystify this technology, and help to evaluate its real value.

An application area which could potentially benefit greatly from the success of agent technology, but is also among the most challenging application fields possible, is the field of telecommunications [4][5][6][7][8]. The rationale behind this viewpoint is the assumption that intelligent

---

1. MAGNA is a project in the R&D programme of DeTeBerkom GmbH, a subsidiary of Deutsche Telekom AG.

† The work of this author is partially supported by Conselho Nacional de Desenvolvimento Científico e Tecnológico (CNPq), Brazil

mobile agents may enable new and much more flexible and efficient ways for the distribution of control and management in telecommunication systems. By supporting the concept of „Remote Programming", mobile agents allow control and management tasks to be carried out „locally" at the resources to be controlled or managed. This concept represents an alternative to the traditional „client/server computing" paradigm, where statically installed clients and servers communicate remotely through RPC (Remote Procedure Call). Remote programming may result in considerable flexibility and efficiency gains. However, it has to be stated, that the value of intelligent mobile agents for the realization of services depends strongly on the application scenario. For some application scenarios the usage of agents provides no benefits and the service modeling according to the traditional client/server paradigm is more adequate.

In order to enable agent-based applications, such as telecommunications and management services to be realized efficiently without giving up the existing RPC-based concepts and mechanisms, it is necessary to develop an integrated platform which is capable of supporting both service modeling paradigms.

This paper introduces the *MAGNA (Mobile AGeNt Architecture)* platform, which is currently under development by GMD FOKUS, sponsored by DeTeBerkom, a subsidiary of Deutsche Telekom AG[1]. The MAGNA platform defines a *distributed agent environment* which consists of a set of distributed agent support components, so-called *agencies*. Each agency provides a set of services addressing various aspects of agent support, such as agent execution, agent communication, agent transport, etc. The MAGNA platform is based on current distributed processing environment (DPE) concepts and technologies, and therefore enables implementations of applications based on both traditional client/server computing and agent-based paradigms to coexist and to be integrated. The MAGNA platform can therefore be seen as an extension of existing DPEs, enabling agent support.

The remainder of this paper is structured as follows: Chapter 2 describes the conceptual background of mobile agents, the so-called remote programming concept. Chapter 3 describes the proposed application domain from an enterprise point of view, identifying the domain structure, the actors and roles, and their particular agent platform requirements. Chapter 4 describes the architectural concepts of MAGNA, covering the agent model, the distributed agent environment, and the services provided. Chapter 5 is concerned with realization aspects of the proposed agent environment. It describes how the proposed distributed

agent environment (DAE) and its services may be mapped to a DPE, and more specifically to a CORBA DPE. Finally, Chapter 6 summarizes the paper.

2. The Mobile Agent Paradigm

The notion of agents and agent functionality differs considerably within the various different application domains where the term *agent* is used [1][3][2][14][5][9]. A commonly accepted detailed definition of agenthood does not exist yet.

Agent can still generally be defined as self-contained software elements which are responsible for autonomously carrying out one or multiple programmatic tasks. They typically act in place or on behalf of a user or a process thereby enabling task automation. Ideally, agents may exhibit anthropomorphic characteristics and behavior, i.e. they may behave just as the corresponding human role players would behave in the same setting, thereby allowing existing problem domains to be easily mapped to an agent-supported environment. To enable their autonomous behavior, agents contain some level of intelligence, which may be realized by arbitrary means ranging from simple script-based mechanisms to complex AI-based mechanisms. Despite their autonomy, most agents will communicate with the user, system resources and other agents to perform their designated task(s). More advanced agents may engage in complex co-operations with other agents to carry out tasks which exceed the capabilities of a single agent. Finally, as transportable or even active objects, they may move from one system to another to access remote resources or to meet or cooperate with other agents.

RPC vs. Remote Programming

In contrast to *static* agents, which remain in a single location throughout their entire life time, *mobile agents* roam the network to perform their tasks. All aspects of an agent's behavior including execution, communication and agent mobility are supported by respective support services of an agent run-time environment, the so-called *Distributed Agent Environment (DAE)*. Allowing agents to move around enables them to perform their tasks locally, i.e. at the location where the involved resources/entities are located. This concept is often referred to as „Remote Programming". Remote programming represents an alternative to the traditional client/server interaction paradigm, where static components located at different computers communicate remotely via a Remote Procedure Call (RPC) mechanism.

Generally, two levels of agent mobility can be identified:

- *Remote Execution*: An agent (i.e. program code and

---

1. Parts of the platform are implemented in cooperation with the PAGE Project of PUCCAMP, Brazil [10]

data) is transferred to a remote system (see Step 1a in Figure 1), where it is activated and executed in its entirety. In the context of its remote execution the agent will then interact with resource objects located at the receiving server host (see Step 2a in Figure 1). After performing its task(s), the agent may either be destroyed or it may stay at the server host in a dormant state in order to be re-activated at a later time.

- *Agent Migration*: In the course of its execution, an „active" agent may move from one node to another node in order to accomplish its task progressively. An agent environment supporting agent migration allows active agents to suspend their execution, to move to another node in the network, and to resume execution from the point where it was suspended. This requires the preservation of the agent's execution state. Migrating agents may return to the originating host in order to deliver specific results (see Step 3a in Figure 1).

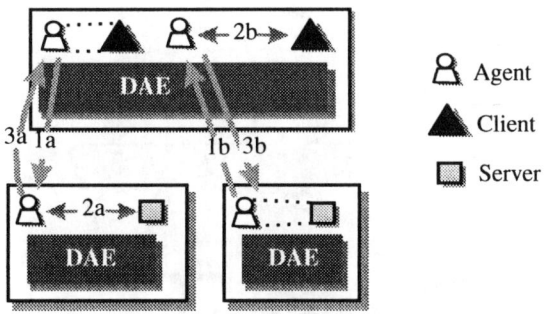

Figure 1: Remote Programming

As depicted in Figure 1, mobile agents may be used to implement client functionality as an agent and to upload such client agents to remote server hosts where client (agent)/server interaction will be handled locally. However, server functionality could also be integrated into mobile agents (see Step 1b in Figure 1). These mobile server agents could then „travel" to the client host to deliver server functionality locally (see Step 2b in Figure 1), effectively downloading the server to the client host. After performing its task, the server agent may migrate to another client or back to the originating server host (see Step 3b in Figure 1). This concept also supports the distribution of customized or tailored server agents.

Rather than considering Remote Programming and the traditional form of client/server interaction as mutually exclusive approaches, they should be seen as potentially complementary. Both concepts can be integrated in a meaningful way. The benefits of this integration are tremendous:

- Existing services can be customized on demand by installing corresponding customized server agents at existing server nodes.
- New services can be provided instantly (and customized) by installing new server agents at server nodes and by uploading service client agents to potential service customers.
- Formerly centralized services can be implemented in a distributed manner by downloading customized server agents to the clients, i.e. customer end systems.

Mobile service agents thus enable services to be deployed and provisioned more flexible than it is possible in existing telecommunication architectures.

3. Electronic Service Markets

The MAGNA enterprise model for electronic service markets is strongly influenced by the enterprise model implications of the TINA-C service architecture, and the „Information Services Market" concepts described in [15]. The basic assumption of this model is, that, enabled by technological developments and changes in the regulatory environment, the previously largely separate domains of information and telecommunication services will converge, creating a common market for electronic services in the process, characterized by complex cooperation and competition relationships between a growing number of network and service providers.

In order for the MAGNA system architecture to adequately reflect the environmental conditions and the resulting service requirements of future open service markets, the MAGNA enterprise model identifies and describes organizational domains, actors, roles, and their respective agent support requirements.

3.1. Domains, Actors and Roles

Concentrating on service provisioning aspects, the MAGNA enterprise model considers the existence of the four types of domains described below. In general, such domains are characterized by strong technological heterogeneity.

Consumer Domain

The consumer domain which is home to at least the end user and customer roles is mainly interested in locating, subscribing and using the required information and telecommunication services. Actors playing these roles require functionality for creating application agents and customizing them, describing the specific activity they have to execute. Consumers must also be able to operate/administer application agents.

If agents are to be used in the service provisioning process, agent support components will have to be provided in the customer domain. However these may be installed, operated and administered by the respective service providers.

Service Provider Domain

The service provider domain may use agents for the following purposes:

- Advertise service availability to retailers via advertising agents
- Locate retailers with the help of search agents
- Distribute service clients/ server agents (supported by up/down-loadable service agents)
- Receive retailer agents requesting services/service modifications

Clearly, the service provider will have to operate its own agent support components to achieve these goals. The service provider therefore may be both, user and agent system operator.

Service Retailer Domain

The service retailer may use agents in the following context:

- Advertise service availability via advertising agents to the consumer
- Distribute service clients/ server agents (supported by up-loadable/down-loadable service agents)
- Receive consumer agents requesting services/service modifications

Assuming that the retailer is only reseller of services provided by other service providers it may not have to operate/administer its own agent components to distribute service clients/service agents, but may alternatively use the respective agent infrastructure/services provided for this purpose by the service provider (e.g. it could trigger the creation of service provisioning agents in the service providers domain). The retailer will be agent system user, and may also operate its own agent system.

Service Broker Domain

The service broker may use agents in the following context:

- Receive consumer agents requesting services/service modifications
- Locate suitable service offerings via search agents

Service brokers will greatly benefit from the availability of a dedicated agent infrastructure/services. Since serv-

ice brokering is inherently concerned with mediation between different domains, service brokers may heavily rely on mobile brokering agents. They should therefore operate/administer and use their own agent infrastructure.

3.2. Implications for Agent Systems

As described in the previous chapter all the role players in the open service market may benefit from the use of agents in performing their respective roles. The mapping between the open service market enterprise model and the agent system supporting the respective interactions can be seen as shown in Figure 2, below.

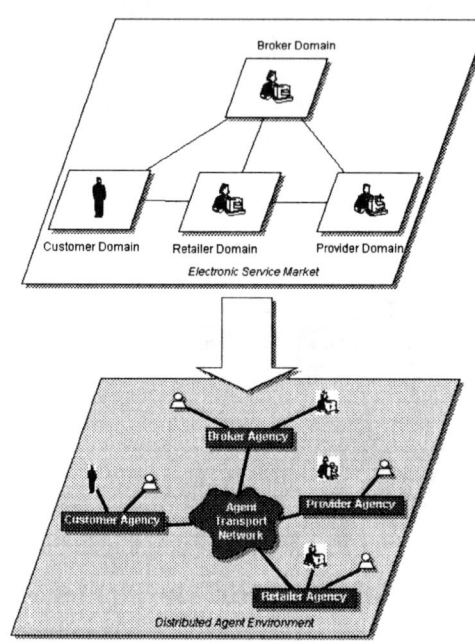

Figure 2: Mapping the Open Service Market to an Electronic Service Market

According to this proposal, each domain will be represented in the agent-based electronic service market by at least one local agent system (or agency). The role players will access their respective agent systems to instantiate mobile agents performing the required service market interactions on their behalf. All role players will thus be users of the distributed agent system infrastructure. In general, each domain will also be responsible for the operation of its local agent system(s) in the context of the distributed infrastructure.

4. MAGNA System Architecture

The design of the platform was governed by the following goals:

- Hardware/software independence
- Flexibility
- Scalability
- Support for Heterogeneity
- Integration of legacy concepts/platforms
- Extensibility

Due to these goals the architecture provides only a conceptual framework, leaving many realization aspects to the implementor.

The architecture consists of an agent model and a description of the distributed agent environment (DAE) which are described in the following two subchapters.

4.1. Agent Model

The guiding principle behind the MAGNA agent model was the support for an anthropomorphic design of agents and agent systems, i.e. agent should exhibit „human-like" characteristics/behavior to enable real world situations to be mapped to agent scenarios as easily and as directly as possible.

4.1.1. Agent General Characteristics

- **Intelligence:** As mentioned before, all agents exhibit a certain level of intelligence (as a consequence of their autonomous behavior). The level of intelligence may however vary greatly.
- **Asynchronous Operation:** An agent may execute its task(s) totally decoupled from its user or other agents. This means that agents may be triggered by the occurrence of a certain event, or by the time of day. An agent placed within the network may operate totally asynchronous to the user, performing its task by interacting with various system resources and potentially with other agents.
- **Communication:** During their operation, agents may communicate with various system resources and users. From an agent's point of view, resources may be local or remote. There is a wide range of system resources agents may access, for example, application programs, data bases, information systems, and so on.
- **Cooperation:** This attribute indicates that the agent system allows for cooperation between agent entities. The complexity of cooperation may range from a client/server style of interaction to negotiations and cooperation based on AI methods, such as con-

tract nets and protocols derived from the speech act theory. This cooperation may necessitate the exchange of knowledge information and represents the prerequisite for multi-agent systems.

- **Mobility:** In order to perform specific tasks, agents may be transported through a network to remote sites, usually in a specific run-time environment.

4.1.2. Agent Abstract Structure

The support for an anthropomorphic design of agents led to the identification of the requirement for multitasking/ multithreading support at the agent level. Agents should be able to follow multiple tasks at the same time. Execution relationships among tasks may be determined by some control flow specification. Each task is associated with information describing its current state and its priority.

An agent contains also a set of data that it processes during its execution. Such data represent information gathered through interactions with servers and the „knowledge" the agent possesses. Tasks and internal data enable an agent to act in place or on behalf of somebody delegating a task to the agent. They are sufficient to model the planned behavior of an agent.

In order for agents to also be able to deal with unforeseen events it must possess event handling knowledge.

Figure 3 illustrates the abstract structure of a MAGNA-conform agent.

It should be noted that the concrete representation of the task, event and knowledge data structures is outside the scope of the MAGNA architecture. Agent "knowledge" may for instance be realized in the form of procedural or declarative descriptions.

4.2. Distributed Agent Environment

4.2.1. Purpose and Overall Principles

The Distributed Agent Environment (DAE) represents MAGNA's platform for agent applications, i.e. it is the platform on which mobile intelligent agents constructed according to the MAGNA framework principles rely throughout their life-cycle, i.e. during creation, operation, management etc.

The DAE relies on the functionality and concrete services provided by an underlying Distributed Processing Environment (DPE), e.g. a CORBA or TINA-compliant DPE.

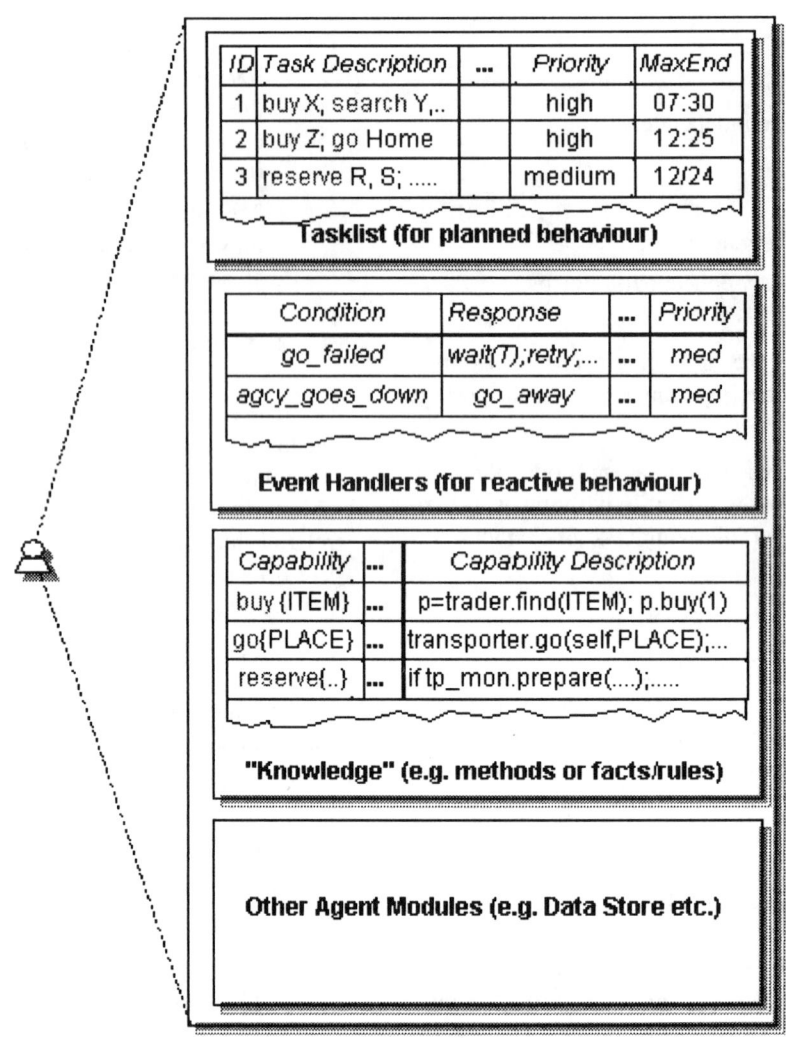

ID	Task Description	...	Priority	MaxEnd
1	buy X; search Y,..		high	07:30
2	buy Z; go Home		high	12:25
3	reserve R, S;		medium	12/24

Tasklist (for planned behaviour)

Condition	Response	...	Priority
go_failed	wait(T);retry;...	...	med
agcy_goes_down	go_away	...	med

Event Handlers (for reactive behaviour)

Capability	...	Capability Description
buy {ITEM}	...	p=trader.find(ITEM); p.buy(1)
go{PLACE}	...	transporter.go(self,PLACE);...
reserve{..}	...	if tp_mon.prepare(....);.....

"Knowledge" (e.g. methods or facts/rules)

Other Agent Modules (e.g. Data Store etc.)

Figure 3: Agent Structure

4.2.2. DAE Structure

Within our architecture, the distributed components of the DAE are called „agencies", thereby referring to the fact that their primary purpose is to host and support agents. In accordance to OMG's CORBA architecture, agencies can be seen as agent facility instances, i.e. as components offering a certain set of agent support services. It should however be noted that the agent facility services proposed in document differ from those proposed in OMG's respective agent proposal [12].

A typical distributed agent environment will consist of a large number of separate agencies (see figure 4). Agencies may be grouped into domains based on some grouping principle, e.g. a common characteristic such as ownership, security policy, direct connectivity or common requirements.

With the exception of „inter-agency" services (e.g. ex-ternal communication, transport, registration services), the services defined for an agency may only be available to agents and facility objects residing within the same, possibly distributed agency. Agencies are thus agent facility service domains. This introduces the concept of places to distributed processing environments, with service availability restricted to agency domains.

Non-distributed agencies allow a direct mapping to physical localities, distributed agencies allow logical agencies to be defined.

Whether or not „internal" facility services should be available outside an agency depends a lot on the desired degree of integration with existing DPE environments. Allowing direct access bears, however, the danger of softening the notion of an agency, and its important role as a means for structuring complex distributed environments. Allowing external accesses may for instance allow configurations where agents rely on services of a „remote" agency, thereby eliminating the very definition of an agency as

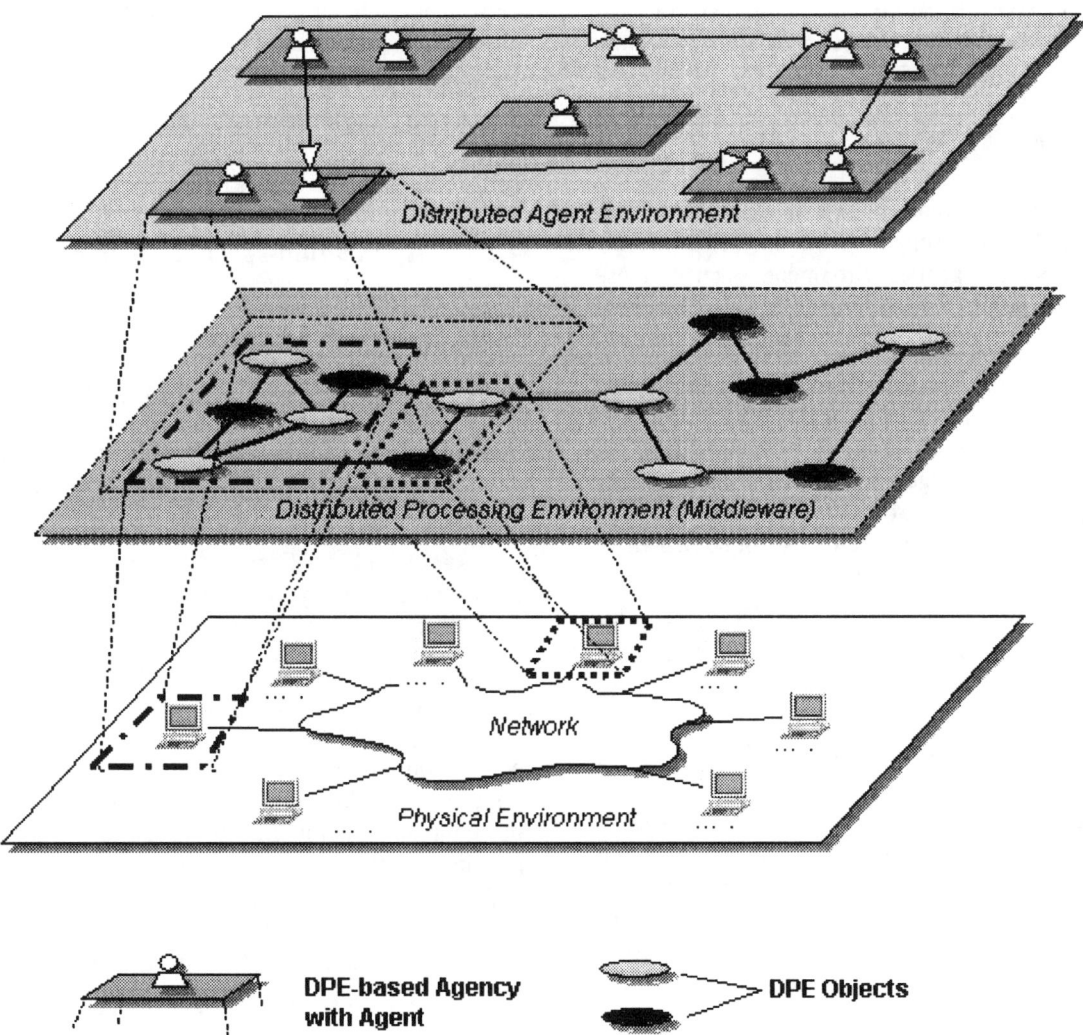

DPE-based Agency
with Agent

DPE Objects

Figure 4: The Mobile Agent Architecture (MAGNA)

a set of interacting objects exclusively providing services to agents co-located with the facility objects.

To avoid this danger, direct interactions between agents or other „internal" facility objects , and objects outside the „local" agency (i.e. objects in other agencies or „legacy objects" outside any agency domain) should be avoided if possible. Interactions should be realized via the provided agency interaction services which are described below.

Although it is assumed that interaction between agencies is realized via DPE-based services, other non-DPE based forms of communication may exist as well, e.g. email, FTP, specific inter-agency protocols, agent communication or transport protocols etc.

4.2.3. Agent Facilities

The primary purpose of the services defined for an agent facility is to provide agents located within an agent facility instance or agency with the necessary functionality to be created, to live, and to operate efficiently within the DAE.

Depending on the type of agent considered, or more specifically depending on its envisioned skills and capabilities (e.g. mobility, ability to communicate or cooperate) agent facilities have to provide a broad range of support functions.

In order to allow the dynamic and flexible realization of these functions, and in order to allow the integration of legacy systems, MAGNA agent facilities are realized in a modular fashion rather than as monolithic entities. This also corresponds to the OMG Common Facility Architecture (CFA) [11], where a facility is seen as an abstract entity which realizes a set of object interfaces (providing certain agent support services).

4.2.4. Agent Facility Services

The agent facility comprises a set of service interfaces which address different support requirements of agents within a Distributed Agent Environment. Each of these service interfaces addresses a different functional aspect of agent support.

Some services (e.g. the agent execution services, or the registration services) are essential and need to be implemented at least in a minimal form for any agency/agent facility to exist. Others (e.g. agent transport services) are only required for particular types of agents (e.g. mobile agents) and may therefore be absent or only be implemented in their most minimal form.

The following sections describe the service categories considered for MAGNA, identifying individual services and their purpose.

4.2.4.1. Common Agent Support Services

This service category comprises all agent facility services which are related to agent programming, creation, and execution, excluding those which require inter-agency interaction (e.g. communication, mobility) or which are highly specific to a particular application field or a specific agent model (e.g. AI-based intelligent or learning agents). The common agent support services provide the basis for the very existence of an agent and must therefore be supported to a varying degree by each agent facility. These services span the entire life-cycle of an agent.

Agent Lifecycle Service: as indicated by its name, this service defines interfaces that enable control of the life-cycle of agent objects. The interfaces defined in this service support agent creation, deletion, copy and movement.

Agent Execution Control Service: an agent may be in one of the following „execution states" after it has been created: idle, active, suspended, waiting. The agent execution control service interfaces allow the state of an agent object to be controlled externally (subject to access control), i.e., the agent execution can be started, suspended, resumed and stopped.

Agent Registration Service: this service enables the registration of information related to an agent (for example, the agent's name and origin information), so that it is known by the agency.

Agent Naming Service: since agents are *per se* only

known by their DPE reference they have to be assigned globally unique names in order to be identifiable within the DAE. This service assigns such names to agents.

Agent Implementation Repository Service: provides functionality for the storage and retrieval of agent code (complete agent code or libraries of agent classes).

4.2.4.2. Application-specific Agent Support Services

These services support agents during their execution by providing them with additional capabilities such as mobility or communication. The number and scope of application-specific agent support services depends largely on the envisaged agent skills and application area, i.e. on the tasks the agent is supposed to carry out.

Agent Mobility Service: this service enables an agent to move from one agency to another. Both mobility levels, remote execution and migration, can be provided.

Agent Communication Service: this service supports communication between agents. Agents can communicate via messages, method invocation or a blackboard mechanism. Communication through messages may be done point-to-point, by multicasting or broadcasting. Furthermore the agent communication service includes also support for semantic communication, where concepts like KQML (Knowledge Query and Manipulation Language), KIF (Knowledge Interchange Format), EDI (Electronic Data Interchange), among others are important.

Agent Event Service: this service enables the management of general types of events. Agents may register themselves to be advised of the occurrence of a certain event. When that event happens, each agent registered to that event receives a notification.

Agent Task Control Service: this service enables control over individual agent tasks. Similar to agent execution control service, the agent task control service enables the requestor to create, delete, enable and disable agent tasks or control their priority.

Agent Security Service: this service provides support for guaranteeing agent privacy and integrity during its travel. This service includes support for agent encryption, decryption, authentication, credential generation, among others.

Agent Persistency Service: this service enables the storage and manipulation of the persistent data state of an agent.

Agent Intelligence Support: during its execution, the agent may use knowledge to analyse and decide what should be its next action. The reasoning mechanism may be offered by the platform.

Agent Transaction Support Service: this service comprehends functionality for supporting correct and reliable execution of agents in presence of concurrency and occurrence of failures. This service supports the execution of reliable workflows over the distributed agent environment.

4.2.4.3. Agency Services

This service category comprises all services which are necessary for the organization and management of a single agency. They differ from the services of the previous categories in that they are not directly concerned with the execution of agents, nor do they directly support them. Their main purpose is to support and maintain the mechanisms on which the other services rely.

Agency Initialization/Startup Service: this service coordinates the installation of all the components that comprise an agency, when it is started up.

Agency Configuration Service: this service allows the management (list, enable, disable, add, delete) of services available at the agency.

Agency Management Service: this service will overlap with other agency services. The notion of an „Agency Management Service" is a container for many other agency-related services. This service comprises support for agency fault, configuration, accounting, performance and security management.

Agency Security Service: this service comprises functionality to protect an agency from attack of malicious agents.

4.2.4.4. Distributed Agent Environment Services

This service category comprises all services which are necessary for establishing and maintaining a distributed agent system consisting of multiple agencies.

Domain Management Service: according to section 4.2.2, agencies may be structured forming domains. This service supports the management of such domains, i.e., insertion and deletion of agencies in a domain, list of agencies pertaining to a domain, etc.

Agent Location and Tracking Service: due to agent autonomy, an agent may roam a network without its user (human-user or application) knowing *a priori* to where it goes. This service allows for finding agents and providing information about the agencies it visited.

Trading Service: as one of the fundamental services

for supporting the management of information in the distributed agent environment, the trading service allows for registration and dynamic discovery of services and servers in the environment. The discovery of services/servers may be done by complex queries to the trader, specifying requirements that the service/servers must fulfill.

4.2.4.5. User Support Services

This category represents the support for the interaction of a human-user with the agent system.

Agent User Console Service: this service supports the human-users in the „pre-service" and „post-service" phase of agents, i.e., providing functionality for editing, testing, debugging, installing and de-installing agents.

Generic Agency Control Service: this service provides a GUI (Graphical User Interface) for users and administrators to control the activities of agents and agencies. Through these GUIs, agent users may get information about the current status of their agents (for example, current execution state or current location) and administrators obtain information about the agents executing at an agency (for example, resources used, statistical information, accounting information, etc).

5. Engineering and Technology Aspects

Architectural basis for the MAGNA agent platform is the Object Management Architecture (OMA) [13], a DPE architecture developed by the Object Management Group. As is shown in Figure 5, the realization of agency services relies on the services provided in the underlying DPE. The distribution of the objects realizing these services is fully transparent at the agency level, i.e. the implementor of a certain agency service is not aware of the physical location or distribution of the objects or components providing the service.

An agency and all the agents within it map to a group of CORBA compliant client and server objects. Via the underlying CORBA DPE service, agents may utilize the services of the agency. Agency service components are always servers of agents running at the agency, however they may also be clients of other agency or external DPE services. This concept is flexible enough to also allow agency services to be realized by agents performing as both, clients and servers.

To allow agents to access other CORBA services, especially other agent facility services, and in order for it to provide its services as part of the agent facility, the respective agent (and thus the agent language) must contain „hooks" allowing agents to access other facilities via the

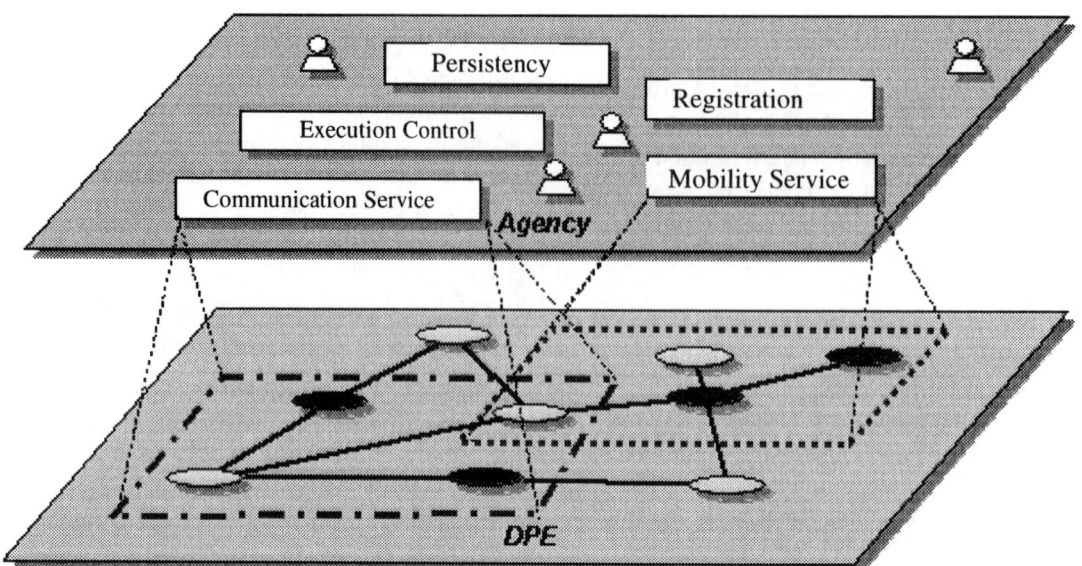

Figure 5: DPE-based Realization of an Agency

underlying DPE, and to be perceived by other facilities/ agents as a DPE-enabled server component. The hooks are realized as system or user defined libraries accessible from within the agent programs. During execution of the agents the hooks are invoked either from within the interpreter (in case the agent execution environment is interpreter based) or from within the compiled agent executable.

6. Summary

The emerging open electronic service market requires the provision of advanced service provisioning platforms, where both RPC-based and agent-based applications have to be supported. This paper has provided an overview of the MAGNA distributed agent environment for mobile intelligent agents, which represents an extension of existing Distributed Processing Environment technology in respect to intelligent mobile agent support.

7. References

[1] Communications of the ACM Journal "Intelligent Agents", Vol.37, No.7, July 1994

[2] N.S. Borenstein: "Computational Mail as Network Infrastructure for Computer-Supported Cooperative Work", CSCW«92 Proceedings, Toronto, November 1992

[3] K. Indermaur: „Baby Steps", BYTE Magazine, pp. 97-104, March 1995

[4] T. Magedanz: „On the Impacts of Intelligent Agent Concepts on Future Telecommunication Environments", in: Lecture Notes on Computer Science 998 - „Bringing Telecommunication Services to the People - IS&N'95", A. Clarke et al. (Eds.), pp.396-414, ISBN: 3-540-60479-0, Springer Verlag, 1995

[5] T. Magedanz, K.Rothermel, S.Krause: „Intelligent Agents: An Emerging Technology for Next Generation Telecommunications?", IEEE INFOCOM, San Francisco, California, USA, March 24-28, 1996

[6] T. Magedanz, T. Eckard: „Mobile Software Agents: A new Paradigm for Telecommunications Management", IEEE NOMS, Kyoto, Japan, April 15-19, 1996

[7] T. Magedanz, S. Krause: "Mobile Service Agents enabling "Intelligence on Demand" in Telecommunications". IEEE Global Telecommunications Conference, London, United Kingdom, November 18-22, 1996

[8] T. Magedanz, R. Popescu-Zeletin: "Towards Intelligence on Demand - On the Impacts of Intelligent Agents on IN". 4th International Conference on Intelligent Networks (ICIN), Bordeaux, France, December 2-5, 1996

[9] M. Mendes , W. Loyolla, T. Magedanz, F.M. Assis Silva, S. Krause: "Agent Skills and their Roles in Mobile Computing a Communications", IFIP World Conference on Mobile Communications, Canberra, Australia, September 2-6, 1996

[10] M.Mendes, C.Mendez, O.Falsarella, P.S.Silva, I.Fontes, C.Tobar, W.Loyolla: „Architectural Considerations about Open Distributed Agent Support Platforms", ISADS'97, Berlin, April, 1997

[11] OMG Document 95-1-2: „Common Facilities Architecture", Revision 4.0, January, 1995

[12] OMG TC Document cf/96-08-01, „Mobile Agent Facility Specification - Submission", August, 1996

[13] OMG TC Document 92.11.1: „Object Management Architecture Guide -", Revision 2.0, September, 1992

[14] C. Guilfoyle, E. Warner: „Intelligent Agents: the New Revolution in Software", Technical Report, OVUM Limited, 1994

[15] TINA-C Overall Stream Deliverable: „The Market for Information Systems and its Demands on TINA-C", Draft Version, April 26, 1995.

A Distributed Programming Platform using Mobile Agents

Djamel H. Sadok

Universidade Federal de Pernambuco
Departamento de Informática
Recife, Pernambuco, Brazil 50732-970

Judith Kelner and R. A. Silva

Universidade Federal de Pernambuco
Departamento de Informática
Recife, Pernambuco, Brazil 50732-970

Abstract

Current distributed computing is based on rigid paradigms such as client server computing, that lack run-time adaptive behaviour. The designer of such systems has to consider all possibilities during the design stage and the services that are offered remain constant until a new re-design is done.

Although there have been attempts to provide different programming languages that consider programming both distributed and non distributed applications the same, the two remain however different. New paradigms are needed that consider factors such as dynamic adaptability, interaction, ability to monitor and react to the environment, mobility, network latency, partial failure and concurrency. In this work, we show that the agent based programming paradigm, agentware, *is closer to fulfilling some of these requirements and describe the proposed Agentware Virtual Machine (AVM), a generic distributed agent-based computing environment.*

With the emergence of intelligent agent-based technology, a number of distributed applications are being re-engineered *taking advantage of this new design methodology.*

1 Introduction

Existing distributed computing is largely based on rigid paradigms such as client server computing, that lack run-time adaptive behaviour. It is during the initial design stage that the designer of such systems has to consider all possibilities. Once this stage is concluded, the offered services remain constant until a new re-design is done. Today's distributed systems architectures often use rigid (static) communication protocols and sometimes even define new ones.

Recent work in the area of distributed computing such as the OMG's CORBA adopts a unified view of objects. In these systems an object, whether remote or local, is defined in terms of its interfaces (declared using an Interface Description Language - IDL) and its methods. The programmer writes the same code when dealing with local or remote objects, it is the responsibility of the underlying mechanisms to take care of the execution aspects.

[JWK94] argues that distributed systems not only failed to paper over the distinction between local and remote objects using their object abstraction, but that they also failed to support basic requirements of robustness and reliability as they do not consider network latency, failures and concurrency. With the need for wider distributed systems, these problems are even greater.

Although ideally the use of new programming languages specifically designed for distributed programming has been recommended throughout the literature, this work is of the view that this alone is not sufficient and that under a new programming paradigm, *agentware*, distributed programming can be made both easier to develop, to understand and more powerful. We chose to use run-time interpretable code as the basis to provide mobile active distributed processing objects or *agents*.

First, client server computing and current distributed systems and platforms are discussed. Next, basic agent concepts and definitions are reviewed. This paper then describes the Agentware Virtual Machine (AVM) which implements active distributed processing services. Finally it ends with some concluding remarks about this new paradigm and the AVM development platform.

2 Distributed Programming

With the rapid growth of networks such as the Internet, a more complex computing scenario has emerged, offering users access to a wide variety of information, products and services.

Initially distributed systems were programmed using existing sequential languages with the addition of new library procedures to handle the distribution aspects. Clearly, this approach soon turned out to be unsatisfactory and new programming languages more adequate for distributed computing were investigated. A review of such languages and their design aspects is given in [HEBT89]. When using a specially designed language, the programmer is shielded from some of these difficulties. Such a language reflects the underlying distributed programming paradigm and allows the programmer to structure the application in a higher level of abstraction easier to understand.

Our definition of distributed systems extends those of [HEBT89] and [Mul89] in that communications between their elements is not limited to passing *data* messages but self-contained independent objects (with programs, "knowledge", state information, data, intermediary results, etc) as well. This opens up an entirely new paradigm for the distributed programming

103

problem as described later on in this document.

The justification of this unified vision is that whether a call is local or remote it has no impact on the correctness of a program. Most of existing distributed platforms adopt an object oriented unified vision of objects all the way down.

2.1 Limitations of Current Approaches to Distributed Programming

Traditional client server applications and distributed applications do not scale well when the network delay increases and where local decisions are required. A typical example is the management of a network fast switch where there is a large volume of highly volatile data to be interpreted at a central point or the manager. Some solutions use cacheing mechanisms to avoid performance problems but this is not always a good solution. Cacheing cannot support highly volatile distributed data.

Other, more infra-structure related problems, emerging from the use existing distributed systems include:

- at least parts of the distributed system must be pre-installed and well configured before the use of applications[GDMH95];

- operating systems have to implement a minimal set of distributed protocols into the kernel which restricts its genericity and adaptability;

- when developing applications, the programmer must explicitly handle efficiently the data exchange;

- installation, testing and debugging problems happen during the installation and configuration processes of different parts, or modules, of a distributed system;

- the low performance and high cost per machine-user of existing distributed platforms;

- work on the standardisation of distributed systems is a lengthy process and generally produces large and complex specifications. For instance, although the OMG effort started in 1991, there is still ongoing work to define new Object Request Brockers (ORBs) and the CORBA specifications continue to evolve;

- with the emergence of different object-based distributed systems, it is unlikely that one of these would dominate and hence there would be no interworking at this level;

3 Objects for Interworking

The use of object abstraction in integrating heterogeneous systems and components characterises recent developments for the integration of distributed architectures. It is not however clear if any of the competing object models will dominate. Some solutions for interoperability have been suggested in [MH93] where a reduced (common) instruction object model is proposed to achieve object level interoperation.

In this work we show that another level of integration made at a special type of objects that are active, or agents, further consolidates interoperability, introduced to model objects with dynamic behaviour and to capture their capabilities and knowledge.

4 Agent-based Distributed Computing

Agentware is not going to replace current efforts such as CORBA, DCE, OLE, etc, but it brings to the spotlight extra capabilities that many distributed applications would benefit from.

Although many researchers are working with agentware, and where some of these even specified entire architectures (e.g. [ea95]), platforms offering well structured interfaces to distributed programming in this new paradigm remain to an extent lacking. This work is a step taken in this direction.

The new paradigm does not impose any protocol, state machine protocol or the like. It provides advantages such as:

- bandwidth conservation by keeping the processing closer to the data;

- situations that were not considered, during the design stage, do not pause *long term* problems and can be avoided;

- dynamic extensibility of distributed applications and resource location;

- initial design can be made to concern itself with failure and performance issues by moving data and processing closer to the resources;

- design of more powerful intelligent applications with adaptability, autonomy, interaction and initiative;

Agentware technology still has a number of drawbacks mainly in the areas of security and efficiency. Although the latter may seem easier to overcome with better optimised hardware based implementations of interpreters, security concerns remain a major preoccupation. The following are some of current limitations of agentware technology:

1. interpreted scripts are not however adequate for real-time applications such as real-time manufacturing control, process control, multimidia applications, etc;

2. the control of the use of resources in distributed environments, accounting and monitoring functions are yet to be addressed;

3. safety against, intentionally or not, ill behaved software;

4. agentware requires the availability of script interpreters throughout the network;

5. for security reasons, these languages offer limited controlled access to existing tools and programs, and hence offer limited integration;

6. although it is desirable that agents may be able to negotiate with their environment, it is not clear yet how this can be done[Maj96].

5 Distributed Processing with AVM

AVM offers a generic distributed environment with the following characteristics:

- it offers a parallel execution environment where sizeable applications can be divided into manageable tasks to be executed in parallel on separate processing resources;

- intelligent configuration management and monitoring in that operations for re-configuration and the management of agents are supported during run-time;

- more fault tolerance as a result of the previous configuration point;

- intelligent distributed processing in that environment resources are programmable in real-time;

- load balancing of distributed processes;

- support for agent migration in that programs may terminate executing in a node and move to another one under certain circumstances such as during a fault or a similar problem;

- support for communications among agents, dynamic loading of remote objects and remote invocation of agents and services;

- support for proxy agents enabling AVM to interact with its environment and mobile agents from other platforms;

- the whole distributed platform may be mobile;

Although many of the above objectives may seem common to many existing distributed systems, AVM takes a radical approach to processes in that they are portable at the code level, self-contained and free to move between nodes in a network.

5.1 AVM Agents

AVM distributed objects, or agents, inherit the concept of division between interface description and implementation as supported in current object oriented languages. AVM uses agents for the mapping and representation of both system and application processes as well as services, offering a high level structuring of distributed programs that may be visualised using a graph.

Unlike the generic agent structure suggested in [GDMH95], an AVM agent is defined as an object containing the parts: header (who), objective (what), body (how) and status. The header tells who this agent is, the objective part shows the nature of its mission, its body part contains information on how it intends to undertake its work and to finally the status part contains intermediary results, processing information, etc, as shown in figure 1. The fields shown in this figure include:

1. a textual description of the agent;

2. data: parameters used at execution;

3. intermediary results;

4. active behaviour: unlike in object oriented languages, AVM objects have active behaviour associated to them allowing them mobility, decision support at run-time, etc;

5. state information: about the environment of the execution of the agent. Such information is useful to keep when agents are migrating and to recover from failures, etc;

6. mailbox: used to receive messages from other agents and its environment.

7. time to live: used by AVM to control to remove old agents from the network;

8. credentiels: include the identity, domain name and certificates of an agent.

9. history information: allows AVM to know agent's origin, itenary etc;

10. header checksum to detect agents information integrity, stronger mechanisms such as hashing algorithms (e.g. Message Digest 2 - MD2 and MD4) may also be used;

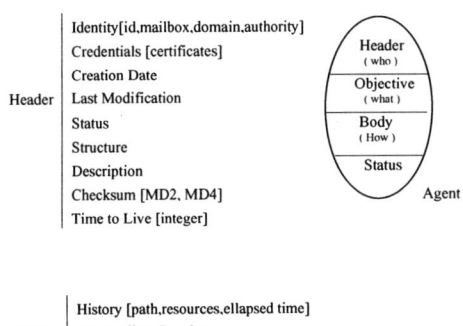

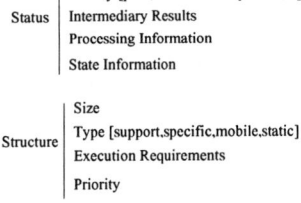

Figure 1: Agent Format

5.2 AVM's Functional Structure

The environment supported by the virtual machine includes services for locating and managing mobile agents, load balancing, garbage collection, interaction with the environment, communications and synchronisation, security and authentication, decision support, etc. AVM allows the construction of an active network of agents to appear as a single machine where agents execute concurrently. This *mobile* network of nodes may be configured dynamically during run-time or in a static manner by the application. The AVM software typically consists of four main parts:

1. the interpreter used for the execution of agents;

2. the receptionist, a special agent that controls access to AVM resources at a node;

3. the scheduler, responsible for agent management;

4. miscellaneous AVM support agents *and* services responsible for the control and management of AVM functions and services such as remote execution, load balancing, etc;

5. AVM user agents which are application specific;

6. proxy agents are used to support AVM interaction with non AVM software including directory services, the operating system, editors, tools, packages, foreign agents, etc;

7. library classes for use by both support and user agents.

A minimal node configuration includes the interpreter, the receptionist, the scheduler and an object loader. Services that are not present within a node, may be loaded on demand from other nodes when necessary.

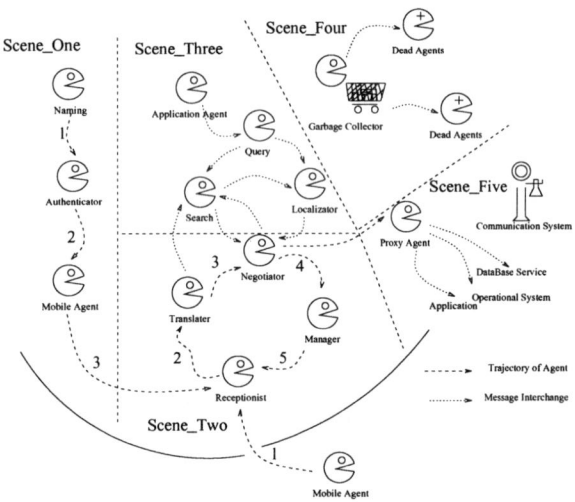

Figure 2: Functional Structure

Although these services seem similar from the structural point of view, in terms of the implementation, some of these such as the configuration manager can be mapped onto agents, whereas others remain services that are implemented just like in existing distributed platforms. Services represent the

static view to distributed system functionality whereas agents provide a more dynamic one. Consequently AVM adopts some of its basic functionality using services whereas more elaborate ones make use of agents.

Next are described the main components of the AVM functional structure as well as the different services this plataform offers.

5.2.1 Naming and AVM Domains

In order for AVM to keep track of its agents there is clearly a need for an agent naming scheme. Naming is also important to services such as security, configuration management, garbage collection, etc. A special naming scheme has been adopted as already shown in figure 1. AVM names are a combination of an agent, authority and domain names.

5.2.2 The Receptionist

A given AVM node must have a reception service, which checks the credentiels of incoming agents and directs them to the appropriate location where they can be offered the services they seek to accomplish their tasks. The receptionist is both the agent who gets first contact with agents (foreign and returning agents) entering the premises and is locally responsible for dispatching visitors.

5.2.3 Security and Access Control

Currently AVM implements basic simple security and access control functions based on the use of AVM domain and authority strings[1]. The assumption is that an AVM agent, including the receptionist and scheduler, trusts agents from its own domain. Domains may also opt to establish trust paths between themselves.

Strong authentication based on the use of agent certificates including the issue of how these certificates are issued to agents during their lifetime are left for further study. For the time being, AVM access control is based on a combination of the use of AVM domain names, Java and the underlying operating system's security support.

5.2.4 The Scheduler

The scheduler is a service that offers flow control operations between concurrent AVM target agents both locally and at the network level. Examples of such operations are the initiating of new agents, remote submission of agents, agent termination and suspension, redirecting, synchronisation, etc.

AVM uses the agent as the unit of parallelism. At this level of granularity, the amount of communications between agents is controlled by the programmer and can be reasonably low. Wihin an agent there may be different threads executing in parallel as shown in figure 3.

---

[1] This is similar to SNMP security based on the use of community strings.

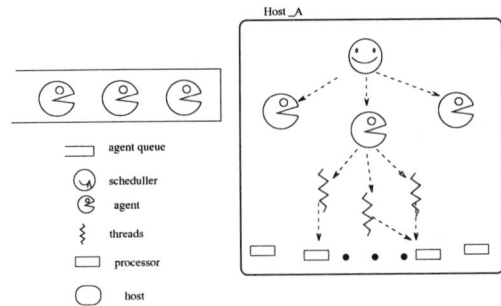

Figure 3: AVM Support for Parallel Execution of Agents and Threads

5.2.5 Communications and Signaling

It is important for AVM to support both the ability to interchange data between agents as well as events (or signaling). A problem to deal with here, is what to do with messages or signals sent to terminated agents. Furthermore, AVM should interact with existing Internet protocols such as http, ftp, telnet, etc through proxy agents and servers, see figure 2. It defines both inter-agent and application defined signaling. AVM presents a loosely coupled distributed system suitable therefore for running large grain parallel applications. The AVM communication model provides blocking and non-blocking send / receive operations as well as broadcast, multicast and ordered multicast modes. The use of asynchronous mode may generate problems due to the loss of order of messages or some of them.

[Hu96] discusses the question of interaction protocols for agents in distributed environments and proposes the use of a two-phase commit protocol, such as the one from[LS76], when dealing with *group* services. First a service probe is used, then the second stage would involve a commit or a rollback action. Note that this approach is also taken in the OSI CCR service element for concurrent services. It is assumed that any agent may send a message to another one. Support for group oriented communications such as a two phase transaction protocol is left for further extensions of the proposed environment.

5.2.6 Performance and Load Balancing

Distributed systems establish different schemes for load balancing. These can be classified into two groups: static application defined load balancing and runtime dynamic load balancing. Some environments such as the Parallel Virtual Machine adopt a *Pool of Tasks* paradigm [ea93] where there is a master that sends out tasks to slaves as they become idle.

AVM takes a different approach to providing runtime adaptive load balancing. AVM agents may query their environment to continuously monitor the state of a node. A load monitoring agent is responsible for determining when agents may be advised to migrate to other nodes if possible or to avoid a given one. The load monitoring agent interacts with the receptionist and scheduler in order to avoid situations where performance is degraded due to the presence of a large number of agents.

5.2.7 Quality of Service Negociation

Agents reaching a given AVM node may negotiate some QoS parameters. Examples of these include resources such CPU time, memory, access to communication channels, etc. Agents may even book some of these resources for themselves or for others. Further work is needed to integrate AVM QoS control with those from underlying subnets such ATM, mobile networks and TCP/IP.

5.2.8 Agent Migration

An agent is an active mobile program and may therefore chose to (through some pre-programmed process of deduction) move to another node. An example would be an agent that is trying to purchase a given number of items.

An interesting aspect of agent migration to be considered is the ability for agents to perform *roaming while maintaining connections* or on-line migration. AVM allows agents to migrate between network nodes to resume processing. This process is a two step one. First the machine must ensure that the agent may be migrated by checking its status and negotiating with the destination AVM node. If all is well, only then that the agent may migrate.

It is also important to have a service whereby an agent may probe the network in order to determine which nodes may offer services that it requires. This would enable the agent not to waste time roaming through the network or simply getting lost. This is also useful for allowing applications to perform self configuration before execution. Furthermore, in time critical applications, it would be important to have such a service avoiding unnecessary agent transfers.

AVM nodes are expected to cache information about the passing of agents. This may be useful in the case users may want to locate an agent for any reason including its removal, as see later in sub-section 5.3 dealing with configuration management.

5.2.9 Fault Tolerance

Ideally the recovery from processor failure should be transparent to the user. A more common approach would be to give the programmer to determine which processes, objects or data are important to the application and therefore must be recovered and the mechanisms for their recovery in a high level abstract language. Examples of languages that adopt this approach include Argus and Aeolus [HEBT89].

Failure of an AVM host should be automatically detected by the AVM machine. Applications may use a *monitoring agent* to identify such conditions. They may then decide to migrate the agent to another network host or leave control of this with the AVM scheduler. In other words it is not only the responsibility of AVM applications to attempt recovering from host failure but the virtual machine itself independently of the application initiates fault tolerance procedures.

5.3 Agent Location and Configuration Management

Agent location and configuration information is very volatile and therefore requires a very exhaustive and elaborate solution. A solution based on the use of static tables is clearly out of consideration as the agents are not persistent objects and are mobile entities. Solutions based on the use of directory services, distributed information services such as the Internet Domain Name Server (DNS), distributed databases, etc, are not sufficient but may be integrated through proxy agents.

5.3.1 A Configuration Platform for AVM

The platform consists of the following entities as shown in figure 4.

- Anode - Agentware Node: is a network host with support to AVM mobile agents;

- Ag - Agents: are software programs together forming an agent based application;

When AVM stations (Anodes) are the only ones that may hold data relevant to agent configuration, we may see, as well as the available resources to deal with requests from managed agents, different alternatives to the type of information maintained by these stations:

1. each Anode maintains a table *home-list* of its own active agents it created. The configuration process consists of updating this list after the moving of an agent. The AVM station which currently hosts the mobile agent (*guest-Anode*), sends a message to the AVM owner of this agent (*Home-Anode*), see figure 4 (a).

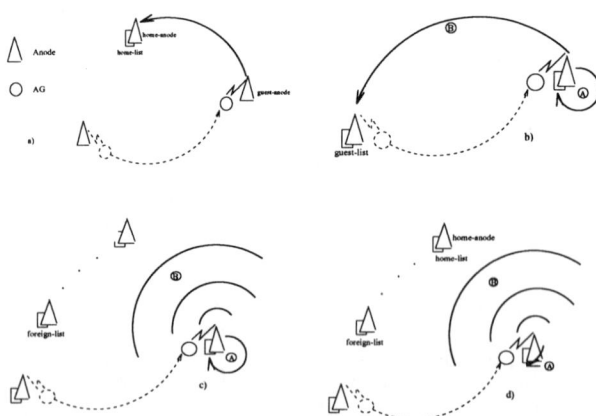

Figure 4: AVM Configuration Structures

2. each of the Anode stations maintains a list (*guest-list*) of the agents currently hosted. The process of configuration, shown in figure 4 (b), is limited to the registration of agents in the current Anode host (step A) and the removal of its record from the previous station (step B). The identification of the more recent of previous Anodes the agent passed through, should be available to the current Anode, in order to allow this to send a removal message of the previous record. This information could be contained in the mobile agent itself or in an another computational entity.

3. each Anode maintains information on the previous hosted agents as well as current ones in a register known as a (*foreign-list*). Here, the process of configuration, shown in figure 4 (c), consists of the agent registration into the current AVM host (step A) and the updating of the records (information) in all the previous agents (step B). In this updating process, the information spreading may take place in the form of a broadcast to all AVM stations and previous hosts of the considered agent; or in the form of a multicast only to the previous AVM nodes (Anodes) that, at a given moment, already hosted the agent.

4. each station (Anode) has data on the agents it never hosted, as well as those it is hosting or has already hosted as shown in figure 4 (d). This happens through the updating of Anode tables each time a broadcast is made to adjust the configuration. The difference of this method from the one above is in the way messages are manipulated or processed by the stations.

Future work would investigate the possibility where only agents hold location information independently of the platform.

5.4 Proxy Agents

Proxy agents may be used to allow AVM agents and services to interwork with external services. For example X.500 directory services, DNS services, WWW support, SNMP network management services may be used through proxy agents.

6 Implementation of AVM

The current implementation of AVM is based on Java, where basic support agents and services such as the receptionist, the scheduler, the loader and configuration manager have been written in the form of java classes reflecting the format shown in figure 1. A shopping application has been developed to hightlight some of the benefits of AVM, see figure 5. The scenario adopted in the following: first the user sends a probe agent asking on where given items are available and their respective prices. Such a request may be broadcast to all nodes in an AVM network or to selective ones. Once replies are returned, the application configures which AVM nodes that would take part into the application, based on the received replies and some selection criteria. It would then submit a shopping agent to the first AVM node in the configuration.

The shopping agent will then be received by the receptionist which would then check the services the coming agent is seeking as well as its credentiels and decide whether it is elligible or not for access as shown in figure 2. In the case where the agent is granted access, the scheduler is then responsible for initiating its execution. Once its task, that of buying a given item is carried out, the agent is issued a bill and allowed to leave the shop (local AVM node). The agent then migrates to the next node where it carries on with its shopping and so on. Finally, the shopper returns back to its user with information of what it bought as well as the receits of purchases it made.

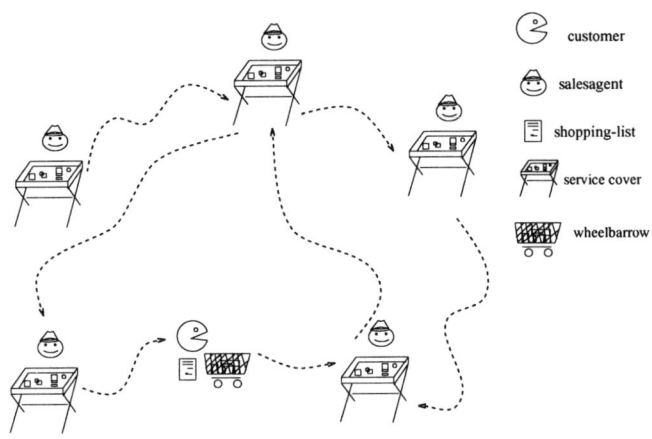

customer

salesagent

shopping-list

service cover

wheelbarrow

Figure 5: Shopping using Software Agents

When we started the implementation of AVM modules, many of the team members had difficulties coming to grasp with the new paradigm. As a result, initial implementations of services and agents where modelled on the client server paradigm. It took some considerable efforts to redirect our coding practices into using the new paradigm instead of thinking in terms of communications between client and server and to restruture accordingly our code. We chose to think of each module or agent as a small independent being. This helped us imagine the widest types of applications and scenarios that may benefit from agentware. For the first time we found ourselves refering to program modules or agents as persons in charge of specific tasks. For example, to illustrate the role of the agent responsible to receiving incoming agents, we used the name receptionist and drew an analogy with the tasks of a company receptionist, and so on.

Unlike debugging serial programs, the debugging of distributed applications remains a difficult task. Under AVM two approaches are taken to solve this problem:

- a debugger agent would run on each the application nodes;

- provide some basic steps or guidelines on how AVM code can be debugged and problems traced.

Developing a fully fledged debugger for AVM is a considerable undertaking. Initially the second approach, i.e. that of providing users with some basic guidelines for identifying problems into their AVM applications. These may also be helped using some visualisation tools for monitoring the execution. A next step is the implementation of a distributed debugger which uses Sun's , *jdb*, used for debugging Java programs. Although a complete debugger will not be available for some time, there are immediate major problems that we need to deal with in the context of debugging AVM applications, namely: preventing deadlocks and livelocks.

7 Summary and Conclusions

In order to isolate AVM programmers from the complexities of programming using script based languages, a more natural meta-language as well as graphical tools and distributed debuggers, are required. This are however the object of future work. We chose the definition of a new meta-language instead of extending the Java language as the former would be easier to use from the user's point of view. In fact we see today many systems that define interpreted meta-languages or scripts to enhance their application programming interface. Some solutions such as IMP[BB95] adopt the use of a script language for device access and interaction. Although the Java language is in itself interpreted it is not as "natural" as these scripts.

7.1 Interworking

The restriction of AVM into interacting with agents within a given domain is not due to technology restrictions but rather as an interim solution to improve security and to re-enforce access control. Once stronger security services are put into AVM based on the use of cryptographic algorithms and strong authentications schemes, it would then be feasable to consider interworking among AVM domains. Some systems such as ANSA make use of delegation to link their traders, ANSA entities similar the locater in terms of their role.

A graphical interface has been built to allow the monitoring of agents through an AVM platform in real time. One advantage of AVM (agentware) is the direct manipulation of objects or more specifically agents as shown in the example. This feature has been identified as an important one by software designers [Shn87].

7.2 Conclusions and Future Work

A new paradigm for distributed processing has been discussed and implemented into a virtual distributed agent based machine. Problems inherent to distributed processing such process control and management, process mobility (migration), location and configuration have been discussed. Although functions such as security have not yet been dealt with, AVM has shown that agentware technology brings a new horizon of benefits for distributed computing, that it is much more flexible, requires much less re-design efforts, and provides simpler solutions to some complex distribution problems such as fault tolerance, configuration management and application migration. Agentware is not going to replace current efforts such as CORBA, DCE, OLE, etc, but it brings to the spotlight extra capabilities that many distributed applications would benefit from.

The machine has been designed to be generic enough to be used with different programming languages and hardware environments. Further studies of other scripting and interpreted languages for mobile agents is required as well as their integration into AVM.

The Java language cannot be used by itself to convey high level semantics that are specific to given applications. Further work is needed to define application oriented languages that can capture and represent application *knowledge*. A given application generally tends to use its own specialised vocabulary. For example database access clients will use a different language semantics to that of a distributed shopping application. Whereas the first one will have constructs for querying and locating the information base, the second application on the other hand will have constructs that allow agents to make decisions about products for sale, compare prices, order items, make payments, etc;

A future extension to AVM agents under consideration is the inclusion of declarative agents with a rule based language. One way we are considering in doing this is the implementation of declaractive agents using Java based procedural agents. Similarly other high level languages, more specific to applications, such as transaction processing may also be implemented in the form of declarative agents.

8 Acknowledgements

This work has been undertaken as part of the Project SAM (Sistemas de Autoria Multimidia), sponsorship by the CNPq/Brasil of the ProTem-II program.

References

[BB95] John Bates and Jean Beacon. Supporting interactive presentation for distributed multimedia applications. *Multimedia Tools and Applications - An International Journal*, 1(1):47–78, March 1995.

[ea93] Al Gueist et al. Pvm 3 user's guide and reference manual. Technical Report ORNL/TM-12187, Oak Ridge National Laboratory, May 1993.

[ea95] D. Chess et al. Itenerant agents for mobile computing. *IEEE Personal Communications*, 2(5):34–49, October 1995.

[FTM94] H. Sunahara F. Teraoka, K. Uehara and J. Murai. Vip: A protocol providing host mobility. Technical report, Sony Computer Science Laboratory Inc., 1994.

[GDMH95] C. Tschdin G. Di Marzo, M. Muhugusa and J. Harms. The messenger paradigm and its implications on distributed systems. In *Proceedings of ICC'95 workshop on Intelligent Computer Communication*, 1995.

[HEBT89] Jennifer G. Steiner Henri E. Bal and Andrew S. Tanenbaum. Programming languages for distributed computing systems. *ACM Computing Surveys*, 21(3):260–322, September 1989.

[Hu96] Yuh-Jong Hu. Interaction protocols for intelligent agents in distributed services environment. In *IFIP/IEE International Conference on Distributed Platforms*, pages 252–256, February 1996.

[JWK94] A. Wprath J. Waldo, G. Wyant and Sam Kendall. A note on distributed computing. Technical Report SMLI TR-94-29, Sun Labs., November 1994.

[LS76] B. Lamson and H. Sturgis. Crash recovery in a distributed data storage system. Technical report, Computer Science Laboratory, Xerox, Palo Alto research Center, CA, 1976.

[Maj96] Steven D. Majewski. Distributed programming: Agentware, componentware, distributed objects. Technical report, University of Virginia, 1996. Draft Document under construction.

[MH93] F. Manola and S. Heiler. A RISC object model for object system interoperation: Concepts and applications. Technical Report TR-0231-08-93-165, GTE Labs Inc., August 1993.

[Mul89] Sape Mullender. *Distributed Systems*. Addison Wesley, 1989.

[Shn87] Ben Shneiderman. *Design the user Interface: Strategies for Effective Human-Computer Interaction*. Addison-Wesley Publishers Ltd., 1987.

Session 2B

Service Development

Chair

Chair

Jung W. Cho

A Framework for ADS Application Software Development Based on CORBA

Stephen S. Yau and Shaowei Mao
Computer Science and Engineering Department
Arizona State University
Tempe, AZ 85281-5406 USA
email: {yau, mao}@asu.edu

Abstract

Autonomous Decentralized System (ADS) which has the characteristics of on-line maintainability, on-line expandability and fault-tolerance has been successfully used in many distributed computing domains, such as factory automation, traffic control, office automation, nuclear power plants. In order to realize many benefits of object-oriented software development, a framework for ADS application software development based on Common Object Request Broker Architecture (CORBA), which is a set of standards for object systems in heterogeneous distributed environments, is presented. In this framework, CORBA is extended and built over ADS system software. A CASE environment for ADS application software development based on CORBA is also presented.

Keyword: Autonomous Decentralized System, application software development, CASE environment, CORBA, framework.

1. Introduction

Autonomous Decentralized System (ADS) [1] [2], which has the characteristics of on-line expandability, on-line maintainability and fault-tolerance, has been successfully used in many distributed computing domains, such as factory automation, traffic control, office automation and nuclear power plants. In order to have effective ADS application software development, Distributed Object Computing (DOC) [3] [4] seems to be a very promising approach because it provides a much better way to capture the inherently decentralized nature of

distributed computing and many benefits of object-oriented technology (encapsulation, reuse, portability, and expandability) for distributed application software as for stand-alone application software. Object-oriented middleware, such as Object Management Group's (OMG's) Common Object Request Broker Architecture (CORBA) [5] [6] and Microsoft's Distributed Component Object Model (DCOM) [7], is an enabling technology for DOC. CORBA is an emerging industry standard for distributed object systems. DCOM is an application-level protocol for object-oriented remote procedure calls for distributed, component-based systems. In order to use DOC in ADS application software development, ADS system software should be enhanced to support distributed objects. One of the approaches to supporting distributed objects in ADS system software is to add CORBA over ADS system software. Since CORBA is a common distributed infrastructure supported by many vendors, this approach also guarantees the interoperability with other CORBA-compliant distributed computing systems.

In this paper, we will present a framework for ADS application software development based on CORBA. In our framework, we will build CORBA over ADS system software, extend CORBA to retain ADS characteristics, and then provide a CASE environment for ADS application software development based on CORBA. We will use an Automatic Teller Machine (ATM) as an example to illustrate how our framework will work.

2. Background

113

In this section, we will provide an overview of CORBA and ADS system software architecture for the sake of completeness.

2.1. Overview of CORBA

CORBA [5][6] is a set of standards which enables objects to transparently make and receive requests and responses in a distributed environment. It is based on OMG object model, where a client can send a message to an object cross address space. The client accesses the services of the object by a well-defined encapsulating interface which isolates the client from the implementation of the services and the object interprets the message to decide what service to perform. The object model describes object semantics and object implementation. Object semantics is related to the client, and includes such concepts as object and object reference, requests and operations, types and signatures. Object implementation includes such concepts as methods, execution engines, and activation.

The interface of an object in CORBA is defined by OMG Interface Definition Language (IDL). IDL is a declarative language which describes the services of the objects and needs to be mapped into particular programming languages. IDL mappings to C, C++, and Smalltalk have been specified by OMG. An IDL compiler is needed to bind IDL to a particular programming language.

The message passing between a client and an object implementation is performed by the Object Request Broker (ORB). The ORB, together with object adapters, provides all the mechanisms required to find the object implementation for the request, to prepare the object implementation to receive the request, to transfer the request, to activate and deactivate the object implementation, and to create and manage object references. The ORB functionality is defined by the ORB interface using pseudo-IDL and its binding to C++ is specified by OMG.

An object adapter specifies how an object implementation access services provided by the ORB. There are several object adapters with interfaces that are appropriate for specific kinds of objects. Basic Object Adapter is an object adapter specified by OMG, and can be used for most ORB objects with conventional implementation.

OMG also specifies a set of services, called *Common Object Services*, to provide the basic functions for using and implementing objects and a set of services, called *Common Object Facilities*, to provide general purpose capabilities useful in many applications.

2.2. ADS System Software Architecture

Conceptually, ADS has the feature that every software subsystem has autonomy to manage itself and coordinate with other software subsystems [1] [2]. Coordination is achieved by communicating with other software subsystems through Data Field (DF), in which the data circulates and software subsystems select the data according to the content code. The software subsystem in ADS is called *Atom*. Every Atom is connected to Data Field. Data also can circulate among the software modules in Atom. Data Field in Atom is called *Atom Data Field*.

Atom consists of not only the application software, but also its own management system software called *Autonomous Control Processor* (ACP). Each ACP is self-contained, operates according to its local information and communicates asynchronously with other ACPs by message broadcast in Data Field. Data Field Management Module in ACP is responsible for receiving the data from Data Field and sending the data into Data Field. The application software module is driven by the data from Data Field according to its content code. It is activated by Execution Management Module in ACP, receives the data from Data Field, processes the data and sends the resultant data to Data Field. Therefore, each ACP can operate even when other ACPs fail, and fault tolerance at system level is achieved. Fault tolerance at the application software level is supported by replicating the application software modules in different Atoms with a threshold-voting mechanism. The replica application software modules process the same data from Data Field and send the resultant data to Data Field. Data Consistency Management Module in each ACP selects the correct resultant data from the replica application software modules by the threshold-voting mechanism. On-line expandability is supported by the construction management module as an application software module. The construction management module can independently install the application software module to an Atom without interrupting other

Atoms. On-line maintainability is supported by the Built-In Test module (BIT) in each ACP and the EXternal Tester module (EXT) as application software modules. The BIT and EXT can independently test the application software modules and decide to start the operation of the application software modules according to the test result. The ACP software architecture is shown in Fig. 1.

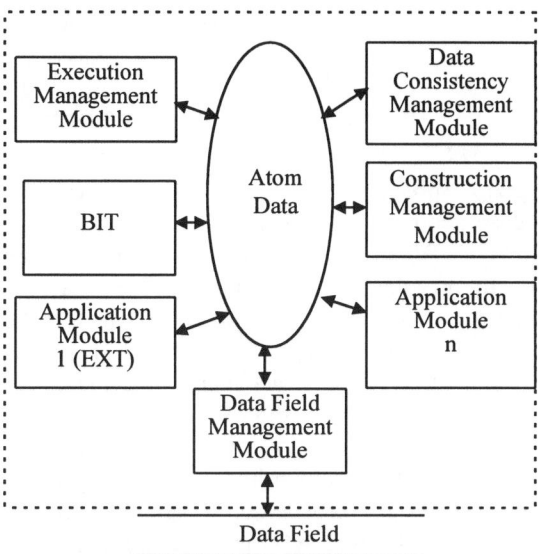

Fig. 1. The ACP software architecture

3. Building CORBA over ADS System Software

Since ADS system software provides a communication layer and the mechanisms to locate, activate and deactivate an application software module [1] [2], we decide to implement CORBA over ADS system software. Specifically, we will implement the ORB functionality by a pair of libraries, one for client application, one for server application, and ORB daemon. ORB daemon is implemented by a few system software modules from *ACP*, and it is responsible for locating, activating and deactivating objects, initiating and receiving remote object request. The client library can initiate the remote object request by forwarding it to ORB daemon and the server library can initiate and receive remote object request through ORB daemon. We will also implement Interface Repository (IR) as a system-resident ADS application. The software architecture of our CORBA implementation is shown in Fig. 2.

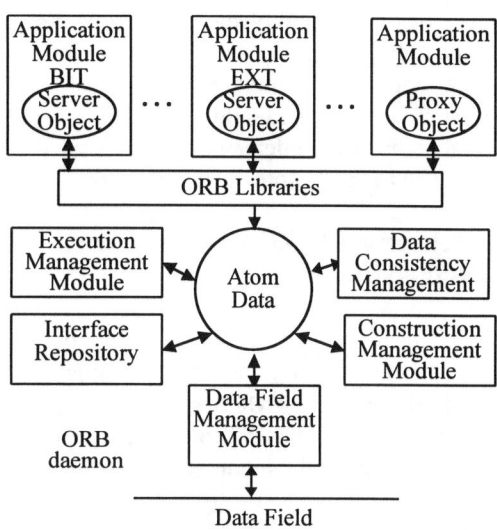

Fig.2. Software architecture of our CORBA

Our implementation also includes an IDL compiler. The IDL compiler provides a language binding from OMG IDL to C++. It generates C++ stub code for client application, implementation skeleton for server application and type information for IR. In our implementation, a client corresponds to an ADS application software module. The default behavior of C++ stub code generated by the IDL compiler is to marshal the remote object request, forward it to the ORB daemon, receive the result from the ORB daemon, and unmarshal it. All these functions are encapsulated in a proxy object generated by the IDL compiler for every IDL interface. The proxy object provides the same methods as the remote server object so that the client invokes these methods of the proxy object just the same as the remote server object. The proxy object also provides other functions in our implementation, such as data consistency check discussed in the next section. The default behavior of the implementation skeleton is to register implementation definition to the ORB daemon, to create or destroy the object reference, and to prepare to receive the requests from the ORB daemon and send the result to the ORB daemon. In our implementation, a server corresponds to an ADS application software module. Each application software module can contain multiple active objects of a given implementation. The Interface Repository provides type information for other application modules to check type dynamically.

Execution Management Module in *ACP* is modified to support part of BOA's functionality, which is to maintain implementation repository, bind the client to the server object in the software application module and activate the application software module according to the information in the implementation repository in the response to the remote object requests.

4. Extending CORBA to support ADS characteristics

In order to maintain ADS characteristics of on-line maintainability, on-line expandability and fault-tolerance, CORBA needs to be extended to support object group, state transfer, data consistency check and object migration.

4.1. Object Group

Fault-tolerance can be achieved by replicating the server object in different *Atom*s to form an object group to respond to the requests from a client. Basically, there are two kinds of replica. One is Active Replica, in which every server object in the object group responds to a request. If at least one object in the object group works, the client gets the services. However, in Active Replica, each object in the object group needs to be consistent with each other. If one object in the object group fails, it needs to be restored to the same state of the other objects. Another is Passive Replica in which only one object as the primary object responds to the requests from a client, other objects act as the backup objects. Once the object fails to provide the services, one of the backup objects takes over and continues to provide the services. In Passive Replica, it takes time for the backup object to get to the same state as the primary object. Active Replica Object Group is correspondent to replica application software modules in ADS. In our implementation, we select Active Replica Object Group, and proxy object in a client is bound to an active replica server object group instead of a single object implementation shown in Fig. 3.

Group object communication protocol in our CORBA implementation is a broadcast protocol since ADS is built over a reliable LAN which provides broadcast. In our group object communication, each message is broadcasted with a header, in which there is a message identifier containing the identity of the broadcasting ORB daemon, a message sequence number, and content code. The operation of

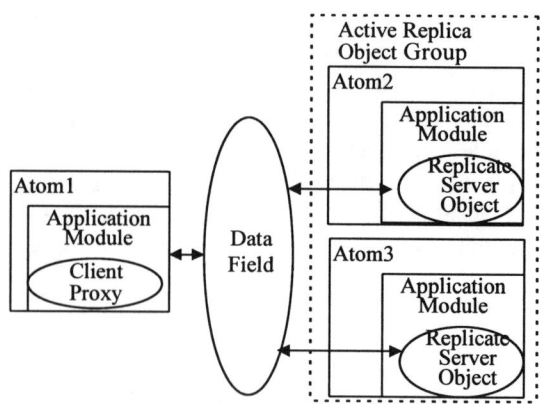

Fig. 3. Active Replica Object Group

group object communication protocol is illustrated with the following scenarios:

- When a client calls an object's binding operation, the ORB daemon A broadcasts a message with reserved content code for the server object binding. The data field of the message includes interface name, object name.
- The ORB daemon B which receives the A's message checks if there is the required server object in its own host which can provide the services and broadcasts an acknowledgment message indicating whether it can provide the services. The ORB daemon C on receiving the B's acknowledgment message will check if it has received A's message. If so, it acknowledges A's message by broadcasting a message indicating whether it can provide the services. If not, it broadcasts a negative acknowledgment message indicating that it has not received A's message. Any ORB daemon which receives C's negative acknowledgment message will rebroadcast A's message if it has received A's message.
- The ORB daemon A collects all acknowledgment messages within some time interval, builds an object group for this client, assigns an In content code and Out content code for the object group and broadcasts a message to inform all the ORB daemons of the creation of an object group. The message includes the assigned content codes and the group member list. Each member in the object group will use the same content codes in the following communication.
- All ORB daemons which can provide services activate ADS application software module in which the server object in the object group resides to prepare for receiving the requests from the client.

116

- The client initiates a request by asking the ORB daemon A to broadcast the request.

- An ORB daemon D in the object group receives the A's request, and checks the sequence number in the message to see if it is more than one greater than the largest sequence number it received from source A. If not, it processes the request, records the sequence number in the message, calls the method of the server object, and broadcasts the acknowledgment message in which the result is packed. Otherwise, it has missed some messages from source A, and broadcasts an acknowledgment message indicating that it cannot provide the service and asks for state transfer. State transfer is equivalent to method invocation on one of the other objects in the same object group to transfer to that object's state. After receiving D's acknowledgment message, the ORB daemon E will respond to A's request in the same way as D by broadcasting acknowledgment message for A's request if it has received A's request. If it has not received A's request, it will broadcast a negative acknowledgment message for A's request. Any ORB daemon which received E's negative acknowledgment message (including A) will rebroadcast A's request if it has received A's request.

- The ORB daemon collects all acknowledgment messages within some time interval, uses threshold-voting mechanism to select the most-likely correct result, and forwards it to the client.

- When the client exits, it asks the ORB daemon to broadcast a message to inform each member in the object group to release the resource for the object group.

The group object communication protocol described above can tolerate transient transmission failure and provides the total order of the messages and atomic delivery of the messages to server-client group communication over LAN in our CORBA implementation. Hence, it guarantees that the state of the object group is consistent if the server object does not fail.

4.2. State Transfer

If a new object joins the object group or a object recovers from failure, the object needs to get to the current state of the object group. The oldest object in the group will send its state to the new object. During this period, the ORB daemons will delay all the requests to the object group until the state transfer is completed.

4.3. Object Migration

Sometimes, a member of the server object group needs to migrate to some other machine in order to achieve load-balancing and fault-tolerance. The migration must not affect the services of the server object group. A possible solution will work as follows.

The server object leaves the server object group and the executable image of the server object is transferred to another Atom by Construction Management Module in ACP. The ORB daemon in that Atom will activate the server object application module. The server object will rejoin the server object group again and restore to the current state of the server object group by state transfer.

5. ADS Application Software Development Framework based on CORBA

Our framework for ADS application software development based on CORBA is shown in Fig. 4. It is modified from our object-oriented software development framework for ADS [8] [9].

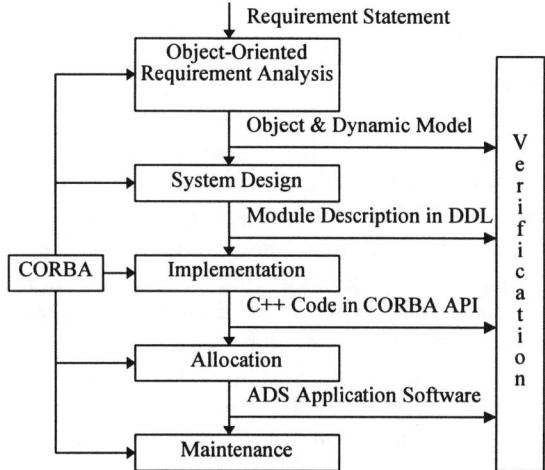

Fig. 4. Our object-oriented framework for ADS application software development

Our framework has the following phases: object-oriented requirements analysis, system design, implementation, allocation, verification and maintenance. We start ADS software development with a set of requirement statements, which is transformed into the object model and dynamic model using object-oriented requirements analysis (OORA) technique. The object model shows classes and their hierarchical structures derived from the knowledge about application domain and the requirement statements. The dynamic model shows a finite-state machine model for each class in the object model. The object model and dynamic model are represented by the module description in Design Description Language (DDL) [8] in the system design phase. Then, each module design in DDL is implemented in C++ and the implemented modules are allocated to processors. At the end of each phase, the adequacy of the results of each phase is verified. We have a CASE environment to support all these phases. Our CASE environment consists of CASE tools for object-oriented analysis and system design [9], object clustering [10] and module allocation [11].

In our new framework, we add the support for CORBA-based application software development in all the phases and our CASE environment. In the OORA phase, we will emphasize the role of objects, persistency of objects, object life cycle, association relationship among objects, and data flow between the producer object and the consumer object in order to provide enough information for object clustering [10] and organize class inheritance hierarchy to make full use of CORBA common object services [6]. In the design phase, our object-clustering CASE tool [10] will cluster the objects to the modules, identify the objects which export their interfaces to other modules, generate IDL interfaces according to the object information in the object model, and use our IDL compiler to generate IDL stubs and implementation skeleton for the implementation phase. In the implementation phase, each class is implemented in C++ using IDL stubs and implementation skeleton generated in the previous phase, and then each module is implemented in C++ using CORBA APIs. In the allocation phase, our CASE tool will emulate the real distributed systems by the workstation cluster, allocate the modules to the workstations according to our allocation algorithm [11], replicate the same module to different Atoms to achieve fault-tolerance, verify the consistency between the modules and the Atoms, run the

application to tune our allocation. During the maintenance phase, the object implementation in each module can be enhanced without affecting the clients which use the services of the object because CORBA completely separates the interface from the object implementation.

6. An example

In this section, we will use an Automatic Teller Machine (ATM) example [9] to illustrate how our framework supports ADS application software development based on CORBA. The software requirements for the ATM system are specified as follows:

Develop the software to support a computerized banking system with automatic teller machines (ATMs) to be shared by a consortium of banks. Each bank has its own computer to maintain its accounts and make updates to accounts. ATMs communicate with a central computer of the consortium. An ATM accepts a cash card, interacts with the user, communicate with the central computer to process transactions, and dispenses cash. For simplicity purpose, the system is assumed to have two ATMs, two Banks and one consortium.

For the object-oriented requirements analysis, we obtain the object and dynamic models of the ATM system according to the procedures in our framework [9]. The object model is shown in Fig. 5.

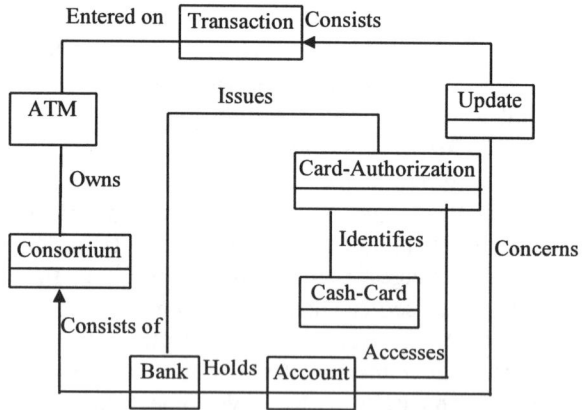

Fig. 5. The object diagram for the ATM system

For the system design, we first identify objects and inter-object communications. In this example, the objects are ATM1, Cash_Card1, ATM2, Cash_Card2, Consortium, Bank1, Transaction1, Bank2, Transaction2, Account1,

Update1, Card_Authorization1, Account2, Update2, and Card_Authorization2 as shown in Fig. 6. The inter-object communications are showed by the lines with arrows in Fig. 6. In the next step, we cluster the objects into several modules shown in Fig. 6, where the objects in the gray boxes are identified as the objects which export their interfaces to other modules and IDL interfaces are generated by IDL generator according to the information in the object model. Then, IDL stubs and implementation skeletons are generated using our IDL compiler.

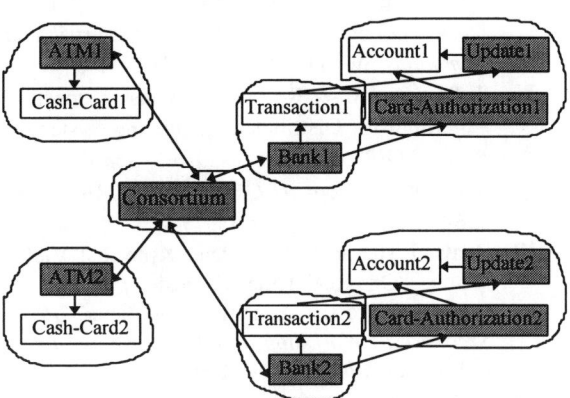

Fig. 6. The object clusters of the ATM example

For the implementation phase, the IDL stubs are used by client objects to invoke the methods of the remote objects, and implementation skeletons are used to implement the remote objects. For the allocation phase, we allocate the modules to different *Atoms* as shown in Fig. 7. The server object Consortium is replicated to two different *Atoms* to form an active replica object group to provide the services to other modules to achieve fault tolerance. Bank1 module and Update1 module are allocated to the same *Atom* since communication between these two modules are heavy. Similarly, Bank2 module and Update2 module are allocated to the same *Atom*. The two ATM modules are allocated to two different *Atoms*.

7. Discussion and Future Work

In this paper, we have presented a framework for ADS application software development based on CORBA. We are implementing the CORBA over ADS system software, extending the CORBA to achieve ADS characteristics, and modifying our CASE environment for ADS application software

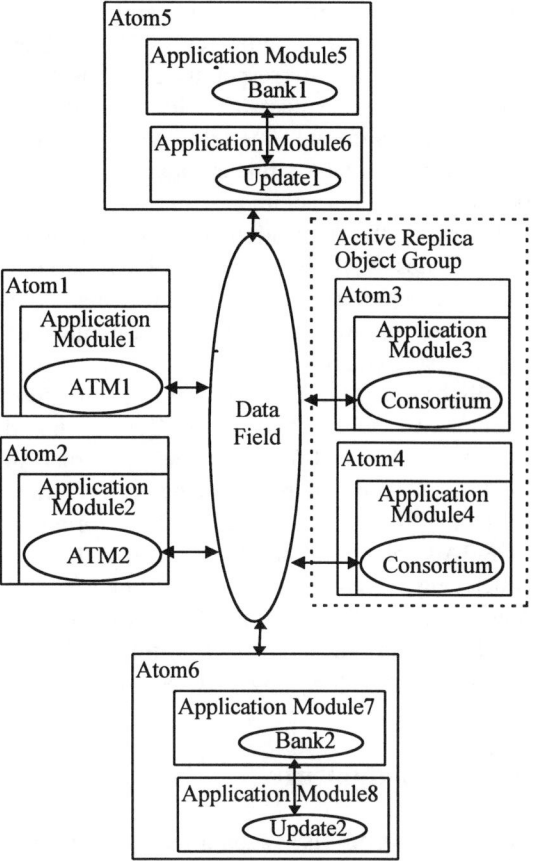

Fig. 7. Module Allocation

development based on CORBA. We also plan to incorporate a performance monitor into the ORB daemon in our implementation to obtain communication traffic information among different modules to dynamically support load-balancing among the hosts and reconfiguration of the application software modules by object migration and achieve interoperability with other CORBA implementations in the near future.

Since ADS has been successfully used in a number of different domains, we plan to analyze existing ADS applications in the different domains using our framework, migrate existing ADS applications to our CORBA-compliant development and running environment, and build domain-specific frameworks for ADS applications.

In order to migrating existing ADS application software modules to our CORBA-compliant development and running environment, we must encapsulate ADS application software modules by IDL interface. Basically, there are two ways for encapsulation. One is coarse encapsulation, which specifies the functionality of every ADS application software module by IDL interface. The object implementation just responds a request by executing the executable image of the application software module with different parameters. Another is fine encapsulation, which factors out the common functions among different ADS application software modules, encapsulates it as the common object services available to every object, and rewrites ADS application software modules as the application objects using common object services.

Since many ADS applications are time critical, we will extend our CORBA-based framework for real-time systems. In order to meet rigid temporal requirements in real time systems, we must have high-performance CORBA implementation which makes full use of underlying communication link, like Asynchronous Transfer Mode (ATM), and provide Quality of Services(QoS) for distributed objects accessible by ADS application developers.

Acknowledgment

This work is supported under the collaborative research agreement between Arizona State University and Hitachi, Ltd.

References

1. K. Kawano, M. Orimo and K. Mori, "Autonomous Decentralized Systems: Concept, Data Field Architecture and Future Trend", *Proc . First Int'l Symp. on Autonomous Decentralized Systems*, 1993, pp. 28-34.
2. K. Mori, et. al., "Autonomous Decentralized Software Structure and its Application", *Proc. FJCC'86*, 1986, pp. 1056-1063.
3. Douglas C. Schmidt and Steve Vinoski, "Introduction to Distributed Object Computing", *C++ Report*, January 1995.
4. Douglas C. Schmidt and Steve Vinoski, "Modeling Distributed Object Computing", *C++ Report*, February 1995.
5. OMG, *The Common Object Request Broker: Architecture and Specification*, 2.0 edition, July 1995.
6. OMG, *CORBAService: Common Object Service Specification*, 95-3-31 edition, March 1994.
7. Microsoft, *Distributed Component Object Model Protocol – DCOM/1.0*, http://ds1.internic.net/internet-drafts/draft-brown-dcom-v1-spec-00.txt.
8. S. S. Yau and G.-H. Oh, "An Object-Oriented Approach to Software Development for Autonomous Decentralized Systems", *Proc . First Int'l Symp. on Autonomous Decentralized Systems*, 1993, pp. 37-43.
9. S. S. Yau, et. al., "An Object-Oriented Approach to Software Development for Autonomous Decentralized Systems", *Proc . Second Int'l Symp. on Autonomous Decentralized Systems*, 1995, pp. 405-411.
10. S. S. Yau and H. Ying, "A Clustering Algorithm for Object-Oriented Development of Distributed Computing System Software", *Proc. 5th IEEE Workshop on Future Trends of Distributed Computing Systems*, 1995, pp. 274-281.
11. S. S. Yau and V. R. Satish, "A task Allocation Algorithm for Distributed Computing Systems", *Proc. 17th Int'l Computer Software & Applications Conf. (COMPSAC 93)*, 1993, pp. 336-342.
12. Bhavani Thuraisingham, Peter Krup and Victor Wolfe, "On Real-Time Extensions to Object Request Brokers: A Panel Position Paper", *Proc. Second Int'l Workshop on Object-oriented Real-time Dependable Systems*, 1996.
13. Gotter Sean, *Inside Taligent Technology*, Addison-Wesley, 1995.
14. INOA Technologies Ltd, "The Orbix Architecture", January 1995, http: //www-usa.iona.com/www/Obix/arch/Summary.html.
15. John A. Zinky, David E. Bakken and Richard Schantz, "Overview of Quality of Service for Distributed Objects", http: //www.bbn.com/offerings/dcutu/duduse/ Dualuse-final.html.
16. Jon Siegel, *CORBA Fundamentals and Programming*, John Wiley & Sons, 1996
17. Kenneth P. Birman and Robbert Van Renesse, "Reliable Distributed Computing with the Isis Toolkit", IEEE Computer Society Press, 1994.
18. Paul D. Ezhilchelvan, Raimundo A. Macedo and Santosh K. Shrivastava, "Newtop: A Fault-Tolerant Group Communication Protocol", http: //arjuna.ncl.ac.uk/arjuna/papers.html
19. P. M. Melliar-Smith, et. al, "Broadcast Protocols for Distributed Systems", *IEEE Transactions on Parallel and Distributed Systems*, Vol. 1, No. 1, pp. 17-25, January, 1990.

An Autonomous Decentralized System Architecture and Techniques for On-line Development and Maintenance

Shigetoshi Sameshima*, Katsumi Kawano*, Jirou Kumayama**, Toshihiko Ito***
Ken Inoue****, Satoru Fujishiro****

*Systems Development Laboratory, Hitachi Ltd.,
1099 Ohzenji, Asao, Kawasaki, Kanagawa 215, Japan
**Omika Works, Hitachi Ltd.,
2-1 Omika-cho 5-chome, Hitachi, Ibaraki 319-12, Japan
***Industrial Computer and System Dept., Hitachi Ltd.,
6 Kanda-Surugadai 4-chome, Chiyoda-ku, Tokyo, 101, Japan
****Kashima System Planning Dept., Sumitomo Metal Industries, Ltd.
3, Hikari, Kashima, Kashima-gun, Ibaraki 314, Japan

Abstract

The recent movement to customer-driven production has resulted in the needs for flexible production systems, that are compatible with the localization, the concurrency, and the cooperation of development and maintenance processes under on-line operation.

This paper proposes an architecture and techniques to make on-line development possible. This architecture enables the on-line development of frequently reconstructed systems and is based on the Autonomous Decentralized System concept, which is tolerant of heterogeneous subsystems in a system. Techniques for on-line development are described, with which development and maintenance subsystems cooperate with on-line subsystems, share mutual data and on-line data synthetically, and are prevented from disturbing on-line operation. These features make it possible to develop or maintain systems locally and quickly even when specifications of the whole system are ambiguous.

The proposed architecture and techniques have been applied to many industrial production systems, and their effectiveness has been verified in a steel production system.

Keywords: ADS (Autonomous Decentralized System), Data Field, ATP (Autonomous Terminal Processor), EXT (External Tester), On-line Development and Maintenance, Data Field, development, maintenance

1 Introduction

The worldwide economic recession and frequent changes in consumer demand recently has caused the production of an increasing variety of products to shift away from mass production to varied production. User's business process reengineering (BPR) has been in progress under this shift, and the advances being made in internetworking and intranetworking technologies are accelerating this trend. A worldwide coordination of production is about to be achieved. A new production system, therefore, is required that can easily be reconstructed frequently to provide an increasing variety of products.

In such a production system, it is often necessary for developers to tune, debug, or test application programs by monitoring and checking the controlled machine's actual motion and the actual data used for control. Faults also need to be analyzed in this way, and the application programs have to be both developed and maintained partially and concurrently without stopping development or on-line operation. In other words, a development and maintenance subsystem must cooperate with the on-line subsystems.

The Autonomous Decentralized System (ADS) has therefore been proposed [Refs. 1, 2, 3], which ensures such on-line properties as on-line expansion, on-line maintenance, and fault tolerance. Several tools based on the ADS architecture, such as the monitoring tool BIT/EXT (Built-in Tester/External Tester), have been proposed and applied in industry [Ref. 4]. A software development framework based on ADS has also been proposed [Ref. 5]. The need for frequent system reconstruction is making it increasingly important to have a mechanism for developing and maintaining the production system in a distributed environment.

The software development and maintenance environment has already been addressed, for example, in [Refs. 6, 7, 8] which discuss the CASE (Computer Aided Software Engineering) platform intended to provide an integrated development environment. This platform integrates several development support tools — requirements analysis tool, editor, and so on — and enables these tools to share the data they generate. The platform, however, only helps to share the data generated in the development phases and is not able to acquire on-line data or to communicate with target application programs. When trying to get on-line data in such an environment, a specific process must be installed in the system under on-line operation. Consequently, its operation is stopped.

This paper, therefore proposes an architecture and techniques that make on-line development and maintenance possible. This architecture is based on the ADS concept, and it provides for the interaction of development, maintenance, and on-line operation.

Section 2 of the paper will consider manufacturer needs in conjunction with system requirements related to software development and maintenance. Section 3 will then propose on-line development as a new paradigm. Section 4 proposes a new ADS architecture for on-line development and maintenance, and Section 5 proposes some techniques used in this architecture . Section 6 describes an application example to a steel production system, and section 7 concludes with a brief summary of this paper and a few words on the future prospects for on-line development and maintenance.

2 Background and Requirements

2.1 Background and User Requirements

Recent severe competition has resulted in manufactures having the following production requirements [Ref. 9, 10, 11].

(1) Customer-driven production

In recent years, manufactures have been changing their production style from making a few lots of many kinds of products to making products according to customer demands for higher quality and more variety. When producing a new product, it becomes necessary to change software, to reconstruct a subsystem, or to expand a production line, and these modifications should be made without stopping the production of other products. In a customer-driven production system, it should be possible

to reconstruct a subsystem quickly and easily while other subsystems are operating.

(2) Low cost production

Production costs must be kept low in order to keep the prices of products low. This can be done by dispersing production to those parts of the world where personnel expenses are lower and by reducing system costs. Reducing system costs means reducing maintenance costs and keeping the initial cost of the production system low by adding and removing components in a step-wise fashion.

2.2 System Requirements

In order to satisfy the requirements mentioned in Section 2.1, the flexible production systems which can be constructed and modified partially are required. From the viewpoints of software development and maintenance, the requirements for flexible production systems are broken down as follows.

(1) Individual development of subsystems

The specification of the total system is changed as the system is expanded/reconstructed. It becomes very difficult for developers to understand the specification of all subsystems in detail.

Each subsystem - hardware component / software - should be developed individually, even if the totality of the system is ambiguous.

(2) Detection of influenced subsystems

When a developed subsystem is going to be added to the target system, it is necessary to check the relation between the developed subsystem and others - the linkages with input/output data - and examine the areas the expansion or reconstruction influence from several viewpoints. The reexamination of linkages is also necessary for diagnosis and trouble-shooting when an error occurs. As the linkages between subsystems are increased and changed as the system grows, it should be possible to specify the subsystems which are affected by the developed subsystem or which affect the developed subsystem.

(3) Cooperation between subsystems

When a developed subsystem is tested, it is necessary to check the correctness of its behavior without suspending the operation of other subsystems. For example, it is necessary to examine the input data from other subsystems, output data to them, the timing of execution cooperating with other subsystems.

If the developed subsystem is not tuned well, it may disturb the process of on-line — often real-time — subsystems. In order to prevent this from happening and not to make the developer/maintainer too cautious, it is important to have an on-line protective mechanism.

122

3 On-line Development

Conventionally, it is thought that development, maintenance, and on-line operation are separate from each other. Development subsystems configure on-line subsystems by making application software for on-line subsystems and defining parameters. Maintenance subsystems monitor the state of on-line subsystems and developed subsystems. This concept, however, cannot meet the requirements of a flexible production system. According to requirements (3) mentioned in Section 2.2, the development of a software module has to be performed with on-line data such as state, timing, and a machine's motion. For requirements (1) and (2), if some design specifications of a target system are not complete because of the complexity of the target system or because the revision records are missing, the developer has to recover them from the maintenance information himself. Maintenance also cannot be performed adequately under the conventional concept, if the system is complicated. Therefore, the separated development and maintenance are not suitable for the required production system; these subsystems need to cooperate.

On-line Development is a new paradigm making development, maintenance, and on-line operation harmonious. On-line development requires development and maintenance subsystems to meet the following three conditions.

(1) Localization

Each subsystem must be localized functionally. This means that each subsystem can manage its own work by itself without being directed by other subsystems.

(2) Concurrency

Each subsystem must manage its work concurrently. One development subsystem can progress without waiting for the development of another subsystem to finish. The same is true of maintenance subsystems.

(3) Cooperation

Each subsystem must cooperate with on-line subsystems. Development and maintenance subsystems

can get on-line data without disturbing on-line operation.

Figure 1 shows the model of these conditions. Each circle represents a subsystem. Localization is the individual operability of each subsystem. For example, the development subsystem can develop its target by itself. The cooperation between development and maintenance subsystems facilitates concurrency. Each development subsystem can recognize the structure and state of the current system from maintenance information (i.e., without depending on other development subsystems), and a maintenance subsystem can understand its target hardware/software by getting information from development subsystems. Cooperation integrates and isolates an on-line subsystem: it ensures transparency of on-line data and protection of the on-line process. Conventional separated development and maintenance subsystems are decomposed into smaller functions and local data that are embedded into on-line subsystems.

On-line development requires development and maintenance subsystems to cooperate, in other words, it requires the total system to be built regarding each development, maintenance, and on-line subsystem as an autonomous unit. The proposed architecture and mechanism for on-line development is therefore based on the ADS concept.

4 ADS Architecture for On-line Development

4.1 ADS Architecture

Before proposing the mechanism for on-line development, the cooperating Autonomous Decentralized System is reviewed briefly [Ref. 12]. This architecture has been proposed as a means to the flexible system described before. In this architecture (Fig. 2), there are two information spaces: a Data Field (DF), where the data is broadcasted and circulates logically; and a Multicast Group (MCG), which is a subset of node computers and can restrict transparency of the data in a DF to only those node computers that belong to the MCG. This architecture has two distinctive functions; the inter-DF communication function and the traffic control function.

The inter-DF communication function enables data to be sent to several DFs without changing application software. It helps to communicate with other subsystems in a different DF without stopping on-line operation.

The traffic control function controls the timing of sending data according to the priority of data, the state of its internal buffer, and so on. It is also applied to the data to be received. Each piece of data is passed to application software modules according to its priority. This function helps to prevent a subsystem from disturbing others.

The data sent to DFs is attached with its content

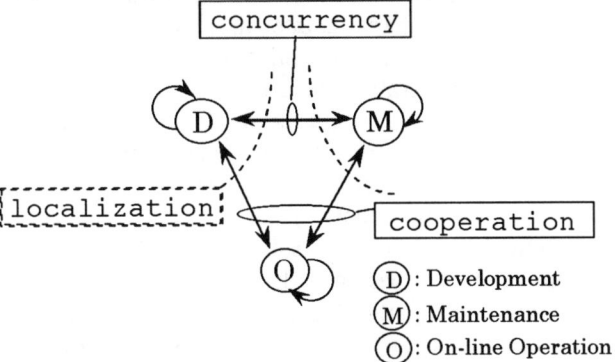

Fig. 1 On-line Development.

(D): Development
(M): Maintenance
(O): On-line Operation

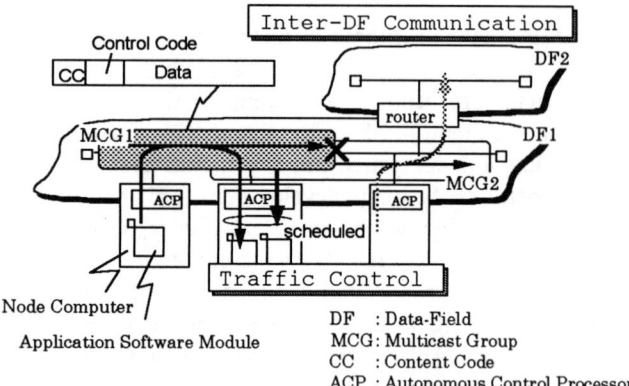

Fig. 2 Autonomous Decentralized System Architecture.

DF : Data-Field
MCG: Multicast Group
CC : Content Code
ACP : Autonomous Control Processor

code (CC), which is uniquely defined in the DF. Each application software module can receive the data simply by registering the CC of necessary data to the configuration table of an ACP (Autonomous Control Processor) (Fig. 3). The ACP judges whether to accept the data on the basis of its CC. The CC is attached to the data sent to DFs by the ACP according to its configuration table. Each software module exchanges data in this "message-passing" way. No software module directly control the others, but controls itself autonomously.

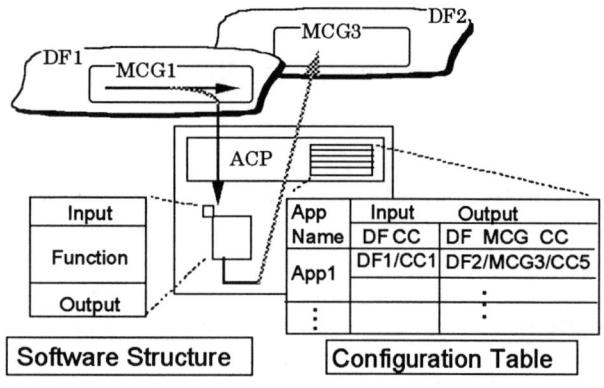

App: Application Software Module

Fig. 3 Software Structure on ADS.

As the interfaces between the software are unified, software modules can be added to the system or removed from the system, without stopping the system's on-line operation, by changing the configuration table of an ACP. However, the construction method (that is, how to register or unregister a software module) and the maintenance method are not presented, and there is no cooperation between the development, maintenance, and on-line operation.

4.2 An On-line Development and Maintenance Mechanism based on ADS Architecture

A mechanism is proposed which is characterized by its structured DF based on the ADS architecture. Structured DF is a policy to organize necessary information from on-line, development, and maintenance data even when the system is wholly ambiguous. It is formed from three elements: the DF and the MCG for development and maintenance in that DF, the configuration table in each node computer, and the design information about the system (Fig. 4). In this architecture, each node computer stores the development information — the application definition in the node computer, the input/output CC of it, and so on—about itself in the configuration table. The development tools and maintenance tools join the system via this DF and MCG. Development tools develop their targets by using development information, and maintenance tools monitor their target by getting maintenance information. Moreover, this architecture and mechanism provides the following three properties.

(1) Transparency of heterogeneous data

Development and maintenance tools can acquire necessary information selectively used in on-line operation, development, and maintenance. For example, it is easy to get on-line data, the states of application software, or such supplementary data of application software as specification documents, on demand. It is not necessary to make a special program to catch the data used in on-line operation or to stop on-line operation to install the program.

(2) Autonomy of development and maintenance subsystems

Each tool with an ACP is regarded as autonomous. They gather and select the necessary data themselves from DFs/MCGs. They never give orders to other tools or receive orders from other tools. In addition, since the data flowing through the development and maintenance MCG has a lower priority than that in on-line MCGs, the on-line data is controlled to take precedence of the development or maintenance data in operation by the traffic control function of the ADS. In this way, there is no interference with on-line operation from the development and maintenance tools.

(3) Recognition of the target application system

The configuration table in an ACP contains the information about itself and the application module in each node computer. Each tool can restore the tables of node computers if necessary. It is noteworthy that this mechanism lets the tools get the true application information of the on-line subsystem. Even if the development tool does not have the whole system specification, it can gather the necessary information form the on-line subsystem. Even if there is a discrepancy between the on-line system structure and the design documents because of changes by another developer or

124

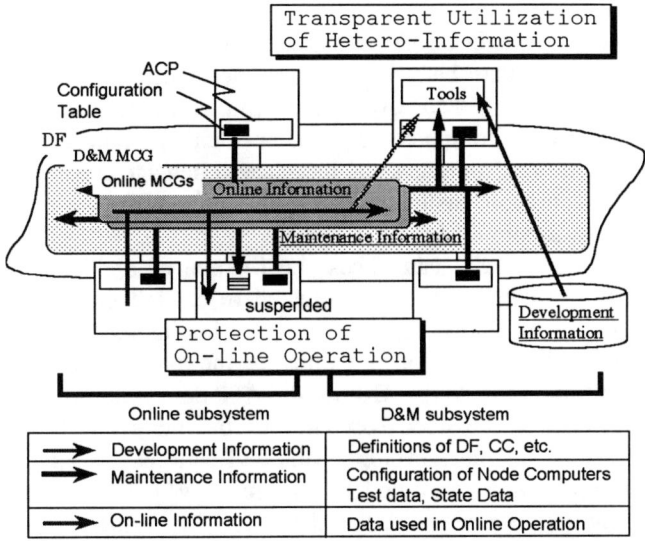

Fig. 4 Structured DF for On-line Development
and Maintenance

irregular modification such as patch-putting, it is possible to get the true definition used in on-line operation. What is defined in the configuration tables, how to construct the tables, and how to get the necessary information with the tools will be discussed in Section 5.

This architecture and mechanism makes it possible for development and maintenance subsystems to cooperate with on-line subsystems. The developed software module can be downloaded and registered without waiting for the development of other software modules to be completed. The software module is tested with on-line data [Ref. 4]. The result and related on-line data are gathered and analyzed with supplementary data defined by development tools. This mechanism for on-line development has enabled software modules to be developed while other modules are in operation and has enabled development to be completed by trial-and-error with on-line data. This is not possible in the conventional systems, and this mechanism can meet the requirements which could not be met by the conventional systems or tools.

5 Techniques for On-line Development and Maintenance

5.1 Conventional ADS Development/Maintenance Support Tools

In the conventional ADS, there is an Autonomous Terminal Processor (ATP) that works as a design support tool and an External Tester (EXT) that works as a test and monitoring support tool [Refs. 3, 4] (Fig. 5).

The ATP has the following two functions.

Data formatting function supports configuring the data format for each content code from several data items.

Data flow generation and flow correctness check function generates the diagram showing the data flow among application software modules. The ATP recognizes the relation between software modules on the basis of the input/output CC of software modules, and it shows the data flow diagram. In this process, the ATP detects the incorrectness of the flow, unused data, infinite loop, and so on.

The ATP assists the process of designing the data flow (application linkage), but does not support the configuration of the system. It is just an off-line tool. The configuration has to be input to the ATP and then it is registered on each node computer. This configuration information and other information defined by designers is used only for design, not for configuring node computers or for maintenance.

The EXT supports on-line testing and state monitoring. For on-line testing, it gathers on-line data and test data, checks the test result, and notifies the result to other node computers or operators. The state-monitoring function supports monitoring the state (sound or faulty) of node computers and application software modules by watching the alive-message cyclically sent by each node computer.

The information generated or gathered by the EXT is not open to the development subsystem. It must be analyzed with the know-how or knowledge about the target system. For instance, without the knowledge about the whole system, a maintainer does not know the meaning of the CC sent from a software module designed by others or

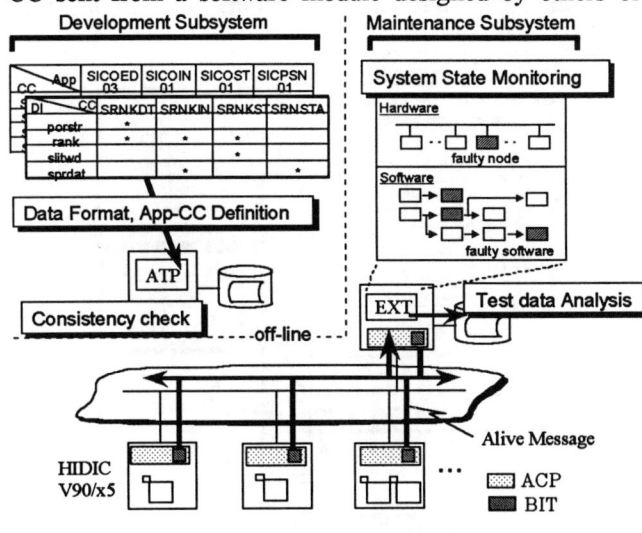

Fig. 5 ATP and EXT.

125

whether or not the sequence or the timing of software modules' execution is normal. This makes it difficult for an unskilled system engineer to maintain the frequently modified system with only EXT.

The ATP and EXT are independent tools from the viewpoint of cooperation between development and maintenance explained before; hence, they have to be extended to meet the requirements discussed in Section 2.

5.2 On-line Development Tools: ATP/EXT

Some techniques and tools based on the proposed architecture are described. In the rest of this paper, the term ATP refers to the extended ATP as an on-line development tool, including the conventional functions. Also, EXT is the extended EXT.

5.2.1 Construction and Restoration: ATP (Fig. 6)

The ATP can cooperate with on-line subsystems and make it easy to construct the system consequently.

(1) Construction

The developers input the following two kinds of definitions in order to construct the system.

One is system definition which consists of DF definition, MCG definition, logical node definition. And another is application definition which consists of CC and its composing data items with format, software modules and its input/output CC with DF.

The definitions are checked for consistency immediately by showing data flow diagram to developers, by checking re-definition, and so on. Then the ATP generates the configuration table and sends it to the target node computer.

Since the ATP can reflect the input definition to the target

system, it enables developers to construct the system much more easily than before. The designers need only input the definition. Moreover, as it can construct various types of node computers, such as personal computers and workstations, from the same interface, it helps to construct the system more easily.

(2) Restoration

The ATP can check the linkage of application software modules that are executed in on-line operation currently by gathering the configuration table of each node computer. It also compares the table with the definition in the ATP. Even if the ATP does not have the definition of the node computer or the table has been revised by another developer, this tool can get the actual definition and examine the discrepancy between the configuration table and the definition by restoring the configuration table. A warning message is issued if there is any difference. This technique helps to prevent registration errors and inconsistency between the definition and the on-line system.

5.2.2 DF Analyzer: EXT (Fig. 7)

It is necessary to check the behavior of the developed software module in on-line subsystems, for example:

· whether or not the content of input/output data of the module is correct,

· whether or not the sequence of application modules is executed as designed, and

· which data caused the fault.

This technique helps to recognize the sequence and contents of data and shows them as well as the linkage between software modules which the ATP conventionally showed. The sequence or timing is especially important in the software controlling machines like those in PLC (see Section 6), so the debugging and testing of such software will be improved by this technique. In other words, this technique helps to analyze the "dynamic" relation among software modules in addition to the "static" relation between them. How to get data and how to show it is described below.

(1) Data collection

The EXT collects the following three kinds of data. System structure is collected in the same way as the ATP's restoration function. Additional information is the data other than the on-line data and the on-line components, such as the names and descriptions of node computers and application software modules, the formats of CCs, and so on. These are shared with the ATP. Messages (the data with CC) are received by opening the interface to the MCG where the necessary data are sent.

(2) Analysis

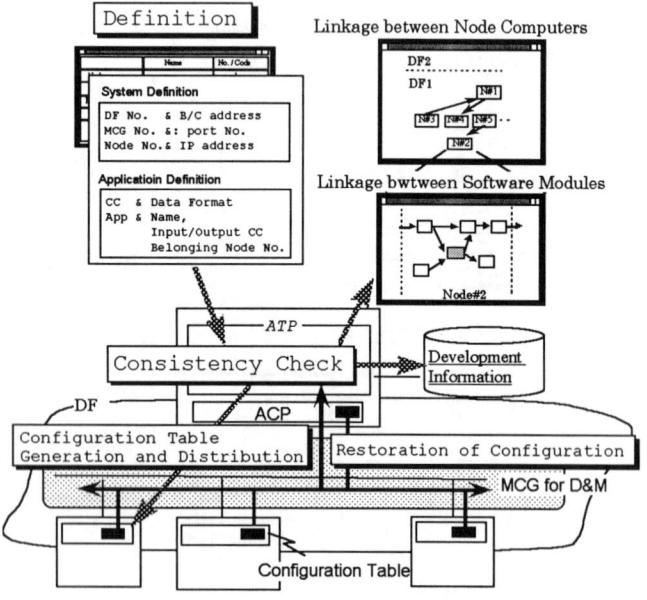

Fig. 6 Construction and Restoration of Configuration

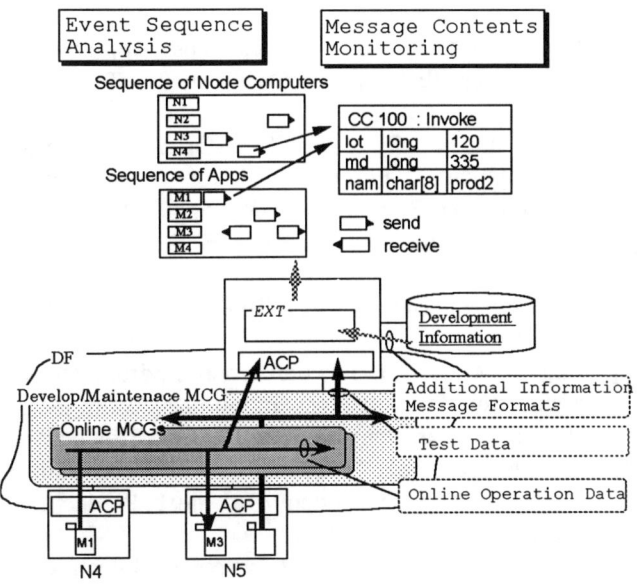

Fig. 7 DF Analyzer.

and it is not necessary to take the trouble of decomposing and analyzing the data.

This DF analyzer technique enables the maintainer and developer to easily analyze mutual relations and behavior of the distributed system as if it were a single system.

6 Application

The proposed architecture and tools—ATP/EXT—have already been applied to a lot of production systems. Figure 8 shows an example.

This is a steel-pipe production system which consists of a press line, a welding line, inspection lines, and a shipping line. Each line is controlled by the PLC (programmable logic controller) in charge. PLCs are connected to the network (Ethernet) to which a process computer for process control and PCs (personal computers) for system monitoring and developing are also connected. The process computer distributes scheduling data and processing parameters to the PLCs, gathers the processing results sent from the PLCs, and stores them. PLCs control the processing machine according to the schedule and processing parameters and send the processing results data, state of processing, tracking data, and state of the PLCs themselves (sound/faulty). This PLC control is a real-time process. The monitoring PCs gather the state data from PLCs and display the state of production or the state of computers. It is also used to maintain the data held in the process computer.

This production system was constructed and reconstructed in several phases. The addition or modification of processing machines and their controller was performed step by step and caused functions to be extended to application software modules which were executed. New

The DF analyzer technique shows the following information in addition to the definition and supplementary data managed by the ATP.

(i) sequence generation

The EXT shows the matrix of sender node computers or software modules and the event data received. This sequence makes it easy to confirm the sequence of execution.

(ii) content of data

The content of the data is recognized according to its format definition, and the EXT shows the values of its organizing data items. The designer or maintainer can take a look at the on-line data in the form it was defined,

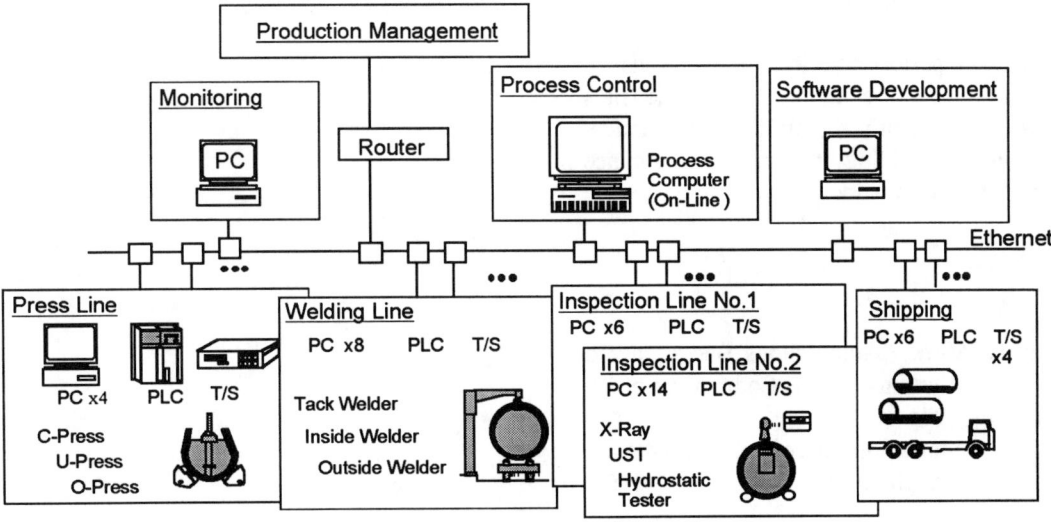

Fig. 8 Application Example
- Large Welded Pipe Mill Process Control System -

127

data, functions, and linkages were generated as the system grew.

Each computer has an ACP and an ATP/EXT is installed in a development PC. The configuration table generated by an ATP is downloaded to each node computer several times for debugging and testing in each phase. Test and behavior analysis was carried out with an EXT. When an error occurs, the operators monitoring the system use an EXT to find out which node computer or software module is faulty, which module can affect the faulty element, and the content of the data sent to the faulty element.

The proposed architecture, which makes on-line development possible, has already been applied to more than several score of systems and its effectiveness has been verified.

7 Conclusions

In this paper, as a solution to the new trend and user requirements, a new paradigm and a new architecture for on-line development and maintenance has been proposed based on ADS architecture. This architecture is characterized by its structured DF and has enabled on-line development providing localization and concurrency of development and maintenance as well as cooperation with on-line operation. Several techniques and tools based on this architecture have also been proposed. This architecture has been applied to various production systems and its effectiveness has been verified in a steel production system.

The following issues are under discussion and are subjects of future work.

(1) Cooperation between developers and source code management in the ATP.

(2) These tools are not perfect solutions to maintenance problems because how easy the maintenance is depends on how simply or independently the application is made. Therefore the design methodology to make maintenance easy is also a subject of future work.

References

[1] H. Ihara and K. Mori., "Autonomous Decentralized Computer Control Systems", IEEE Computer, vol.17, no.8, 57-66, Aug. 1984.

[2] K. Mori, et al., "Autonomous Decentralized Software Structure and Its Application", FJCC, 1056-1063, 1986.

[3] K. Mori, "Autonomous Decentralized System : Concept, Data Field Architecture and Future Trends", Proc. ISADS 93, 28-34, 1993

[4] K. Kawano, et al., "Autonomous Decentralized System Test Technique", COMPSAC 89, 52-57, Sept. 1989.

[5] S.S. Yau et al., "An Object-Oriented Software Development Framework for Autonomous Decentralized Systems", ISADS 95, 405-411, 1995.

[6] "Feature on Integrated CASE", IEEE Software Mar. 1992.

[7] "Feature on Development Tools", IEEE Software Mar. 1995.

[8] Cagan, M.R., "The HP SoftBench environment: an architecture for a new generation of software tools", Hewlett-Packard Journal, June 1990, pp.35-47.

[9] Y. Mashino et al., "An Autonomous Decentralized Process Computer System for Steel Production", ISADS 93, 390-397, 1993.

[10] M. Omura el al., "Hi-Cell System Architecture for Manufacturing Systems", ISADS 95, 154-161, 1995.

[11] J. Mori et al., "Flexible Manufacturing and High Performance System for Steel Plants", FTDCS, 112-119, 1995.

[12] H. Wataya, et al., "Cooperating Autonomous Decentralized System Architecture", Proc. ISADS 95, 40-47, 1995.

Service Accelerator (SEA) System for Supplying Demand Oriented Information Services

Kinji Mori*, Shin'ichiro Yamashita**, Hiroaki Nakanishi***,
Kazuyuki Hayashi****, Kazuhiko Ohmachi****, Yuutarou Hori*****

* Systems Development Laboratory, Hitachi, Ltd.
1099 Ohzenji, Asao, Kawasaki, Kanagawa 215, Japan
** Software Development Center, Hitachi, Ltd.
5030 Totsuka, Totsuka, Yokohama, Kanagawa 245, Japan
*** Business Development Office, Hitachi, Ltd.
4-6 Kanda-Surugadai, Chiyoda, Tokyo 101, Japan
**** Telecommunications Division, Hitachi, Ltd.
216 Totsuka, Totsuka, Yokohama 244, Japan
***** Omika Works, Hitachi, Ltd.
5-2-1 Omika, Hitachi, Ibaraki 319-12, Japan

Abstract

Information service system is proposed to satisfy the requirements for use, provision and guarantee of service from the standpoints of end users, service providers and network providers. Here the service is defined as three views of customization (who is requesting information service), situation (where and when they request it) and quality (how important the request is). These requirements are achieved by three basic techniques of discernment of service from three views, of broad action to access the related service providers or end users on the basis of the discernment, and of synchronization to integrate the multiple information provided from the broad action. This information service system SEA(Service Accelerator) is realized as a service platform in the open environment to coexist with the conventional systems and to attain the flexible services.

1. Introduction

Telecommunication services have been liberalized and globalized. Almost all telephone companies have become involved in delivering non-telephone services to end users. Most cable television (CATV) providers are interested in telephone and Internet services as well as video-on-demand (VoD) services. These companies compete with each other and most believe that cost savings are possible through value-added services. Also, the number of end users is increasing daily and these users have unique interests such as stock market, gardening and cooking, and because of these interests, they require different services from service providers.

Recently, many end users want access to services not only from their homes, but also outside the homes, such as while commuting or traveling. This has been led by the development of portable PCs and personal digital assistants (PDAs) [1]. To meet this increased demand, the number of information service providers (SPs) has also increased. SPs are able to provide various information services to their users, but because there are so many services available, users tend to become overwhelmed and have a hard time choosing a SP that will best meet their needs. SPs are eager to increase the amount of information they provide, but, they typically provide information that does not meet their users' demands. Another important point is the complexity of information services. Many people who do not have much technical knowledge have begun to utilize information services. These services are often too difficult for the user who does not have much computer knowledge or skill. This general lack of skill results in end users not being able to efficiently use the services available to them.

Based on the above, this paper concludes that end users have three specific needs regarding information services. First, end users need services that are customized to users' requirements. SPs should provide contents that reflect individual preferences. Second, end users do not want services that are inflexible or do not reflect time. Different services should be available in the morning, such as weather information, or on the weekend, such as leisure information. They should also be able to respond to the users' needs in terms of his location. For example, a user may require access to a service at home, at a hotel, or some other location. End users need services that reflect these different situations. Third, services

129

should vary according to levels of importance of users' request. For example, if one user uses a service to earn 100 pounds while a second user uses the service to earn 1 million pounds, the second user would like to handle his service more carefully and the SP must do so.

To meet these end users' needs, a system called Service Accelerator (SEA) is provided, which is a user-oriented system that operates between end users and SPs, and is able to provide appropriate contents for each end user.

In information service systems, each component of the system should be autonomous [2]. Therefore, the information selecting mechanism of the SEA is based on the content code communication technique used in Autonomous Decentralized System (ADS) [3,4]. The SEA thus has on-line expandability of services, and can handle the addition of new services flexibly amid the increasing competition in the information service industry .

This paper discusses concepts and architecture of the SEA system and shows an example application.

2. SEA Concept

2.1. Conventional System

This section reviews characteristics of conventional computing systems used as information service providing systems. Conventional computing systems are constructed from the SPs' point of view. End users request to have contents sent to them by SPs by specifying identifiers of particular contents that they want. SPs can then provide contents which match the end users' requests [5]. For example, in Object Request Broker (ORB), end users (client) send requests to SPs (server objects) via ORB and receive responses from the SPs [6,7]. In this system, end users are aware of contents that they want. In other words, end users know in advance what contents will satisfy their demands. These conventional systems are said to be "access-oriented" system.

A service flow is the chain of procedures for providing a service to an end user. In conventional systems, the service flow has the following characteristics (Fig. 1):

- The same flow is applied regardless of the end user. For example, user A and user B get the same contents even if different contents would be more appropriate.
- The same flow is applied regardless of the situation. There is no discernment between differences in time and place; end users in any situation receive the same contents with the same flow.
- The same flows is applied regardless of the level of

importance of demand. The level of importance of demand is different for each end user. There may be a case that an end user wants some contents sent very carefully.

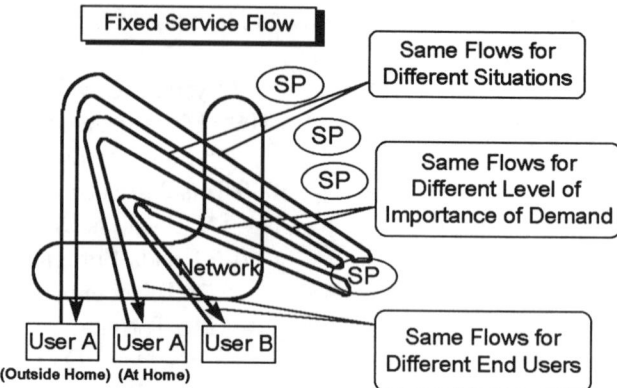

Fig. 1 Conventional Systems

2.2. SEA Concept

Hospitality of service is defined as the degree of satisfaction an end user feels towards services provided. End users feel hospitality of service when the systems provide services appropriately according to when, where, whom and how to provide the services. To make services more hospitable, information service systems should take the following factors into consideration,

- To whom will the service be provided?
  Who will provide the service?
  (customization to each end users and SPs)
- When will the service be provided? (the time) Where will the service be provided? (the place of end users)
- How will the service be provided? (the level of importance of request)

Service Accelerator (SEA) is introduced in order to manage these factors and improve hospitality of service. The SEA is a third party which selects appropriate contents provided by SPs and provides them to end users. Prior to using the system end users need to specify the three factors listed above to the SEA, and SPs need to specify contents that they provide. The SEA then can select appropriate contents according to the factors. End users are not aware of the contents, but specify their demands. Thus, the SEA can be said to be a "demand-oriented" system.

To manage the three factors, the SEA defines the following three views.

- **Customization**: the differences of end users in terms of interests, jobs, family-life, etc. and the differences of SPs in terms of the provided services.
- **Situation**: situation (time, place, schedule, machine environment) of end users and SPs.

- **Quality**: level of importance of the service for end users and SPs. For example, an end user wants contents to be sent by express or registered mail.

In this section these are defined as the basic views that determine services in the SEA system. These three basic views are characterized by end users and SPs in the SEA. To select appropriate contents for the end users, the SEA considers demands given by end users and strategies given by SPs on the basis of these basic views (Fig. 2). The strategies relate to properties of contents which SPs provide as class, type, components, etc. For example, information about an apple has property of fruit as class, sweet as type and apple juice as component.

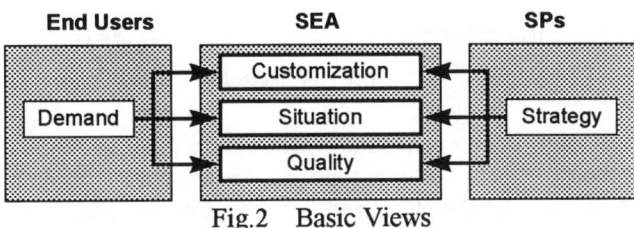

Fig.2　Basic Views

Services can be determined by end users' demands and SPs' strategies from the three views (Fig. 3). There are relationships between these three views and services. The three views influence services as follows. Customization are utilized to select actual contents that are appropriate. Situation is utilized to determine time and place for service provision. Quality specifies the level of importance of services.

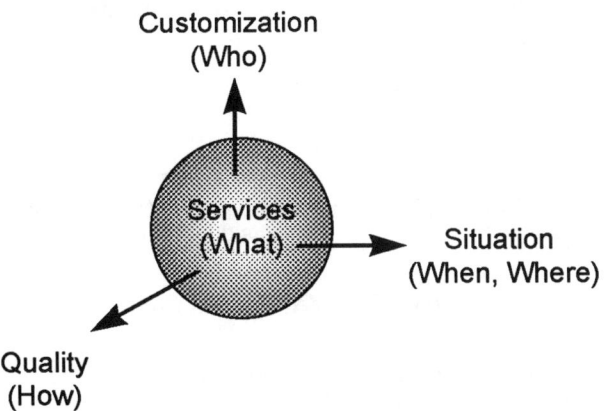

Fig.3 Three Views of Services

To provide service it is necessary to define these basic views and relationships between views and services. These relationships are (1) relationship between customization and services, (2) relationship between situation and services, (3) relationship between quality and services and other relationships among the three views.

(1) Relationship between customization and services:

There are two kinds of relationships between end users and services. One is the relationship between properties of end users and properties of contents. This is a general relationship between two properties and can be applied to all end users and contents in order to select suitable contents from a particular end user or suitable end user from particular contents indirectly. One example is that end users who are interested in health have strong connection with contents about sports.

The other kind of relationships is the direct relationship between end users and contents. This is based on end users' history of access to particular contents or specifications by end users. This relationship is utilized to select appropriate contents from particular end users or appropriate end users from particular contents directly. An example of this would be end users that access information frequently.

(2) Relationship between situation and services

This is a relationship between situation and properties of contents. This is utilized to distinguish the kinds of properties of contents that are suitable for particular situation. One example is that contents of travel guides are suitable for end users who are traveling.

(3) Relationship between quality and services

This relationship determines default levels of importance for different demands on the basis of contents. For example, contents "stock information" would be given a default level of importance valued at "express". If an end user does not specify quality of demand, the default level is adopted. This relationship also determines appropriate contents for different qualities of demands.

There are other relationships among customization, quality and situation. Finally, end user can obtain the services as the integration of results specified by these relationships.

In the SEA system, service flow has following characteristics (Fig. 4):

- Different flows may be applied to similar requests by different end users. Customized contents for each end user are provided.
- Different flows may be applied to similar requests under different situations. For example, different contents are provided to the same end user for different situations.
- Different flows can be applied based on quality. Some requests may be treated more carefully when the level of importance of the requests is high.

The SEA system enables flexible service flows that

increase hospitality of service according to differences in customization, situation and quality of end users and SPs.

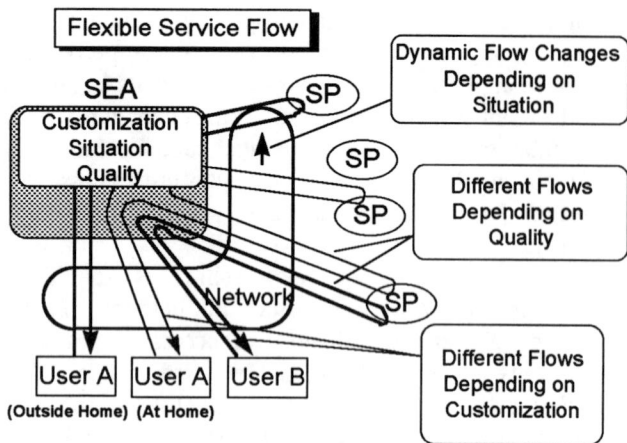

Fig.4   SEA Systems

2.3. Use, Provision and Guarantee of Service

Hospitality of service can be attained by service use, service provision and service guarantee. Service use is the operation of end users based on their demands. Service provision is the operation of SPs in which contents are selected and provided to end users according to the factors indicating what contents, to whom, when, where and how they should be provided. Service guarantee is the operation of network providers that directs the service provision function and coordinates these five factors. *What contents* refers to which contents are selected. If there are no suitable contents, the SEA selects alternatives. *To whom, when and where* refers to the contents being chosen depending on the end users and the end users' situations. The SEA also decides *how* to send the contents to end users. Important contents are sent to end users carefully and reliably, and other contents are sent in the usual way.

3. SEA Architecture

3.1. Service Data Field

In conventional systems, an end user specifies a SP directly and requests information, then the SP provides the information to the end user. In the SEA system, end users and SPs are connected to each other through Service Data Field (SDF) where information on customization, situation and quality circulate (Fig. 5). Before using the SEA system, end users must specify demands on the three views to the SDF, and SPs must specify strategies on the three views to SDF. Then, the relationships among the

three views and services are checked and appropriate contents are selected. Finally, the contents are provided to end users.

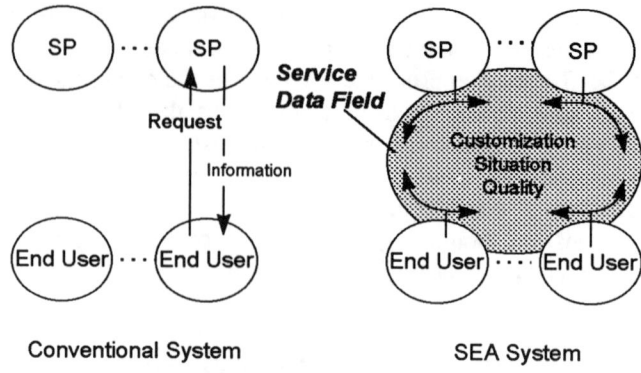

Fig.5   Service Data Field

SDF is an expansion of Data Field (DF) which is a concept of Autonomous Decentralized System. Autonomous Decentralized System Concept has been proposed to attain on-line expandability, on-line maintainability and fault tolerance of not only the hardware but also the software system. The basic feature introduced in the Autonomous Decentralized Software Structure is the DF. In the DF, each data has a content code which indicates its meaning. The software module broadcasts all data with its content code into the DF and judges whether to receive the data from the DF on the basis of this code. DF uses only one identifier (content code), but SDF uses an extended content code. The code has multiple identifiers which are Customization ID, Situation ID and Quality ID (Fig. 6).

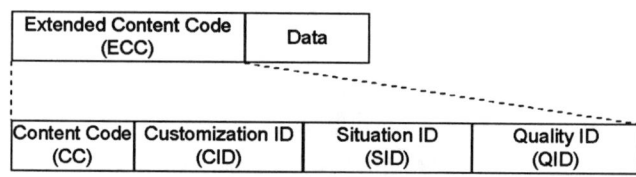

Fig.6   Extended Content Code

The following are examples in which the SDF concept is implemented (Fig. 7).
(1) An independent server (SEA server) is located between end users and SPs. End users register their basic information from the three basic views to the SEA server and SPs also register their basic information from the three basic views to the server in advance. The SEA server manages these information and checks the relationships among them. Then the SEA server selects appropriate contents, gathers them from SPs and sends them to end users.

132

(2) The SDF is located at each end user. The end users registers their basic information from the three basic views in the SDF in advance. SPs send their basic information from the three basic views to all end users. After receiving the information from SPs, SDF at the end user checks the relationships among the basic views and selects appropriate contents. Then the end user gathers the contents from the SPs.

(3) SDF is located in the SPs' servers. The SP registers their basic information from the three basic views in the SDF in advance. The end user sends their basic information from the three basic views to all SPs. SDF at the SP's server checks the relationships among the basic views and selects appropriate contents. Then the SP sends the contents to the end user.

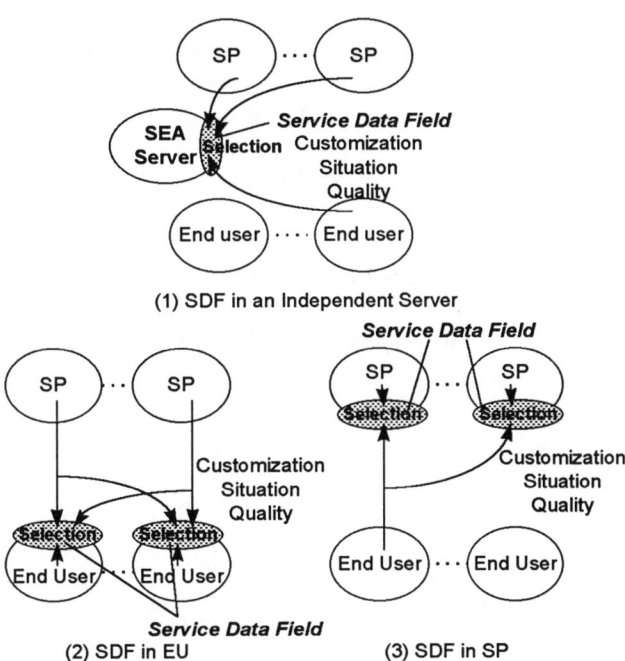

Fig.7   Implementation Examples of SDF

3.2. Basic Functions of SEA

In the following discussion, it is assumed that case (1) in Fig. 6 is adopted as the implementation of SDF. However, almost all of the discussion applies to case (2) and (3) as well.

The service provision mechanism of the SEA system is discussed here in more detail. The standard flow of information provision using the SEA server is as follows (Fig. 8).

(1) SPs register the strategies to the SEA server in advance.

(2) End users register their information from the three basic views to the SEA server in advance.

(3) End users send a demand to the SEA server.

(4) The SEA server selects suitable contents according to the relationships between the basic views ("Discernment of service" function).

(5) The identifiers of selected contents are sent to "Broad action" function.

(6) The SEA server gathers the selected contents from SPs ("Broad action" function).

(7) The SEA synchronizes and integrates gathered contents according to the relationships among basic views ("Synchronization" function).

(8) If the contents are not available, then the SEA server tries to select alternatives.

(9) The SEA server sends integrated contents to end users.

Discernment of service, broad action and synchronization function are the three basic functions of the SEA.

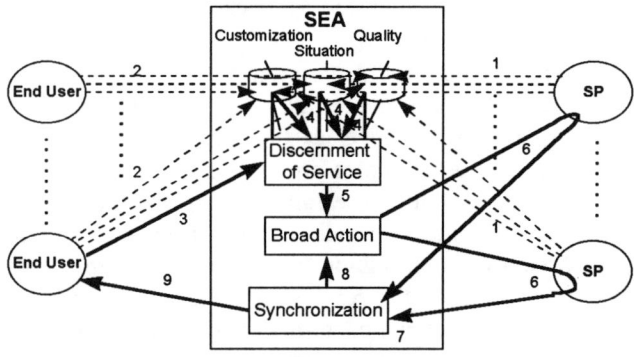

Fig.8 Basic Functions of SEA

- **Discernment of Service (DoS)**

The purpose of DoS is to select appropriate contents for end users. DoS uses the three basic views and the relationships among them. The Basic views and their relationships are stored in a local disk in advance. When the SEA server receives a demand for information from an end user, DoS discerns services (or contents) using the relationships, and determines a value that indicates the strength of the relationship between basic views and services. Then DoS compares the values and appropriate services are selected.

The DoS function is based on the content code communication technique in ADS. It is an expansion of this communication technique in the sense that they both select data by checking the date's identifiers. Content code communication technique uses only one identifier (content code), but DoS uses extended content code. The code has multiple identifiers which are Customization ID, Situation ID and Quality ID.

- **Broad Action**

Since three views are used to select services (or contents), so it may not be possible to narrow the selection down to only one service. In this case, the SEA has to investigate multiple services in different SPs and compare them. Broad action function executes and manages these multiple processes. These processes have the same action ID and are managed collectively. They are executed in sequence or in parallel. Broad action may stop current processing according to the condition of received data from SPs in Synchronization function. If the data received from SPs at the first request in a sequential processing is good enough, the rest of processing may not be executed. Broad action may also resend the request if there are no responses from SPs because of a delay or fault. The timing and times of resend is decided according to the three basic views.

In the above flow, broad action obtains identifiers of selected services from DoS and sends requests to appropriate SPs to obtain the services.

- **Synchronization**

Synchronization function have two sub-functions, that is, waiting sub-function and integrating sub-function. Waiting sub-function decides the way how to wait the responses. Some examples are at-most-one, waiting for fixed numbers of responses and waiting for fixed time. Waiting sub-function sends a request of completion, interruption or re-execution of the processing to Broad action according to the result of the waiting. Integrating sub-function specifies how to integrate the received data. There are various way to integrate such as majority.

3.3. SEA Architecture Based on TINA

This section shows the implementation of SEA function on TINA, the standard model for service management of telecommunication[8]. TINA defines four roles which are end users, SP, Retailer and Network Provider in the business model. Corresponded with SEA, TINA defines the Retailer which mediates between SPs and end users and provides the services. By adding the Retailer to SEA functions, the TINA system can provide the services more flexibly and manage communication service like the QoS control on ATM network on the basis of users' demand.

TINA divides into three parts for provision of the services from end user to SP. It describes about each parts and class of components supported each parts.

- **Access Session**

This part covers the interactions required for end user and SP to establish the Service Session parts. It consists of three components, PA(Provider Agent) which manage the information of SP, UA(User Agent) which manage the information of end users, and SF(Service Factory) which creates the service session components for a service.

- **Service Session**

This part represents information and functionality related to capabilities to execute, control and manage services. It consists of two components, SSM(Service Session Manager) which comprises the service-specific and generic session control segments like a network routing information, or a service path between end users and SP, USM(User Session Manager) which manage session information between end user and Retailer.

- **Communication Session**

This part represents a general, service view of stream connections and a network technology independent view of the communication resources. It is constructed by CSM and TCSM.

In order to realize the SEA functions, expanded components are required. By using these components the Retailer can decide appropriate SP on the basis of three views of end users' demand, customization, situation and quality in Access Session part and Service Session part. Fig.9 shows expanded components for realizing SEA, and it is as follows;

(A) UA, which manage the information about three views for end users and SPs.

(B) SSM, which manage the information about three views for SPs.

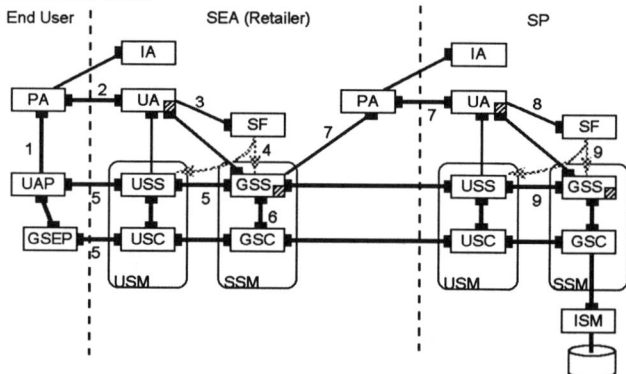

Fig. 9   SEA Architecture Based on TINA

The standard flow of information provision using expanded TINA architecture is as follows;

(1) End user application requests the service to PA.

(2) UA in Retailer receives the request from PA in end user. UA in Retailer registers three views for requested end user, and analyzes properties of demand. The three views using by SEA function add to User profile in UA specified by TINA.

(3) UA in Retailer requests the demand of connecting service session to SF in Retailer.

(4) SF creates SSM and USM for a demand for the service from end user.

134

(5) USM creates communication session between USM in Retailer and end user Application.

(6) SSM analyzes three views for each SPs registered in SSM, and judge the SP satisfied the demand of EU. If SSM finds a suitable SP, it send the information of network resources (QoS Parameter etc.) about this service to TSM.

(7) SSM requests a demand of service connection to PA in Retailer judged in step (6). PA in Retailer requests the session to UA in SP.

(8) UA in SP analyzes properties of demand by using three views of the SP. The three views using by SEA function add to Service profile in UA specified by TINA. And UA in SP requests the demand of connecting service session to SF in SP.

(9) SF creates SSM and USM for a demand for the service from Retailer, and SSM requests a demand of service session connection to SSM in Retailer. At last Service session creates by SSM in SP and SSM,

4. Application

An example application of the SEA system used for shopping is shown (Fig. 10). The system consists of end users, the SEA server and the SPs, all of which are connected via the Internet. Service providers consist of a credit card company, a delivery company and a retailer. The SEA manages the contents provided by the retailer. The retailer in this example sells food and daily goods. The scenario is explained along with the service flow.

(1) John, who is a gold member of this retailer which means that he is considered an important customer, is at home and logs in on his desktop PC to the SEA in order to buy some food.

(2) The SEA accepts John with his password and determines that he is at home and uses a desktop PC by the information included in the login message.

John at home

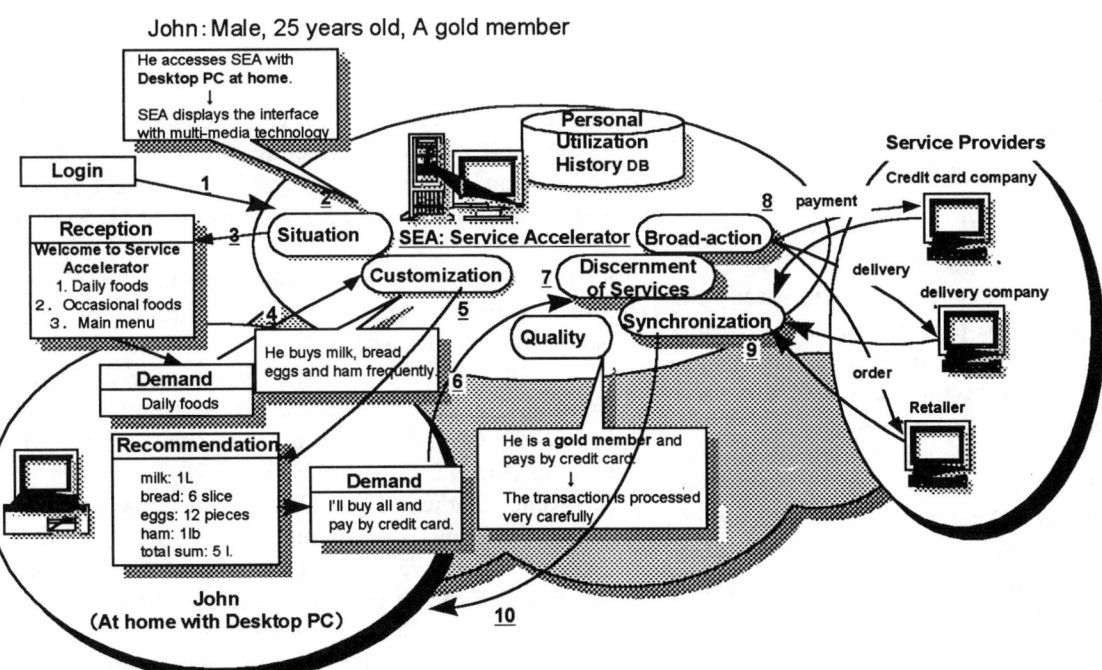

Fig. 10 An Example of Application

USM in Retailer.

The rest of steps making communication session is compliant with TINA documents. The expanded functions of UA and SSM for SEA are based on section 3.2. A special feature of SEA implemented on TINA architecture is a point that Retailer searches for the most suitable SP automatically for demand from end user by using three service views, customization, situation and quality of end users and SPs.

(3) The SEA sends the first menu for John. This menu is suited for John's current situation. That is, the menu is constructed with multi-media technology of audio and pictures and the contents are for daily shopping. (Situation)

(4) John selects "Daily food" to buy daily food.

(5) Since the SEA has a record in a personal utilization history DB that John often buys milk, bread, eggs and ham frequently, the SEA recommends these foods to

John.(Customization)

(6) John decides to buy the recommended foods as usual and specifies to pay by credit card.

(7) The SEA determines the level of importance of this transaction according to the fact that John is a gold member and pays by credit card. (Quality)

(8) The SEA sends transaction messages to the credit card company to pay for the items, to the delivery company to ask them to deliver the items, and to the retailer to order the items.

(9) The SEA receives responses from SPs. If there are problems with the responses, the SEA tries to solve them.

(10) If the responses are normal, the SEA sends a transaction completed message to John.

In the above example, customization, situation and quality of end user are all considered when the service is provided.

5. Effects of SEA

The effects of the SEA system can be summarized as follows:

(1) End users do not need to know what contents are available, they only need to specify their demands.

(2) The SEA selects appropriate contents for each end user according to his/her customization, situation and quality.

Effects of the SEA are discussed as a realization of Autonomous Decentralized System (ADS). As above, in conventional information systems, end users specify the name of a particular SP and particular contents. In this case, end users need to know what SPs and contents are available. Even if a new SP becomes available, end users can not utilize them unless they are aware of them.

In the SEA system, end users do not specify any SPs or contents. End users specify demand based on customization, situation and quality, then the SEA selects appropriate contents by using registered strategies. When new SPs or contents are added, the SEA can select contents from these sources as well. Thus end users do not need to know about the addition of new contents.

The advanced nature of this system that comes from the fact that the mechanism of DoS is an expansion of the content code communication technique of the Autonomous Decentralized System. The on-line expandability of services allows the SEA system to keep up with competition in the information service industry.

6. Conclusions

This paper presented the Service Accelerator concept

and its architecture. The SEA selects suitable contents autonomously according to customization , quality and situation. By using the SEA, end users do not have to know what contents are available, they only need to specify their demands. If new contents are added, end users can have access to the new contents through the autonomous selection function that the SEA provides. Furthermore, the SEA also manages the level of importance of the contents. This point, however, is still being studied to make it more effective for the end user.

This paper only discusses communication initiated by end users trigger for obtaining contents from SPs. There are other cases in which the SPs initiate communication and send contents to end users, or end users send data to other end users. The SEA can be easily extended to cope with this case.

References

[1] T. Eckardt, T. Magedanz and R. Popescu-Zeletin, "Application of X.500 and X.700 Standards for Supporting Personal Communications in Distributed Computing Environment", Proc. IEEE CS Workshop on Future Trend of Distributed Computing Systems, Aug. 1995, pp. 232-241

[2] R. Popescu-Zeletin, V. Tschammer and M. Tschichholz, "Y - Architecture for the Integration of Autonomous Components", Proc. IEEE CS Int'l Symp. on Autonomous Decentralized Systems (ISADS '93), Mar. 1993, pp. 21-27

[3] K. Mori, et. al., "Autonomous Decentralized Software Structure and its Application", Proc. of FJCC '86, Dallas, Nov. 1986, pp. 1056-1063

[4] K. Mori, "Autonomous Decentralized Systems: Concept, Data Field Architecture, and Future Trends", Proc. IEEE CS Int'l Symp. on Autonomous Decentralized Systems (ISADS 93), Mar. 1993, pp. 28-34

[5] Anne Clarke, Mario Campolargo, Nikos Karatzas (Eds.) Bringing Telecommunication Services to the People - IS&N '95 Third International Conference on Intelligence in Broadband Services and Networks, Heraklion, Crete, Greece, October 1995

[6] Robert Orfali, Dan Harkey, Jeri Edwards: The Essential Distributed Objects Suvaival Guide, John Wiley & Sons, Inc(1995)

[7] Thomas J. Mowbray, Ron Zahavi: The Essential CORBA, John Wiley & Sons, Inc(1995)

[8] H. Berndt, etc.: Service Architecture, Version 2.0, TINA-C, March 1995

Session 2C

Distributed System Paradigms

Chair

Jean-Pierre Banatre

Fearful Symmetry in System Structures

Reiko TANAKA and Seiichi SHIN
Course of Mathematical Engineering and Information Physics,
Graduate School of Engineering,
The University of Tokyo
Hongo, Bunkyo-ku, Tokyo 113, JAPAN

Abstract

This paper discusses the relationship between the level of symmetry in a system structure and loss of controllability. For a system consisting of a large number of modules, symmetrical structure is considered as the identity of modules and their interconnections. We especially dealt with three basic structures, such as ring-type, bus-type, and star-type structures. If the entire system becomes uncontrollable because of failures in a set of k submodules, this number k indicates the level of fault tolerance. On the other hand, the level of symmetry is evaluated by the number of transformations that keep the structure invariant. Through the evaluation of fault tolerance of three basic structures, we considered the relationship between lowering the level of symmetry and loss of controllability because of failures.

1 Background

1.1 "Symmetry", why?

An object is said to be symmetric if it holds the original figure after some transformations. With the number of such transformations that keep the subject invariant, the level of the subjects' symmetry can be compared one another. For example, a circle is invariant under any rotations, whereas a square is not invariant under the rotations other than multiple of $\pi/2$ radians. The figures which are invariant only by the rotations of certain angles are said to have symmetry of lower level than circles.

Lowering of the level of symmetry is often seen in natural phenomena. When you let fall a drop of milk in your morning cup which is full of milk, the form of rebound looks like a crown: a crown with 24 bumps[1]. Before the drop, a cup of milk had a perfect symmetry as a planar circle, i.e., symmetry for 360° rotation. Then a drop brakes it up, and $360/24 = 15°$ rotation symmetry appears, which has lower level of symmetry than a circle.

Various phenomena show this sort of lowering the level of symmetry at the moment of transition, such as condensation of gas, where Landau theory [2] mentions about it explicitly. The bifurcation theory intensively studies how the characteristics of the system transfers according to the change of the parameters that describe the system. Moreover, bifurcation theory goes well with group theory[3] to deal with sym-

metry possessed in the natural phenomena. It is said that lowering the level of symmetry, or a break-down of symmetry may bring the variety of transitions in the characteristics of the object.

1.2 What for "system structures"?

In the engineering systems, various studies have been carried out concerning large-scale systems consisting of a large number of modules, such as autonomous decentralized systems (ADSs) and hierarchical multilevel systems [4]. For several years, ADSs have shown a striking diffusion in real systems of various fields, for example, FA systems[5], and railway systems[6]. The main idea of ADSs is that cooperation among modules enables systems to realize some functions that are difficult to be realized by a single module alone. In other words, addition of a certain number of modules to a system can cause a transition of its functions. Conversely, failures of some modules may cause a negative transition, such as loss of controllability.

This paper focuses on lowering the level of symmetry in system structure at the moment of loss of controllability. We especially consider the case where a system loses its controllability as the result of increase in the number of modules that are in the outage. This transition of controllability is defined by the authors[7] as autonomous controllability. A system is said to be autonomously controllable at level l if it is controllable even with failures in arbitrary l modules. Conversely, if a system becomes uncontrollable because of failures in at least a set of k modules, this level k serves for a quantitative evaluation of fault-tolerance. As a result, fault tolerance of the system can be evaluated for various types of structures.

Symmetrical structure is realized by the identity of modules and their interconnections. From the practical view, assemblage of identical modules has the advantage to facilitate analysis and design of the entire system, and enables to get a variety of systems. Replacements of the failed modules are also be available by their mass-production. It is therefore suitable to the recent demands on the production of small quantity with more varieties. This paper considers three basic and widely-used structures shown in Fig. 1: (a) ring-type, (b) string-type, and (c) star-type structures.

However, systems with symmetrical structure tend

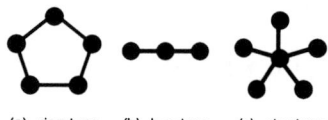

(a) ring-type (b) bus-type (c) star-type

Fig. 1. Three basic structures of interconnections among identical modules. (a) Ring-type, (b) string-type, (c) star-type.

to become uncontrollable by the failures of the modules distributed symmetrically[8]. Therefore, arrangement of the failed modules with which the system becomes uncontrollable can be obtained merely by its structural information. The following example describes well the problem considered in this paper.

Example 1. A ring-type structured system becomes uncontrollable because of failures in some control channels distributed symmetrically. In Fig. 2, the arrows represent effective inputs, so that the modules without the arrow are in the outage. Even though such failures, all the states of the system containing the states of failed modules are desired to be controllable. However, the failures shown in Fig.2 (b) cause the system to be uncontrollable. It is also observed that the level of symmetry lowers from (a) to (b) in Fig. 2: (a) is invariant under $\frac{2\pi i}{9}$ rotation, whereas (b) is invariant under $\frac{2\pi i}{3}$ rotation.

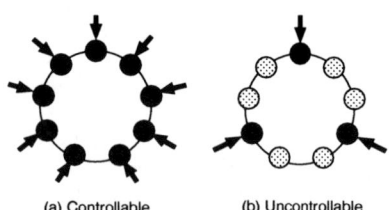

(a) Controllable (b) Uncontrollable

Fig. 2. (a) A ring-type structured system that is controllable. (b) When failures occur in the modules represented by dotted bullets, the system becomes uncontrollable.

Through the evaluation of fault tolerance of these three structures, we consider the relation between lowering the level of symmetry and loss of controllability because of failures. Moreover, this evaluation is carried out for various types of structures seen in the real systems, such as railway systems and air conditioning systems.

1.3 Organization of the paper

This paper is organized as follows. At first, Section 2 defines the system structures considered in this paper. It also clarifies the notion of controllability, which plays a primary role in this paper. After a notion of failure in a module and autonomous controllability are defined in Section 3, this section provides the results about loss of controllability in three basic structures. With this result, Section 4 discusses lowering the level

of symmetry defined for system structures. In Section 5, analysis of fault tolerance is extended to polyhedral structures, and hierarchical structures which are often seen in real systems. Finally, Section 6 gives the prospects for the future.

2 System structures

This section defines the system structure considered in this paper. In order to focus on the characteristics determined merely by the structure of interconnections among the modules, a module structure matrix will be defined.

2.1 System description

A system considered in this paper consists of identical module, say, $\{\mathcal{S}_1, \mathcal{S}_2, \cdots, \mathcal{S}_m\}$ (Fig. 3). Each module has its own control channel. In Fig. 3, control inputs from the channels are described by arrows. By virtue of the interconnections among the modules, the system functions as a whole.

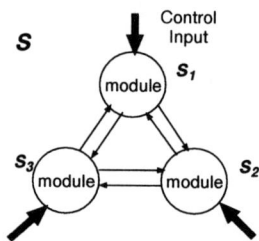

Fig. 3. A system consisting of three modules ($m=3$).

Each module $\mathcal{S}_i\,(i=1,2,\cdots,m)$ is described by a state transition equation

$$\dot{x}_i(t) = A_{ii}x_i(t) + B_{ii}u_i(t) + \sum_{j\neq i} A_{ij}x_j(t) + \sum_{j\neq i} B_{ij}u_j(t),$$

where $x_i(t) \in \mathbf{R}^{n_0}$ and $u_i(t) \in \mathbf{R}^{r_0}$ denote the state of $\mathcal{S}_i$ and the input from its control channel, respectively. The entire system $\mathcal{S}$ is described in the standard form of a state transition equation

$$\dot{x}(t) = Ax(t) + Bu(t) \qquad (1)$$

$$= \begin{bmatrix} A_{11} & A_{12} & \cdots & A_{1m} \\ A_{21} & A_{22} & \cdots & A_{2m} \\ \vdots & \vdots & \ddots & \vdots \\ A_{m1} & A_{m2} & \cdots & A_{mm} \end{bmatrix} \begin{bmatrix} x_1 \\ x_2 \\ \vdots \\ x_m \end{bmatrix} + \begin{bmatrix} B_{11} & B_{12} & \cdots & B_{1m} \\ B_{21} & B_{22} & \cdots & B_{2m} \\ \vdots & \vdots & \ddots & \vdots \\ B_{m1} & B_{m2} & \cdots & B_{mm} \end{bmatrix} \begin{bmatrix} u_1 \\ u_2 \\ \vdots \\ u_m \end{bmatrix},$$

$$(2)$$

where the state and the input of $\mathcal{S}$ are denoted as

$$x = \begin{bmatrix} x_1 \\ x_2 \\ \vdots \\ x_m \end{bmatrix} \in \mathbf{R}^n, u = \begin{bmatrix} u_1 \\ u_2 \\ \vdots \\ u_m \end{bmatrix} \in \mathbf{R}^r, \qquad (3)$$

respectively. The diagonal blocks of A and B stand for the effects of each module $\mathcal{S}_i$ on its state x_i. On the other hand, the off-diagonal blocks imply the interactions among modules. Since all components and the existing interconnections among them are supposed to be identical, interconnections between modules are described as

$$A_{ii} = K, \quad A_{ij} = \begin{cases} L & \text{(if } S_i \text{ and } S_j \text{ are connected)} \\ O & \text{(otherwise)} \end{cases},$$

$$B_{ii} = P, \quad B_{ij} = \begin{cases} Q & \text{(if } S_i \text{ and } S_j \text{ are connected)} \\ O & \text{(otherwise)} \end{cases},$$

with nonzero matrices K, L, P and Q of adequate size.

2.2 Controllability of a system

In order to design controllers, it is natural to ask how we can find a suitable input $u(t)$ that will take the system to any desired state in a finite time. If we can always find such inputs, we shall say that the system (A, B) is controllable[9]. It is a fundamental property of a realization. Therefore, a system (1) is said to be controllable if every state x_i of S_i can attain arbitrary points independently of one another, with adequate control inputs. Mathematically, controllability can be judged by the controllability matrix[9] of (A, B), i.e., $R(A, B) = [B\ AB\ A^2B \cdots A^{n-1}B]$. The system (A, B) is controllable if and only if $R(A, B)$ is of full-rank. Throughout this paper, the system (1) is supposed to be controllable. Fault tolerance of the system is then discussed by the number of failed modules, which cause the system to be uncontrollable. The loss of controllability called "autonomous controllability" will be defined in the next section.

2.3 Module structure matrix

This paper discusses the characteristics determined merely by the structure of interconnections among modules and their identities, and independently of numerical information of parameters K, L, P and Q. We then define the module structure matrix A_m^{type} (type=ring,bus,star) as follows.

Definition 1. Module structure matrix A_m^{type} of $m \times m$ is obtained by replacing K and L by scalar parameters k and l, respectively. Thus, the module structure matrix A_m^{type} describes the structural information about the system.

Module structure matrices for each type of structure $A_m^{\text{ring}}, A_m^{\text{bus}}, A_m^{\text{star}}$ are defined as

$$\begin{bmatrix} k & l & 0 & \cdots & 0 & l \\ l & k & & & & 0 \\ 0 & & & & & \vdots \\ \vdots & & & & & 0 \\ 0 & & & & k & l \\ l & 0 & \cdots & 0 & l & k \end{bmatrix}, \begin{bmatrix} k & l & 0 & \cdots & 0 \\ l & k & & & \vdots \\ 0 & & & & 0 \\ \vdots & & & & l \\ 0 & \cdots & 0 & l & k \end{bmatrix}, \begin{bmatrix} k & 0 & \cdots & 0 & l \\ 0 & k & & & \vdots \\ \vdots & & & & 0 & l \\ 0 & \cdots & 0 & k & l \\ l & \cdots & l & l & k \end{bmatrix},$$

respectively.

3 Fault tolerance of systems

This section discusses fault tolerance of systems with three basic structures defined in the previous section.

3.1 Failure of a module

Failure of a module is restricted to that of its control channel [10]. In other words, to say a failure occurred in a module S_i at time t_0 means

$$u_i(t) = 0 \qquad \text{for} \quad \forall t \geq t_0. \tag{4}$$

Systems for practical use may be equipped to make the controller stop when any kind of doubtful signals are detected. Then, the suspected parts should be inspected with their control inputs off. Therefore, this definition of failures as (4) can deal with the case of emergency stops, or, maintenance of modules, as well. Even in such cases, fault-tolerant systems are desired to keep functioning as a whole.

3.2 Autonomous controllability

As a characteristic desired to fault-tolerant systems, the authors defined autonomous controllability of the entire system[7]. This property guarantees the controllability of the system even with the failures in some modules. That is, the state of the entire system including those of failed modules can achieve arbitrary points by cooperation of active control inputs. Moreover, the number of failed modules is called "level". A system is said to be autonomously controllable at level l if it is controllable even with failures in arbitrary l modules. Conversely, if a system becomes uncontrollable because of failures in at least a set of k modules, this level k serves for a quantitative evaluation of fault-tolerance.

Remark.
Autonomous controllability defined by Mori *et al.* [10] is useful if it is possible to remove the modules in the outage, for example, in computer network systems [11]. Their autonomous controllability is the controllability of the states of the functioning modules with all the failures in some modules, so that it makes no mention of the states of modules in the outage.

3.3 A measure of fault tolerance

Suppose that it is given the module structure matrix A_m^{type} and the knowledge about the failures of each module.

In order to describe the measure of fault tolerance, we adopt the notion of "cut" in the study of the reliability[12]. We first associate with each module S_i a variable f_i depending on if it is in the outage or not, such that

$$f_i = \begin{cases} 0 & \text{(if } S_i \text{ is faulty)}, \\ 1 & \text{(otherwise)}. \end{cases} \tag{5}$$

Consider a pattern $(f_1, f_2, \cdots, f_m)$ with which all the systems of the structure A_m^{type} is uncontrollable.

Definition 2. Let $\bar{S}$ be the subset of modules whose indices belong to J. If the entire system is not controllable when all the modules in $\bar{S}$ have failed, i.e.,

$$\left. \begin{array}{l} f_j = 0, \ j \in J \\ f_j = 1, \ j \notin J \end{array} \right\} \Rightarrow S \text{ is uncontrollable,}$$

one says that $\bar{S}$ is a *cut* of the structure A_m^{type}.

All the systems with the structure described by A_m^{type} lose controllability because of failures in all modules in $\bar{S}$. Therefore, the number of the modules in a cut $\bar{S}$ may manifest the level of fault tolerance of the systems.

3.4 Structural analysis for fault tolerance of three basic structures

The authors have conducted structural analysis for fault tolerance of three basic structures[13]. The results are reviewed here, and their fault tolerance is discussed.

Corresponding to a pattern of failures, a failure matrix is defined by $F = \text{diag}\{f_1, f_2, \cdots, f_m\}$. The matrix $\bar{B}F_{r_0}$, where $F_{r_0} = F \otimes I_{r_0}$, after the failure has zero column blocks that correspond to all the subsystems in the outage. We then investigate the cut $\bar{S}$ described above, where the rank of $(A, \bar{B}F_{r_0})$ becomes less than n.

Since the rank of the controllability matrix $R(A, B)$ is invariant under state transformations, we shall consider $\text{rank}R(\tilde{A}, \tilde{B})$, where $\tilde{A} = Z_{n_0}^{-1}AZ_{n_0}$ and $\tilde{B} = Z_{n_0}^{-1}B$ after state transformation by $Z_{n_0} = Z_m^{\text{type}} \otimes I_{n_0}$. The matrix Z_m^{type} is found as such that $(Z_m^{\text{type}})^{-1}A_m^{\text{type}}Z_m^{\text{type}}$ becomes diagonal, independently of parameters. Consequently, $\tilde{A}$ becomes block diagonal, say, $\text{diag}(A_1, A_2, \cdots, A_m)$, and $\tilde{B}Z_{r_0}$ is also block diagonal as $\tilde{B}Z_{r_0} = Z_{n_0}^{-1}BZ_{r_0} = \text{diag}\{B_1, B_2, \cdots, B_m\}$. Therefore, $\tilde{A}^i\tilde{B} = \text{diag}\{A_1^iB_1\ A_2^iB_2 \cdots A_m^iB_i\}Z_{r_0}^{-1}$. By permutating some rows and columns,

$$\text{rank}R(\tilde{A}, \tilde{B}) = \text{rank}\left\{\begin{bmatrix} R_1 & & & \\ & R_2 & & \\ & & \ddots & \\ & & & R_m \end{bmatrix}(Z_m^{\text{type}})^{-1} \otimes I_{nr_0}\right\},$$

where each R_j $(j = 1, \cdots, m)$ is (6)

$$R_j = \begin{bmatrix} B_j\ A_jB_j\ A_j^2B_j \cdots A_j^{n-1}B_j \end{bmatrix}. \quad (7)$$

In (6), we suppose that R_j is of full-rank, i.e., $R(A_i, B_i)$ is of full-rank because the order of A_i is n_0. Under this condition, uncontrollability after failures can occur by the existence of a zero row in $(Z_m^{\text{type}})^{-1}F$.

Cuts for three basic structures are obtained by the procedure above. In the following, we show the matrices that diagonalize each module structure matrix A_m^{type}, and let them denote by $Z_m^{\text{ring}}, Z_m^{\text{bus}}, Z_m^{\text{star}}$, respectively. Then we clarify the cuts of each structure.

(a) The matrix Z_m^{ring} for ring-type structures is obtained as follows by a group theoretic approach. This procedure shall be explained in the next section. If m is an odd number,

$$\begin{bmatrix} \frac{1}{\sqrt{2}} & \cos\theta & \sin\theta & \cdots & \cos\lfloor\frac{m-1}{2}\rfloor\theta & \sin\lfloor\frac{m-1}{2}\rfloor\theta \\ \frac{1}{\sqrt{2}} & \cos 2\theta & \sin 2\theta & \cdots & \cos 2\lfloor\frac{m-1}{2}\rfloor\theta & \sin 2\lfloor\frac{m-1}{2}\rfloor\theta \\ \vdots & \vdots & \vdots & \vdots & \vdots & \vdots \\ \frac{1}{\sqrt{2}} & \cos m\theta & \sin m\theta & \cdots & \cos m\lfloor\frac{m-1}{2}\rfloor\theta & \sin m\lfloor\frac{m-1}{2}\rfloor\theta \end{bmatrix}, \quad (8)$$

and if m is an even number,

$$\begin{bmatrix} \frac{1}{\sqrt{2}} & \cos\theta & \sin\theta & \cdots & \cos\lfloor\frac{m-1}{2}\rfloor\theta & \sin\lfloor\frac{m-1}{2}\rfloor\theta & -\frac{1}{\sqrt{2}} \\ \frac{1}{\sqrt{2}} & \cos 2\theta & \sin 2\theta & \cdots & \cos 2\lfloor\frac{m-1}{2}\rfloor\theta & \sin 2\lfloor\frac{m-1}{2}\rfloor\theta & \frac{1}{\sqrt{2}} \\ \vdots & & \vdots & & \vdots & \vdots & \vdots \\ \frac{1}{\sqrt{2}} & \cos m\theta & \sin m\theta & \cdots & \cos m\lfloor\frac{m-1}{2}\rfloor\theta & \sin m\lfloor\frac{m-1}{2}\rfloor\theta & -\frac{1}{\sqrt{2}} \end{bmatrix}. \quad (9)$$

where $\theta = \frac{2\pi}{m}$.

(b) The matrix Z_m^{bus} for string-type structures is explicitly obtained for tridiagonal matrices as follows.

$$Z = \begin{bmatrix} \sin\theta & \sin 2\theta & \cdots & \sin m\theta \\ \sin 2\theta & \sin 4\theta & \cdots & \sin 2m\theta \\ \vdots & \vdots & \vdots & \vdots \\ \sin m\theta & \sin 2m\theta & \cdots & \sin m^2\theta \end{bmatrix},$$

where $\theta = \frac{\pi}{m+1}$.

(c) The matrix Z_m^{star} for star-type structures is calculated as follows by the aid of Z_{m-1}^{ring} for ring-type structures.

If $m' = m - 1$ is an odd number, Z_m^{star} is obtained as

$$\begin{bmatrix} \cos\theta & \sin\theta & \cdots & \cos\lfloor\frac{m'-1}{2}\rfloor\theta & \sin\lfloor\frac{m'-1}{2}\rfloor\theta & 1/2 & 1/2 \\ \cos 2\theta & \sin 2\theta & \cdots & \cos 2\lfloor\frac{m'-1}{2}\rfloor\theta & \sin 2\lfloor\frac{m'-1}{2}\rfloor\theta & 1/2 & 1/2 \\ \vdots & & & \vdots & \vdots & \vdots & \vdots \\ \cos m'\theta\ \sin m'\theta & \cdots & \cos m'\lfloor\frac{m'-1}{2}\rfloor\theta & \sin m'\lfloor\frac{m'-1}{2}\rfloor\theta & 1/2 & 1/2 \\ 0 & 0 & \cdots & 0 & 0 & \frac{\sqrt{m'}}{2} & -\frac{\sqrt{m'}}{2} \end{bmatrix}$$

and if $m' = m - 1$ is an even number, Z_m^{star} is obtained as

$$\begin{bmatrix} \cos\theta & \sin\theta & \cdots & \cos\lfloor\frac{m'-1}{2}\rfloor\theta & \sin\lfloor\frac{m'-1}{2}\rfloor\theta & 1/2 & 1/2 & -\frac{1}{\sqrt{2}} \\ \cos 2\theta & \sin 2\theta & \cdots & \cos 2\lfloor\frac{m'-1}{2}\rfloor\theta & \sin 2\lfloor\frac{m'-1}{2}\rfloor\theta & 1/2 & 1/2 & \frac{1}{\sqrt{2}} \\ \vdots & \vdots & & \vdots & \vdots & \vdots & \vdots & \vdots \\ \cos m'\theta\ \sin m'\theta & \cdots & \cos m'\lfloor\frac{m'-1}{2}\rfloor\theta & \sin m'\lfloor\frac{m'-1}{2}\rfloor\theta & 1/2 & 1/2 & \frac{1}{\sqrt{2}} \\ 0 & 0 & \cdots & 0 & 0 & \frac{\sqrt{m'}}{2} & -\frac{\sqrt{m'}}{2} & 0 \end{bmatrix}$$

where $\theta = \frac{2\pi}{m'}$.

Examination of zero and nonzero pattern of Z^{-1} for each of them leads us to the following theorem.

Theorem 1.
The cuts of three basic structures to be generically uncontrollable are described as follows.

(a) The cuts of the ring-type structure are $\mathcal{S} - \{\mathcal{S}_q, \mathcal{S}_{2q}, \mathcal{S}_{3q}, \cdots, \mathcal{S}_m\}$, for all the divisors q of m, except for the case $m \equiv 2 \pmod 4$ where $q = 2$ is not included.

(b) The cuts of the string-type structure are $\mathcal{S} - \{\mathcal{S}_q, \mathcal{S}_{2q}, \mathcal{S}_{3q}, \cdots, \mathcal{S}_{m+1-q}\}$, for all the divisors $q > 1$ of $(m+1)$.

(c) The cuts of the star-type structure are arbitrary $(m-1)(1-\frac{1}{q})$ subsystems in $\{\mathcal{S}_1, \mathcal{S}_2, \cdots, \mathcal{S}_{m-1}\}$, for all the divisors q of $(m-1)$, except for the case $(m-1) \equiv 2 \pmod 4$ where $q = 2$ is not included.

Theorem 1 shows definitive failure patterns of modules to cause the system to be uncontrollable, merely by its structures. From analysis for (c), the module $\mathcal{S}_m$ is not necessarily identical with the others.

Example 2.
Figure 4 gives examples of Theorem 1. Failures occur in the modules represented by dotted bullets. The arrows denote the effective control inputs to functioning modules.

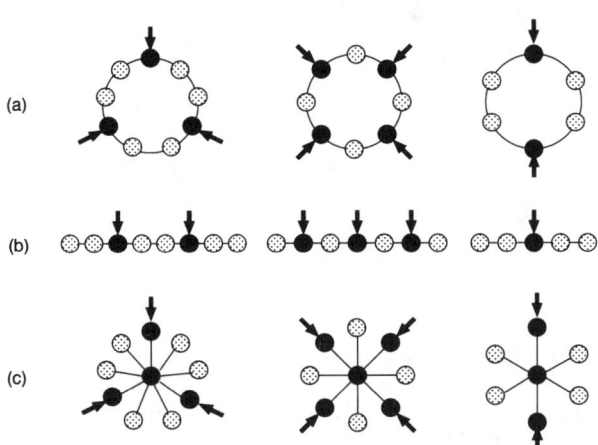

Fig. 4. Examples of the systems consisting of identical modules. They are uncontrollable independently of parameters. Failures occur in the modules represented by dotted bullets. (a) Ring-type, (b) string-type, (c) star-type.

In the failure patterns that induce uncontrollability as shown in Theorem 1, the functioning modules are distributed in periodic ways (see Fig. 4). This periodicity derives from two kinds of symmetry. Namely, if a system has structural symmetry, the system is not tolerant for failures of modules distributed symmetrically. That is, uncontrollability is induced by symmetry. In this context, it might be worth mentioning that ring-type structure with a prime number

m, string-type structure with a prime number $m + 1$, and star-type structure with a prime number $m - 1$ of modules connected to the module at the center, are superior in fault tolerance. In these cases, the system never becomes uncontrollable by symmetry as far as arbitrary two modules remain functioning.

4 Lowering of symmetry and uncontrollability

In the discussion above, when a system with symmetrical structure becomes uncontrollable because of some failures, another symmetrical structure is found again in the system. The group theory[14] is known to serve as a powerful tool to consider symmetrical structures. In fact, the primary point in the procedure to derive Theorem 1 is the diagonalization of A_m^{type}, for which the finite group theory exhibits efficiency for the systems with symmetrical structures.

4.1 Preliminaries

Before moving on to the definition of symmetrical structures, we introduce some fundamental facts about group representations (see, for example [14]) utilized in the following discussion.

Let G be a finite group of order $n(G)$. Let V be a finite-dimensional vector space over $\mathbf{K} = \mathbf{R}$ or $\mathbf{C}$, where $\mathbf{R}$ is the field of real numbers and $\mathbf{C}$ is the field of complex numbers, and denote by $GL(V)$ the group of all nonsingular linear transformations of V onto itself. A *representation* of G on *representation space* V is a homomorphism $\mathbf{T} : G \to GL(V)$. The *dimension* of the representation is $n = \dim V$. A subspace W of V is *invariant* under τ if $\tau(g)\boldsymbol{w} \in W$ for every $g \in G, \boldsymbol{w} \in W$. The representation τ is *irreducible* if the only invariant subspaces of V are $\{0\}$ and V itself. An n–dimensional matrix representation of G is a homomorphism $T : G \to GL(n, K)$, where $GL(n, K)$ denotes the group of all nonsingular matrices over $\mathbf{K}$ of order n. If a basis $\{\boldsymbol{v}_1, \boldsymbol{v}_2, \cdots, \boldsymbol{v}_n\}$ is fixed for V, we obtain a matrix representation T of G. The character $\chi : G \to \mathbf{K}$ of τ is defined by $\chi(g) = \text{Tr}\tau(g) = \text{Tr}T(g)$. Note that the character χ is independent of the choice of basis vectors for V. A subgroup H of G is a subset which is itself a group in G. A representation T_H of subgroup H of G is obtained by restricting T to H, $T_H(h) = T(h), h \in H$. We write $T_H = T|H$.

We denote by $\{\tau^\mu \mid \mu \in R(G)\}$ a complete list of nonequivalent irreducible representations of G, where $R(G)$ denotes an index set for the irreducible representations of G. A complete list of nonequivalent irreducible matrix representations of G is denoted by $\{T^\mu \mid \mu \in R(G)\}$. We denote the dimension of T^μ by n^μ. Every finite-dimensional representation of a finite group can be decomposed into a direct sum of irreducible representations. The direct sum decomposition is obtained by

$$V = \oplus_{\mu \in R(G)} \oplus_{i=1}^{a^\mu} V_i^\mu, \tag{10}$$

where V_i^μ are invariant subspaces of V which transform irreducibly under the restrictions τ^μ of τ to V_i^μ, and the *multiplicity* a^μ of τ^μ in τ is uniquely determined. Then the matrix representation can be put into a block-diagonal form

$$T(g) = \oplus_{\mu \in R(G)} \oplus_{i=1}^{a^\mu} T_i^\mu(g), \qquad g \in G, \qquad (11)$$

where T_i^μ is irreducible, if we first choose a basis $\{v_{ij}^\mu | j = 1, \cdots, n^\mu\}$ for each V_i^μ and adopt their union as a basis of V. Moreover, by choosing a basis adequately, we can have $T_i^\mu(g) = T^\mu(g)$ $(1 \le i \le a^\mu)$. Namely, the decomposition is as

$$T(g) = \oplus_{\mu \in R(G)} \oplus_{i=1}^{a^\mu} T^\mu(g). \qquad (12)$$

The following is known as Schur's lemma, see, e.g., [14].

Lemma 1.
Let T_1 and T_2 be irreducible matrix representations of G over $\mathbf{K}$, and H be a matrix over $\mathbf{K}$. Assume that

$$T_1(g)H = HT_2(g), \qquad g \in G. \qquad (13)$$

1. If T_1 and T_2 are not equivalent, then $H = O$.
2. If $\mathbf{K} = \mathbf{C}$ and $T_1 = T_2$, then $H = \alpha I$ for some $\alpha \in \mathbf{C}$.

4.2 Symmetrical structures

We now define a symmetrical structure as follows. Then, it is shown how we obtain the matrix Z_m^{ring} by the use of its structural symmetry.

Definition 3. A structure A_m^{type} is said to be *symmetric* for a specified group G if

$$T(g^{-1})A_m^{\mathrm{type}}T(g) = A_m^{\mathrm{type}}, \quad g \in G, \qquad (14)$$

where $T(g)$ is a matrix representation of G.

This equation (14) reflects the underlying geometric symmetry in the system structure, such as the homogeneity of the modules and that of the interconnections among the modules.

Lemma 1, together with the equations (12) and (14), shows that $Z^{-1}A_m^{\mathrm{type}}Z$ becomes block-diagonal with Z made by arranging $\{v_{ij}^\mu\}$:

$$Z^{-1}A_m^{\mathrm{type}}Z = \oplus_{\mu \in R(G)}(A_{\mu,m}^{\mathrm{type}} \otimes I_{n^\mu}), \qquad (15)$$

where A_μ^{type} is of order a_μ. The basis $\{v_{ij}^\mu\}$ can be determined from $\{T^\mu(g)|\mu \in R(G)\}$ by means of the projection methods, see, e.g., [14].

Ring-type system is symmetric for the dihedral group D_m of order $2m$, which is defined as

$$D_m = \{1, r, \cdots, r^{(m-1)}; s, sr, \cdots, sr^{(m-1)}\}, \qquad (16)$$

where $r^m = s^2 = (sr)^2 = 1$. The planar regular polygons are invariant under D_m. The irreducible representation decomposition is described as

$$T = \begin{cases} A_1 \oplus B_1 \oplus E_1 \oplus E_2 \oplus \cdots \oplus E_{\frac{m}{2}-1} & (m \text{ is even}), \\ A_1 \oplus E_1 \oplus E_2 \oplus \cdots \oplus E_{\frac{m-1}{2}} & (m \text{ is odd}), \end{cases}$$

where one-dimensional irreducible matrix representations are given as

$$T^{A_1}(g) = 1 (g \in D_m)$$

and

$$\begin{cases} T^{B_1}(g) = 1 (g \in \{1, sr^{2j}, r^{2j}, r^{m-2j}\}), \\ T^{B_1}(g) = -1 (g \in \{sr^{2j+1}, r^{2j+1}, r^{m-2j-1}\}), \end{cases}$$

with $0 \le j \le \frac{m}{2} - 1$, and two-dimensional irreducible matrix representations are given as

$$T^{E_k}(r^l) = \begin{bmatrix} \cos kl\theta & -\sin kl\theta \\ \sin kl\theta & \cos kl\theta \end{bmatrix}$$

and

$$T^{E_k}(sr^l) = \begin{bmatrix} \cos kl\theta & -\sin kl\theta \\ -\sin kl\theta & -\cos kl\theta \end{bmatrix},$$

with $0 \le l \le (m-1)$. The first column of Z_m^{ring} in (8) and (9) correpsonds to A_1 and the last of (9) to B_1. A pair of columns whose first entries are $\cos k\theta$ and $\sin k\theta$ corresponds to E_k. The matrix Z_m^{ring} is thus obtained from the irreducible decomposition of the representation.

4.3 Emergence of irreducible representation and loss of controllability

The ring-type structure is symmetric for the dihedral group D_m. According to Theorem 1 (a), uncontrollable systems with some failures are also symmetric for the dihedral group D_p, where p is a divisor of m except for the case $m \equiv 2 \pmod{4}$ where $p = m/2$ is not included (Fig. 4(a)). Since D_p as above is a subgroup of D_m, the level of symmetry lowers at the loss of controllability. Moreover, when the loss of controllability occurs, $D_m|D_p$ shows the emergence of some irreducible representations that are not seen in the decomposition for D_m.

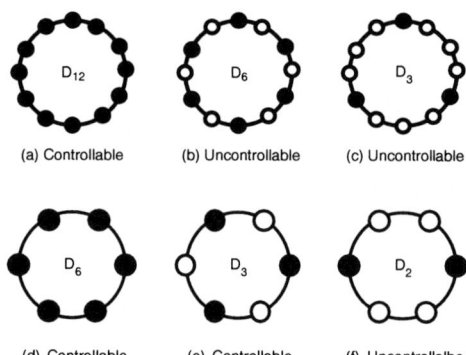

Fig. 5. Ring-type structured systems. Failures occur in modules represented by circles (o). Each system is symmetric under the group indicated in the figure. Failures as shown in (b), (c), and (f) cause the system to be uncontrollable.

Example 3. Consider a ring-type structured system with 12 modules (Fig.5(a)). The irreducible representation decomposition for D_{12} is described as

$$A_1 \oplus B_1 \oplus E_1 \oplus E_2 \oplus E_3 \oplus E_4 \oplus E_5. \qquad (17)$$

The system after the failures in $\{S_2, S_4, S_6, S_8, S_{10}, S_{12}\}$ (Fig.5(b)) is symmetric under D_6, and falures in $S - \{S_4, S_8, S_{12}\}$ cause the system to be symmetric under D_3 (Fig.5(c)). In both cases, the system becomes uncontrollable. And the irreducible representation decomposition of $D_{12}|D_6$ and $D_{12}|D_3$ are described as

$$2A_1 \oplus B_1 \oplus \mathbf{B_2} \oplus 2E_1 \oplus 2E_2, \qquad (18)$$
$$3A_1 \oplus \mathbf{A_2} \oplus 4E_1, \qquad (19)$$

respectively. The emergence of irreducible representation is seen as B_2 in (18) and A_2 in (19).

Example 4. Consider a ring-type structured system with 6 modules (Fig.5(d)). The irreducible representation decomposition for D_6 is described as

$$A_1 \oplus B_1 \oplus E_1 \oplus E_2. \qquad (20)$$

The system after the failures in $\{\mathcal{S}_2, \mathcal{S}_4, \mathcal{S}_6\}$ (Fig.5(e)) is symmetric under D_3, but keep the system controllable. On the other hand, falures in $\mathcal{S} - \{\mathcal{S}_3, \mathcal{S}_6\}$ (Fig.5(f)) cause the system to be symmetric under D_2 and uncontrollable. The irreducible representation decomposition of $D_6|D_3$ and $D_6|D_2$ are described as

$$2A_1 \oplus \oplus 2E_1 \qquad (21)$$
$$2A_1 \oplus \mathbf{A_2} \oplus 2B_1 \oplus \mathbf{B_2}, \qquad (22)$$

respectively. The emergence of irreducible representation is only seen as A_2 and B_2 in (22), where the system loses its controllability.

5 Extention for a variety of structures

This section carries out evaluation of fault tolerance for the structure of regular polyhedron and hierarchical systems obtained by combination of three basic structures discussed above.

5.1 Regular Polyhedron

It is known that there exist only five types of regular polyhedrons. There also exist the groups that keep each of them invariant: tetrahedral group, hexahedral group, octahedral group, dodecahedral group and icosahedral group.

Example 5. Fault-tolerance of the structure of tetrahedrons and hexahedrons are investigated similarly. As the result, failures as shown in Fig. 6 cause the system to be uncontrollable.

The other types are to be analyzed similarly, and the level down of symmetry can be also seen in the uncontrollable systems after failures. Discussion for three-dimensional structures may have relations with crystallography and quantum mechanical systems.

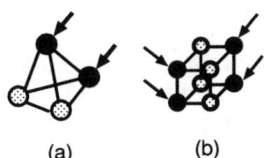

Fig. 6. Polyhedral structures: (a) tetrahedral and (b) hexahedral structures. We can see these systems are generically uncontrollable with failures in the modules represented by bullets(•).

5.2 Hierarchical systems

A variety of structures can be obtained by combination of three basic structures discussed above. For example, hierarchical systems are often seen in FA systems and electric power plants[4]. This section discusses fault tolerance of hierarchical systems as shown in Fig. 7. A hierarchical system is supposed to consist of a number of identical modules, each of which consists again of a number of identical submodules. The fault-tolerance of these systems can be analyzed in the similar way[15].

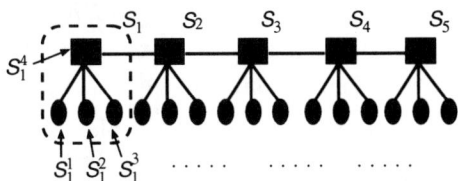

Fig. 7. A hierarchical system with bus-type interconnections among modules and star-type interconnections among submodules.

Suppose that each module $\mathcal{S}_i$ consists of h submodules $\{\mathcal{S}_i^1, \cdots, \mathcal{S}_i^h\}$. Therefore,

$$x_i = \begin{bmatrix} x_i^1 \\ x_i^2 \\ \vdots \\ x_i^h \end{bmatrix}, u_i = \begin{bmatrix} u_i^1 \\ u_i^2 \\ \vdots \\ u_i^h \end{bmatrix}, \qquad (23)$$

where $x_i^j(t) \in \mathbf{R}^{n_1}$ and $u_i^j(t) \in \mathbf{R}^{r_1}$ denotes the state of $\mathcal{S}_i^j$ and the control input from its own channel, respectively. If matrices K and P are divided as

$$K = \begin{bmatrix} K_{11} \cdots K_{1h} \\ \vdots \ddots \vdots \\ K_{h1} \cdots K_{hh} \end{bmatrix}, P = \begin{bmatrix} P_{11} \cdots P_{1h} \\ \vdots \ddots \vdots \\ P_{h1} \cdots P_{hh} \end{bmatrix}, \qquad (24)$$

according to (23), interconnections among submodules are denoted as

$$K_{jj} = K^\bullet, \quad K_{jl} = \begin{cases} K^\circ & \text{(if } \mathcal{S}_i^j \text{ and } \mathcal{S}_i^l \text{ are connected)} \\ O & \text{(otherwise)} \end{cases},$$
$$P_{jj} = P^\bullet, \quad P_{jl} = \begin{cases} P^\circ & \text{(if } \mathcal{S}_i^j \text{ and } \mathcal{S}_i^l \text{ are connected)} \\ O & \text{(otherwise)} \end{cases},$$

with nonzero matrices $K^\bullet, K^\circ, P^\bullet$ and P° of adequate size. The same is true of L and Q, which are defined as $L^\bullet, Q^\bullet$, respectively if corresponding submodules are connected over the different modules, and otherwise they are defined to be O.

Submodule structure matrix K_h^{type} is defined similarly to module structure matrix.

Definition 4. Submodule structure matrix K_h^{type} of $h \times h$ is obtained by replacing $K^\bullet$ and K° by scalar parameters $k^\bullet$ and k°, respectively.

The structure of a hierarchical systems is then represented by a pair $\left[A_m^{\text{type}}, K_h^{\text{type}}\right]$ of the module structure matrix A^{type} and the submodule structure matrix K_h^{type}.

Failure in a submodule is defined similarly to that of a module. Then, to say a failure occurred in a module S_i^j at time t_0 means

$$\boldsymbol{u}_i^j(t) = 0 \qquad \text{for} \quad \forall t \geq t_0. \qquad (25)$$

The similar calculation as in Section 3 leads to the following theorem for the cut of $\left[A_m^{\text{type}}, K_h^{\text{type}}\right]$. It clarifies that the cut is obtained by combination of the cut of the both structure matrices.

Theorem 2. Let the cut of A_m^{type} denote by $\bar{S}^A$, and I represents the index set of $\bar{S}^A$. Similarly, J represents the index set of $\bar{S}^K$, that is the cut of K_h^{type}. Then, the cut of the hierarchical structure $\left[A_m^{\text{type}}, K_h^{\text{type}}\right]$ is given by a set of submodules $\{S_i^j | \forall i \in I, \forall j \in J\}$.

Example 6. Fig. 8 shows some examples of hierarchical systems which are uncontrollable because of failures in some submodules. The structure shown in (a) and (b) are found in railway systems, where each module corresponds to a station in which the control system adopts the star-type structure. This structure is found also in computer networks. The structure shown in (c) and (d) is found in air conditioning systems in a building, where each module corresponds to the jets of a story. The structure shown as (g)–(i) is found in FA systems, for example.

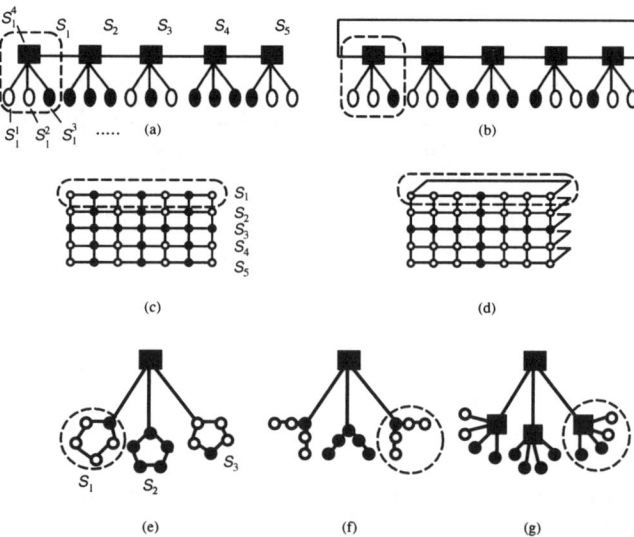

Fig. 8. Failure patterns that cause the hierarchical system to be uncontrollable. Failures occur in the submodules represented by circles (o).

6 Prospects for the future

This paper evaluated the fault tolerance of the system consisting of a large number of modules, quantitatively by the number of failed modules to cause the system to be uncontrollable. And we mentioned that the level of symmetry in the system structure lowers when the loss of controllability occurs. We also showed the change of the irreducible representations for the symmetry of the system structure at the transition point. Theoretical clarification for this emergence of irreducible representations is for future research.

Acknowledgement

The authors wish to express our gratitude to Dr. S. Iwata of Kyoto University for his effective suggestions and guidance in the course of this research and to Dr. K. Suyama of Tokyo Univ. of Mercantile Marine for helpful discussions.

References

[1] I. Stewart and M. Golubitsky (1992). *Fearful Symmetry.* Penguin Books Ltd., London.

[2] L. D. Landau and E. M. Lifshitz (1980). *Statistical Physics.* Oxford, Pergamon.

[3] M. Golubitsky, I. Stewart and D. G. Schaeffer (1988). *Singularities and Groups in Bifurcation Theory.* Springer-Verlag.

[4] M. D. Mesarovic, D. Macko and Y. Takahara (1970). *Theory of Hierarchical Multilevel Systems.* Academic Press, Inc.

[5] M. Omura and M. Oku (1995). Hi-Cell System Architecture for Manufacturing Systems. *Proc. Int. Sympo. Autonomous Decentralized Systems,* pp. 154–161.

[6] S. Harashima, Y. Taniyama, S. Ogawa, K. Udono and T. Seki (1995). *Proc. Int. Sympo. Autonomous Decentralized Systems,* pp. 80–85.

[7] R. Tanaka, S. Shin and N. Sebe (1994). Controllability of Autonomous Decentralized Systems. *Proc. 3rd Seiken/IEEE Symposium on Emerging Technologies and Factory Automation,* pp. 265–272.

[8] R. Tanaka, S. Shin and N. Sebe (1995). Invitation to Dissymmetry —— System Invariance and Basic Properties of Linear Systems. *Proc. IFAC Conference on System Structure and Control,* pp. 449–454.

[9] T. Kailath: Linear Systems, Prince-Hall, Inc. (1980)

[10] K. Mori, K. Sano and H. Ihara (1981). Autonomous Controllability of Decentralized System Aiming at Fault-Tolerance. *Proc. 8th IFAC World Congress,* pp. 1833–1839.

[11] H. Ihara and K. Mori (1984). Autonomous Decentralized Computer Control Systems. *IEEE COMPUTER,* **17**, 8, pp. 57–66.

[12] A. Kaufmann, D. Grouchko and R. Cruon (1977). Mathematical Models for the Study of the Reliability of Systems. Academic Press.

[13] R. Tanaka, S. Iwata and S. Shin (1996). Structural Analysis of Fault-tolerance for Homogeneous Systems. *Proc. 35th IEEE Conference on Decision and Control* (to be presented).

[14] W. Miller (1972). Symmetry Groups and Their Applications. Academic Press, New York and London.

[15] R. Tanaka and S. Shin (1997). Autonomous controllability of hierarchical systems (in Japanese). *Proc. 9th SICE Symposium on Decentralized Autonomous Systems* (to be presented)

Integration of Control & Information Systems by Open Autonomous Decentralized System Architecture and its Application for Distributed Manufacturing System

Hiroshi Wataya*,　Keijirou Hayashi*,　Junichi Toyouchi**,　Takeiki Aizono**,
Tamio IIzuka***,　Satoru Shibao****,　Masaru Omura****,　Masaharu Oku****

* Omika Works, Hitachi Ltd.,
5-2-1 Omika-cho, Hitachi, Ibaraki 319-12 Japan, wataya@omika.hitachi.co.jp
** Systems Development Laboratory, Hitachi Ltd.,
1099 Ohzenji Aso-Ku, Kawasaki, Kanagawa 215 Japan
*** Hitachi Process Computer Engineering. Inc.,
5-2-1 Omika-cho, Hitachi, Ibaraki 319-12 Japan
**** Automation Controls & System Development Dept., BRIDGESTONE CORP.
3-1-1 Ogawahigashi-Cho, Kodaira, Tokyo 187 JAPAN

Abstract

Under the recent severe economic situation, computer system of companies have to produce new products or to supply new services flexibly to meet changing customers' demands. To realize flexible computer system, the system should be able to expand and modify its software as well as hardware rapidly. Moreover the companies have been tackling Business Process Re-engineering (BPR) to share various kinds of information among different departments, for quick response to the consumers' needs and cost reduction.

Open ADS architecture that is a new control & information system architecture, is proposed to fulfill the flexibility of system. This paper introduces practical manufacturing system realized on the platform. These system are developed based on Hi-Cell concept that is defined as an assembly of autonomies called Homogeneous Max Intelligent Cells (HIC). In the system, techniques for resumption have been applied, and the effectiveness of it has been evidenced.

Keywords: *Autonomous Decentralized System, Control & Information System, Open ADS, Hi-Cell concept, Promotion Activity, Expansion, Maintenance, Resumption*

1. New Trend of Control & Information Systems

The control & information systems which integrate control systems and information systems have been widespread as social infrastructure such as power plant control systems, traffic control systems, water utilities management systems, factory automation systems, logistics control systems and so on.

In conventional business environment, main trend was Mass Production: which means large amount of products was manufactured cheaply, quickly and without fail. On the other hand, in recent severe economic situation, as the business has been changing drastically, the success key of business is rapid development of new products and services to satisfy individual and changeful customer's demands in addition to the traditional requirements for low cost and high quality.

For example, in the tire market, the manufacturers have been competing with one another in providing custom products produced in small lots to meet consumers' taste quickly. One major tire company are producing more than 20,000 kinds of tires for automobile, motorbike, airplane, and so on [1].

In the new trend, BPR, the requirements of manufacturing system has also changed [2]. The flow-through operation among manufacturing sub-system, production management sub system, logistics sub-system, and sales management sub-system based on the integration of each systems is eagerly life or death in the business. If the sales information is reflected to the production management system and manufacturing system as soon as possible, the wasteful stock of product can be minimized or the possibility of dead stock of popular products also can be minimized on the contrary. The new control & information system also strongly requires flexibility for easy integrating other information sub-systems and extensibility in proportion to the increase of business volume.

2. Realization of New Control & Information System Architecture

2.1 Concept of Open ADS

In BPR environment, the role of control & information systems can be said that it is not sufficient to automate only the closed manufacturing floor but to easily cooperate with and easily integrate other information sub-systems flexibly.

Recently, many Japanese manufacturing companies have tried to move the factories, and sales and service bases from Japan to world-wide countries because of strong yen and global competition. In this business situation, it is a desirable solution to construct such kind of huge and wide-spread system to utilize the most suitable hardware and software with the best system architecture from multi-vendors in the recent open computing environment.

We has presented the Open ADS architecture that have the following two features.
(1) Global system integration architecture

Flow through operation among global manufacturing, logistics, sales, service sub-systems can be possible by this architecture.
(2) Multi-vendors system integration architecture

Integrated middle software platform for multi-vendors component can be possible by this architecture.

2.2 Promotion activities for Open ADS

Aiming for realization of global control & information system based on Open ADS, the following scenario have been taken.
(1) Implementation of Open ADS platform on multi-vendor component

Open ADS platform that is ADS-support middle software has been already developed and released on Windows 3.1/NT/95 based IBM PC/AT compatible computers and UNIX based workstations. Especially in information world, the word "Multi-Vendor" has been very common and easy to realize recently, since the computer market has been already matured by several powerful computer vendors' architectures such as Intel*1, Microsoft*1, SUN*1, HP*1 and so on.
(2) Promotion of defacto standardization by information openness.

Especially in the control world, there are many kinds of control components which are suitably developed for specified control systems such as power plant control, factory automation, railway traffic control and so on.

Since in the control & information system, the control components are key ones, Open ADS protocol support on these components is indispensable for realization of global system control & information system. The 'Open ADS' protocol on the top of the defacto standard protocol: UDP/IP has been opened to other companies to support ADS function on these various kinds of multi-vendor controllers in addition to the other kind of computers as stated above. As of today July, 1996, more than fifty companies including American, Japanese and European powerful control vendors has already got Open ADS license. Some vendors have already supported Open ADS protocol, some vendors are now planning to support Open ADS protocols. Moreover, "The FA Open Promotion Committee" by International Robotics and Factory Automation Center(IROFA) supported by the Ministry of International Trade and Industry (MITI) in Japan has decided to promote 'Open ADS' and the promotion activity has started on June, 1996.
(3) Participation in device level LAN organization of defacto standard

In the conventional control systems, devices such as sensors and actuators are directly connected to the specified controllers by way of peer to peer communication line. The device level LAN which is developed for aiming at saving wiring cost of huge number of devices in the system and standardization of devices is considered to be the new key component in control & information systems in the near future. Pursuing new Open ADS architecture including device level integration, we have participated in ODVA*1 (Open Devicenet Vendors Association supported by Allen Bradley*1) as a founder who can present the new function on Devicenet. New function for device level autonomy has been already presented and decided to be standard in ODVA.

3. Open ADS Architecture for Highly Flexible and Extensible Systems

3.1 Conventional Software Development in Control & Information Systems

The disturbing factor from flexible and expandable systems can be said   the complexity of software structure. In case of system modification, the exact portion of software which required modification might not be clearly limited as the software structure is highly complicated and each software module is tightly locked together:

which is generally called 'Spaghetti Code'. (See Fig.1)

Especially in the control & information systems, since the software state has to be changed dynamically and timely based on the plant state and operator's instruction, the structure of the application software is apt to be highly complicated and disturb from flexible and expandable systems. It can be said that the simpler the software structure, the more flexible and expandable the system is.

On the other hand, the consuming hour for making document related to the software such as function specification, design document and software test specification and so on is much larger than the hour of writing source code and testing software. The maintenance cost of these document is also much in case of that the application software is frequently modified. There exists also another following problem: these document has been inconsistent with the real software structure since the maintenance of these document has been missed sometime for long years after the system was started to operate. It also can be said that the less the software document and the simpler the description of the software structure is, the higher the productivity of application software is.

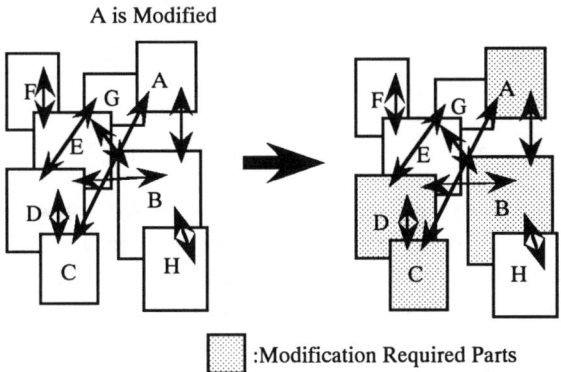

A is Modified

☐ :Modification Required Parts

Fig.1 Disturbing Factor for Flexible Expandable System

As the above discussion is more conceptual, the more detailed one is next from the view point of software engineering. The specification of the software can be described as the unit of processing (module) such as called process, task, thread and so on and the linkage (communication method) among other units of processing. There are several kinds of linkage among modules generally. For example, in computer the linkages might be shared file, shared memory, sub-routine call and outside the computer by way of network, the linkage might be socket interface on TCP/IP network. From the point of the complexity of the software structure, the more

kinds of these linkages are used at the same time, the more complicated the structure of the application software.

3.2 Open ADS Architecture for Flexible and Extensible Control & Information Systems

As Open ADS architecture exclude the above complexity of application software structure, flexible and extensible control & information systems can be realized in global operation [3][4].
(1) The concept from molecular biology

In molecular biology, it is well known that each cell with the same DNA information is grown to various organs by continuous cell division and at last to one living body. In Open ADS concept, same as a cell in molecular biology, each node can receive and keep the same information from other software module in and outside of each node and autonomously do the processing.

To realize the above mechanism, in Open ADS contend code communication is implemented as below.
(2) Content code communication for excluding dependency among software modules

In conventional communication method, the sender module specifies the destination address and sends the data. However, in content code communication method, the sender does not specify the destination address but only broadcasts the data with Content Code which is uniquely defined with respect to the content of the data. The receiver module selectively receive the data by Content Code and do the operation autonomously (See Fig.2). The data is not sent to the specified software module, but only sent out to the data pool which is called data filed in Open ADS.

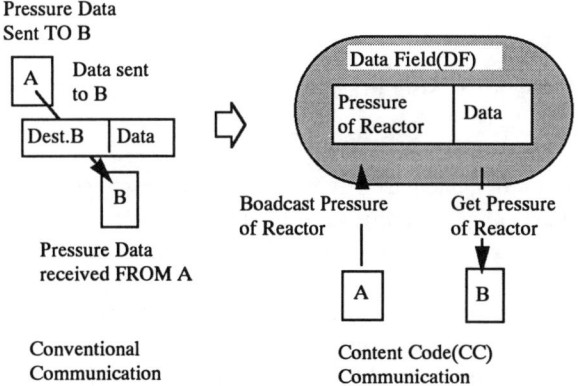

Fig.2 Conventional and Content Code Communication

In Open ADS, as the connection among software

modules are replaced by the connection between each software module and data pool, the mutual dependency among software modules can be definitely excluded. The receiver module can receive the data selectively by content code not by source address or destination address. Even if the sender module is replaced by other software module on the other node, there is also no side-effect for receiver software module.

(3) Data driven software module execution with no influence from other modules

In the receiver node, the data with specified Content Code is received, the specified module is executed: this execution mechanism is called 'Data Driven Software Module Execution'. In this mechanism, the software module is automatically executed only when the required sets of data with the specified Content Code are ready in the receivers data pool.

In Open ADS, the linkage of software module in and inter node is only Content Code communication. The execution method of software module is also restricted to use only data driven software module execution mechanism for the simple software structure. By applying the Content Code communication and data driven software module execution, the software structure can be drastically simplified even in the large scale application software in control & information systems. (See Fig.3)

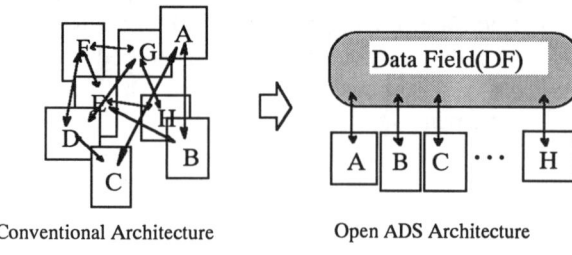

Conventional Architecture   Open ADS Architecture

Figure 3. Simplification of Software Structure

(4) Content Code base system design for bottom up design

The linkage among software modules in and out side of nodes is unified to Content Code, the software linkage can be defined partially. In case of the shared file and memory, only after the whole the file and memory design is finished, the following system design is started. There might be a lot of backtrack in the process of the system design.

For example, guess that there are 5 sub-systems in one Data Field. Since the usable Content Code number is 63000, sub-system A uses 1 to 10000 and sub-system B uses 10001 to 20000, sub-system C uses 20001 to 30000 and so on. In this style of system design, as there is no relation among sub-systems, the bottom up system design can be easily possible in Open ADS. The backtrack work in the process of system and application software design can be minimized by this metrology.

(5) Less documentation of application software in Open ADS

The description of relation among software modules in the whole system can be also drastically simplified in Open ADS, since only the Content Code communication is used as a inter module communication both in the node and by way of networks. The description of relation among modules is only that some module broadcasts the data with specified Content Code, and that some other modules receive data with the same Content Code. This diagram is called Data Flow Chart. (See Fig.4) By the Data Flow Chart, user can easily get the relation among software module in the whole system at a glance.

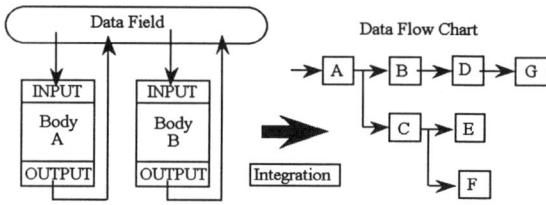

Fig.4 Data Flow Chart

The volume of documentation of application software can be also surprisingly decreased in Open ADS compared with the conventional design methods since the system structure is much simplified as stated above. In Open ADS, only next four kinds of documents are sufficient for seize whole the system structure of control & information systems.

(a) Hardware components structure including network configuration
(b) Data Flow Chart both in each node and inter nodes
(c) The Content Code data structure
   (message specification of Content Code data)
(d) Source code of software modules (or flow chart)

In our Open ADS design support middle software package, the function of automatically generating the above 3 types of documents ((a) to (c)) based on the system configuration information is supported, aiming for reducing the cost of making software design documents.
(6) The components of Open ADS

Open ADS has been realized on TCP,UDP/IP based defacto standard network such as Ethernet, FDDI and

ATM, and also on our proprietary control network. The hardware components which can support Open ADS on theses network are PC, WS, multi-vendor PLCs and other control components for such as power generation, train control and so on in addition to the computers and controllers of our products. Theses multi-vendor hardware composed Open ADS control & information systems.

3.3 The integration of control and information sub-systems by Open ADS

In Open ADS concept, all the components in the system are assumed to have the same information. By Content Code communication based on the broadcast communication, all the nodes in the network can share the information from other nodes. However, in the real control & information systems, whole the system is divided into several sub-systems such as control sub-system, production management sub-system and so on. These sub-systems are connected each other by way of router or wide area network such as Internet and composed global control & information systems.

Since the quality of data required for the floor control sub-systems and that of production management sub-systems are quit different, it is effective that one Data Field is also assigned for each sub-system and each Data Field is cooperated with each other to do the global operation. Although the prior discussion in section 3.2 is mainly for intra-sub-systems, all the advantages in it are furthermore practicable for inter-sub-systems.

The relation among the sub-systems can be said to be just equal to the relation among software modules which reside in different Data Fields. Each sub-systems might be developed by different vendors and different period or might be in the separate countries such as in Japan and China. Consequently, the superior points of Open ADS such as the minimization of mutual dependency among software modules, bottom up design method, the simplification of design steps and program structure and so on are also advantageous for integrating sub-systems: integration of control & information systems.

In Open ADS, the node which is connected with plural Data Fields is called 'gate way node'. In this node, a data received from one Data Field can be relayed to another Data Field. Since the gate way node can know the content of the data by Content Code, only required data for the other Data Fields can be relayed. The gate way node is also possible to translate suitably for the other Data Fields or add information to the data for the other Data Fields.

As illustrated in Fig.5, the output data from the production control Data Filed can be easily injected to the

newly added sales department Data Filed. Moreover, the estimate of sales data in sales department Data Field can easily relayed to the production control Data Field as the data of production schedule.

In Open ADS, these kind of integration in sub-system level can be flexibly and extensibility realized. Since the gate way node can change the quality of the data in the process of relaying the data, flow-through operation among heterogeneous divisions such as factory floor, production control, sales department and so on can be acquired.

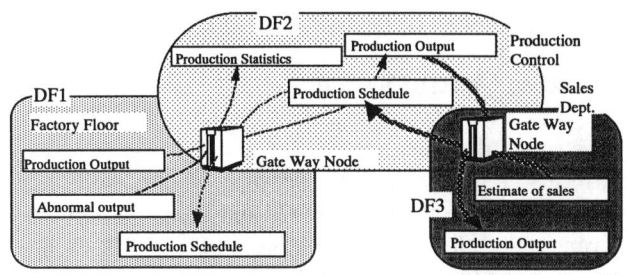

Fig.5 Flow Through Operation Base on Open ADS

4. Application

4.1 Manufacturing System based on Hi-Cell Concept

Practical tire manufacturing systems are introduced in which two kinds of system function recovery methods are applied in case of partial failure or modification of system elements on Open ADS platform. These applications are developed based on Hi-Cell Concept that is proposed to fulfill the flexibility of manufacturing system. 'Building Block' design and 'Restructuring' system are indispensable factors in this architecture [5].

Hi-Cell automation system is defined as an assembly of autonomies called Homogeneous Max Intelligent Cells (HIC). Each HIC works autonomously by its own judgment without being affected by other HICs and cooperates with other HICs so that it can accomplish the object of whole system. HIC is connected to Information Field, and it can communicate with the other HICs using broadcasted messages. HIC selects and receives messages within the Information Field and secures autonomy by using data-driven mechanism. All HICs have Hi-Cell Kernel (HCK) which manages Information Field. HCK secures autonomy of each HIC and realizes coordination among HICs.

Common purposes and rules in the system is called

Global Order. Under Global Order, all HICs work on coordinating each other. Local purposes and rules in one HIC do not affect other HICs, and they are called Local Order. Common purposes and rules in a certain group of HICs are called Regional Order [5].

In Hi-Cell Concept, the continuity of function in case of cut separation and reconnection of partial element can be improved by duplication of function block. However, in practical system, duplication of function blocks are difficult because the number of functions is huge and many functions' software depend on hardware structure. Therefore we introduce data back-up and recovery method without duplication of function block [6].

4.2 Process Control System

(1) System Outline

Process control system for one whole works is connected to more than 1,000 PCs and PLCs by network and runs online for 24 hours. These PCs and PLCs are partially stopped several hundreds times per year for partial failure or modification.

Therefore we adopt Open-ADS structure and Open-ADS platform for this system. There exists a process control controller per one production process, and every process control controller receive a global production schedule from business host computer. Each controller create local production schedule for own production process based on the global schedule, production status of own and other process dynamically. This local production schedule contains numbers of products, production timing, parameters for manufacturing controller, and so on. The schedule is transferred to equipment control system as a production order, and modified by receiving actual production number from control system. Since the equipment control system have a function to store the production order from process control system and actual production number in it, it can continue to manufacture products for about one hour in case of the process control system is down or separated. (See Fig.6)

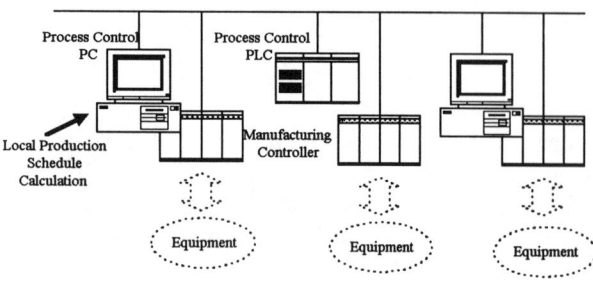

Fig.6 Process Control System

(2) Needs for separation and reconnection of elements

When one or several process control controller is separated and then reconnected to a whole system, the process control controller requires a history of process that is not only present status but all log while disconnected to resume to its normal condition. It's not impossible that operator input all history data by hand, but is very hard because the amount of the data is very large.

(3) Solution and Method

To recover autonomously after reconnection, the process control system must have functions to store necessary backup data and replay it to disconnected controller exactly.

<Backup Method>

There are alternative techniques to store backup data as follows;

(i) External Device for backup : An extra machine pools historical data.

(ii) Local backup : Every controller store history data in own local memory.

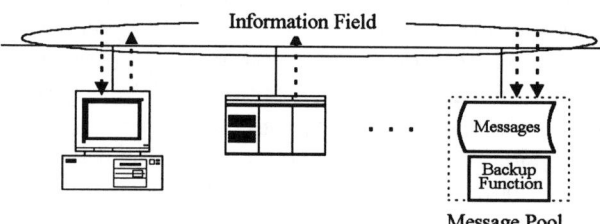

Fig.7 Message Pool

In the system, the process control controller don't have enough memory to store historical data of one hour. Therefore technique (i) was applied to the system. Message Pool is a PC to store history data for several hours and reply it (See Fig 7).

<Resumption Procedure>

To resume to controller's normal condition, all history data while disconnected is necessary, however the order of data is not important to modify production schedule and redundancy of data is allowed because the reconnect controller can distinguish redundant data. Here we introduce resumption procedure.

(i) Reconnected controller send a reconnect announce with 'Last Received Data' to Information Field (IF: In the system IF is realized by Ethernet).

(ii) When Message Pool receives this reconnect announce, it start to replay history data from the second previous data to the 'Last Received Data'.

(iii) The reconnected controller receives the replay data and ongoing actual production number data concurrently to resume to its normal condition step by step.

In the actual system, Message Pool sends one history data per 30 seconds. In a practical case, a controller disconnected for one hour must receive about 20 stored data, therefore it was necessary for about 10 minutes to resume completely. And 2 or 3 ongoing data was received by the controller during the resumption term.

(4) Evaluation

In an early stage of installation of the process control system, several tens failure and modification occurred in each local process control system. If the system was developed based on conventional system structure and had no Message Pool, whole system should be stopped to fix these problems and it would be very inefficient development. However, in the process control system, partial elements could be separates, reconnected and resumed automatically, therefore modification of the system can be easily operated.

4.3 Manufacturing Process System

(1) System Outline

Manufacturing process consists of three kinds of elements, two assembling machines, ten processing machines and one AGV (autonomous grand vehicle) that carries WORKs between the assembling machines and the processing equipments (See Fig.8). A controller controls an equipment one by one. A WORK is a processing equipment prepared for one type of product, and it is settled on a processing machines to work. This manufacturing process was designed to meet a small lot production efficiently.

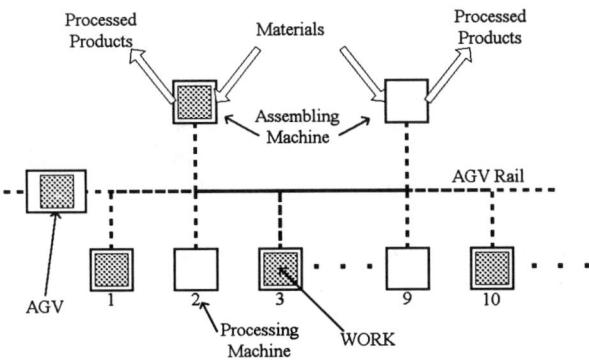

Fig.8 Manufacturing Process System

Manufacturing procedure is as follows;
(i) A AGV demounts a WORK from a processing machine.
(ii) The AGV carries a WORK to either of two assembling machines.
(iii) The assembling machine takes a processed product out of the WORK.
(iv) The assembling machine put a processing material into the empty WORK.
(v) The AGV carries and mount the WORK to the processing machine.
(vi) The WORK processes the material on the processing machine according to the manufacturing parameters called RECIPE.
(vii) When the WORK finishes processing, the AGV demounts a WORK from a processing machine again, then the same procedure is repeated.

Ten processing machines run according to this procedure using own parameter in parallel, and at most ten types of products are processed continuously. For efficient operation of these equipments, the system must have a function to monitor the whole status of the system to control AGV. Master-Slave method in which the master controller gathers all status from controller(PLC)s for processing machines and assembling machines, then decides and orders AGV controller, was one solution to realize the system. However, this method would have the following problems.

(i) The whole manufacturing process will stop in case of the master controller fails.

(ii) The cost of the system become expensive because the master controller is indispensable even for a small process like the manufacturing process system.

Therefore we adopt Open-ADS structure and Open-ADS platform. In the system, each PLC has sufficient data to control local equipment by itself, and a PLC for processing machine can call AGV in the ordinary order considering the whole status of the system.

Equipment	Time Limit of Disconnection	Necessary Data for Resumption
AGV (1)	5 ~ 10 min.	Present Status of Assembling Machine and Processing Machine
Assembling Machine (2)	None	Present Status of Assembling Machine and AGV
Processing Machine (10)	None	Processing Parameter only for Re-connected Machine

Table 1   Restriction for Resumption of Equipment

(2) Needs for separation and reconnection of partial elements

Three types of equipment composing the manufacturing process system have different restriction for separation and reconnection to the whole system. (See Table 1)

AGV and Assembling Machines decide the best next course based on the whole status of the system by themselves. The whole status mean present data but historical one as process control system above described. Therefore, AGV and Assembling Machines must have a function to get the present whole status for quick resumption.

(3) Solution and Method
<Backup Method>

In the system, the necessary data for resumption of reconnected equipment is only the 'newest' status, so technique (ii) Local backup was applied to the system.

When each controller's status changes, it sends the present status to network. Since the frequency of change of status is only once per about thirty minutes, each controller can keep the newest data to send to a reconnected controller.

<Resumption Procedure>

Here we introduce resumption procedure.

(i) Reconnected controller sends a reconnect announce to Information Field.

(ii) When each controller receives this reconnect announce, it sends the 'newest' status stored in it.

(iii) The reconnected controller receives only necessary data and resume to its normal condition. (See Fig.9)
(4) Evaluation

Resumption function was easily implemented because the system was constructed based on Open-ADS structure and Open-ADS platform. The resumption method is independent of the numbers of autonomous elements.

5. Conclusions and Discussion

This paper has presented the concept, and advantages of Open ADS architecture which integrate control & information systems globally. The unified inter module communication method: Content Code communication with data driven module execution mechanism makes whole the system much simplified, flexible and expandable. Since, we pursue the "real" multi-vendor control & information system by Open ADS architecture, the protocol has been opened widely and the standardization activities of Open ADS has been promoted aggressively. We wish more vendors and users join our activities and give continuous support.

We had proposed two kinds of implementations of resumption methods for reconnected equipments on Open ADS platform, and proved their availability. As Message Pool method allows redundancy of data and does not care if data is transferred in wrong order, it have some restrictions. However the method is available for almost all system of tire manufacturing plant. On the other hands, local backup method will be more reasonable if memory for controller become cheaper.

*1 All brand or product names are trademarks or registered trademarks of their respective holders.

References
[1] M.Oku, "paradigm Shift Towards The 21 Century", Bridgestone Firestone, 1990.
[2]M.Hammer, et.al.," Reengineering The Corporation A Manifesto for Business Revolution" Nicholas Brealey Publishng
[3]K.Mori et. al., " Autonomous Decentralized Software Structure and its Application", FJCC, 1986
[4]H. Wataya, K. Kawano and K. Hayashi, "The Cooperating Autonomous Decentralized System Architecture", Proc. of the 2nd Int. Symposium on Autonomous Decentralized Systems, Arizona, Apr. 1995, pp. 40-47
[5] M.Oku, M.Omura, J.Kann, M.Perrone and M.Roth, "Hi-Cell Architecture and the Project Model for Manufacturing ADS", ISADS 93, 1993.
[6] M. Omura and M. Oku, "Hi-Cell System Architecture for Manufacturing Systems", Proc. of the 2nd Int. Symposium on Autonomous Decentralized Systems, Arizona, Apr. 1995, pp.154-161.

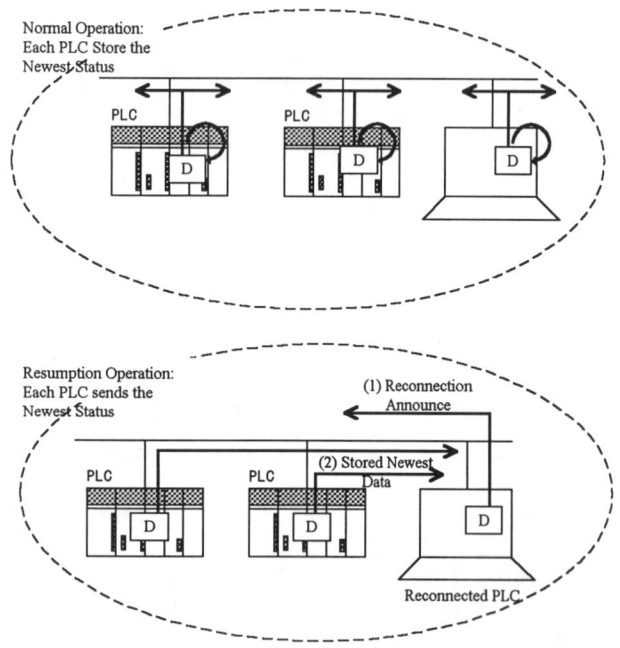

Fig.9 Resumption of Reconnected Controller

Network Computers – Ubiquitous Computing or Dumb Multimedia?

Ralf G. Herrtwich and Thomas Käppner
RWE Telekommunikation
Hollestraße 7b
45127 Essen
email: {rgh, thomas.kaeppner}@rwe-telliance.de

Abstract

In May '96 a group of major hardware and software manufacturers jointly announced a set of guidelines forming an industry standard for a new class of devices called network computer (NC). The new industry standard – called the Network Computer Reference Profile 1 *– is designed to accelerate the development of low-cost computing.*

The group of firms consisting of Apple, IBM, Oracle, Sun Microsystems and Netscape Communications unveiled plans to provide network computer operating software and a variety of NC hardware models by end of '96. The NC Reference Profile 1 considerably builds upon existing Internet and World Wide Web technology. This paper gives a brief overview of the goals pursued by the introduction of NC Reference Profile 1. It surveys the technologies employed for NCs and discusses their application and market potential in different environments.

1 The perspective

In May '96 a group of major hardware and software manufacturers jointly announced a set of guidelines forming an industry standard for a new class of devices called network computer (NC). The industry standard is called *Network Computer Reference Profile 1* and its target is, according to the announcement, to „make multimedia Internet computing as affordable and ubiquitous as telephone and television services".

The group of firms consisting of Apple, IBM, Oracle, Sun Microsystems and Netscape Communications unveiled plans to provide network computer operating software and a variety of NC hardware models by end of '96. Looking at the NC Reference Profile 1 it is obvious that network computers are not a totally new invention, but rather a utilization of existing technology according to a new concept. This paper gives a brief overview of the goals pursued by the introduction of NC Reference Profile 1. It surveys the technologies employed for NCs and discusses their application and market potential in different environments.

The introduction of NC Reference Profile 1 has one major intention: It is designed to accelerate the development of low-cost computing devices. Industry-wide compatibility will serve both customers and providers of systems and services by protecting investments, opening a mass-market thereby reducing costs. For that purpose, it is important that the profile is an open standard that, in principle, can be implemented by anybody.

The guidelines can be illustrated as a common denominator of popular and widely used features and functions across a broad range of scaleable network computing devices. NCs are envisaged as entertainment devices in combination with a TV monitor, as palmtops, laptops, desktops, video phones, pagers, or conventional PCs. Common to this wide range of devices is their link to the Internet or an intranet which enables them to interoperate with other network devices or central nodes.

Each network computer has the capability to run one or a range of basic applications. End-user ease of use and simplified system administration are important attributes of NCs. Devices are to be developed for communications and commerce to be usable in homes, schools, businesses and institutions. Multimedia technology will be employed for communications, security features are prerequisites for commercial applications.

To relate them to NCs, multimedia PCs are well introduced since years and in fact, today's personal computers are fully capable of supporting the draft NC Reference Profile. Unlike PCs, however, NCs are designed from the very beginning as devices that are attached to the network, that are fed by the network, and that provide network-centric services.

2 The technology

Network computers are embedded in an IP-based network where they communicate with other network devices and servers via the Internet Protocol. The programming environment consists of Java which enables network-resident as well as stand-alone applications to be executed. Both of these technology choices reflect the objective to leverage the installed base of systems and software for network computers: The number of networks based on the Internet Protocol (IP) is doubling almost every year and

155

Java is a technology studied and utilized extensively in context with small World Wide Web (WWW) applications called applets.

Java is an object-oriented, interpreted language designed to be architecture-neutral, robust, and secure. A Java compiler delivers a special byte-code that is interpreted on any target machine on which the Java Virtual Machine has been implemented. While the Java compiler is strict in its compile-time static checking, the language and run-time system are dynamic in their linking stages. Classes are linked only as needed. New code modules can be linked in on demand from a variety of sources even from sources across a network. Therefore software, especially thinking of small WWW applications, can be updated easily and transparently via network connections, applets can extend their capabilities as needed by linking additional or newer modules.

Besides IP and Java the NC Reference Profile 1 will include a whole set of open standards covering general resource guidelines, World Wide Web standards, e-mail protocols, common multimedia formats, boot protocols, and security features. Based on the announcement in May '96 and ongoing discussions the following is expected to form the first industry standard for network computing devices:

Internet Protocols: General support for IP as mentioned above is required, so that the Internet Protocol can be for the exchange of messages with other devices. Specific hardware attachment, i.e., specifications below IP will not be part of the profile. Yet support for several higher-layer internet protocols plays an important role for the architecture of network computers. The Transmission Control Protocol (TCP) adds reliable transport to the basic services of IP, whereas the User Datagram Protocol (UDP) offers plain end-to-end message transport. The File Transfer Protocol (FTP) is a service on top of TCP, allowing to transmit data files across networks. FTP is introduced as an option for devices on the network that support the notion of file systems and their distribution. Telnet is a higher layer service which is used to log-on to remote hosts via terminal emulation. Telnet, as FTP, is considered optionally, being necessary only for NCs that support a character based terminal access to remote hosts. The Network File System (NFS) supports distributed file systems for network computers and it is required for NCs that will support the notion of distributed file systems.

World Wide Web standards: The Hyper Text Markup Language (HTML) is the format in which documents are published within the WWW. The Hyper Text Transfer Protocol (HTTP) is the means of communication between browsers and servers of the WWW. For the reasons introduced above, the Java application environment including the Java Virtual Machine and runtime environment and

Java class libraries provide a platform for the development of portable, flexible, and extensible applications.

Configuration and administration: The Dynamic Host Configuration Protocol enables NCs to boot themselves over the network, to dynamically acquire an IP address and to transmit configuration information over the network. Alternatively the simpler Bootstrap Protocol (BootP) can be employed for booting an NC over the network in statically-addressed IP environments. The Simple Network Management Protocol (SNMP) offers services which could relieve end-users from the task of administrating NCs and offer them remote administration and automatic updates.

General equipment: The minimum screen resolution is equivalent to standard VGA, i.e. 640 x 480 pixels. A pointing device, such as trackball, mouse or touchscreen is required as is the capability for the input of text via keyboard or a similar device. Audio output is considered as necessity, whereas local storage is regarded as an option.

Mail communications: The Simple Message Transfer Protocol (SMTP) ensures that network computing devices are able to participate in the exchange of email messages. The Internet Message Access Protocol Version 4 (IMAP4) and the Post Office Protocol Version 3 (POP3) are client/server protocols for the access of messages stored at mail servers.

Multimedia formats: Several media formats, such as JPEG, GIF WAV, AU, will be adopted for the Reference Profile, as they already play an important role for the interchange of multimedia information.

Security features: The ways in which security will be provided are still under discussion. It is envisaged, however, that emerging APIs, such as the ISO 7816 (Smart Cards) or the Europay/Master Card/Visa specifications are candidates for security support in network computers.

Printing: Connecting a printer to a network computer will be specified rather as an option than as a requirement.

Some manufacturers have already presented prototypes that support most of the standards listed above. The original prototype by Oracle, which led to proposing the idea of NCs, has been built upon the Advanced RISC Machines (ARM) 7500 processor, a multimedia, 32-bit RISC silicon with performance equivalent to a 66 MHz Intel 486. The ARM 7500 is a highly integrated chip incorporating many functions, such as video and I/O subsystems, keyboard input, audio capability, etc., which are typically handled separately.

IBM showed a prototype of a network computer at the announcement, dubbed an „application-centric terminal" to be announced in the 3rd quarter '96. The machine will run low-power embedded Power PC 403 processors and let corporate users surf the Internet and run mainframe, PC server, and AS/400 and groupware applications tuned for intranets. The software architecture envisaged for such a device is illustrated in Figure 1.

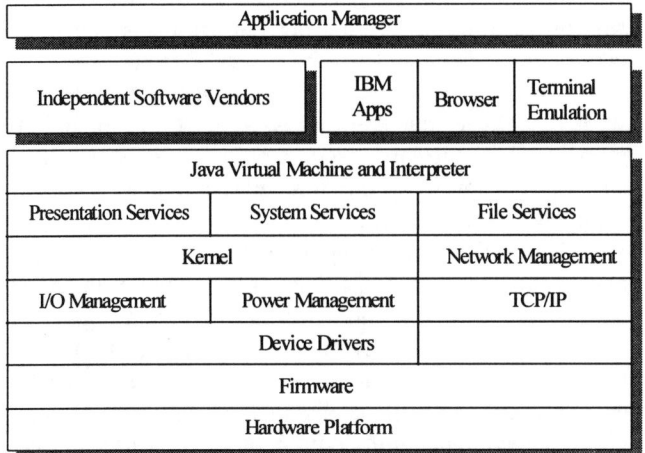

Figure 1: Network computer system architecture

These application-centric terminals will adhere to the NC Reference Profile, their pricing is expected to be about $700 for the device without a monitor, keyboard, and mouse. According to IBM officials the whole point of such products is to have PC applications on a desktop without the cost and complexity of a PC.

Oracle's NC System Software Suite will also support the set of network computing standards adopted by the five original standard bearers. This suite will include a microkernel operating system, multimedia extensions and additional system software, it is designed to download all software from servers over a network to any one of a range of inexpensive, diskless devices with minimal memory, a low-power CPU and no hard drive. This procedure promises to relieve users from software configuration and other administrative tasks and decrease maintenance costs. Oracle plans to leverage Object Request Broker technology to ensure adequate connections between service providers and NC users. Data of individual NCs will be stored in an Oracle Universal Database.

3 The rules

Oracle can be seen as the driving force in defining the Network Computer Reference Profile 1. Oracle itself will however, except for demonstration purposes, not develop any network computing devices, but it has created a subsidiary company to devote its entire resources to the development of software for the NC. Oracle's NC System Software Suite has already been licenced to over 30 manufacturers. In turn Oracle licensed Sun's Java, Adobe's Bravo and Macromedia Inc.'s Director.

It is important to note that the NC Reference Profile 1 sets guidelines for standard functionality but does not impose limitations. Manufacturers may add more functionality as the profile does not limit designs to a specific set of features. Compliance with the profile is defined as support for every standard or technology included as a requirement by the profile. It is unlikely that every implementation in the future will cover all standards defined by the NC Reference Profile 1, instead it is envisaged that new profiles will be recommended as unique requirements or technologies for particular markets and applications emerge. The vision is that NCs support a base level of standards and that classes of devices that may provide unique features and functions specific to their particular market do so consistently and based on open standards.

An NC brand will be made available to manufacturers that provide NC products complying to the reference profile, yet conformance tests are still to be defined. Oracle will promote the concept of NCs by an NC branding campaign. In addition, the group of announcing firms will establish an *NC Friendly* logo which makes web sites recognizable as providing content that is compatible to network computers, i.e., it is accessible to NC devices. While the web site may contain content outside the scope of the NC Reference Profile, and certain profile-compliant devices may support such content, the site should be capable of providing alternative content which is *NC Friendly*.

4 The applications

NCs are designed as potential replacements for PCs in most situations in which PCs are used today, e.g. offices, homes, or schools. In offices most users have PCs on their desk to perform tasks like database and WWW access, word processing, spreadsheets, and e-mail. Looking at the history of office computing it can be established that the process of downsizing and decentralizing computing equipment has almost been finished and that corporations begin to realize the costs associated with supporting and maintaining a network of PCs.

Whereas during the phases of host computing high expenses were mainly due to expensive mainframes, after installing PCs in every office the costs shifted to hardware and software configuration, maintenance, and support for this large number of desktops. Studies by Gartner Group have found the five-year cost of owning a Windows PC for a business to be about $40,000, a major part being support costs for operating systems and applications software products. Network-centric computing promises to enable lightweight hardware and software products, which could ease administration and maintenance and reduce support costs. Since network computers load their software from a central repository at boot time, software updates and configuration could be eased significantly.

In internal corporate networks based on the Internet Protocols, so-called intranets, the links of NCs usually have enough bandwidth so that performance is not that much degraded when loading applications from a central server. Most businesses already store crucial data on servers, so that a fast reliable network connection is not an issue. Where this is not the case yet, as in small offices, additional investments for storage and applications servers are needed.

On the other hand the guidelines and prototypes demonstrated so far suggest, that NCs, compared to regular PCs are going to limit the end-user. The low screen resolution and memory equipment let expect that layered windows and menus will be restricted if available at all. NCs are made for tasks as browsing databases and the Web or simple word processing, in other words: NCs are rather for consuming than for creating content. Power users, such as project leaders or designers, will not be satisfied with these restrictions.

In the past there has been a clear tendency to put more storage and computing power on every desktop, regarding for the ever-growing demands of office applications. The proponents of network computers have yet to prove that it is possible to provide similar capabilities to office users with a totally opposite approach. The key to perceptibly higher performance could lie in Java's dynamic linking feature. Monolithic office applications can be broken down into smaller dynamically loaded modules, in which case just the parts of software that are currently needed are loaded from the central server into memory of NCs. This would counteract the fact that current PC software is lavish with resources, especially memory. On the other hand the new application environment is a major problem for the dissemination of network computers, as every single productivity tool has to be programmed again, if not accidentally the NC is based on the same software architecture as a PC. This lead of PC software development can only be caught up with, if the mass of Java software that is currently becoming available free of charge all over the world will be leveraged.

The mass market has a somewhat different starting point: The importance of access to the Internet and the WWW is growing dramatically. Since 1993 the number of users increased from 18 to 60 million worldwide; it is expected to double again by the end of 1997. Network computers are ideally suited to serve this mass market, since they are designed for operation within the Internet. For instance the dedicated online device Minitel generates a revenue of $1,5 billion annually serving 14 million users in France. The PC penetration is about 25% in Germany (48% in the US), 4% of households have access to online services or the Internet (20% in the US). These numbers show that, in principle, there are millions of potential users for a product at the cost of a terminal with the functionality of a PC, especially if it comes with online or Internet access. Microsoft, however is ignoring those aspects and is criticizing the announcements in a way that led analysts to compare them with Detroit's car industry. As they didn't want to believe that Americans would trade in their fuel guzzlers for cheap Japanese compacts 20 years ago, Microsoft today ignores that the mass market demands less powerful, cheaper computers.

Yet the price advantage of NCs is not very large and comes at a cost: Using a TV set as monitor for an NC cannot provide the quality home users know from regular PCs. Although techniques have been developed to improve the readability, viewing text-rich Web pages with regular fonts on a TV set can still be dissatisfactory. Since prices for disk storage are declining continuously, sparing a hard disk does not really yield a big margin. But without persistent storage users are left dependent on the availability of network connectivity and storage servers.

Performance appears to be another problem: Oracle is proposing 8 MB and a 486 or similar processor, which compared to desktops' standards slows down tasks considerably. In contrast to office environments users are restricted to considerably less bandwidth at home so that the time for downloading applications might yield a high performance penalty. Moreover, current Internet traffic jams and today's typical 14.4 Kbit/second modems could put off many users since almost all homes have only standard telephone lines as their communication pipeline. Cable modems, e.g. as announced by Motorola's Multimedia Group, will allow data transfer rates of 30 Mbits/s downstream and 768 Kbits/s upstream in the future. This technology requires investments, however, in order to enable current cable networks for bidirectional communication. It is still questionable that there is enough bandwidth available to put every home online.

With the dissemination of NCs the business of online service providers will experience a considerable shift. Since most NCs will not be equipped with hard disks, persistent storage must be offered in addition to the services provided today, such that users can retrieve software and store their personal data. An issue is the privacy of end-users, whether they will accept to store their tax files, resumes, expense records, etc. at sites of the service provider is questionable. In addition service providers who invest in expensive equipment and applets will have to put a costly premium on their services. Tariff models, such as nickel per click can become very expensive in the long run.

Simplicity and ease of use might prove as substantial advantages of NCs over traditional PCs, especially for opening the mass market. The fact that users are relieved from backups, installation problems, or driver conflicts, that administrative tasks are automated and software support is available at the service provider might attract many new users who have been deterred by the complexity of PCs in the past. Future versions of NCs might come as appliances that are even more specialized and simplified, targeted at people who do not need the broad capabilities of personal computers but who nevertheless want access to specific functions that require some sort of underlying computer hardware.

Home users do not have such high demand for productivity applications, like spreadsheets and word processing, as office users do. Many of them are looking for up-to-date information in the Web, which is readily available. Moreover the Web is evolving rapidly and applets for home shopping, or home banking could become killer applications, especially if NCs could dispense electronic cash turning them into Automated Teller Machines. Other mass market applications like games might even be to demanding for NCs.

5 The outlook

Opinions about the Java application environment are varying from 'insecure' and 'will take years to mature' to 'the ultimate solution to software distribution problems'. Fact is that current speed problems will be mitigated by Java chips manufactured by Sun and that just-in-time compilers, available from third parties will accelerate application development. Although the mainstream productivity applications are not yet available, there are many useful applications already found in the Web. Data delivery is the current Java killer application. The success of NCs seems to be coupled directly to the success of Java which is very promising today.

Another key factor is whether users will accept that they have to log on for the simplest task to be performed by their NC. Some analysts view the capability to store data locally without retrieving every single piece of information from a central server as a necessity. We believe it is sufficient to automate the process of dialing-in or to make it transparent to the user to be successful. Ideally NCs would be staying online and users would only be charged for the services they demanded.

The change in the value chain of the desktop/online market are also important to consider: The current role of software vendors may completely be taken over by online service providers, who deliver software and updates up to the desktop. Software manufacturers could cooperate with the service providers leaving vendors to some extent superfluous. For the success of network computers it will be important how fast service providers adapt to the changing market conditions. Only if storage and software services are available in time, network computers can become a mass market product.

In the future the information accessible via NCs will play a major role. In order to attract the mass market the information must be actual and of direct benefit to the users. Technically, it will be important that text documents can be easily read via high resolution displays. Appliances that can be plugged in and left on are the ones that promise the largest success for the mass market. The announcement of the NC Reference Profile 1 can be considered as one step further towards the development of such devices.

Panel Session P1

The Role of Autonomous
Decentralized Systems in Network Computing

Panel Chair

Liba Svobodova

Panelists

Hendrik Berndt

Thomas Magedanz

Hiroaki Nakanishi

Stefano Zatti

The role of ADS in Network Computing

Liba Svobodova
Panel Chair

IBM Research Division, Zurich Research Laboratory, CH-8803 Rueschlikon, Switzerland
Tel: +41 1 724 8274 , Fax: +41 1 710 3608, Email: svo@zurich.ibm.com

Hendrik Berndt
Global-One, 12490 Sunrise Valley Drive, Reston, VA 20196, USA
Tel: +1 703 689-6304, Fax: +1 703 689-6724, Email: Hendrik.Berndt@Global-One.net

Thomas Magedanz
GMD FOKUS, Hardenbergplatz 2, D-10623 Berlin, Germany
Tel: +49 30 254 99 229, Fax: +49 30 254 99 202, Email: magedanz@fokus.gmd.de

Hiroaki Nakanishi
Hitachi Ltd., Kanda-Surugadai 4-chome, Chiyoda-ku, Tokyo, 101 Japan
Tel: +81-3-5295-5403, Fax: +81-3-3258-2086, Email: nakanish@cm.head.hitachi.co.jp

Stefano Zatti
European Space Agency - ESRIN, Via Galileo Galilei, I-00044 Frascati, Italy
Tel: +39-6-9418 0445, Fax: +39-6-9418 0442, Email: szatti@esrin.esa.it

Liba Svobodova (Panel Chair), Technical Strategy, Communication Systems, IBM Zurich Research Laboratory, Switzerland.

Network computing is an emerging paradigm that promises to extend computing and information access capabilities transparently beyond the domain of a specific organization and enable unprecedented collaboration and business transactions in a global open network environment. It also challenges the present client-server paradigm, which has proven to be rather expensive, owing to the high cost of maintaining and managing individual workstations and PCs.

Network computing is closely coupled with the Internet and World Wide Web explosion. The HTTP-based browser is becoming a standard open universal interface not just for WWW browsing, but for access to corporate databases, for remote system/network management, communication and collaboration, workflow management, and for other services and applications. The Network Computer (NC) is viewed by many as the successor of the familiar PC, that will get most of the needed software and services on demand from the network. To realize the vision of network computing, powerful, robust, and efficient "backend" infrastructure is required to provide such applications and services within an organization, across domains, and on the global scale.

What kind of role does the autonomous distributed system (ADS) paradigm play in network computing? Are these two different worlds? Are they complementary? Or, is ADS an inseparable part of or a base for the fledging network computing?

The individual panelists address these questions from the point of view of different environments and applications. They discuss technologies that support these paradigms and form a bridge between them.

163

Hendrik Berndt, Executive Director for Advanced Technology, Global-One, USA.

Focus: The integration of computing and telecommunications approaches to deliver sophisticated flexible customized multi-media services to the customer in a global environment.

Even so telecommunications systems have always been decentralized systems, the drive is now for an autonomous, object oriented solution suitable to meet telecommunications requirements (e.g. continuous information flow, stream interfaces, stream channels). Moreover there is a need for a platform (DPE) and platform services (e.g. notification services) that fit the telecommunications environment. For various reasons there is a need for uncoupling the service itself from the connectivity provided by the network. It is a tremendous challenge to find the common modeling principle and mechanisms to provide the service, and to control and manage it independently from the existing native computing environment.

The reference model of Open Distributed Processing (ODP) serves as a basic framework for addressing the above issues. The platforms and platform service specifications will most likely be shaped by an enhanced CORBA-compliant product family. The service issues are strongly influenced by the next generation IN and more importantly by the TINA service architecture. Performance issues are key for success. The most prominent contenders here are best effort service delivery and guaranteed quality of service approaches. An interesting question is whether some "in between" approach is feasible and could be the best solution.

Thomas Magedanz, Head of Intelligent Communications Environment Department (ICE), GMD FOKUS, Germany.

Focus: Integrated service platform combining Distributed Processing Environment (DPE) and Mobile Agent technologies.

The general trend in the telecommunications environment is toward object-orientation, where multiple services are realized by a common set of reusable service components (i.e. objects). Distributed Processing Environment (DPE) technology allows for the transparent distribution of service components and supports the communication between these components. On the other hand, Mobile Agents (MAs) and Network Computing can be regarded as enabling technologies for dynamic and on-demand combining and downloading of service modules to specific nodes (both network servers and customer systems) and for distributed realization of new value-added services, aimed at meeting specific requirements such as performance, cost, asynchronous operation, etc.

The future service platform has to provide the necessary flexibility to cope with a multitude of customer systems and service provisioning scenarios, e.g. centralized or distributed intelligence in the network, intelligence outside the network, on-demand service provisioning, etc. The combination of DPE technology and mobile agent technology is regarded as the most powerful and flexible base for an integrated service platform. Taking into account that DPE technology such as OMG's Common Object Request Broker Architecture (CORBA) represents the state of the art in object-oriented middleware technology, the most appropriate approach is to enhance it in the direction of mobile agents. The current specification of a Mobile Agent Facility within OMG is a strong indicator of a global support of this direction. TINA, which is also based on CORBA, defines a flexible service architecture for future information networking services, with primary focus on the provider domain. Mobile agent technology could enhance TINA for the dynamic provisioning of new service client modules. However, an important aspect in the introduction of such a new service platform is its applicability in different application domains and the migration and interworking with existing platforms (i.e. legacy systems).

Hiroaki Nakanishi, Executive Staff, Business Development Office, Hitachi, Japan.

Focus: Application-oriented network computing from the viewpoint of an application planner and business developer.

The Personal Computers have spread out vastly into the consumer market. This generates new business opportunities in deploying inter-media environments, which combine the functions of the Internet, digital satellite broadcasting and multi-media storage systems such as DVD devices. Various applications utilizing inter-media environments are planned and being tried right now, such as electronic newspapers, electronic publishing, home education, and home shopping.

In these systems, the broadcasting of data is one of the key source of information; this is one of the typical mechanisms of ADS. The major technical issues are how to make the applications and the supporting system functions flexible and extensible, since the final required functions cannot be pre-designed. That is also one of the requirements in the ADS technology.

Stefano Zatti, Network and security manager, European Space Agency - ESRIN, Italy.

Focus: The computing issues and tradeoffs at ESA, a largely distributed organization, in which the network plays a vital role.

The current computing paradigm at ESA is granting a high degree of local autonomy to the different sites, whereas a number of corporate applications (financial, purchase, personnel) are centralized. On-going trend is for client-server computing and downsizing, with progressive phase-out of centralized systems. The major motor of this trend is cost reduction. As such, our feeling is clearly in favor of autonomous decentralized systems, with minimalist clients supported by a number of distributed, cooperating servers.

The reaction to the different opportunities offered by network computing is twofold:

- Positive, for the possibility of further cost reductions (cheaper equipment) and of easier control of where and how the information is stored;
- Diffident, with respect to the possibility (such as with Java) of having (possibly) uncontrolled software running on proprietary machines, with potential damage. The experience acquired with dumb and X- terminals does not favorably encourage a return to such devices, unless their problems are overcome by the new paradigm.

Keynote Address

Experiences Teach Us the Future of Autonomous Decentralized Systems

Speaker

Hiroshi Kuwahara

Experiences Teach Us the Future of Autonomous Decentralized Systems

Hiroshi Kuwahara
Hitachi, Ltd.
6-27-8 Minim Oi, Shinagawa, Tokyo 140, Japan

Abstract

In these 20 years, the Autonomous Decentralized Systems (ADS) have been developed and now several hundred systems have been in operation effectively. As the ADS innovates the system concept, the new system architecture and technologies have been induced to develop for growing up the merits and down the demerits of ADS. In this paper, two typical application systems are shown to describe the problems, the solutions and the further works for realizing ADS.

These experiences teach us the future of ADS towards the telecommunications-based information service systems in which the end users autonomously select to utilize the service. The research and development for the next generation ADS technologies in telecommunications-based information services will be accomplished through the cooperation in the world.

1. Introduction

Autonomous Decentralized Systems (ADS) were proposed 20 years ago as a revolutionary concept [1] whose practical use brought with it plenty of risks. However, the development of computer and communications technologies and the expanding needs assured the future of ADS, and the development and application of these systems have been continued. This paper describes our research and development experiences in the field and examines the future prospects for ADS.

Over the past 20 years, R&D into ADS spread in Japan and around the world.   In 1993, the first ISADS [2] was held under the sponsorship of IEEE and Japan's two societies, IPSJ (Information Processing Society of Japan) and SICE (Society of Instrument and Control Engineers of Japan).   The second symposium [3] was held in Phoenix, U.S.A., and now this is the third time for the symposium to be held.   In Japan, in 1991, the Ministry of Education started concentrated research in ADS which continued for a period of three years[4].   An ADS Committee was formed at SICE and an ADS symposium is held every year.

Autonomous decentralized systems is an important topic in the international IMS (Intelligent Manufacturing System) project under the auspices of the Ministry of International Trade and Industry of Japan[5].   In 1995 the international consortium in the field of factory automation, ODVA (Open Device-Net Vendor Association) approved ADS architecture as a de-facto standard [6].   Work to standardize activities in the factory automation (FA) and building control fields then began in Japan with international firms taking part.

In telecommunications, ADS has also been starting to be used and this is the special topic in this year's ISADS.   The experiences of R&D will be surely able to be put to valid use in future technologies and applications in the field of telecommunications.

2. Experiences

2.1 Manufacturing Control System

Fueled by harsh competition on an international scale and changing user needs, manufacturing systems have developed enormously in the past 20 years. Changes in quality have been more drastic than changes in production volume.   Here the experience in ADS development for a steel production system and new technical advancement in the field are discussed.

2.1.1 Requirements and Solutions
(1) Requirements
These days, steel production system requires "variable-types and variable-volume production" so that a certain type of product can be produced in a certain quantity according to user needs.   This differs from conventional "mass production" as well as from planned "multiple type and small-lot" production.   The industry is now in an environment where it is unpredictable for changes in user needs and production costs through production materials and logistics, constantly fluctuating owing to changes in social and economic conditions.   In such a situation, it

169

is necessary to respond to user needs rather than increasing the variety of products and to provide only the volume and type of products required. And, it is also necessary to be vigilant about cutting production costs and improving product quality.

To respond to such requirements, computing systems need to be modified and expanded during system operation. To effect these modifications and expansions it should be possible to test and subsequently maintain systems while they are in operation.

(2) Solutions

Autonomous decentralized systems are based on the notion that basically subsystems exist and then system is formed as integration of them. Each subsystem is autonomous in the sense that it can control its own task and rather cooperate with existing subsystems while failures, maintenance, or construction of the other subsystems. This autonomy means that on-line properties (expansion and maintenance during operation, and fault-tolerance) can be obtained in system.

This autonomous decentralized system is realized from data field system architecture [2] (Fig. 2.1). Data field system architecture consists of a data field (DF) where data is circulating and subsystems which are connected to it. The subsystems select to receive data based on its content, and once all the required data has been received, begin execution of the process. A content code that corresponds to the data content is tagged to the data and this is broadcast in the data field as a message. None of the subsystems receive instructions from each other to send/receive data and to begin processing. Communication and processing is only carried out as judged by each subsystem while mutual coordination is possible via the data field.

The needs of steel production systems are consistent with the aims of the autonomous decentralized system, and this enabled practical application of this data field architecture. In particular, there are two main reasons why application of the system is so successful.

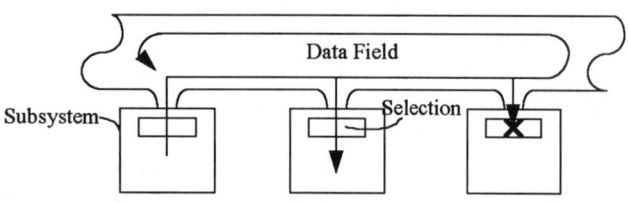

Fig. 2.1 Data Field Architecture

First, in a control system, the actuators works by using sensor data from the plant and management data. This actuation modifies the process, the modification is monitored, and the results is fed back to the control system. While the sensor, actuator, and monitor devices include smart intelligence, they are still device-dependent and their functions can be capsulated. These capsules correspond to the subsystems of an autonomous decentralized system [2] (Fig. 2.2).

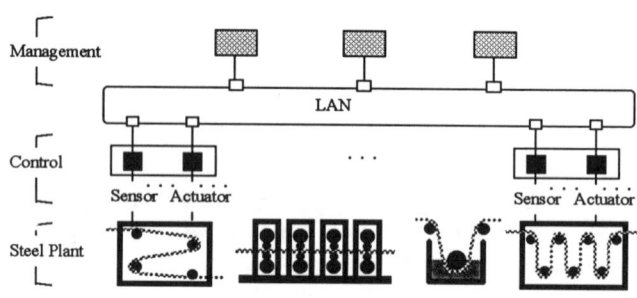

Fig. 2. 2 Steel Production Computer Control System

Second, the input/output data of the capsulated functions can be defined clearly and uniformly. The capsules are mutually connected not according to the processing circumstances but via the data itself. In the ADS, autonomous subsystems are initiated via the data in the data field, making this the ideal structure for the steel production system.

2.1.2 Problems and Solutions

As mentioned in 2.1.1, ADS answered the steel production system needs. But to do so, another problem had to be solved, namely, system cost. Because ADS differs from conventional systems, it took more time to design application programs, and to carry out development, tests and maintenance.

This problem has been solved thanks to development of support tools for ADS design, development, and testing. In data field architecture, the application software modules are data-driven. Therefore, tools are developed that allow bottom-up design of application software modules for distributed development. However, testing of the developed application software is desirable while the system is operating in on-line mode. The data field architecture is designed so that even if on-line data and test data co-exist in the data field, the difference between both would be identifiable by flagging them in the message. This means that even when tests are carried out on test modules using on-line data, the resulting data of those tests is identifiable as test data and will not be used by other operating on-line modules in their operations.

170

Neither will on-line databases or files be affected. This allows the system's on-line operations to be carried out at the same time during the test. The other advantage of using data field is that, when connecting modules via data field, any data can be receive anywhere so that the test monitor can operate without interfering with any other system.

Therefore, these on-line test support tools and design and development support tools can be used to boost software productivity by 30%. However, although the tools had been developed, since the data-drive software design differed from conventional design methods, extensive training of designers and engineers is required to familiarize them with the tools. A full solution to this problem has not yet been found and the future research that will provide with a design policy is necessary.

2.1.3 Further experience

Cost-cutting of system has been consistently improved more under another technical trend. In the field of control systems, sensors, actuators, and switch devices are becoming smarter and then advances are made in ultra-distributed systems. As the example of steel production systems shows, control systems functions can be capsulated. Once each device has been made intelligent, autonomous capsulation becomes even easier. If each device becomes autonomous, this will allow the configuration of an ultra-distributed system to which thousands or even 100,000's of devices can be connected. Once the number of built-in devices is increased thanks to their autonomy, their prices drop, thereby also lowering system costs. It is with goals such as these that the international ODVA (Open Device-Net Vendor Association)—a vendor consortium in the field of factory automation — was established in 1995. The ODVA approved ADS as a system architecture and has already determined to treat the content code communications functions for the common interfaces of each devices as standard specifications. This has been made available as ODVA Version 2.0[6]. In this respect, Hitachi which holds the patents on some of ADS technologies has made the license freely available to promote its spread.

These ultra-distributed systems are appearing not only in the field of factory automation but are also spreading to other control fields in countries around the world while the trend towards more open systems, and standardization is likely to pick up pace. Within these trends, R&D into ADS still continues with international cooperation.

2.2 Transportation Traffic Control and Information Systems

As demands for safety, speed, economy, and ecology in transportation increase, many countries have been in the way to systematize the field. In Japan, in particular, train transportation is important. The country's train transportation systems employ cutting-edge computer communication technologies and represent Japan's biggest large-scale systems. The following section describes the process that led to the development of the ADS in Japan's largest transportation system, the Tokyo Metropolitan Train Traffic Control System, and the new technical developments that is emerged from this.

2.2.1 Requirements and Solutions
(1) Requirements
The Tokyo metropolitan area comprises 19 train lines and 300 railway stations. Until now, each of these train lines has been capable of providing a safe, and high-density transportation service. However, factors such as diversification of transport means and the resulting intensity of competition, and more flexible service needs call for the development of a new train traffic computing system. Particular demands are the running of trains in line with passenger demands, and passenger information services, including the operational state of the trains. Conventionally, trains were to run exactly according to a predetermined timetable for each line, and if the timetable was upset, efforts were made to return it to normal as soon as possible. While these needs have not changed, recent response to passenger needs requires running trains over a wide range of lines, or running special types of trains for specific events. At the same time, passengers now look for information related to travel time so they can select a complex combination of routes to reach their destination according to circumstances, or whether trains have been delayed or not. These days, it is necessary to fulfill these rather different demands for reliable real-time train transportation control and flexible information services. To answer these needs for such a gigantic railway network requires the step-by-step construction of a system over a period of eight years that would not interrupt train operations while it has been built. This, in turn, requires the establishment of a computing system that covered a total line length of 1,200 km, and computerized each station, ensuring safety and fault tolerance and allowing construction, expansion, and maintenance without disrupting operations.
(2) Solutions
The needs of this railway system matches the goals of autonomous decentralized systems allowing neat

traffic control is managed by a successive cooperative linkage between neighboring stations, which also allows control of all lines. The localized station control and coordination between neighboring stations form the basis for the system. As is the case in the manufacturing control system mentioned in 2.1 above, the sensor, actuator, monitoring functions could be capsulated. Moreover, these functions could be further capsulated per station, allowing autonomy of the station as subsystem [3] (Fig. 2.3).

The second reason is that data for passenger information services has to be available everywhere and at any time. Train management data is broadcast in the data field where current information on all trains and stations is kept as data flow. Therefore, any information under any circumstances is available to passengers in real-time at any time. While train management is based on local control, the information service is based on shared data. The capsulation of stations and the inter-station data field have enabled simultaneously these two localized control and large-scale sharing of data.

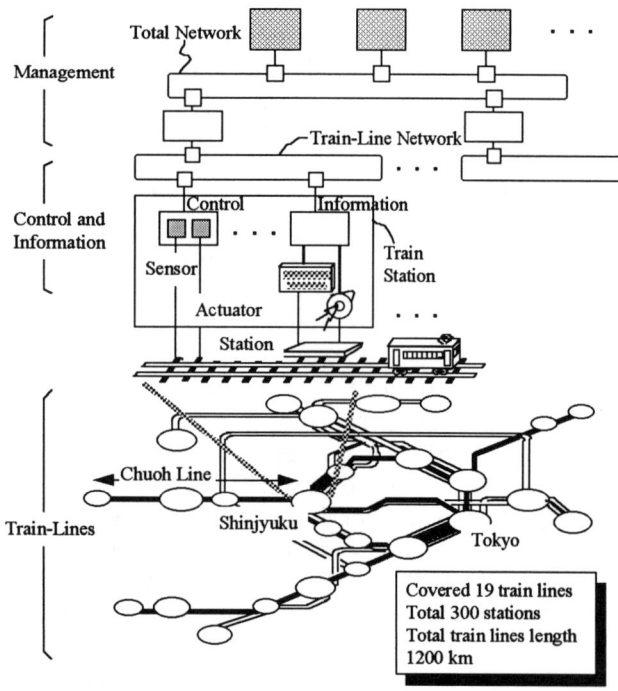

Fig. 2.3 Tokyo Metropolitan Train Traffic Control and Information System

2.2.2 Problems and Solutions

While expert use of the advantages of ADS has fulfilled the railway system's needs, new problems has arisen. These problems has emerged when the train traffic control system and the passenger information service system are integrated. A control system has to be highly reliable, and able to maintain a real-time condition. On the other hand, an information system aims for flexible shared information amongst users. The information system must not be allowed to adversely affect the reliability and performance of the control system. Conversely, the control system must be able to actively provide the required data to the information system. Filtering technology is developed in order to simultaneously maintain separation and integration of the control system and information system. This filtering technology identifies the data in the data field and enables or disables the control or information system from collecting the data there. In general, while data from the control system may enter the information system, data from the information may not enter the control system. This filtering technology is effective in providing a clear separation between the control system and the information system. However it does not fully solve problems such as classifying the vague range of either control or information system or when the role of the subsystems changes according to circumstances.

2.2.3 Further experience

This railway computing system is gigantic, and as it is being developed over a lengthy period, it is necessary to use products from many different vendors. And, since the control system has to be integrated with the information system, the environment of the heterogeneous systems has o be harmonized. Therefore, ADS has to be realized with standard products such as personal computers, workstations, Ethernet[*1], FDDI and software including operating systems (Windows[*2], Unix[*3]) and protocols such as TCP/IP and UDP/IP.

The ADS is constructed by using an ACP (autonomous control processor) as a distributed OS on Windows and UNIX. The ADS content code communications protocol is created using UDP/IP while ADS database management systems as database support are constructed on such as ACCESS[*4].

[*1] Registered trademark of Xerox Corp.
[*2] Registered trademark of Microsoft Corp.
[*3] Registered trademark in the United States and other countries, licensed exclusively through X/Open Company Limited.
[*4] Registered trademark of Microsoft Corp.

In Japan, a consortium has been established to standardize Ethernet-based ADS. This activity should not be confined to Japan and in the near future, it is likely that standardization bodies around the world will cooperate in this task.

3. The Future of ADS

3.1 Information services-oriented society and its systems

Towards the 21st century, an information-oriented society has been increasingly emerged. It is expected that the advent of more sophisticated systems particularly in telecommunications, will create infrastructures to support advanced information needs. These functions will do more than simply provide reasonable and global access to multimedia information. A systematized telecommunications network will provide a wide range of services. In other words, the source of this social revolution lies in telecommunications. Service providers based on telecommunications networks offer a wide variety of services, from electric commerce, to home education services, or telemedicine services. These services can be expected to proliferate via networks in response to user needs.

The following three points are the basic attributes these telecommunications-based information service must offer.

(1) <u>Diversity and flexibility</u>: There are many different kinds of information service that can be offered over the network and these vary according to social, economic, and individual conditions. For these reasons, the service must be functionally diverse and flexible. And, beyond this, the network that supports these services must itself be flexible in areas such as structure, transmission volume and quality, and in providing value-added information.
(2) <u>Dependable provision/guarantee</u>: Because of the communal and widespread use of these services, they must be dependable on a global scale. Safety and the maintenance of security must be guaranteed despite any breakdowns or overloading that may occur on the network.
(3) <u>Open access to and standardization of services</u>: While advances in international standardization of telecommunications is a given, service transactions should also be possible on a global scale. Therefore, the service functions and transaction methods should be made open and standardized.

Regarding the above three demands, ADS research and development has until now been concerned with ways to build computer systems and to apply ADS in

the fields of plant control systems and information systems. It seems that the same system techniques can be applied to the telecommunications-based information service systems. The vast experience can be exploited in the field of system development as the world of telecommunications expands and becomes increasingly intelligent, forming the basis for a wide range of information service.

3.2 Telecommunications-based information services

The world of telecommunications that forms the basis for a diversity of service functions is made up of the following three types of participants (Fig. 3.1).

(1) End Users (EU): There are consumers or corporate bodies who "use" services to obtain practical benefits. Depending on the services offered the end users make practical use of various functions such as shopping, making inquiries, and personal communications.
(2) Service Providers (SP): These "provide" services to the end user. As the provider depends on the service prices, it offers services the end user wants and receives payment for this.
(3) Network Providers (NP): These come between end users and service providers and "guarantee" the transmission of the information required by both of end users and service providers. The basic function of telecommunications as derived from telephone companies is to connect end users to one another and to ensure that this can happen in all circumstances by striving to provide the appropriate technologies, equipment, and investment. Now these providers are creating network functions that go beyond this by providing network infrastructures for the growing telecommunications-based service and anticipating the continually diversifying and changing nature of these services.

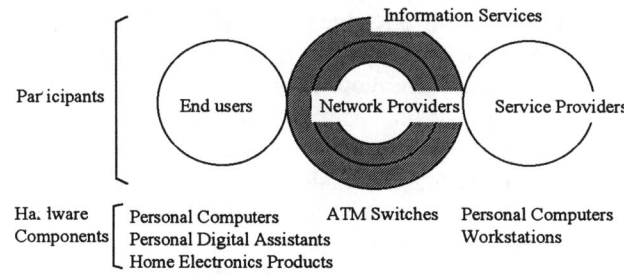

Fig. 3.1 Telecommunications Participants

The cost of actual services is borne by companies and ordinary consumers and it is only when he benefits of these services are felt that they are given social significance. A variety of needs are dependent on each service and they are defined from the point of view of service functions as follows.

(1) Customization: Use, provision, and guarantees of the service should be specified in accordance with the end user of the service. Use, provision, and guarantee of optimum service should be in accordance with the tendencies of the end users, the history of their needs, and the organizations they belong to.

(2) Situation: Use, provision, and guarantees of the service should be in accordance with the situation at the time the end user requires it. The conditions that determine the end user situation include the type of I/O terminal, location, time, and degree of urgency.

(3) Quality: Use, provision, and guarantee must be specified with respect to the quality of information required by the end user of the service. Concretely, this means information response (high-speed or normal), security, visibility (multimedia, screen quality, etc.).

Based on these three views (customization, situation and quality), ADS can be concretely applied to constructing service platform in a telecommunications system that consists of end users, service providers and network providers.

3.3 Service platform

The age of production and distribution systems for mass production and mass consumption is past. These days, manufacturers and distributors, and most industries are preoccupied with defining subjects such as constantly changing consumer needs and how to respond to them, specifically, the individual wants of consumers and how carry out appropriate marketing. As services develop over the networks, the previously mentioned three views of customization, situation and quality are inevitable. Service platforms must allow autonomous management of service use, provision, and guarantees by end-users, service providers, and network providers. Services are continually changing, being added or removed and at the same time, end-users and service providers also fluctuate as do network configurations of network providers. This is because in such an environment it is impossible to prevent services from operating. Participants using, providing, or guaranteeing services, are also responsible of making judgments, and selection of services from other participants must be possible. By building a service platform that has an autonomous decentralized structure,

these use, provision, and guarantee needs can be met. One effective way to incorporate ADS in the basic telecommunications service platform structure is to introduce the fundamental ADS mechanism of content code communication. In the content code communication, an autonomously functioning module receives only the necessary broadcasted data and processes it autonomously according to its role. A service platform is formed by incorporating these as functions of the modules that make up the network. The ADS can be expanded, and in order to realize the service functions available on the network, sub codes that adhere to the three views of services can be added to content code communication protocol. One proposal for doing this now under study is to add a customization ID (CID), situation ID (SID), and quality ID (QID) to the content code header (Fig. 3.2). As the customization ID (CID), situation ID (SID), and quality ID (QID) are attached to the message on the network, end-users, service providers and network providers can easily access, collect, and analyze information at any time. This expanded content code data can drive diverse service functions, allowing the generation of a wide diversity of data. As this wide diversity of data driven by various functions can be incorporated, processes such as synchronization for ordering and selecting the data need to be carried out. These processes will determine the diversity and effectiveness of each service.

Implementation of this mechanism allows the creation of a use, provision, and guarantee of services infrastructure in a network that can realize three views of services.

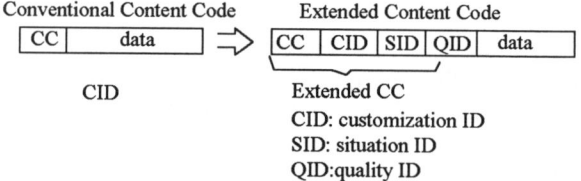

Fig. 3.2 Extended Content Code

The service provider can according to the situation directly examine in detail the demands generated by end-users from information broadcasted on the network. The service platform can enable the following value-added functions demanded by service providers.

(1) Service providers can access and analyze the purchasing history of clients and their tendencies while this function make it possible to improve service.

(2) Service providers can create a database on the results of client data analysis. But these functions of (1) and (2) needs to be handled carefully in the interests of

174

maintaining privacy.

(3) As the service provider subscribes to the network as an autonomous subsystem, it can offer consistently dependable services despite any interruptions to the network due to faults or expansion of the other subsystems.

3.4 Dependable network resources management

In the telecommunications ATM (Asynchronous Transmission Mode) switches have been widely accepted to realize the telecommunications networks. A total system is being developed and expanding day-by-day. The very form of telecommunications network itself is in a state of flux, and the basic requirements determined for information services (diversity, flexibility, dependability, openness) apply equally to telecommunications network. One example of this is the need for network resources management functions. Using the ADS technologies in the network resources management functions for monitoring, analyzing, testing, and recovering of the network allow the establishment of a highly dependable network. Not only the network resources management functions but also the service management functions using ADS technologies will be developed in the future in ATM related devices (switches, routes, and other autonomous intelligent modules).

3.5 International telecommunications

Telecommunications-based information service system will expand based on the creation of more intelligent networks. The personal computer (PC) will no longer be the only end-user device but television, personal digital assistants (PDA), and other devices will also come in play. The more telecommunications equipment develops, the fewer will be the technological limits of information processing. A common wish is for more service provided across national boundaries. This will be made possible by international standardization (such as by TINA-C: Telecommunications Information Networking Architecture-Consortium) and by the continuous progress towards this goal. Services reflect the culture they serve and therefore differ according to region, country, and society. Services that are becoming increasingly available on an international telecommunications network should be developed in consideration with the similarities and differences between cultures, and on standardized infrastructures. ADS based ideas, architectures, technologies and platform surely constitute a powerful telecommunications resources management and service management systems.

From this viewpoint, the telecommunications devices, software and systems have been researched and developed in an effort to concretely provide the kind of highly convenient services mentioned in this paper. International cooperation is essential to meet the goal of information services-oriented society. ISADS '97 will hopefully promote the creation and development of these links.

4. Conclusions

The research and development of Autonomous Decentralized Systems (ADS) over a 20-year period is presented here through two specific application examples: manufacturing control system and train traffic control system. The factors behind the success of these systems are addressed as well as remaining problems. Moreover, in light of the expanding user needs and the advancement of ADS technologies, the information services systems in the field of telecommunications are shown as one important application of the next generation ADS. The role of ADS should certainly expand in the advanced information society. For effectively developing the technologies needed for computers, communications and control, it is essential that continuous work be performed on an international basis in cooperation with the academic, industrial and governmental sectors of society. Now this cooperation has begun in the world.

References

[1] K. Mori et al, "Proposition of Autonomous Decentralized System Concept", Transaction of The Institute of Electrical Engineering of Japan, Vol. 104, No. 12, December, 1984.

[2] Proceedings of ISADS 93: International Symposium on Autonomous Decentralized Systems, March 30 - April 1, 1993.

[3] Proceedings of ISADS 95: Second International Symposium on Autonomous Decentralized Systems, April 25-27, 1995.

[4] "Report on Autonomous Decentralized System Project", sponsored by Ministry of Education of Japan, September, 1990.

[5] "Journal of Robotics and Mechatronics", Vol. 6, No. 6, Dec., 1994, IMS Center under auspice of MITI.

[6] "DeviceNet Specifications", Open DeviceNet Vendor Association, Inc., April 22,1996.

Session 3A

Architecture I

Chair

Richard Soley

Support for Distributed Multimedia Services in the TINA Architecture [*]

Mikael Jørgensen[a], Peter Leydekkers[b], Marcel Mampaey[c] and Henry Yang[d]

[a]Tele Danmark R&D, Lyngsø Allé 2, DK-2970 Hørsholm, mikael@tdr.dk
[b]KPN Research, P.O.Box 15000, 9700 CD Groningen, The Netherlands,
P.Leydekkers@research.kpn.com
[c]Alcatel Telecom Research Division, Francis Wellesplein 1, 2018 Antwerpen, BELGIUM
mmam@rc.bel.alcatel.be
[d]Lucent Technologies - Bell Labs., 67 Whippany Rd, Whippany, NJ 07981, U.S.A.
xyang@bell-labs.com

Abstract

TINA is a world wide initiative with the objective of defining an architecture for the future telecommunications infrastructure, which should be able to handle multimedia services. This paper investigates and discusses the concepts currently available in the TINA architecture to model multimedia services. The focus of the paper is targeted towards the support for describing, requiring and setting up the necessary multimedia connections between the parties involved in distributed multimedia services. By means of an example we illustrate how the TINA modeling concepts can be successfully applied to the description of a multimedia conferencing service scenario.

1. Introduction

The Telecommunications Information Net-working Architecture Consortium (TINA-C) was established in 1993 with the objective to define a general architecture for telecommunication services and the supporting infrastructure. Currently, the TINA architecture consists of four sub-architectures, i.e. the Service architecture, Resource architecture, DPE architecture, and the Management architecture.

The *Service architecture* defines a set of concepts, principles and guidelines for constructing, deploying, operating and withdrawing of tele-communications information services in a TINA compliant environment. A service is provided by a set of interacting objects that are located in an open distributed processing environment (DPE). The service architecture defines the objects that constitute a telecommunication service, and how these objects should be used and interact.

The *Resource architecture* defines a set of generic concepts that describe transport networks in a technology

independent manner and provides the mechanisms for the establishment, modification and release of network connections.

The *DPE architecture* defines the information [2], computational [3] and engineering languages that are used in the other architectures to specify telecommunication services. Additionally, the DPE architecture defines a generic Distributed Processing Environment (DPE) that is used as the platform on which TINA compliant services are built. A (partial) implementation of the DPE could be the Common Object Request Broker (CORBA) as defined by OMG.

The *Management architecture* provides generic management principles and concepts for the management of services, network and computing platform, but as this is outside the scope of this paper it will not be further addressed.

All these sub-architectures define concepts, rules and guidelines that should complement each other in the trajectory of specifying a TINA service. An interesting issue is how well these sub-architectures fit together when focusing on a particular subset of TINA services, i.e., interactive multimedia services which are an important group of TINA applications. The purpose of this paper is to investigate the most important concepts currently available in the TINA sub-architectures to model multimedia services and examine how well the different concepts developed in the sub-architectures fit together.

The paper is organized as follows: Section 2 identifies several important requirements that should be fulfilled by the TINA architecture to model multimedia services in an open distributed environment. Section 3 identifies the most important concepts currently available in the TINA architecture to model multimedia services. Section 4 discusses the correspondences between the requirements expressed in Section 2 and the available concepts presented in Section 3. Section 5 then describes an example how to model a multimedia conferencing service

---

[*] This work was performed during the authors stay at the TINA-C Core Team. However, as this paper presents results from ongoing research activities within the Core Team, it does not reflect consortium wide agreements and reflects the view of the authors.

in a consistent manner using the concepts from Section 3. Finally, Section 6 will provide the conclusions and outline directions for further work on related issues.

2. General multimedia service requirements on the TINA architecture

Multimedia services consist of some combination of software and hardware resources which can be controlled and connected to each other while executing in a heterogeneous distributed environment. An architecture for multimedia services like TINA, should provide the means to describe such services at a sufficient abstraction level. This implies that TINA must provide the abstractions and modeling concepts which allows us to model multimedia services without paying attention to the specific characteristics of the underlying execution platform, communication network and attached multimedia devices.

The emergence of low cost broadband networks opens a market for real-time multimedia services. These services support multimedia traffic, such as audio and video streams, that have additional functional and performance requirements that need to be met by distributed processing environments (DPEs). These multimedia services require for instance the support of a variety of communication protocol stacks suitable for stream communication (e.g. XTP [19]), and support for and manipulation of quality of service (QoS) parameters, allowing performance and real-time guarantees. Multimedia applications have strict QoS requirements and this implies that the TINA architecture should provide the modeling concepts to specify QoS commitments adequately. Also, the underlying infrastructure should support the strict performance requirements demanded by real-time multimedia applications.

Another important issue is interoperability between multimedia services that are located in different administrative domains or in different computing nodes. This requires that multimedia characteristics should be specified in a standardized manner which should be prescribed by the TINA architecture.

Finally, the TINA architecture is meant as an overall architecture which should be used and refined by other consortia. Consortia such as DAVIC [14] and IMA [10] focus on the definition of specific multimedia architectures and it is important to align the concepts developed in TINA with those consortia to ensure that the TINA architecture meets their demands.

	Service Architecture	Resource Architecture	DPE Architecture
Information viewpoint (information exchange between information objects)	- Service session graph - User/data relations between parties	- Logical connection graph (stream interface, flow endpoint, flow connection) - Connectivity layer	
Computational viewpoint (operational and continuous interactions between computational objects through well defined interfaces)	- Service session graph interface - SSM, USM, GSEP - Stream interfaces - Binding of stream interfaces	- Logical connection graph interface - Flow endpoint/access point - CSM,TCSM,CC - Physical and Nodal connection graph	
Engineering viewpoint (operational and continuous interactions between distributed objects using supporting objects)			- Stream interface reference - Channel (stub, binder, protocol) - DPE Kernel

Table 3-1 TINA terminology and modeling concepts related to multimedia in the various sub-architectures

3. Concepts related to multimedia in the various TINA sub-architectures

Table 3-1 depicts the TINA terminologies and modeling concepts that are of interest for the specification of multimedia services. The terminologies and modeling concepts are organized along the ODP information, computational and engineering viewpoints [13] (vertical column) and the TINA sub-architectures (horizontal column). Ideally, the TINA terminologies and modeling concepts should smoothly fit together in the trajectory of specifying a TINA Service.

The following sections provide an overview of each of the concepts shown in Table 3-1.

3.1 Service architecture

The service architecture [4] defines a session model used for the subscription, access, and usage of a service by a number of users. Subscription and access are modeled by the access session, and the usage of a service by the service session. The service session is described by means of information and computational models. The information model uses the concept of a Session Graph which depicts the control and connectivity relations between the members of the service session. A general session information model is shown in Figure 3-1 upper part using the OMT notation. The complete information on the Session Graph model can be found in [7].

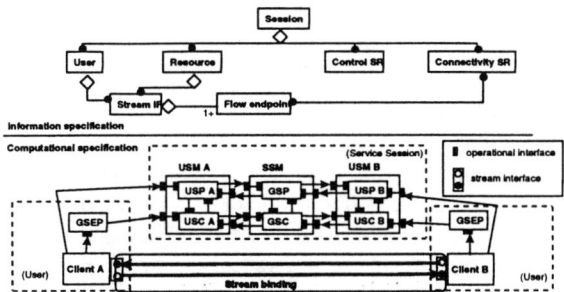

Figure 3-1 Informational and computational model of service session

Each of the classes in the OMT diagram in Figure 3-1 is briefly explained below:

- The *Session* class contains all the information necessary to model the configuration of a session and the relationships between the elements of the session (i.e. users, resources) at a certain point of time.
- The *User* class models a user involved in the session. The user will have control capabilities and can be represented by an address (e.g. E.164)
- The *Resource* class models any resource that can be utilized by users within the session. This includes special resources (e.g. conference bridges, converters) as well as information resources like data files, containing for instance multimedia documents.
- The *User and Resource* class contains *Stream Interfaces* by which they are able to participate in connectivity relations. The Stream Interface class is an aggregation of *flow endpoints*, each with an associated flow type description (e.g. audio, video).
- The *Connectivity SR* models connectivity associations between Session Members (more specifically between the stream interfaces they contain), enabling to model simple point-to-point connection but also complex multipoint-to-multipoint connections.
- The *Control SR* models control associations within the session, used to model e.g. ownership of connections by the users, permissions for a user to add new users in the session or to add (delete) connections, etc.

The Service Session model is very generic and models service session configurations and associations that are valid for a broad range of multimedia services. It defines precisely the relations between the parties involved in a multimedia service session, as well as, the connectivity relations between the multimedia sources and sinks involved by means of the stream interface class.

Figure 3-1 lower part shows a configuration of computational objects which model a service session. For the computational viewpoint the service architecture defines several computational objects that should always be present in a TINA compliant service session. These computational objects are:

- *Service Session Manager (SSM)* which is the central point of a service session. The SSM coordinates and serves all the requests from the members of the

session. The SSM computational object is decomposed into two objects each offering a different type of operational interface. One object is the *Global Session Control (GSC)* which offers a generic session control interface which provides operations on the Session Graph used for both connectivity and control relationships requests. It allows, for instance, users to join or leave the service session. The second computational object is the *Global Service Part (GSP)* which offers an interface providing service specific operations.

- *User Session Manager (USM)* which represents each client in a session. The USM object is decomposed into two computational objects each offering a different type of operational interface. The *User Session Control (USC)* object provides the client with the generic session control interface. Through this interface the client can invoke operations such as adding or removing other session members, thus changing the appearance of the session graph which describes the service session at a certain point of time. The second object is the *User Service Part (USP)* which provides the service specific interface.
- *Generic Session End Point (GSEP)* object provides each client with a generic session control interface which can be used to access and use all services.

In the computational model the connectivity session relationship is represented by the stream interfaces and the binding. The binding is a computational concept which abstracts over the actual connection through the transport network (see also Section 3.3).

3.2 Resource architecture

The TINA Resource architecture [5] [15] is used to describe and implement the needed connections between service session users and resources. The connections are described by means of *connection graphs*. The connection graph concept provides the means to express connectivity between stream interfaces as well as connectivity between resources found in the underlying networks.

Figure 3-2 depicts the basic concepts of connection graphs (information specification) and how they relate to entities in the computational and engineering model.

A *stream interface* contains a number of *flow endpoints* which can be either sinks or sources for continuous dataflows. A *Logical Connection Graph (LCG)* is an information object that represents a stream binding (Figure 3-2 upper part). The LCG contains one or more flow connections. A *flow connection* is the representation of a binding between a source flow endpoint and a number of sink flow end points. A *Nodal Connection Graph (NCG)* is an information object that represents a nodal binding and it contains one or more nodal connections. A nodal connection is the representation of a connection between a flow endpoint and a network access point within a node. A *Physical*

181

Connection Graph (PCG) is an information object that represent a transport network binding and it contains one or more transport connections. A *transport connection* is the representation of a connection between one or more network access points.

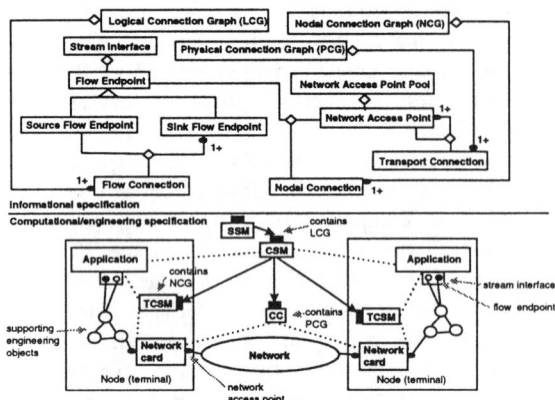

Figure 3-2 Information and combined computational-engineering model to express connectivity in TINA

For the computational viewpoint the TINA Resource architecture defines several computational objects that handle and use the connection graphs as described in the information model. These computational objects are shown in Figure 3-2 lower part which is a combination of an ODP computational and engineering specification:

- The *Communication Session Manager (CSM)* is responsible for the end-to-end application communication, i.e., provide binding between flow endpoints (depicted by the dotted lines in Figure 3-2, lower part). The logical connection graph is used by the Service Session Manager (SSM) to describe and express to the CSM object the characteristics and configuration of the binding between the flow endpoints in the context of a communication session. The CSM will then decompose the requested binding (i.e. LCG) into two parts: the nodal part (i.e. NCG) and transport network part (i.e. PCG). The CSM object requests the Terminal Communication Session Manager (TCSM) to take care of the nodal part while the Connection Coordinator (CC) is responsible for the transport network part.

- The *Terminal Communication Session Manager (TCSM)* is responsible for the nodal binding and binds the flow endpoint to the network access point (Figure 3-2, lower part). The characteristics and configuration of the nodal binding is described by the CSM in the form of NCG within the context of a terminal communication session.

- The *Connection Coordinator (CC)* is responsible for the end-to-end transport connection of a transport network, i.e., from network access point to network access point. The characteristics and configuration of transport connection is described by the CSM in the

form of PCG within the context of a transport connection session.

3.3 DPE architecture

Multimedia concepts in the DPE architecture are mainly reflected in the ODP compliant computational and engineering models. These are the stream interface, binding, stream interface reference, and stream channel.

Stream interface. Stream interfaces are used for describing the points through which an entity may participate in the exchange of continuous data.

A stream interface can contain a number of flow endpoints and each flow endpoint is a unidirectional source or sink of a flow. When stream interfaces are bound, the flow endpoints of the stream interfaces are connected to each other. The information exchange between the connected endpoints occurs in the form of a bit sequence with a certain frame structure (data format and coding) and with a certain quality. The stream interface is described in terms of TINA-ODL as reported in [8] and [17].

Each flow in the stream interface describes its associated QoS by means of the so-called *Service Attribute*. Since the QoS can be different depending on the causality of the flow, we distinguish between *offered QoS* and *required QoS* values for respectively source and sink flows. The QoS described in the stream interface is on the service level, i.e. it uses terms such as "color", coding type (e.g. MPEG1, PAL) and image dimensions. Based on this service level QoS description, requirements from the network and supporting (computer) system can be deduced. For instance, a certain frame rate and coding format for a video flow requires certain bandwidth and jitter characteristics from the network, and certain processing and memory from the computer systems to process the video flow as specified in the service level QoS description.

Figure 3-3 outlines the content of a TINA-ODL object template, including the specification of a stream interface, i.e. the flow endpoints and QoS requirements and offerings. The object template provides the information for creating the object, and the stream interface template gives the information to create stream interfaces and contains the QoS properties of the flow endpoints. The information in the service attribute is used in the binding process to determine whether the object can participate in stream bindings [17].

Binding. Interactions between computational stream interfaces are only possible if a binding, i.e. a relation between the stream interfaces involved, using some communication path, has been established between them (see Figure 3-3). The computational language specifies explicit binding actions for stream interfaces. The binding abstracts over the actual connection established between the involved stream interfaces through a transport network and it can be controlled through the *Connection Graph Interface*. More details on the binding can be found in [6].

A binding is initiated by one of the objects involved in the binding or a third party (e.g. SSM object) and has the effect of establishing at each interface the necessary information for interaction to take place. Thus the identity and QoS properties of the interfaces are exchanged. A precondition for a successful explicit binding action is that the involved flow endpoints can be matched. This means having comparable types (e.g. both are video or audio) and complementary causality (i.e. one is a producer and the other a consumer).

Figure 3-3 shows an example of a binding between two stream interfaces. The binding is shown as the shaded oval in which the flows between the flow endpoints are indicated by thick lines. The binding object which exports an operational interface is the object through which the binding can be controlled. This interface corresponds to the interface exported by the CSM object as shown in Figure 3-2, and it offers operations to initiate and alter the LCG that describes the binding.

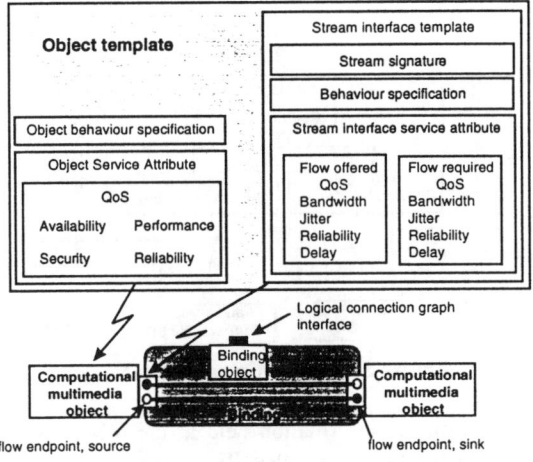

Figure 3-3 Stream interface object template and binding

In TINA the binding only supports point-to-point and point-to-multipoint connection of stream interfaces. In order to handle multipoint-to-multipoint connections the SSM will use a suitable combination of resources, e.g. audio/video bridges to mix the stream flows, and bindings, to connect the involved stream interfaces to the resources. From the perspective of the clients the SSM can be seen as the object providing control of the multipoint-to-multipoint "bindings" whereas the CSM provides the control of point-to-point or point-to-multipoint bindings. This issue is further elaborated in [1].

Stream interface reference. Interface references are unambiguous identifiers for the interface they reference. In a TINA/ ODP system it is assumed that the nucleus is responsible to create an unambiguous reference. Given such a reference it is possible to discover the type of the

interface, and a communication address at which binding to it can be initiated. It also contains other information about the expected behavior of stubs, binder and protocol objects within the channel. A stream interface reference has the same general characteristics as the operational one [17]. Its real-time multimedia characteristics appear mainly in the

- computational interface type that it implements, including the QoS specification (i.e., QoS offered / QoS required),
- channel configuration of stubs, binders, protocol objects, and
- possible interceptors needed that have real-time characteristics.

The information in a stream interface reference is used by several objects in the TINA architecture and it enables, for instance, the SSM object to create a session graph that is in line with the characteristics of the stream interface supported by the client. Also the CSM object uses information conveyed in the interface reference to create a LCG that corresponds to the transmission capabilities supported by the clients.

Stream channel. The *Channel* is the partial engineering implementation of the computational binding. Thus computational objects that are bound through either stream or operational interfaces are connected in the engineering specification by means of stream or operational channels. In TINA-C, channels consists of stub, binder and protocol objects as outlined in Figure 3-4.

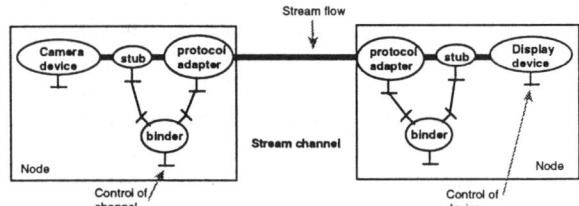

Figure 3-4 Stream channel objects

- The *stub* provides data conversion of data exchanged between objects. It offers a control interface through which the data presentation provided by the stub can be changed.
- The *binder* object manages the integrity and QoS of the stream channel. It offers a control interface through which the structure and configuration of the channel, e.g. the QoS characteristics, can be changed. This interface can be used by the TCSM to establish and manage the channel, i.e. the binder can be viewed as the intra-channel manager.
- The *protocol adapter* provides communication functions which allows the application objects to interact across node boundaries. It offers a control interface through which the communication functions

can be changed. For more details on the actual content of the protocol adapter see [9].

4. Review of the TINA Multimedia concepts

The previous section showed the most important concepts currently available in the TINA architecture to model multimedia characteristics of distributed services. Whether these concepts are sufficient to meet the requirements expressed in Section 2 is discussed below.

The multimedia concepts are mainly situated in the information and computational viewpoint and indeed do not refer to any specific technology. The engineering support for multimedia applications is based on refinements of ODP engineering concepts and supposes a certain (abstract) configuration of the underlying infrastructure. However, it is expected that an underlying object infrastructure as defined by the OMG Common Object Request Broker (CORBA) [16] will be used as an implementation of the DPE. But CORBA does not support the concepts of streams and stream interfaces, although activities within the OMG investigate to which extent, and how this could be done. This implies that currently a gap exists between the information/ computational modeling of multimedia applications and the actual implementation when using a CORBA compliant implementation of the TINA DPE.

Multimedia applications require *QoS commitment* which should be supported by TINA concepts. The binding object and stream interface concept provide hooks (e.g. service attribute) that allow a precise description of the QoS expected and offered by a computational object supporting stream interfaces. These QoS descriptions are used in the connection establishment phase and operational phase to negotiate and monitor the agreed QoS.

The stream interface reference allows one party to express its communication capabilities to other parties, thus making it possible to interoperate cross domain and technology boundaries. However, the TINA concept of the stream interface reference is still immature but other work, e.g. [11], [12], [17] addresses this issue and their results can be incorporated into TINA.

Until now, TINA has not attempted to define the service specific control operations and stream interfaces for different classes of multimedia applications. Instead the focus has been on the generic aspects of the service control as expressed in the generic service session concepts as outlined in Section 3.1. We regard this focus on generic aspects as a strong point since other consortia, such as DAVIC [14] and IMA [10], already work on these kind of specifications. For instance, the specific control operations for video-on-demand services are defined in DAVIC. Furthermore, a wide acceptance of the generic service session concepts as a basis for many kinds of multimedia services will allow greater interoperability between them. However, there is still a need for alignment

between the work done in TINA and these "service specific" consortia.

5. Multimedia Service example

This section outlines how the TINA concepts, as described in section 3, can be applied through the various stages of service development. The service scenario used in this example is a multimedia conferencing service (MMCS). We will show information, computational and engineering models and the correspondence between them.

5.1 Information model

The OMT diagram in Figure 5-1 is an instance diagram according to the model given in Figure 3-1 (upper part) and it describes a MMCS service session. The MMCS service session consists of two users, User1 and User2, whose stream interfaces are connected through a bridge which is a resource controlled by the MMCS session.

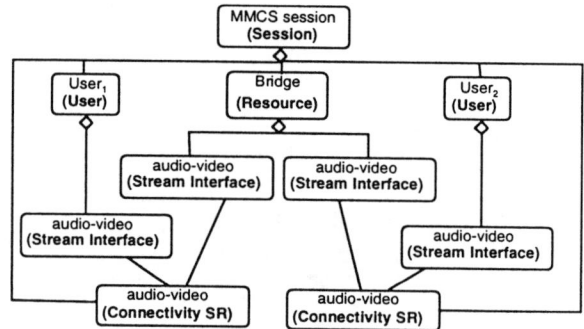

Figure 5-1 Session graph for MMCS

Based on the information held in the MMCS session graph a LCG can be generated. The process for generating the LCG from the information provided by the session graph consists of the following steps:

- For each connectivity session relation it is investigated whether it describes a multipoint-to-multipoint connection or not. If so, further resources need to be introduced to reduce the multipoint-to-multipoint connection to a point-to-multipoint or point-to-point connections which can be described by the LCG. In the case of Figure 5-1 both connectivity session relations describe a point-to-point connection.

- Then, the flow endpoints of the stream interfaces involved in connectivity session relations are investigated. The objective is to match as many flow endpoints as possible. In some cases not all flow endpoints can be matched. For instance, if a user wants to join that only supports audio, then only the audio flow endpoints of the bridge will be used. The matching of the compatibility between two flow endpoints includes:

1. Finding flow endpoints which handle the same flow type but have different directionality, i.e. audio-source to audio-sink or video-source to video-sink;

2. Investigating whether the connection of the flow endpoints uses the same format or requires conversion.

3. Comparing QoS characteristics; when a proper pair of source-endpoint and sink-endpoint are found, the QoS offered by the source must be held against the QoS required by the sink to reveal whether the source satisfies the QoS requirements from the sink.

When these investigations are concluded, a LCG can be build which expresses the desired connections between the flow endpoints involved in the MMC session. The OMT diagram in Figure 5-2 is an instance diagram according to the model in Figure 3-2 (upper part) and describes the LCG for the MMCS session.

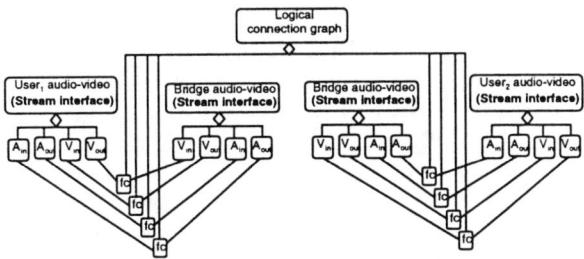

Figure 5-2 LCG describing flow connections (fc) between User₁, User₂ and Bridge

5.2 Computational model

The information provided by the session graph in Figure 5-1 and the LCG in Figure 5-2 corresponds to the computational model in Figure 5-3. This model contains a number of objects reflecting the information provided by the session and connection graphs:

- The MMCS SSM object controls the conferencing session and resources shared by the users. The MMCS SSM object provides a generic service session interface and a service specific interface for interactions with the USM objects.

- The MMCS USM objects (one for each user) communicate with the users involved in the session. Each MMCS USM object provides a generic service session interface and a service specific interface for the interaction between objects in the user domain and the MMCS SSM.

- The binding, (Figure 5-1, shaded area), connects the stream interfaces of the users to the stream interfaces of the bridge as described by the LCG.

- An instance of the CSM object provides control of the binding. Through the CSM control interface, the MMCS SSM object can alter the LCG. Operations are for instance, to add a new stream interface (user) to the

binding, or remove a specific pair of flow endpoints, or change the QoS of the current binding.

The generic service session interfaces provided by the MMCS SSM and MMCS USM are described in detail in [7]. The service specific interfaces are not defined within TINA but they are in many cases defined by consortia working on specific service classes, e.g. DAVIC [14] gives ideas of service specific interfaces that may implement the MMCS SSM and MMCS USM objects.

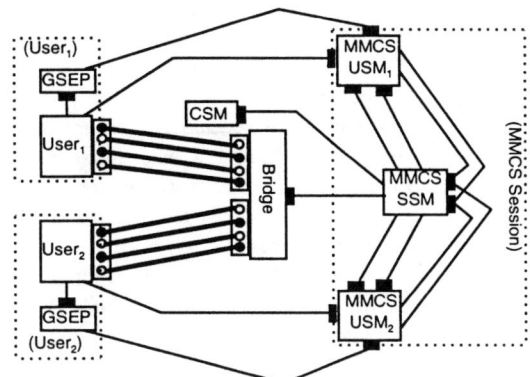

Figure 5-3 Computational model of MMCS session

5.3 Engineering model

When the CSM, TCSM and CC objects have instantiated the connections as described by the various connection graphs an engineering configuration as partly shown in Figure 5.4 is created.

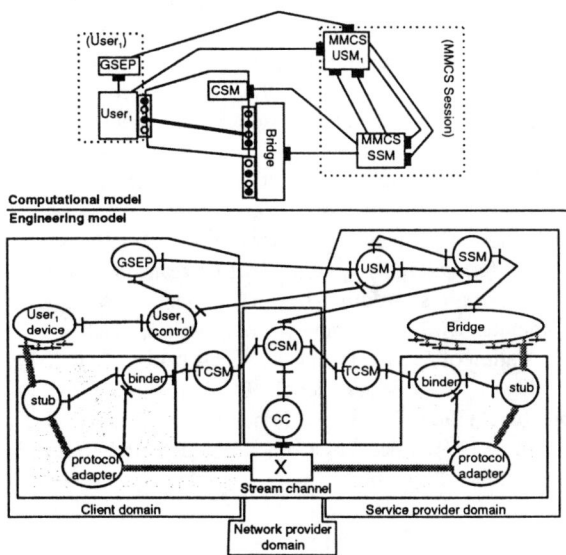

Figure 5-4 Engineering configuration of MMCS

Figure 5-4 shows how a flow endpoint in the stream interface of the user device is connected to a flow endpoint in the stream interface of the bridge through the intra-nodal part of the channel, managed by the TCSM,

185

and the inter-nodal part of the channel, managed by the CC.

6. Conclusion

The purpose of the investigation presented in this paper was to reveal any inconsistencies between the modeling concepts and terminologies in the TINA sub-architectures, and if discovered, to provide suggestions for mending these inconsistencies. Since the TINA sub-architectures are supposed to complement each other in the trajectory of specifying a TINA service, it should be possible to transform one viewpoint service specification (e.g. information specification) to another (e.g. computational specification) using the different TINA sub-architecture concepts and methodologies. These points are crucial if we expect to achieve a consistent framework for specifying multimedia services.

The material presented in this paper incorporates some changes and additions to the current TINA-C baselines, which were necessary to achieve consistency. For example, we do not use the vertex and port concepts, as described in the Resource architecture, since a clear relationship with the DPE architecture and service architecture is not found. Also these concepts are difficult to position in the various viewpoints. Instead, we changed the vertex and port concepts to the stream interface concept which allows a better alignment to the other TINA sub-architectures, as well as, to the ODP standards. Further details on the relations, and mapping between viewpoint descriptions can be found in [1].

Additional work is needed to investigate how the TINA generic service control specifications and the service specific control developed in other consortia relate, and hopefully complement each other. From TINA's perspective this requires that closer alignment with these consortia should be sought.

For multimedia services QoS and the support for continuous dataflows plays an important role. TINA provides the concepts (based on RM-ODP) to associate QoS offerings and requirements to stream interfaces and flow endpoints. Additional work is needed to achieve a complete QoS framework where issues, such as, the translation of service level QoS specifications (e.g. MPEG quality), to network level QoS requirements (e.g. bandwidth and jitter) are detailed. Initial ideas can be found in [17] which describes an ODP based framework for QoS handling.

7. Acknowledgments

The authors would like to thank Hendrik Berndt, Eunho Choi, Emmanuel Darmois, Hannu Flinck, Matthieu Goutet, Hans Hegeman, Hidetsugo Kobayashi, Harm Mulder, Natarayan Narayannan, Tom Handegård and Ajeet Parhar for their comments and encouragements during the work which lead to the results presented in this paper.

References

[1] TINA coreteam, Connection/Session Graph, Stream Interfaces and Channel Model, Document No. TR_PL.001_1.1_95, TINA Consortium, February 1996.

[2] TINA coreteam, Information Modelling Concepts, Document No. TB_EAC.001_3.0_94, TINA-C, December 1994.

[3] TINA coreteam, Computational Modelling Concepts, Document No. TB_A2.HC.012_3.0_95, TINA-C, August 1995.

[4] TINA coreteam, Service Architecture, Document No. TB_MDC.012_2.0_94, TINA-C, March, 1995.

[5] TINA coreteam, Connection Management Architecture, Document No. TB_JJB.005_1.3_94, TINA-C, January 1995.

[6] TINA coreteam, Stream Channel Model, No. TR_MFJ.001_1.3_95, TINA-C, July 1995.

[7] TINA coreteam, Service Session Control Specifications, No. TR_MJM.005_2.1_95, TINA-C, February 1996.

[8] TINA coreteam, TINA Object Definition Language (TINA-ODL) MANUAL, Document No. TR_NM.002_1.3_95, TINA-C, June 1995.

[9] TINA coreteam, TINA Distributed Processing Environment (TINA-DPE), Document No. TB_PL001_1.0_95, TINA-C, July 1995.

[10] IMA, Multimedia Systems Services - Part 2: Multimedia devices and formats, IMA Recommended Practice, Second Draft, 26 September 1994.

[11] *CORBA 2.0/Interoperability Universal Networked Objects, OMG Document No. 95.3.10, March 1995.*

[12] Y. Hoffner and B. Crawford, 'Federation and Interoperability', APM.1514.01, ANSA Phase III, October 1995.

[13] Open Distributed Processing - Reference Model, 'Part 3: Architecture' International Standard 10746-3, ITU-T Recommendation X.903, 1995.

[14] Digital Audio-Visual Council, DAVIC 1.0 specification, Part 1-9, Technical Report, January 1996.

[15] TINA coreteam, Network Resource Information Model Specification, No. TB_LR.001_2.1_95, TINA-C, December 1994.

[16] OMG, The Common Object Request Broker: Architecture and Specification, Revision 2.0, OMG Technical Document PTC/96-03-04, July 1995.

[17] P. Leydekkers, V. Gay, ODP View on Quality of Service for Open Distributed Multimedia Environments, Proceedings of the 4th International IFIP Workshop on Quality of Service (IWQoS 96), J. de Meer and A. Vogel (Eds.), Paris, March 6-8, 1996.

[18] V. Gay and P. Leydekkers, Multimedia Conferencing Services in an Open Distributed Environment, Springer-Verlag, IWACA Workshop, September 1994.

[19] XTP Forum: 'Xpress Transport Protocol Specification - XTP Revision 4.0'; XTP Forum, Santa Barbara, California, USA, March 1995.

A Configurable Protocol Architecture for CORBA Environments

Stephen Crane and Naranker Dulay
Department of Computing
Imperial College of Science, Technology and Medicine
180 Queen's Gate, London SW7 2BZ, UK
{jsc, nd}@doc.ic.ac.uk

Abstract

This paper describes a flexible architecture for building the protocols required to allow interaction between distributed objects in a CORBA environment. A key feature of the architecture is its ability to select the elements of a protocol stack dynamically at bind-time depending on the properties of the interface being accessed. This permits multiple object-invocation protocols to coexist such that a system may support local, intra-ORB and inter-ORB protocols and allows the selection of the most appropriate protocol at run-time. In addition, the architecture is capable of supporting "non-standard" interaction protocols such as multimedia streams. The paper outlines the architectural principles used and describes an efficient implementation of the CORBA Internet Inter-Orb Protocol.

Keywords*: Distributed Objects, Protocol Stacks, Interoperability, CORBA.*

1. Introduction

Modern distributed software development environments are exhibiting an increasing trend towards standardisation and interworking. Notable in this field are the efforts of ISO in its Reference Model for Open Distributed Processing, RM-ODP [1] and the Object Management Group's Common Object Request Broker Architecture, CORBA [2].

Independent of this standardisation work are the demands on distributed systems to carry an increasing variety of data, for example, to carry data with time-dependent characteristics such as multimedia streams and to inter-work with legacy systems such as SQL databases and TP monitors. This has necessitated the development of a plenitude of *ad hoc* transport sub-systems, each dedicated to the maintenance of a particular quality of service, in response to an application's temporal and legacy requirements. It is no longer possible for distributed system designers to supply a single transport which will meet all needs, the semantics of the data are too diverse.

Fundamental to the operation of a distributed system is the establishment of communication paths between its components, or *binding*. Essentially this involves the initialisation of a sender-side endpoint with the distributed address of its receiver-side peer. Since the format of the distributed address is tightly coupled both to the native format of the distributed system *and* to the transport system which recognises it, we believe that a rigorous approach to the act of binding establishment will lead to a flexible communications architecture in which both inter-working *and* the support of multiple transport protocols are elegantly supported.

The structure of the remainder of this paper is as follows. In section 2, we outline the principles which influenced the development of the architecture and the structure of the resulting communications system. Section 3 describes the CORBA inter-ORB protocol and the structure of our implementation of it. It also describes the structure of Regis' native intra-ORB protocol and demonstrates the comparative performance of Regis and a commercial ORB. Section 4 places this work in context with related work and section 5 concludes with future directions in which we intend to take this work.

2. Protocol Architecture

This section describes the salient features of a communications architecture which is capable of supporting multiple styles of binding between different types of interaction classes over multiple transport protocols. The different styles of binding are first defined; then the act of creating a binding is described; finally we present the elements of the communications architecture explicitly designed to support the model of binding.

187

2.1 Binding

We define binding as the establishment of a communication between components. The binding connects the endpoint where a service is *required* to that where it is *provided*.

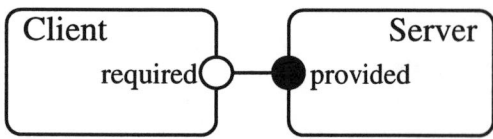

For communication to take place, the endpoints being bound must support the same *interaction style* and take on complementary roles. For example, a function and a function pointer have complementary roles but a function pointer may only be bound to a function with the same signature. Bindings are potentially *many-to-one*, one requirement may only address a single provision, but many requirements may address the same provision, thus preserving the semantics of procedure and variable name binding.

Binding occurs dynamically between endpoints with no prior knowledge of each other. It is a basic requirement of re-useable components that they have as little direct knowledge of their environment as possible. In order to support dynamic binding, there must be some mechanisms of communicating the location of a provided endpoint to where it is required. In our architecture three such mechanisms are supported: first-party binding, third-party binding and back-binding. We use the term *interface reference* to denote an object which is transmissible between address spaces and contains all of the information required to establish a binding to its *referend*.

2.1.1 First-party binding

First-party binding[1] occurs when a client component obtains an interface reference to a provided service and uses it to bind one of its required interfaces. While bindings may be established in other ways [4], in this paper we will concentrate on first-party binding as it is the only style required by the CORBA specification.

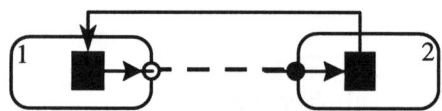

References which support first-party binding are typically typed by the signature of their referend in order to enforce type compatibility. The origin of the interface reference is not defined but the client commonly obtains it from an interface trader or as a result of a previous interaction.

2.1.2 Third-party binding

Third-party binding is performed by an entity (the third-party) which is neither the client nor the server. The third-party obtains a reference to the server's interface, binds to the client's *binding service*, and requests the latter to bind the specific required interface using the reference. Third-party binding is therefore at a conceptually higher level than first-party, since it requires first-party binding to perform its function. Examples of third-party binding are less common than first-party. Environments such as Regis [3] which require programmers to provide an explicit description of program structure use third-party binding to establish an initial component configuration. The idiom is also found in systems which allow controlled modification of their structure by external managers [4]. References involved in third-party binding are not typed by their referends; type checking is typically performed at a higher level, by the configuration language compiler or the management interpreter.

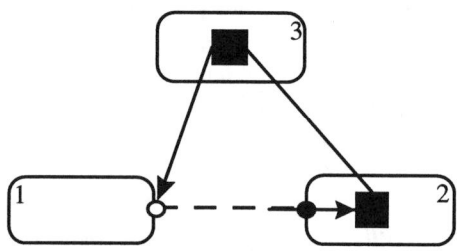

2.1.3 Back-binding

The great majority of interaction in a distributed system is two-way. A client issues a request and at some later time, a server returns one or more replies. Since our model of binding is many-to-one, and many requests can be outstanding at a server's interface, each request must store a return path or *back-binding* to the client. Conceptually this is created from a reference to a private service contained in the client's endpoint.

2.1.4 Interface References

Previous sections have described the motivations for dynamic binding and situations where they occur. This section describes how bindings are created from references.

References in general have opaque structure: it is not possible to infer anything about the information or *names* contained in a reference besides the fact that there must be one or more *contexts* in which some or all of the information makes sense. The act of binding then consists

---

[1] The first-party in a binding relationship is the client, the second-party the server, while a third-party is an external agent which knows, and is trusted by, the first and second parties.

of identifying these contexts and requesting them to create the *elements* of the binding. We name these binding elements *protocols*.

A *strongly-typed reference* is a list of names each of which, by its type, identifies a unique protocol layer in the desired communications system. Construction of a compatible binding is performed by a *factory* which supports the complete type of the reference. Strongly-typed references thus guarantee that a compatible communications sub-system will be created under the client's endpoint. However, since the structure of strongly-typed references is completely known at compile-time, the re-usability of components in which they occur is lessened.

2.2 Configurable Protocol Stacks

In our architecture, a binding is supported by a linear stack of protocols under each of the bound endpoints. Context independence, and hence re-use, is obtained by requiring each protocol to conform to an abstract interface and encouraging 'derived associations' between protocols. At the top and bottom of the stack are the endpoints and devices respectively. Endpoints provide synchronisation with user-level threads while devices manage the interface to the underlying operating system or hardware device itself.

2.2.1 Linear Protocol Stack

The figure shows a bi-directional protocol stack. The interface provided by a particular endpoint defines the *style*, or component programmer's view, of the interaction. Since most interactions are bi-directional, endpoints require a transmission interface (*Session::Client*) and provide an

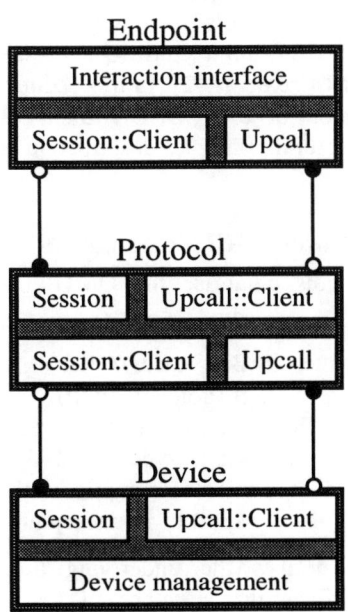

interface (*Upcall*) at which incoming data is delivered.

Protocols are intermediate objects in the stack. On their top side, they provide a transmission interface to their neighbour and require a reception interface, while on their bottom side, they require a transmission interface and provide a reception interface.

Device objects must possess a top-side interface which conforms to that expected by protocols' bottom-sides. Internally, of course, they translate invocations on this interface to device-specific operations.

In our model, protocol stacks are *configurable*: in addition to the end-to-end data path, they provide the ability to issue configuration commands to one or more protocols in the stack, query them for aspects of their state, or receive notification of exceptional conditions.

2.2.2 Protocol implementation

The class structure of the principal elements of a communications system which implements this model is shown in the OMT diagram [5] below. The core abstractions are *Session*, *Upcall* and their accessors *Session::Client* and *Upcall::Client*.

An accessor's role is to manage one or more references to services provided by the neighbouring protocol layer. It provides one operation, `bind`, permitting initialisation of its reference(s).

The `Session` class defines the interface for downward movement of data in the protocol stack. End-to-end data is transmitted using the `transmit` operation, while the protocol layer's state is modified and interrogated by a higher layer via `configure` and `query` respectively.

The `Upcall` class defines the interface for upward movement of data. End-to-end data arrives via the `deliver` operation, while warnings of exceptional conditions arrive via `notify`.

By combining pairs of these core abstractions, we can specify derived abstractions. A `Device` is a combination of a `Session` and an `Upcall` accessor. A transmit-only protocol, or `DProto`, combines a `Session` and its accessor, while a receive-only protocol, or `UProto`, is composed of an `Upcall` and *its* accessor. A normal bi-directional protocol, `Proto`, unsurprisingly combines a `UProto` and a `DProto`. Finally an `Endpoint` comprises a `Session` accessor and an `Upcall`.

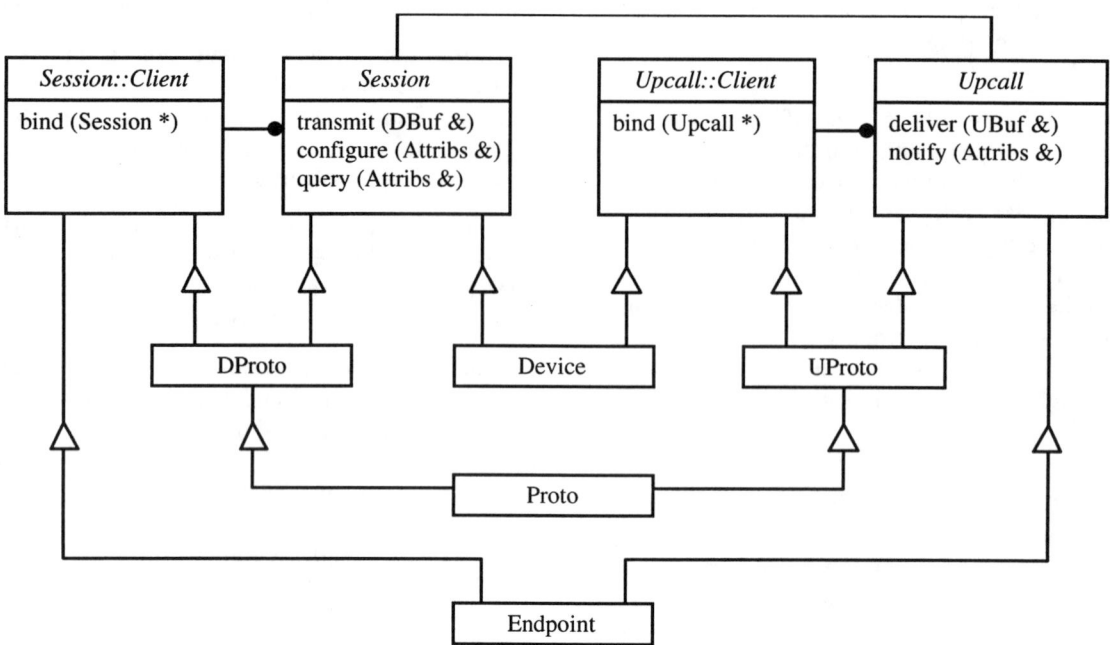

2.2.3 Dynamic Protocol Configuration

Two main principles are reflected in our implementation of protocols stacks:

- that individual protocols should be lightweight, they should not adversely affect overall communications performance.
- that stacks should be dynamic, they should allow protocols to be introduced or removed at any time during the lifetime of the connection.

The dynamic aspect means that the state of the connection is represented by the presence or absence of certain protocols rather than state variables internal to the protocol elements themselves. This has the effect of simplifying other protocol layers in the stacks and making the state of the connection externally visible without breaking the encapsulation provided by each layer.

The combination of the two principles also leads to a greater degree of code re-use since many interactions share the same state transitions; for example all interactions which support third-party binding by a configuration language have the required property that access to them will block the accessor until binding has taken place. This property is provided by a re-useable *bind protocol* element which exists until binding has taken place and is thereafter destroyed, [6]; further, the bind protocol may make configuration decisions based on the locality of the interface endpoints being bound.

2.2.4 Multiplexors (Protocol Trees)

Although the linear stack is an important building-block of the communications system, it cannot alone describe all possible communication patterns. In our model of communication, the most general pattern is the tree. Branching of the tree occurs when a 'naming boundary' is crossed at a multiplexor. Examples of multiplexing are extremely common: UDP and TCP are multiplexed over IP; the TCP layer itself maintains a set of connections each identified by a port minor number. While individual protocol layers may or may not add headers to messages, multiplexors must *always* do so.

3. CORBA Inter-Operability Protocol

The *Object Management Group* has recently formalised a set of protocols to allow implementations of its object request broker architecture to inter-operate, the so-called *Inter-ORB Protocols* [2]. In this section we describe an implementation of the CORBA Inter-ORB protocols using the architectural principles described in section 2.

3.1 CORBA Inter-ORB Protocols

Two protocols comprise the OMG inter-operability specification:

- the General Inter-ORB Protocol (GIOP)
- the Internet Inter-ORB Protocol (IIOP)

The GIOP defines: the format of IDL data on the wire, a small number of message types (and their associated headers) and makes general assumptions about the QoS characteristics of the transport layer underpinning it.

The IIOP defines the underlying transport to be TCP/IP and defines the structure of an Inter-operable

Object Reference, IOR, to consist of, among other things, a TCP/IP endpoint address and an opaque object key to identify a target object within the address space identified by the TCP/IP endpoint.

3.2 GIOP and IIOP Implementation

The diagram below shows the structure of the protocol layers comprising our implementation of the IIOP. At lower layers of the protocol stack, all I/O is non-blocking and reactive. Threading and synchronisation only become concerns at the level of Proxy and Endpoint. Our programming platform, Regis, provides non-pre-emptive user-space threads but also utilises kernel threads where provided by the operating system.

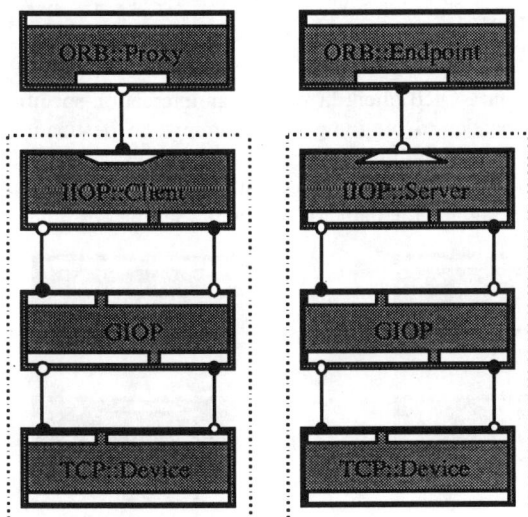

3.2.1 Layer descriptions

We describe the function of each layer in the IIOP protocol stack by tracing the path of an invocation between

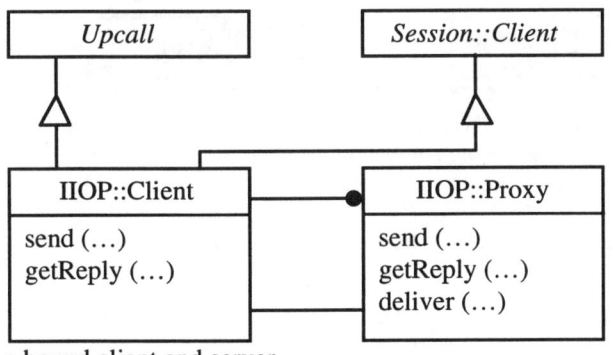

a bound client and server.

The client initiates a remote method invocation by calling the corresponding method of the IDL-generated proxy. This marshalls its arguments into a buffer and calls

the send method of the ORB Proxy class, whose main function is to store the identity of its peer.

Instances of the IIOP Client class are allocated per-connection, multiplexing requests from many proxies to the same server site. Its send method creates a 'RequestHeader' from the opaque target object identifier, the operation name and the marshalled request data. It returns a request identifier in order later to distinguish the reply. The bottom-side interface of Client conforms to the protocol stack combination of Session::Client and Upcall.

The buffer containing the marshalled data and the RequestHeader then arrives at the Session interface of the GIOP layer. This layer is almost identical at either end of the connection but parametrised by its primary role. It builds a MessageHeader, stating the message's type and size, and transmits it via the TCP layer.

On arrival of the data at the server's site, the TCP layer associated with this connection is upcalled. It reads all available data from the stream and passes it up to the GIOP layer. This is not a pure Proto layer because its Upcall interface is implemented by the Packetise class which breaks a stream of data into buffers, one per invocation. This is necessary because data belonging to an invocation might arrive in fragments, depending on the buffering provided by the kernels at either end of the connection.

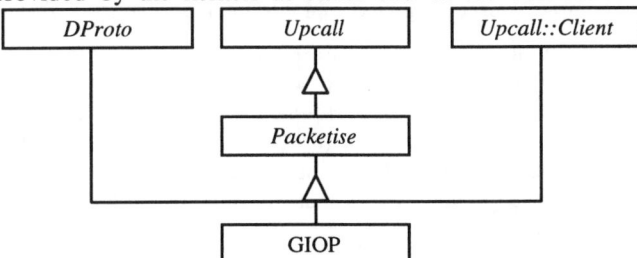

When all of the data has been assembled, the GIOP layer passes it up to the IIOP Server layer which demultiplexes the invocation to its destination Endpoint using the RequestHeader.

An Endpoint is parametrised by a Handler implementing one of several *processing strategies*:

- if the invocation is merely of the 'get attribute' type, it is performed eagerly and the reply transmitted.
- if it requires more processing time, or if the component which owns the interface wishes explicitly to drive the invocation, it is deferred.
- if greater concurrency is required, the invocation is passed to one of a pool of pre-allocated threads for processing.

However it is processed, the invocation eventually completes and a reply returned to the invoker. The Endpoint passes it to the Server layer, to whose

`Session` interface it has retained a reference providing a 'back-binding'. The `Server` adds a 'ReplyHeader' to the data buffer, identifying the request and return status and passes the buffer to the GIOP layer. This adds the MessageHeader and forwards it to the TCP layer for transmission.

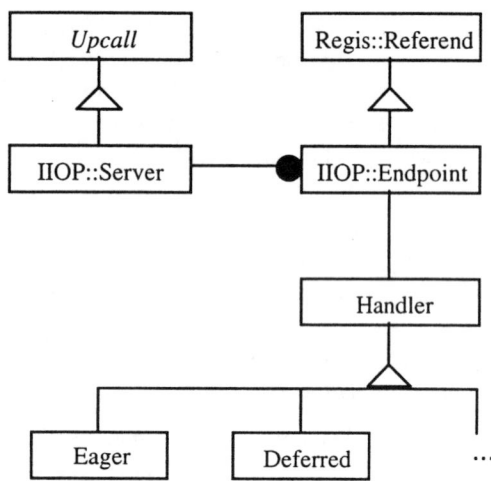

At the client side, the reply filters up through the TCP and GIOP layers to the `Client`. From the request identifier stored in the ReplyHeader, it determines which `Proxy` transmitted the original request and delivers the reply data and return status to it. The `Proxy` then makes any thread blocked on the reply runnable. When it runs, the thread directs its `Proxy` to unmarshall the reply data (or throw an exception) and the invocation is complete.

3.2.2 Binding and Stack creation

Creation of IIOP protocol stacks is managed by an `IIOP` object which is present in every address space. A server creates a reference (an `IOR`) to an interface from the interface's endpoint and its type. The TCP endpoint contained in the IOR is that of the `IIOP` instance at the server's site.

A client obtains the `IOR` and uses it to bind a proxy by asking its local `IIOP Factory` to construct (or re-use) a client-side protocol stack consisting of `Client`, `GIOP` and `TCP` layers. It returns a pointer to the `Client` element at the top of the stack.

When the first invocation between address spaces reaches the server's site, it arrives at the 'listening' TCP endpoint in the server's `IIOP` instance. This TCP sub-layer

creates a new `TCP Device` and passes it the invocation data and a pointer to *its* upcall which is in fact the IIOP `Factory`. When it receives the data buffer, the IIOP factory creates new `Server` and `GIOP` layers and binds them to the `TCP Device`, allowing the next invocation to proceed independently of it.

3.3 Regis Intra-ORB Protocol

We demonstrate the flexibility of our architecture by describing our intra-ORB protocol. This is based on UDP/IP and already supports a wide variety of interaction styles, for example message ports, Ada-style entries and event disseminators. The connectionless nature of UDP/IP results in a slightly different structure at lower levels of the protocol hierarchy. A second level of multiplexing is required now to replace the connection-oriented semantics of TCP/IP. This is supplied by the Regis multiplexor bound to the UDP device.

The intra-ORB client layer adds an interaction-specific protocol header to outgoing data, naming the destination endpoint, and providing a return path for replies. The Regis multiplexor adds a Regis protocol header comprising a byte-order tag and the name of the remote service.

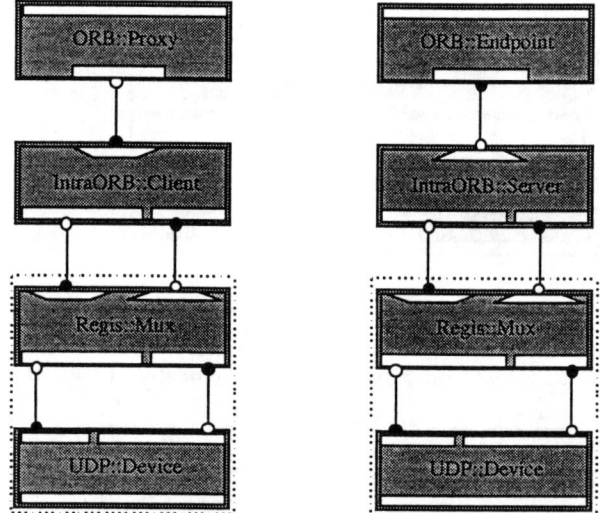

When data arrives at the server's site, the UDP device builds a back-binding containing the client's UDP endpoint, and upcalls the demultiplexor with the data. The

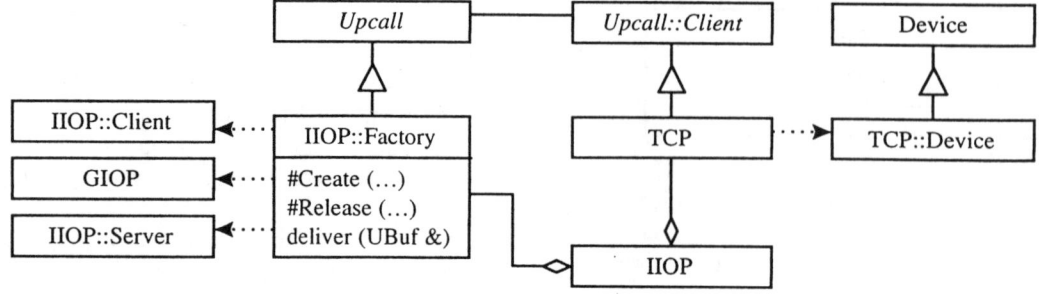

demultiplexor reads the Regis protocol header and uses it to forward the message to the intra-ORB service layer which in turn reads the intra-ORB protocol header and from it, demultiplexes the request to the destination endpoint.

This protocol further differs from the IIOP by using an integer to identify the method to invoke. Two advantages accrue from this: first, the invocation overhead is independent of the number of operations in the interface; and second, the size of the protocol header is constant, allowing efficient marshalling.

While the use of UDP/IP places an upper bound on the size of the data which may be sent between endpoints, it has proved sufficient for the majority of our applications. Furthermore, since it has a well-defined protocol interface, its behaviour may easily be modified by the addition of fragmentation and reliable message delivery layers if required.

3.4 Performance

Preliminary measurements of the performance of our implementation are very encouraging. To test the efficiency of our protocol layers, a client performed several thousand 'null' invocations on a server's interface, passing it a sequence of octets which varied in length.

3.4.1 Inter-ORB protocol

The table shows the minimum time recorded for the different data sizes passed using the IIOP, and for comparison the time taken by a commercial ORB, Orbix-2.0.1, [7], running a similar program with an identical IDL interface. The platform was a lightly-loaded Sun SS-20

Sequence length	Regis	Orbix
1	1.698	3.014
10	1.681	3.037
100	1.714	3.007
1000	1.948	3.103
10000	4.069	4.326

running Solaris 2.4. The Regis program was executed under SunOS binary emulation while the Orbix program was explicitly built for Solaris.

3.4.2 Intra-ORB protocol

The same tests as above were repeated for a program which used a proprietary intra-ORB protocol. The Regis

Sequence length	Regis	Orbix
1	1.275	1.843
10	1.288	1.880
100	1.331	1.852
1000	1.550	1.959

ORB employed the lightweight protocol described in the last section, while we believe Orbix continued to use TCP. The results are shown in the accompanying table. The sequence length was bounded in the Regis case by the maximum length of a UDP packet.

3.4.3 Discussion

One of the features of our implementation which enhances performance is the buffer class. It relies on kernel support for 'scatter-gather' I/O to minimise data copying when a header is to be added to a message. In the common case, this allows an invocation to be assembled from an ordered list of pointers to automatic data. Another contributory feature is the reactive input behaviour which requires no thread switch until the data reaches the endpoint itself and, in this example, none then.

These tests were run on the same machine for several reasons. In the inter-ORB case, nothing would have been gained by distributing the example, it is assumed that the kernel responded to each application identically, so that what is being compared is the relative performances of the two run-time systems. In the intra-ORB case, non-distribution has the effect of ironing out performance differences between UDP/IP and TCP/IP, resulting again in a direct comparison between the two run-times. In passing, it is worth noting the difference between the performance of Orbix's native protocol and its implementation of the IIOP, although they are both out-performed by our IIOP implementation.

4. Related Work

The concept of layered protocols possessing abstract interfaces originates with Ritchie [10] and has also been exploited by the X-kernel [8] and Horus [9] which feature light-weight and reusable communication protocols. Our architecture extends these systems by offering support for configurable protocol stacks and trees and by supporting multiple distributed object-invocation protocols such as those found in CORBA and in Regis.

Our architecture's reactive I/O model also employs *design patterns* [11] for abstracting the key functionality of configurable protocol stacks. Inter-stack operations, such as those supported by the 'configure', 'query' and 'notify' operations, are examples of the *chain-of-responsibility* pattern while the abstract interfaces defined by 'session' and 'upcall' are *facades*. Parameterisation of a remotely-invokable object by an 'invocation handler' is an example of the *strategy* pattern. The work is described in detail in [14]. Our model, although independently developed, is functionally equivalent to Schmidt's *Reactor* pattern [12] while our TCP layer is similar to his *Connector-Acceptor* [13].

193

5. Conclusion

In this paper we have presented an architecture for building a diverse collection of configurable object-invocation protocols for use in a CORBA environment. The architecture can also support "non-standard" protocols such as those needed for time-dependent multimedia streams and legacy access to TP monitors and databases. The abstraction layers of the architecture are design-pattern based and encourage light-weight, layered and re-usable protocol code.

The architecture needs to be enhanced in some areas. The most pressing requirement is to employ the 'short-circuit' binding pattern [14], to support efficient communication between co-located interfaces. Another desirable feature is support for third-party binding and the configuration language Darwin [3]. In a more general scheme of things, distributed bindings between objects, and our view of them, do not differ substantively from bindings between other interactions such as ports and event disseminators. In essence, we are employing the IDL compiler as a factory of configurable RPC-like interaction classes.

Our prototype implementation (in common with Orbix) does not easily allow selection of the desired protocol configuration since references to interfaces are statically typed by the desired transport, with a corresponding effect on component re-usability. Run-time selection of transport protocols will remedy this inflexibility in the future.

In conclusion, we have presented a flexible protocol architecture for CORBA environments. The architecture's flexibility has been demonstrated by its application to the CORBA IIOP. As demonstrated by the prototype's performance figures, this flexibility does not come at a cost.

6. Acknowledgements

The authors acknowledge the financial support of British Telecommunications plc. We also acknowledge the contribution of Halldor Fosså of the back-end to the IDL compiler and many stimulating discussions with our colleagues in the Distributed Software Engineering section, in particular Jeff Kramer, Jeff Magee, Nat Pryce and Morris Sloman.

7. References

1. *ODP Reference Model Part 1, Overview.* Secretariat: ISO/IEC JTC1/SC21/WG7. Standards Association of Australia, PO Box 1055, Strathfield, NSW, Australia 2135, May 1995.

2. The Object Management Group. *The Common Object Request Broker: Architecture and Specification*, Revision 2.0. OMG Headquarters, 492 Old Connecticut Path, Framington, MA 01701, July 1995.

3. J. Magee, N. Dulay and J. Kramer. A Constructive Development Environment for Parallel and Distributed Programs. In *Distributed Systems Engineering*, 1(5): 304-312, September 1994. IEE/IOP/BCS.

4. S. Crane, N. Dulay, H. Fosså, J. Kramer, M. Sloman. Configuration Management for Distributed Software Services. In *Integrated Network Management IV*, Proceedings of ISINM-95, Santa Barbara, USA. Editors A. Sethi, Y. Ranoud, F. Faure-Vincent. May 1995. Chapman and Hall.

5. J. Rumbaugh, *et al. Object-Oriented Modelling and Design.* Prentice-Hall International Inc. New Jersey, 1991. ISBN 0-13-630054-5.

6. S. Crane, J. Magee and N. Pryce. Design Patterns for Binding in Distributed Systems. Presented at the *OOPSLA-95 Workshop on Design Patterns for Concurrent, Parallel and distributed Object-Oriented Systems.* October 1995. Austin, Texas, USA.

7. IONA Technologies Ltd. *The Orbix Architecture.* January 1995. http://www.iona.ie

8. S. W. O'Malley and L. L. Peterson. A Dynamic Network Architecture. *ACM Transactions on Computer Systems*, 10(2): 110-143, May 1992.

9. R. van Renesse, T. M. Hickey and K. P. Birman. *Design and Performance of Horus: a lightweight Group Communication System.* Technical Report, Department of Computer Science, Cornell University, Ithaca, New York, 1994.

10. D. M. Ritchie. A Stream Input-Output System. *AT&T Bell Laboratories Technical Journal*, 63(8): 1897-1910, October 1984.

11. E. Gamma, R. Helm, R. Johnson and J. Vlissides. *Design Patterns: Elements of Reusable Object-Oriented Software.* Addison-Wesley, 1994. ISBN 0-201-63361-2.

12. D. C. Schmidt and T. Suda. The Service Configurator Framework: An Extensible Architecture for Dynamically Configuring Concurrent, Multi-Service Network Daemons. In *Proceedings of the Second International Conference on Configurable Distributed Systems*, March 21-23 1994, Pittsburgh Pennsylvania, USA. IEEE Computer Society.

13. D. C. Schmidt. Acceptor and Connector: Object Creational Design Patterns for Actively and Passively Initialising Network Services. Presented at *The European Pattern Language of Programs Conference*, July 10-14, 1996, Kloster Irsee, Germany.

14. N. Pryce and S. Crane. A Uniform Approach to Configuration and Communication in Distributed Systems. In *Proceedings of the Third International Conference on Configurable Distributed Systems*, May 6-8 1996, Annapolis Maryland, USA. IEEE Computer Society.

A Mechanism to Provide Interoperability between ORBs with Relocation Transparency

Nuccio M. S. Zuquello*

Teleprocessing Division
Bureau of Navy Telecommunication
20090-070 Rio de Janeiro - RJ - Brazil
E-mail: zuquello@rigel.mar.br

Edmundo R. M. Madeira

IC - Institute of Computing
UNICAMP - University of Campinas
13083-970 Campinas - SP - Brazil
E-mail: edmundo@dcc.unicamp.br

Abstract

Ability of heterogeneous systems cooperate to problem's solutions is a real necessity that has been growing, due to the increasing development of the distributed processing technology. This necessity is driving the development of more suitable solutions. CORBA (Common Object Request Broker Architecture) specification presents an architecture which allows interoperability among heterogeneous distributed environment applications. However, because of the great flexibility offered in implementation decisions, many problems to extend interoperability between two ORBs developed by different technologies arise. This paper discusses these problems and their solutions and presents a set of operations to support a bridging mechanism extending access and location transparencies, considered fundamental for a distributed environment, to work beyond an ORB scope. A scheme to provide relocation transparency, extended to support interoperability is also proposed. Finally, an implementation of these mechanisms over an ORB's prototype, developed at UNICAMP, and ORBeline, a commercial product, is described.

1 Introduction

Advances in communication technology with high transmission rates and new cooperative applications have led to growth of distributed processing. In this context, object oriented paradigm is a suitable approach for technology development which supports distributed computing and offers adequate solutions for heterogeneity, reusability and management problems. This approach is used by ISO and ITU-T in RM-ODP (Reference Model for Open Distributed Processing) [4], APM (Architecture Projects management) in ANSAware platform and OMG (Object Management Group) in CORBA (Common Object Request Broker Architecture).

The use of ORB platform as infrastructure is suitable for open distributed environments because it is simple and it has the RM-ODP main concepts. However, CORBA specification permits a high level of freedom at its implementation. This characteristic causes

interoperability problems regarding ORBs from different vendors. CORBA 2.0 partially supports interoperability.

This work describes the necessity of interoperability support, comments the CORBA 2.0 solution and presents a solution that maintains access and location transparencies; besides it proposes a mechanism to support relocation transparency between ORBs. The implemented prototype uses ORBeline [18] and an ORB developed at Unicamp [3]. This work is part of Multiware Project [5, 1, 6], where a platform based on CORBA is being developed to support distributed applications. Multiware also considers ODP concepts.

This paper is organized as following. Section 2 presents the necessity of interoperability support, comments some solutions and proposes a mechanism to implement this support. Section 3 presents a Bridge Model to allow interoperability and Section 4 describes a manner to provide relocation transparency between ORBs. An implementation of these mechanisms is presented in Section 5. Section 6 analyses related work and Section 7 presents conclusion.

2 Interoperability

[13] defines interoperability as "the ability of a client in an ORB A to invoke an operation, defined in IDL, in an object in an ORB B, where ORB A and ORB B are independently developed". CORBA 2.0 defines how the interoperability can be reached.

ORBs from different vendors have particular solutions due to the use of different mechanisms to obtain the same preconized functionalities and due to new functions defined as important to user. For instance, objref, which is a powerful concept in this architecture, is implemented with different compositions. Therefore, it would be a complex task to use an objref developed for an ORB in another ORB. Interoperability is basically related to a transparent change of **domains**, which is a scope where some characteristics and/or rules are preserved (e.g., scopes of an objref or a security policy). Interoperability will only be possible if conversions between domains are possible.

---

*This work was developped at IC - UNICAMP.

2.1 Bridging

The basic problem to reach interoperability is how an object Y, in ORB B, appears as an object X, in ORB A, so that the last ORB can use X in the same manner as if it were using any object implemented by itself. A request to X must be transformed to a request to Y. The object X is the proxy of Y in ORB A [12].

An object, called Bridge, is placed between the ORBs (with different implementations) and allows for an invocation initiated in an ORB to be received by an object in the other ORB. Interception mechanism could be: a) immediate bridging: if conversions are performed from representation of a domain directly to representation of the other domain; or b) mediated bridging: if conversions are initially performed to an intermediary representation (canonic) and then to the new representation. Considering its components location: a) in-line: if conversions are performed inside ORB; or b) request-level: if conversions are performed outside ORB. Request-level Bridge could be: a) specific: if only predetermined interfaces can be converted; or b) generic: if conversion of interfaces known at run time is possible.

This work deals with mediated generic request-level Bridges, because this mechanism is more flexible and it is not necessary to know implementation aspects of ORBs. Bridge is composed of two parts, one in each ORB, that communicate in a private manner. Bridge in the client ORB needs to obtain necessary information to retransmit the request to the object in server ORB, that is, it must dynamically know operation name and parameter types. Dynamic Skeleton Interface (DSI) is used to obtain, at run time, the implemented interface.

2.2 Bridge Functions

In order to support several functionalities claimed to permit interoperability and based on ideas in [10, 13, 14], we propose some operations.

Create_proxy - a **proxy** creation requires creation of a representative of this object on the same ORB where the client stays in order to permit that an object installed at ORB B of being accessed by a client in ORB A. Satisfying this requirement, a certain object must exist in B and be responsible for sending information necessary to proxy creation (LifeCycle, for instance, is an Object Service which defines services and conventions to create, destroy, copy and move objects [9]). When Bridge receives this information, it prepares objref mapping and puts Object Adaptor in action to register the object (its proxy) and generates an objref which is compatible to ORB where it is at this moment. An objref is associated with interface and implementation of object being referred definitions and an opaque datum (inserted by object implementation). During reference mapping, interface definition conversions are not necessary, because of the specification of an unique identifier to the interfaces and because of the maintenance of consistent Interface Repository [11]. Possibility of using Bridge implementation to substitute the one from object which a proxy is being created (suggested in [12]), also avoids

implementation description conversion. Opaque data by default are not mapped.

In fact, an objref to Bridge itself is created and it contains necessary information to permit "real" object identification, whose reference has been mapped by Bridge. For purpose of making the mapping, Bridge, which has Object Implementation properties, associates to objref (of proxy) (created by Object Adaptor of ORB where proxy is being created) an identifier corresponding to original objref on mapping table.

A simple proposal considering a Bridge between only two ORBs only demands objref passing. Half bridge invocation, on ORB A, clarifies the existence of intention to create a proxy on ORB B foregoing source and destination ORB information. Similarly, Object Adaptor does not need to be indicated. It can always be the same (probably a BOA (*Basic Object Adaptor*)), because Object Implementation will ever be Bridge, in other words, implementation style does not affect Object Adaptor choice by ORB which receives the proxy. Characteristic necessities to each implementation will be sent and attempted by Object Adaptor where the "real" object has been registered, having been chosen exactly by the fact it supports these requirements.

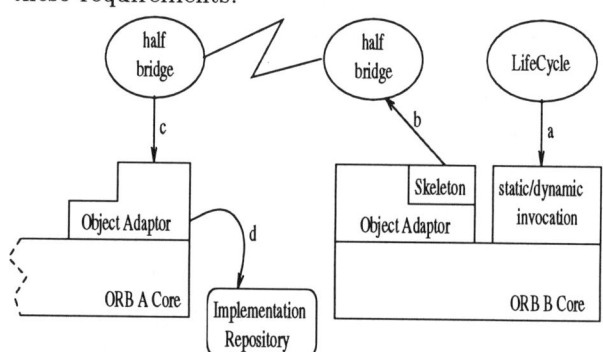

Figure 1: A proxy creation

In Figure 1, following steps are identified: **1)**Any object (on the example, LifeCycle) invokes Bridge to create a proxy of a certain object on ORB A. In spite of figure presenting possibility of static or dynamic invocation, it is supposed that this kind of invocation is mostly static considering that Bridge interface is known at run time; **2)**A skeleton makes Bridge method in action to create a proxy. Similarly to former item, most of invocations are supposed to be static; **3)**Bridge registers proxy through Object Adaptor; **4)**Object Adaptor creates an objref and introduces into Implementation Repository the implementation description (represented here by Bridge description).

In case of Bridge serving more than two ORBs, destination ORB information is necessary to permit identification of which platform agrees creating a proxy. When receiving an invocation applied to a proxy, Object Adaptor sends to dynamic skeleton (able to deal with any kind of interface) a ServerRequest, which is equivalent to Request on DSI and made of information contained on request and Interface Repository.

In order to satisfy interoperability requirements

Bridge will use the following operations, defined on CORBA or proposed at this work:

1)Defined by interface *Object* (of CORBA): get_interface and get_implementation - returns respectively an object which represents object interface definition and an object which describes object implementation where it is invoked.

2)Defined by Dynamic Invocation Interface [8]: a)create_operation_list - defined by interface ORB, returns a NVList (request parameters structured list [8]), initiated with arguments descriptions to the operation to be invoked. It receives as input parameter an operation definition which can be got from Interface Repository if the operation name and interface definition that provides it are known; b)create_request - creates a pseudo_object (a Request) to the object to be invoked. It is invoked on destination ORB with data already converted by bridge as input arguments: context, operation identifier and parameter list; c)invoke - defined by interface Request, which effectively executes invocation after request had been "mounted".

3 - Defined by bridge itself: a)create_array - operation proposed to provide mapping between objrefs. It creates an array with indexes indentical to **ids**, corresponding to respective objrefs; b)convert_ref - operation proposed to convert an objref from one ORB to another one. It receives a proxy (or "real"), obtains its **id** (used as index on an array or as input into a mapping table), searches and returns the "real" (or proxy) corresponding object. For this purpose, it can use: - get_id - operation defined by interface BOA [8], that returns an object **id** and having as input parameter its objref (in this case, proxy's); - search_objref - operation proposed that returns a reference to the object whose **id** is got, searching on a table, for example, which stores mapping between **ids** and theirs objrefs; c)convert_parameters - parameters are passed as a (*NVList*) structure by *NamedValues*, which must have their typecodes transformed into a form understood by new ORB, making it possible constructing a new *NVList*. In case of existence of return parameters (inout,out) it will be also needed its conversion to source ORB, the same fact occurs when an operation result exists; d)convert_context - context contains information about the client, environment or about its request considered inappropriate to be passed as parameters [8].

Conversion functions (Figure 2) must be able to map operation parameters (including the result) and context into a standardized intermediate form. In order to do that, Common Data Representation (CDR), which is presented in [14], can be used. In addition, in a more restricted way but potentially more efficient, a mapping specific of these data can be made directly between representations of two particular ORBs which intend to interact.

CORBA's specification, review 1.2, does not define completely some system parts, looking forward its specialization to different applications and technologies. In this way, in order to make two ORBs being able to cooperate, it should be possible conversion not only of objrefs but typecodes (used to describe an interface parameters), *principals* (contain information about clients responsible for an invocation), contexts (contain general information) and service contexts (contain information about ORB inner services) [8, 14]. First three are defined as opaque data; context does not permit access to all information on it; there is not an interface defined to *principal*; and service context, also undefined, turned out to exist only on architecture review 2.0 [20].

4 - Defined by Object Adaptor: - create_modified - Creates an objref on destination ORB (a proxy), registering it, however, as a dynamic skeleton. Receives as parameters interface and implementation definitions and also opaque datum **id**.

5 - Defined by ServerRequest object: This object is supposed to support operations that permit extraction of all necessary data (interface and operation names, parameters, etc) which will allow Bridge to construct a request on destination ORB.

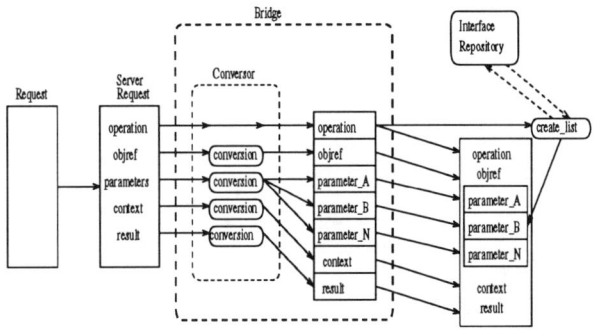

Figure 2: Mechanism of providing interoperability

6 - Modifications on Object Adaptor (OA): when OA receives an invocation on an object recognized as proxy, it directs request to a dynamic object implementation, providing a ServerRequest. Operation invocation will be responsible for putting all abovementioned operations in action which collaborate with goal of making a transmission of a request from an ORB to another and providing necessary data mapping. As input datum it receives a ServerRequest, from where all information necessary to compose a new request through destination ORB is obtained.

3 Bridge Model

Figure 3 shows the modules that may exist in a Bridge. Objref tracking function obeys the interface reference tracking function concepts presented in [4]. Its domain, otherwise, becomes one from an installed ORB, because it defines, in CORBA, objref scope. Because all references must pass through the Bridge, it is the best place to make this kind of control. Before making an object removal, Bridge can be asked in order to verify whether the object had become available in another ORB, when the proxy of this object must be also removed from the other ORB. Obviously, policies can be created to refuse an object removal when the other ORB is interested in it. Another situation that must be considered is when the reference of the object that will be removed had been passed as an argument to other ORB, creating automatically a proxy. This information can also be stored in the Bridge, indicating that there is a potencial Client, probably in the ORB that intends to move the object.

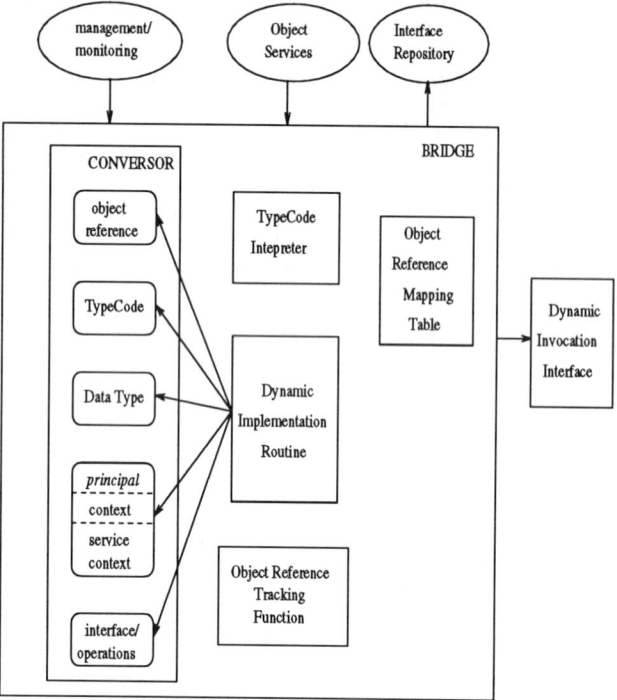

Figure 3: Bridge and its "components"

Dynamic Invocation Routine corresponds to "generic" method that receives all requests to proxies and then send theirs data to the Conversor. Typecode Interpreter is suggested in [12] and has as goal facilitating navigation on arbitrary data structures searching objrefs which should be replaced by proxies before be sent to DII on the other ORB. It is necessary when it occurs the passing of an objref as an argument on an invocation. Objrefs mapping table stores proxies and theirs identifiers in the context of Bridge. It should be also noted Object Services, which can use Bridge information and/or which Bridge itself can use: LifeCycle, Persistency, Security, etc.

4 Relocation Transparency

Through location transparency, objects services are applied without the necessity of client having any indication of their position. Relocation transparency extends this concept to objects moved between invocations [2]. ORB guarantees location and access transparency. Then, we are going to present a mechanism that permits relocation transparency, which is not completely contemplated in CORBA's specification. Latter we will extend it between ORBs. GIOP supports some object migration between ORBs, not inter-ORB, depending on ägent" at the ORBs, maybe a Half-Bridge [13].

4.1 Relocation at the same ORB scope

In this context, we introduce an object called Relocator. It is responsible for mapping between service provider object identifiers and locations to where they have been moved. This object obeys the basic concepts defined in RM-ODP [4]. Object responsible for the relocation (LifeCycle, Externalization, Installation and Activation, or any other object that turns out to be developed with this functionality) also will be responsible for updating Relocator information. Object Service Architecture (OSA) 8.1 [15] suggests the creation of a Relocation specific Object Service.

Relocator manages a table that stores moved object identifiers (objref or name, for example) and their corresponding new locations. Relocator proposed has the following operations:a)**locate** - receives as input parameter an identifier corresponding to the object that has done a movement and returns one of the parameters: **1)** the new location of the moved object, in the format used by platform to locate an object: its new IP address and file (executable or data one) path, for example; **2)**a message containing the information that the above-mentioned object is still moving, whether Relocator had been notified that the object which was initiating a relocation hasn't received its updated location; or **3)** a null value whether the object is not stored in the Relocator;b)**insert_beginning_movement** - receives as input parameter the identifier of the object which has initiated the movement. Firstly, its applies **locate** in order to verify whether there is a new location corresponding to this object. This situation can occur because of successive movements in the context of a single ORB, thus only the existing location should be replaced by a movement beginning signal. Operation **locate** returning a null value means that an object can be inserted jointly with movement beginning signal; **3)insert_new_location** - receives as input parameters a moved object identifier and its actual location and then executes a search using this identifier to introduce a new location of the above-mentioned object; and **4)remove** - receives as input parameter a moved object identifier taking it away from Relocator together with its corresponding information. This function can be used, for example, at the updating of Implementation Repository or of information contained in any device used by ORB to locate the object (Name Service, Trading, etc). It must be performed obeying a coherent policy with the necessity which had initially caused relocation of objects.

For purpose of being moved an object must "collaborate". In this case, it means that it is supposed to inherit LifeCycle interface [15]. For purpose of supporting relocation transparency we propose extending this object functionality which will be responsible for putting the Relocator in action. Its interface provides an operation that moves an object to the context of a **factory locator**. Factory is defined as an object with the ability to create another object, providing the client operations needed to create and initiate new instances. Creating a new object, a client must own an objref to a factory which attends moving object characteristics. Either implementation and definition of factory interfaces are considered as a part of the application development [9].

Following operations are supposed to be applied in Relocator through the

LifeCycle: **a)insert_beginning_movement** - using as input parameter the identifier of the object that will be moved; **b) insert_new_location** - new location can be extracted from the location of the factory that will be used at the object movement.

4.2 Relocation between ORBs with different implementations

Supporting relocation transparency, the following extensions are necessary:

a) Object which has made relocation - must send jointly with relocation necessary information a key which identifies unique and trustly the object being moved (arrow **a**, Figure 4). This identifier does not need to be interpreted by the destination ORB, only the ORB which has caused it is worried about, thus it can be sent as an opaque datum. Guaranteeing type integrity, key must contain the interface global identifier [11] and, whether there is more than one ORB, an ORB identifier (its name or any identifier registered at OMG, for example) can be added. In despite of being omitted in Figure 4, this information also should intercepted and the references to the factories should already be available to ORB A, in this case. This key is sent to Bridge (arrow **b**, Figure 4), where it will be stored for future verification, during the objects mapping. This key is also introduced in the Relocator (arrow **c**, Figure 4), where it will be interpreted as a signal representing that the object to which it corresponds is in movement process.

When it is moved, it can be considered that the object is being installed on the new ORB, thus it must be registered through the Object Adaptor in order to becoming available to be used by the system and in order to generate a new objref which is compatible to the new platform. Providing the migrating object becoming available to its source ORB is responsibility of the object which has made relocation. In so doing, the Object which made relocation must be able to provide resources to a **Bridge** invocation, as of movement destination ORB, making possible mapping between references from objects of platforms under discussion. Key is supposed to be sent together with reference.

We can aggregate this functionality to the object **factory** (LifeCycle), which has duty of creating the object being moved on its new ORB. Objref due to this creation together with the key is sent to Bridge (through create_proxy operation). Key must be inserted in objref data information.

Case of crossing administrations domains it will be necessary a formerly agreement between Administrators.

b) Bridge - being invoked to provide mapping between references of objects used by an ORB and another, Bridge verifies if there is a key aggregated to the reference to be mapped, what indicates that the object being presented, in fact, has been sometime owned by ORB itself to which it is having its reference converted. Noting this fact, Bridge may:

1 - introduce in Relocator together with objref indicated by the key the Bridge location itself added to a signal showing it is a proxy. It can be done in this way, because Bridge works like a local representative of the

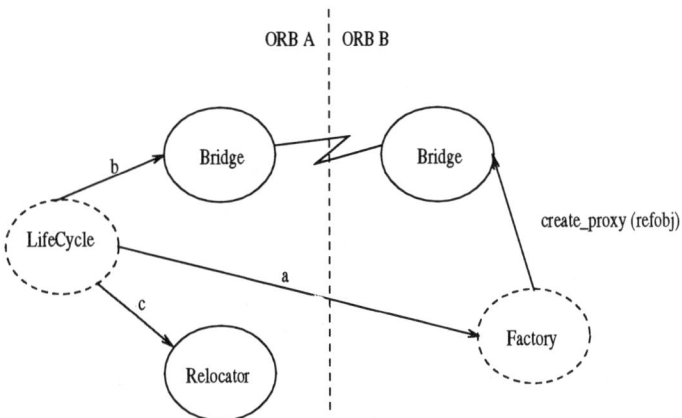

Figure 4: Mechanism for supporting relocation transparency

remote object to which it repasses messages referring to an invocation from which it receives responses, if there exist, delivering them to the client (section 2.2). Operation **insert_new_objref** is responsible for this insertion.

Thus request sending schema remains the same, but invocation is switched to respective Bridge and, when arriving at Object Adaptor, it receives, yet on client ORB, treatment of a proxy.

2 - using objref contained in Relocator to change object implementation definition indicated by this reference by Bridge itself definition. This can be done using Object Adaptor operation, defined from CORBA, **change_implementation**, which can generally be called by Bridge. This procedure is more generic because it can use the same reference resolution mechanism adopted by the ORB being used. Another difference is that the functionality of operation create_proxy, when detecting as relocated object through presence of the key, does not create another object (that would be a proxy), instead of it, modifies the definition of implementation associated to objref. Bridge must then make Relocator to know that its objref remains valid. In this way it is necessary a new reference resolution to make new request. In the same way, request is sent to Bridge.

c) Relocator - yonder functions already described it must support the following operations:**1) insert_new_objref** - used to insert object identifiers. It is put in action by Bridge, which is the object that "knows" which objects should have their new locations inserted in Relocator; **2) substitute_objref** - exchanges the objref that was used to invocation and another one, provided by Relocator, representing a proxy of moved object. It will be able to access request which is being sent towards executing this modification and initiating the "mounting" of a **Server Requisition** [13].

d) Relationship Object - existency of an object that makes reference to other ones is a common situation in a distributed environment. Thus when considering objects movement to other ORBs it arises preoccupation with the consequencies of this fact, because a service provider can use services of other objects, which remains at source ORB and, thus, are inaccessible to the object now dislocated to other ORB.This object manages a repository which contains a reference object of every object of an ORB as well as a report of all objects each one refers. It may support the following operations: **a) insert_object** - inserts an object with respectives referred objects;**b) insert_ref** - inserts new references to an existing object (in case of any modification in the object); **c) remove_object** - takes object and its references away from table; **d) search_ref** - searches object and returns all its references; and **e) remove_ref** - takes away only determinated references of a specific object. Applied when some references used by object do not exist any more, what happens when the object does not use all referred objects and some of them have functionality aggregated to other ones.

When determined object is moved Relocator verifies, at the Relationship object, which will be its references and thus send them to Bridge. Bridge is supposed to create a proxy to each referred object (through operation **create_proxy**). With this procedure, these objects can be invoked by the moved object as if all of them were part of the same ORB.

This work used for the implementation: ORBeline [18] (Sun C++ compiler) from PostModern and an ORB prototype developed at Unicamp [3] (Free Software Foundation C++ compiler and Sun RPC). Today, most of the ORB products are based on RPC [7, 17].

ORB prototype only has the basic functionalities of the CORBA specification. Dynamic Invocation Interface (DII), some features of the Interface Repository and some IDL resources, like exceptions and contexts, were not implemented. Basic framework is shown in Figure 5 in full lines. Dashed objects were added to support interoperability and relocation transparency. Bridge is an object implementation.

Portmapper receives as parameters class identifier and server location and it identifies the port in which the server is associated. Implementation Repository is used to locate a server. Parameter is the object class. This repository is implemented as a file, that contains a record for each server composed of the class identifiers and IP address of the machine. Class is the service interface. Object Repository is a list that mapping objref and C++ pointers. Object Adaptor is called by the implementation due to create an objref for an object. Receiving a call to a service, OA uses objref in the request to identify the solicited object and execute the suitable method calling. RPC Stub prepares parameters to be used by the RPC support routine, which converts the parameters using the XDR library and which transmits them. In the server side, parameters are converted from the XDR format to another one understood by the RPC skeleton.

Each method of the class that is being implemented is related to a stub and a skeleton, generated in C++, which are integrated to the client and server processes, respectively. Objref was implemented as a structure containing the IP address, a service identifier and an identifier of the object of the class that implements it.

ORBeline, version 1.0, implements CORBA review

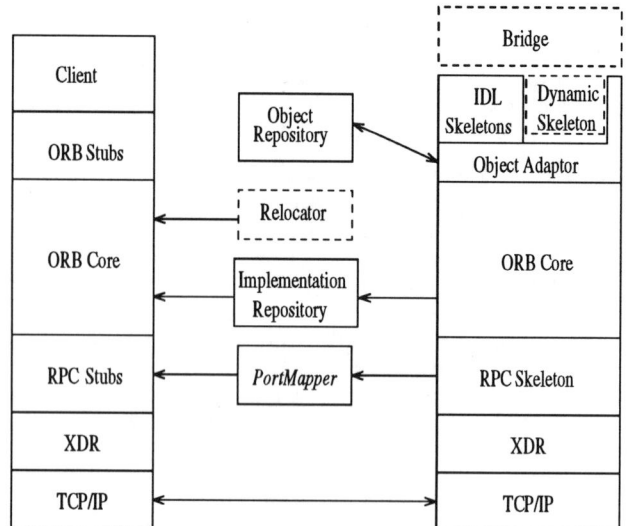

Figure 5: System Architecture

1.1, with all its functionalities.

Our implementation provides, with some constrains: **a)**relocation transparency in the prototype of [3];**b)**invocation, transparently, initiating at the prototype of services that have been registered at ORBeline and put available to the prototype;**c)**relocation transparency of an object that has been moved from the prototype to ORBeline; and **d)**- access and location transparencies were extended to services at the prototype and put available for ORBeline.

4.3 Relocation Transparency Implementation

Relocator was implemented as an object that could access a file containing: an objref (it means, hostname of object originally, object interface identifier, object identifier) and hostname to where object has been moved. When a query is not necessary, a verification uses movement origin hostname. All requests are sent to Object Adaptor, which has been implemented together with skeletons and Bridge at the same process. During an invocation, the following steps are obeyed:**a)**Firstly the location of a server which provides the desired interface is searched on implementation repository; **b)**once found this location, it is verified departing from object identifier if this object is not a proxy. In case of "real" object, an invocation to respective BOA is made. This BOA determines the skeleton to which invocation will be sent. Otherwise an invocation is made normally to BOA which delivers the request to dynamic skeleton which makes its "unpacking" extracting necessary information to repass this invocation to the other ORB; **c)**when trying a connection to service provider object (here called **server**) and server is not responding a calling, a query is made to Relocator in order to seek object new location, whether it has, in fact, been moved and not only been failed;**d)**once found this new location, the procedure is like the one above described, verifying if object is a proxy or not, and if it is acting correctly.

Object movement is not at this work scope and it is realized through the destruction of the process which provides the solicited service, at location where it is being invoked, followed by its instanciation at another location. During instanciation process, object implementation itself will register the object on new ORB, through Object Adaptor, and on Bridge, making it able to be accessed by source ORB. Relocator functions were implemented in accordance to section 4.

4.4 Bridge Implementation

Bridge is made of two distinct parts: one as Object Implementation in [3] prototype called ServerBridge and the other one as a client in ORBeline called Client-Bridge. Information transmission between these half bridges can be made in various ways [12], such as sockets, TLI, RPC or even a third ORB. Our implementation uses sockets. Operations defined to Bridge in section 2.2 were implemented, excepting parameters and context conversions.

On Client ORB, Bridge is implemented together with Object Adaptor and dynamic skeleton. When receiving an invocation, Adaptor using the key detects if it is dealing with a proxy and then puts dynamic skeleton in action. Dynamic skeleton will extract information of coming request and will provide them to Bridge. Because base prototype used does not implement Interface Repository neither DII, request was implemented as a structure where typecodes of each parameter are sent and from this structure are extracted the arguments to be converted and sent to ORB B Bridge. In the same way, it was in fact implemented only conversion to objrefs.

On Server ORB, Bridge, after receiving data sent by Bridge on ORB A, and like any ORB client, must ask Interface Repository to extract necessary data to dynamically make an invocation. Firstly Client-Bridge queries a table containing key received from Server-Bridge and its objref. Therefore, client-Bridge will be responsible for string conversion into an objref (through _string_to_object ORB operation). This operation returns an object of ultimate class (Object) which will be used as parameter for dynamic request creating operation. Objref, operation name and Interface Repository (containing the desired interface) name will be used as input parameters to this operation. Dynamic creation operation, at ORBeline, makes applications programmers work easier, querying automatically an Interface Repository.

Objref is used to get the interface name, which will be used as a search parameter in the Interface Repository. Furthermore, this operation already constructs a template, initializing the interface name, operation identification, mode (in, out, inout) and typecode of each parameter. Thence, programmer must fill out respectives parameters values (only with mode in or out). The values of received parameters, after conversion into ORBeline format, are introduced into the template. Though, in order to make it possible each parameter must be initialized into a class, specific for each type, ultimate class Value derived. This class is a generic form for data representation. For each data type there is a Value derived class that has data access operations. Proper instantiations are constructed through typecodes. Through a factory, that receives a typecode as a parameter (by reference or pointer) and returns a pointer to an object of Value class, for a specified type, already allocating necessary storage size.

Possessing typecodes, retrieved from Interface Repository, we can identify the parameter type and then use received value of the client-Bridge and converted into ORBeline form, to initialize an class Value instance. Encapsulated in this way the received value can be introduced into the request. For instance, from a request, we can get a pointer to a derived Value class, constructed for a type specified by "parameter_name". For each basic data type there is an specific operation to insert its value into respective classes. Also, it would be possible by getting typecode of "parameter_name" (through arg_type operation), to be used to construct a Value to the specified type (factory(CORBA::TypeCode& typecode)). The access operation to be used can be found through kind() operation. Thence the request is "mounted", it can be sent to the service provider object, obeying all dynamic invocation normal proceeddings.

5 Related Work

Recently OMG has made a great effort to completely define solutions obtained to provide interoperability between ORBs. This effort resulted in UNO (Universal Networked Objects) specification [14], that defines GIOP (General Inter-ORB Protocol) to support interoperability at protocol level. GIOP specifies a common transfer syntax (CDR) and standardizes the messages exchanged between ORBs. As a specialization of this specification appeared IIOP (Internet IOP) that is TCP/IP transport mapping of a GIOP. OMG interoperability architecture also defines ESIOPs (Environment Specific IOPs) that are specific protocols for a determined environment. Up to now, only DCE/ESIOP is defined. UNO also preconizes the necessity of a bridging mechanism between ORBs but don't specify them. CORBA 2.0 specification, that basically aggregated UNO definitions, has some modifications by [16].

[20] describes a framework to provide a bridging mechanism that follows OMG concepts and provides interoperability between Orbix from IONA Technologies and DOME from Object Oriented Technologies.

6 Conclusion

This paper has presented the description of a mechanism to provide interoperability between ORBs implemented by differents technologies and a way of providing support to relocation transparency, extended to objects moved across ORB's boundaries. Yet a way of implementing those mechanisms, using a commercial product and a prototype implemented over RPC has been described. Some conclusions of this work are: a)presented scheme extends naturally location and access transparencies to interoperating ORBs context. It can mainly be reached due to objref characteristics maintenance, locally representing service provider object (which is local or remote to client) and allowing, together with stubs and skeletons, an identical mechanism for both cases; **b)**It is possible to map RM-ODP Relocator funcionalities into a similar object of CORBA; **c)**in spite of Relocator becomes a potential bottleneck, too many movements between interoperating ORBs are not supposed to occur, neither do them on a single scope; **d)**mapping of requests and parameters between ORBs, or from an ORB into an intermediate form, depends completely on the ORBs implementation. So, it seems to be a good idea implementing it as a separate module out of the Bridge [20]. Conversion of context and principal (a client identifier), though simple, requires a high level agreement (e.g., between administrators) to guarantee the right semantic; **e)**mapping of objrefs is essential to interoperability and can be reached by a mechanism not so complex, since we can guarantee Interface Repositories consistency;and **f)**a browser embedded in ORBeline DII and the structure of Value class has aided implementation of Bridge.

Due to great flexibility provided for a generic bridge its complexity is very increased resulting in performance dropping, because of each invocation results in various queries (and invocations) by ORB and by Bridge, similarly to what happens with DII [19]. We can reduced this dropping using a specific bridge, though services provided are constrained to those previously "known". It's necessary an analysis of each situation to make the best option, noting actual and future necessities of the ORBs in question. Thinking in this way, this work could be extended to describe an hybrid bridge, initially specific but extensible (due evolution of the goals of working together ORBs) to handle run time known interfaces. This extension would be flexible to allow a dynamic or static invocation initiating at ORB A and, not depending on this decision, a dynamic or static invocation at a final ORB B.

Acknowledgments

This work was in part supported by CNPq (project PROTEM GEOTEC 680061-94-0) and FAPESP (Grant 92/3507-0).

References

[1] F.Costa and E.Madeira. *Distributed Platforms: An Object Group Model and its Implementation to Support Cooperative Applications on CORBA.* Chapman & Hall, pp. 213-229, March 1996.

[2] N. Davies, G.S. Blair and J.A. Mariani. Supporting persistent re-locateble objects in the ANSA architecture. Technical report, Computing Department, Lancaster University, 1992. MPG-92-04.

[3] A. M. V. de Mello, An Object Request Broker Prototype. Master's thesis, University of Campinas - FEE - (in Portuguese), 1995.

[4] ISO/IEC — ITU-T. *Draft Recommendation X.903: ODP Reference Model - Part 3: Architecture*, February 1995.

[5] L. Lima Jr. and E. Madeira. *Open Distributed Processing: Experiences with Distributed Environment - A Model for a Federative Trader.* Chapman & Hall, pp. 173-184, 1995.

[6] F. Lima and E. Madeira. ODP-based QoS for Multiware Platform. In *Proceedings of 4th International Workshop on Quality Service (IFIP IWQoS'96) - Paris-France*, pp. 45-54, March 1996.

[7] J.R. Nicol and C.T. Wilkes and F.A. Manola. Object Orientation in Heterogeneous Distributed Computing Systems. *IEEE Computer*, pages 57-67, June 1993.

[8] OMG. *Common Object Request Broker: Architecture and Specification*, December 1994. OMG TC Document 93.12.29.

[9] OMG. *Common Object Services Specification,vol. 1*, March 1994. OMG TC Document 94.1.3.

[10] OMG. *OMG ORB2.0 Interoperability and Initialisation RFP Response*, March 1994. BNR Submission - OMG TC Document 94.3.4.

[11] OMG. *OMG RFP Submission - Interface Repository*, November 1994. OMG TC Document 94.11.7.

[12] OMG. *ORB 2.0 RFP Submission - ORB Interoperability*, March 1994. OMG 94.3.1.

[13] OMG. *ORB 2.0 RFP Submission - Universal Networked Objects*, September 1994. OMG 94.9.32.

[14] OMG. CORBA2.0/ Interoperability - Universal Networked Objects, March 1995. OMG TC Document 95.3.xx[REVISED 1.8jm].

[15] OMG. *Object Services Architecture*, January 1995. OMG TC Document 95.1.47 rev.8.1.

[16] OMG. *CORBA Interoperability Revision*, May 1996. OMG TC Document 96.5.1.

[17] R. Orfali and D. Harkey. Client/Server with Distributed Objects. *Byte*, pages 151-162, April 1995.

[18] PostModern Computing Technologies, Inc. *ORBeline User's Guide*,1994.

[19] D. C. Schmidt and S. Vinoski. Object Interconnections - Comparing Alternative Client-side Distributed Programming Techniques. *C++ Report*, 7(4), May 1995.

[20] M. Steinder, A. Uszok and G.K. Zieliński. *Distributed Platforms - A Framework for Inter-ORB Request Level Bridge Construction.* Chapman & Hall, pp. 86-89, first edition, February 1996.

Session 3B

Interoperability and Workflow Management

Chair

Randy Chow

CodAlf: A Decentralized Workflow Management System on Top of OSF DCE and DC++

Alexander Schill[a] and Christian Mittasch[b]

[a]Fakultät Informatik, TU Dresden D-01062 Dresden, Germany, schill@ibdr.inf.tu-dresden.de
[b]Institut für Informatik, TU Bergakademie Freiberg D-09596 Freiberg, Germany,
chris@informatik.tu-freiberg.de

Abstract:

This paper describes a new workflow management system (WfMS) named CodAlf. It is based on a completely decentralized system architecture. The approach especially enables the support of distributed organizations as well as dynamic growth and reconfiguration. CodAlf supports collaborative work and decentralization with the help of a comfortable type management. The implementation uses the OSF Distributed Computing Environment (DCE) as a basis. All components are modeled and implemented as objects of a distributed system on top of DC++, an object-oriented extension of DCE. In addition to runtime support, a workflow specification language and a workflow design tool are offered. The system is also compared to commercially available WfMS and to decentralized object-oriented approaches.

1 Introduction

Workflows control business processes. Workflows consist of a number of tasks to be performed periodically by several participants within organizations. While workflows can be executed manually, partial automation can enhance productivity and can enable the use of computerized tools for the various tasks. A workflow management system (WfMS) supports this goal: it coordinates the user and system participants, together with the appropriate data resources to achieve defined objectives by set deadlines. The coordination involves passing tasks from participant to participant in correct sequence, ensuring that all fulfill their required contributions [7]. It lays in the nature of WfMS, that its success depends on a matching presentation of both, the organizational resources and business processes, as a basic expression of collaborative distributed and frequently repeated work.

A lot of earlier research projects worked on this problem; examples are Exotica [10], Meteor2 [13], EuroCoop [5], IPSO [16] and Mentor [20]. These systems resulted in basic prototypes but hardly provided language and tool support. More recently, several workflow management products emerged; some examples are COSA [1], FlowMark [2], WorkParty [19] and Prominand [15]. These systems support comfortable workflow specification techniques and enable the integration of existing standard applications such as text processing or spreadsheets. However, they typically use a centralized database approach for implementing workflow management at runtime. This leads to well-known problems of single points of failure, to potential performance bottlenecks, and also to a lack of scalability and dynamic reconfigurability. Most systems also depend on proprietary communication protocols.

With most existing approaches, control aspects for workflow management users (e.g. employees) are not sufficiently supported. There is a lack of control by the system itself: Most of the WfMS are only able to distinguish if a task is finished successfully or not. They are not able to make out details, like "How long does it probably take the employee to finish the task", etc. And, on the other hand, users have to be supported by a comprehensive user-interaction-server. Furthermore, the applications of WfMS can be extended to more than one department of an enterprise by the exclusive use of a trading instance [11] or by co-operating instances of WfMS. Third, the application of such WfMS improves step by step the understanding of collaboration and of information flow within the organization. Thus, a general object model of the overall enterprise results considering workflows and resources in a balanced manner [17]. In the meantime, standardization efforts have been initiated and are conducted by the Workflow Management Coalition (WfMC) [18], a consortium of hardware vendors, software developers and end users. This organization has proposed a general workflow system architecture that consists of process definition tools, administration and monitoring tools, workflow enactment services (runtime services) and application software. Moreover, a workflow specification language is being standardized. However, the standardization is still under discussion and has not been completed yet. Conceptual work has been done, also considering a framework approach to WfM [8]. Altogether, we come to the conclusion that WfM in environments of limited scope is already well-supported by today's products, but that future applications in larger, widely distributed settings require more advanced, decentralized solutions. Significant efforts are still necessary to integrate the benefits of both areas, i.e. to integrate distribution, transactions, legacy applications, security and additional features.

Based on this background, this paper introduces CodAlf („**Code** Name **Alfa**"), a new, decentralized system towards distributed workflow management. As opposed to existing solutions, this especially enables the support of distributed organizations as well as dynamic growth and scalability. The basic concepts together with an application scenario are described in section 2. Section 3 details the design and implementation of CodAlf with its architectural components, its workflow specification language, its workflow design approach and its decentralized workflow control support. Section 4 presents experiences with this approach and with its applications. We also compare the system with other solutions in more detail, particularly with our CORBA-based very recent development BPAFrame („Framework

for Business Process Automation"). Finally, section 5 concludes with an outlook to future work.

2 CodAlf: Foundations of the WfMS and the use of an open platform

2.1 Impacts of Application Scenarios

A typical compounded workflow application scenario of the environmental administration area is shown in fig.1. Before an extensive construction on a building site can begin, an analysis concerning environmental drags has to be performed. It starts with an official inquiry. This leads to an analysis of suspicious facts by a dedicated department. Depending on the result, a conditional in-depth analysis concerning potential ground pollution, air pollution, and water pollution is being performed by specialized engineers. These tasks can be performed in parallel. Not all tasks have to be performed in any case; their execution depends on special context information of the given project. Each in-depth analysis can consult an existing database about environmental drags. Typically, additional measurements at the very location are performed. In special cases, simulations of expected future pollution for special industrial areas are another option of advance analysis. These three tasks can again be executed in parallel and are mutually optional.

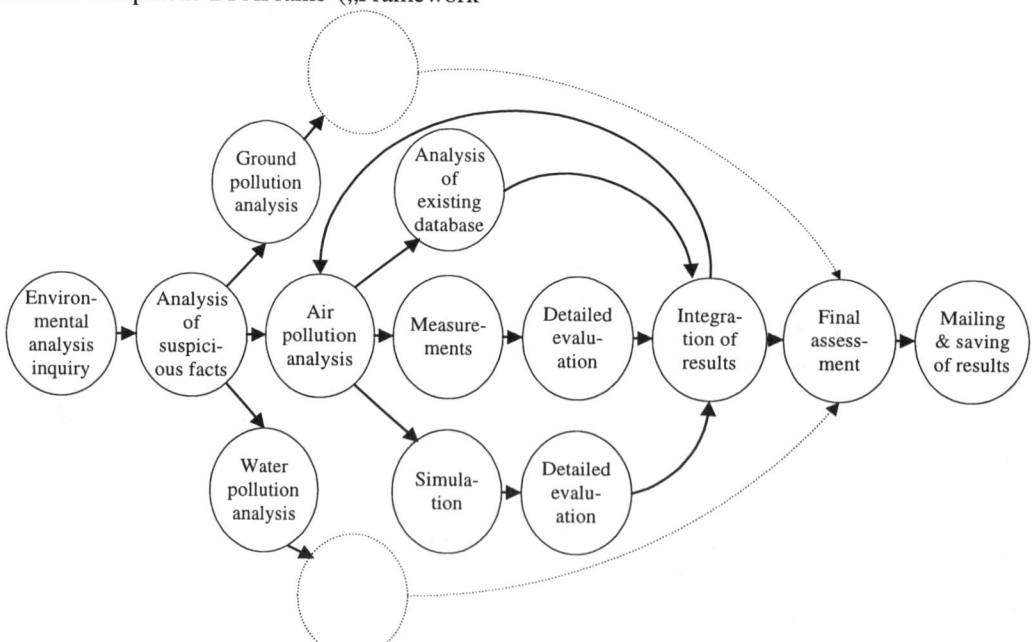

Figure 1: Workflow application scenario (Compounded Workflow)

A detailed evaluation of measurements and of simulation results follows (within sub-workflows).

Thereafter, a synchronization of the various tasks is required in order to collect and integrate the results of one category of analysis (air, ground or water). If the results are still too vague, the analysis has to be repeated in more detail, for example by comparison with similar situations in other cities. Finally, a global synchro-nization and a final assessment are performed and results are mailed to the client and stored. While the air pollution analysis is shown in more detail in the figure, the other two categories are modeled as complex tasks that are refined separately by another team of the department. Basically, they look similar so that the details are not shown here. From this example, several basic characteristics and requirements of workflows can be abstracted:

- *Workflow structure:* Workflows typically have a well-defined structure (sequences and time dependencies) and occur repeatedly in practice. The tasks can be executed sequentially (e.g. mailing of results after final assessment) or in parallel (e.g. analysis of database, measurements and simulation). Execution can also be conditional (e.g. simulation of pollution instead of detailed measurements), and tasks can be optional. Moreover, loops are possible (e.g. for repeating the air pollution analysis). The workflow structure is typically given by a directed *execution graph* (see section 3.1).

- *Workflow types:* Because many workflows represent collections of routine tasks, they are potentially instantiated many times. Therefore, it is important to make a distinction between workflow types and associated instances. This enables reuse of existing workflow types (mainly a collection of basic workflow types) during instantiation and also during workflow specification: An existing workflow type with its execution graph can be embedded into a new, more complex workflow. Workflow instances have a specific owner (initiator) with general access rights; access of others is usually more restricted. Specialized workflows represent the real business of the organization, then.

- *Roles:* The tasks of the given execution graph have to be mapped to actual people, computer systems, tools and applications, i.e. to *execution instances*. For example, the final assessment of results may be performed by an employee using a spreadsheet program. It is important to enable a flexible mapping from tasks to execution instances; for example, several people may be able to perform an air pollution analysis. Therefore, *roles* are introduced: A task is then mapped to a role hiding employees as resources.

Based on these assumptions, a formal workflow specification technique and execution model can be introduced as described in section 3. However, additional design goals concerning the implementation of a distributed workflow management system have to be discussed first.

2.2 Distributed Object-Oriented Workflow Management Systems: Design Goals

Distributed execution of workflows is required due to the physical distribution of execution instances in typical organizations. As a consequence, it is realized by a decentralized architecture of the WfMS. Additionally, a dynamic mapping of tasks and associated roles onto execution instances must be supported. This way, cost optimizations and means for achieving fault tolerance are enabled. As a basis, tool support such as a trader [11] and a type-manager must be part of the WfMS. The decentralized workflow management system must be extendible. It has to be possible to integrate new execution instances at runtime. To achieve this, at least a dynamically extensible directory service is required. In addition, different kinds of application services of execution instances - such as word processing or spreadsheet calculation - should be integrateable via a uniform interface.

We use DC++, a distributed object-oriented system on top of DCE. It provides a C++ class hierarchy and runtime system for location independent remote object invocation and object mobility. Objects have globally unique identifiers, and invocations are forwarded to the target location by a forward addressing scheme. Object migrations are performed upon explicit request, for example to collocate cooperating objects. A migration includes the transfer of all object instance data, the transparent update of inter-object references, and the reinstallation of the object at the target site. Details are found in [3]. An object-oriented basic architecture can facilitate the implementation of workflow management: Workflows can themselves be modeled as (mobile) objects, execution instances can naturally be implemented as objects, too, and inheritance and encapsulation support reuse and structural clarity.

Based on these considerations, we decided to take two directions: (1) We designed and implemented a workflow management system named CodAlf on top of DC++/DCE. This work has largely been completed. We therefore describe this approach in detail in this paper. (2) Recently, we also decided to use CORBA as an alternative basis for implementing a workflow management system named BPAframe [12]. In spite of the limitations, this has been proven to be viable.

3 Design and Implementation of CodAlf

3.1 User Services for the Design of Workflows

Complex applications require a structured description in event lists, graphs, etc., representing control flow and structured tasks.

Our own approach is a Petri-Net-based workflow description language. In fig. 2 a workflow execution graph is specified containing various tasks (nodes) and their interaction structure (edges), including a specification of quality requirements at nodes and edges. It describes the mentioned structural elements, (sequences, parallel edges and alternative edges, loops and sub-workflows).

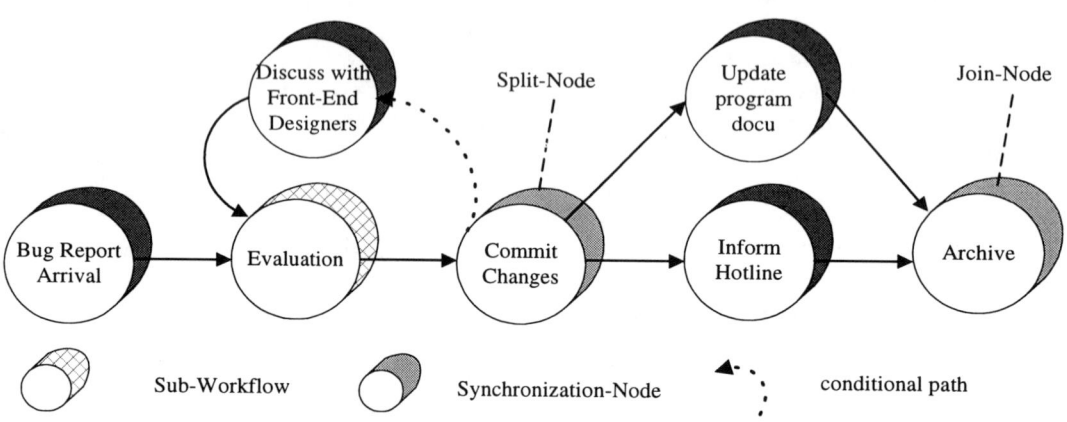

Figure 2: Example of a graphical workflow description

The associated linguistic workflow description consists of two parts: First, the execution graph with general attributes is given. Second, the detailed description of all task nodes is required. The definition of an individual task node contains the following information: The specification of the service with requested quality parameters, the kind of server allocation, with optional resource reservation requests, the description of the subsequent edges (communication services) with requested QoS-parameters of communication links (e.g. transmission rate), control mechanisms (e.g. parallel paths or the condition-dependent splitting of the workflow with subsequent synchronization points in both cases), determination of the behavior in the case of errors (currently, rollback to the last synchronization-point or restart is supported) and a privilege ticket of the workflow initiator (security). To illustrate that, a part of the workflow of fig. 2 is shown in fig. 3.

The workflow design phase is supported by an editor with a graphical user interface, including the additional information (attributes) mentioned above as well as syntactic and semantic tests. A generator translates the workflow description (graphical or textual mode) into an internal type description - the input to the runtime system. When a workflow is started, a workflow object is generated and migrated to the first application server.

```
WORKFLOW bug_report_process_frame;
ATTRIBUTES
            COMPLETION_TIME=5;
            PRIORITY_LEVEL=Medium;
    DATA
            request_message_type report_form;
            catalogue_type software_project;
    INITIATOR
            NOTIFY hotline_server WITH
                    STATUS procedure_started;
    TERMINATION
            MAIL report_form TO r_server;

START_NODE bug_report_arrival;
    SERVICE do_archive_report;
            do_select_developer_expert;
            PARAMETERS report_form [IN];
                    task_message [OUT];
                    analysis_form [OUT]
    NODE_ATTRIBUTES
            time_out = 2;   # 2 hours
    BINDINGS
            SELECT WITH software_project =xx;
    NODE_ACTIONS
            RESERVATION AT_NODE
                    commit_changes
                RESOURCE responsible.name =
                        "Lehmann";
                NOTIFY INITIATOR WITH
                        COMPLETION;
            EXCEPTION
            CAUSE responsible.available
                =false REACTION
                responsible.name="Meier";
    NEXT_NODES
            LINK_TO evaluation;
NODE_END
```

```
# further nodes...

FINAL_NODE inform_hotline;
        SERVICE mail_results to hotline_team;
                PARAMETERS analysis_form [IN];
                            inform_message [OUT];
        NODE_ACTIONS
                # none
NODE_END
```

Figure 3: Example of a textual workflow description

3.2 Runtime System

The overall system structure of CodAlf is shown in fig. 4. The runtime system is installed on any participating node and encloses the application servers. All workflow instances are modeled as mobile objects based on DC++ and move through the distributed system during workflow execution. When a workflow task is to be executed at a specific node, its entrance module receives the workflow object. It is then passed to the runtime system and forwarded again by the exit module after completing the task. A control module inside the runtime system communicates with additional tools and middleware services (naming service, trading service, data type-manager, error recovery, monitor, and system management tool) in order to control workflow execution.

The directory service maps the names of required execution instances onto addresses of matching servers within the distributed environment. It is possible to consider static attributes describing the server facilities (for example, the level of measurement quality in our environmental application). The service is implemented by using the DCE Cell Directory Service (CDS). It is augmented with a trading service (see section 3.3); this service also evaluates dynamically changing attributes of servers to enable an optimized mapping from tasks to servers. The data type manager module (see section 3.3) provides type descriptions of all workflow data objects at runtime. It is required for dynamic access to arbitrary data objects associated with a workflow. The error recovery component is able to write synchronization-points (including the associated data objects) after each task execution and to rollback execution to such a synchronization-point. The monitor keeps track of the execution history of a workflow and monitors the executed tasks with the visited servers. In particular, it provides remote control access to a workflow object. This way, the initiator can dynamically query for the current location, execution status, and associated data objects of a workflow. Exception handling can also be controlled remotely, and workflows can be canceled on explicit demand. A special dialogue and control window is offered for that. The system management tool enables the remote start, shutdown, and status control of execution instances in the sense of remote system configuration management.

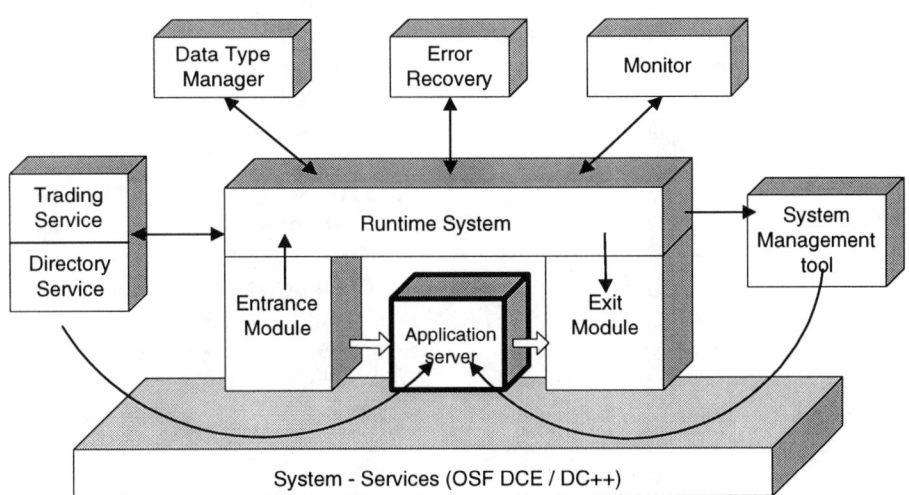

Figure 4: System components at any participating node

General application servers exist at fixed locations for executing the tasks of the workflow application. A workflow is started by an initiator by selecting an associated type description and by specifying the required input data (for example, an application form for an environmental analysis). The workflow executes while the workflow object moves between application servers.

Every workflow object possesses a copy of the workflow type description. The present state of processing is represented by a state pointer within the execution graph of the type description. After completing a task, status information can be passed back to the initiator via a control protocol of the monitor component. Workflow processing at all nodes comprises the following major steps and components:

- *The entrance module.* This module receives the workflow object, separates the data and passes it to the application server. The remaining part of the workflow object, i.e. the workflow type description with the execution graph and the status control information, is passed to the control module. For access to workflow data, the data-type-manager module (DTMM) is used as discussed in section 3.3.

- *The control module.* This module prepares the continuation of workflow processing based on the enclosed execution graph and the knowledge of the actual state during data processing by the application server. In particular, conditional execution paths are evaluated and matching servers are selected as execution instances via the trading and directory service for all paths to be actually executed.

- *The exit module.* This module receives the data produced or manipulated by the application server, adds them to the modified workflow object and causes the continuation of the workflow in accordance with the execution graph by migrating the workflow object to the next selected server(s). If several execution paths have to be joined before executing a subsequent task, a specific synchronization protocol is performed by the exit module: Basically, all other involved execution instances have to send a synchronization message to a selected instance on the primary execution path that is given within the workflow type definition.

The runtime system communicates with application servers via the application-server interface. Application servers consist of an interface for data exchange and an interface for management purposes. Thus, all internal details of application servers are hidden for CodAlf.

3.3 Required Additional Components

First, an additional management system is necessary. It consists of agents on each participating node and allows administrative intervention during workflow execution. The runtime system uses the ability of the management component to start and terminate servers. All application server processes are started initially via the management component. Moreover, the monitor component stores and manages basic events occurred in the distributed system and creates log files of workflow execution. Workflow processing is described in the faultless case here. For error handling, several different kinds of error semantics are supported by the CodAlf system, for instance for the restart of the application by the initiator or the restart at defined synchronization points (see also [11]).

The embedding of the application server into the runtime components requires explicit data management

facilities. The runtime system should be usable for all application servers without changing its source code. On the other hand, it should be possible to pass any data type from the runtime modules to the application server module. We resolve this problem by using a common unique data type which can represent any specific data type using an additional type description. It consists of a byte array with variable length, the byte counter belonging to it and an identifier (UUID) which specifies the real data-type. The conversion from a special data type into its common representation and vice versa is carried out by the data type management module (DTMM).

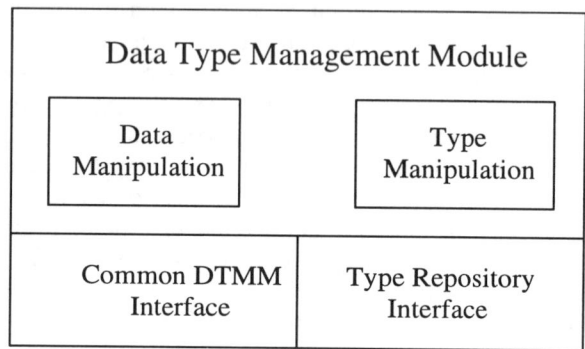

Figure 5: Data Type Management Module

For each data type used by a workflow, the DTMM generates a type description from a description file with an ASN.1-like syntax and exports it to a type repository. If an application server receives a common data type it passes it to the DTMM. This module obtains the matching type description and decodes it. The communication with the type repository takes place at workflow instantiation time; later on, the descriptions are available locally. The functionality of the DTMM (see fig. 5) is useful for almost all components of CodAlf. The workflow initiator can carry out additional semantic tests. The runtime system can evaluate data-based conditions for deciding about alternative execution paths. It can extract any component by element name from a common data type without knowledge of the real data structure implementation. Furthermore, a developer of application servers for CodAlf can generate the necessary C type definitions by using a management interface to the DTMM.

The runtime-system of CodAlf as well as the initiator of a workflow use naming and trading services. The trading-service is realized by our own trader prototype, the X*-Trader [11]. The application gets information about matching server objects and their location. Moreover, the trader offers a comprehensive server selection method - it includes a variety of server attributes and also dynamically changing attribute values. The X*-Trader realizes server and location transparency (according to Boolean expressions of requirements and constraints, respectively). Furthermore, it contains an interface for application development support delivering service type and implementation repositories and statistics of successful and unsuccessful service negotiations. In detail, we are using the X*-Trader as follows:

- *Static service mapping:* At the beginning, a workflow description is compiled in order to generate an executable form. The trader is used in order to check whether the service of the employee is available and whether the static quality of service requirements can be met at this stage. This is environment-dependent and time-dependent, of course.

- *Dynamic service mapping:* The control module of each server is responsible for dynamic server mapping. At runtime, it contacts the trader via a specific trader interface and migrates the workflow object in accordance with the associated execution graph. The trader performs static attribute mapping and selects potential server candidates as described above. In addition, it is able to perform an optimized selection of a specific server based on dynamic information.

- *Mid-term optimization:* The knowledge of the trader about successful and failed service binding will be used for a better translation of the service description, optionally. This feature is not implemented yet.

4. Experiences

After having described the concepts and implementation of CodAlf, we summarize our experiences with this system below. We also compare it with other systems in more detail.

The presented system has been used for designing and implementing several workflow applications. In addition to the environmental administration workflow, a complex image processing workflow and a typical office application (insurance form processing) have been realized. Major experiences are as follows:

- *Genericity and abstraction:* Basically, it makes sense to provide the abstraction of a generic distributed workflow solution. Many example applications have similar characteristics, i.e. well-structured tasks, relatively well-known execution instances, and a distributed environment. Therefore, a generic solution facilitates the implementation of specific workflows significantly.

- *Separation of concerns:* For such a generic solution, it is very important to separate the general roles of system administration (definition and configuration of execution instances and network structure etc.), workflow type definition (specification of workflows and installation within the system), and workflow usage (instantiation and execution of workflows). These roles require different levels of qualification of personnel and

should be supported by different tools as found in CodAlf.

- *Object orientation:* The use of an object-oriented middleware (DC++) facilitated the implementation significantly. Experiences with former, purely client/server-based solutions confirm this observation; much more effort was necessary there. In particular, object mobility and location independent object invocation enabled a rather straightforward implementation of workflow forwarding and status tracking.

- *Standards:* Finally, the use of standards at the middleware level has proven to be very important. The use of DCE facilitated porting the system from IBM AIX to Digital Unix. Other ports are under consideration. Moreover, a standardized platform also provides a significant level of stability that is not achieved with other ad-hoc solutions.

On the negative side, the integration of existing tools and applications (such as standard text processing) is rather difficult with CodAlf. A higher-level interface for application integration is required. Moreover, the system is currently not yet available for Microsoft or Macintosh platforms; this would be very important for typical business applications.

Currently, all DCE/DC++ communication is realized without secure RPCs. It will be necessary to develop solutions for the integration of all DCE security levels for the application of CodAlf. A user should determine which security level is required for which kind of workflow in order to establish a high confidentiality of customers, employees and the management of an enterprise. Based on our described system architecture, security features can be integrated without significant conceptual modifications: It is mainly required to use encryption for privacy and integrity of RPC communication, and to authenticate and authorize execution instances, workflow initiators, and management components. Login of all users is based on DCE-Security-Login so that the basic security control data are already established.

Up to now, only rather simple checkpointing error recovery facilities are implemented. For a consistent rollback of several parallel execution paths, the use of transactional RPCs [6] would be required and would simplify the design and use of workflows. Based on existing products such as Encina, such a facility could also be integrated into our basic architecture.

5. Conclusions and Future Work

This paper presented a decentralized, object-oriented approach towards distributed workflow management. We have illustrated that it is possible to enable a dynamic mapping of tasks onto execution instances, to forward and synchronize workflows in a completely decentralized way, and to provide adequate tool support. It has also been argued that the implementation of such an approach is further facilitated by the use of CORBA, and that existing applications can more easily be integrated this way. Based on the comparison with other approaches, it has been outlined that the integration of various features still requires additional research and development effort.

Our future work will focus on the implementation and evaluation of additional workflow scenarios, on the extension of the BPAFrame-system, and on the integration with other results from our distributed multimedia research [4]. In particular, we are planning to integrate more detailed quality of service specifications into existing workflow specifications, for example to control communication within image processing workflows.

Acknowledgments

This work has been supported by the Deutsche Forschungsgemeinschaft within SFB 358. We would like to thank all colleagues and students who contributed to the design and implementation of the presented approaches, namely Elke Haubold, Wolfgang König, Jörg Schreiter, Birgit Simsch, Timo Welker, Gritta Wolf and Jan Zöllner.

References

[1] Software Ley GmbH: COSA Documentation, 1994

[2] IBM: FlowMark - Modeling Workflows, 1994

[3] Heuser, L.; Schill, A.: DC++. - Bonn: Int. Thomson Publ., 1995 (TAT 15)

[4] Hess, R., Hutschenreuther, T., Schill, A.: Video Communication and Media Scaling System „Xnetvideo": Design and Implementation; European Workshop IDMS'96, Berlin, 1996

[5] Hennessy, P., Kreifelts, T., Ehrlich, U.: Distributed Work Management: Activity Coordination within the EuroCoOp Project; Computer Communications, Vol. 15, No. 8, Oct. 1992, pp. 477-488

[6] Houston, P.J.: Extending DCE for Business-Critical Computing with the Encina Monitor. Transarc White paper 1995

[7] Jablonski, S.: MOBILE: A Modular Workflow Management Model and Architecture; Proc. Int. Working Conf. on Dynamic Modelling and Information Systems, Nordwijkerhout, 1994

[8] Joosten, S.; Brinkkemper, S.: Fundamental Concepts for Workflow Automation in Practice. Submitted Paper to ICIS `95, Amsterdam 1995

[9] Lockhart, H.W.: OSF DCE: Guide to developing distributed applications. Mc Graw Hill, Inc. 1994.

[10] Mohan, C, et al..: An Overview of the Exotica Research Project on Workflow Management Systems. Proc. 6th Int'l. Ws. on High Transaction Syst., Asilomar, Sept. 1995

[11] Mittasch, Ch.; König, W.; Funke, R.: Trader supported Distributed Office Applications. - Int. Conf. On Distr. Platf. Dresden, 1996, In: Schill, et. al. (Eds.): Distributed Plattforms Conf. Proc. pp. 230-244, 1996

[12] Mittasch et al.: Design and Use of BPAFrame - a Decentralized CORBA-based WfMS. IFIP World Computer Congress, Canberra, Sept. 1996, In: Terashima, N.; Altman, E. (Eds.): Advanced IT Tools, Chapman & Hall, S. 303-310, 1996

[13] Miller, J.A.;et al.: CORBA-Based Runtime Architectures for Workflow Management Systems. J. of Database Management, Special Issue on Multidatabases, vol. 7 (1996), No. 1, pp.16-27

[14] Nicol, J.R., et al.: Object-Orientation in Heterogeneous Distributed Computing Systems; IEEE Computer, Vol. 26, No. 6, 1993, pp. 57-67

[15] IABG: Prominand Documentation, 1995

[16] Schill, A., Gütter, D.: Extending Group Communication Facilities to Support Complex Distributed Office Procedures; Int. J. of ICIS vol. 3, No. 2, 1994, pp. 203-223

[17] Tailor, D.A: Business Engineering with Object Technology. John Wiley & Sons, 1995

[18] Workflow Management Coalition, Work Group 1/3: Interface 1: Process Definition Interchange, WfMC, Doc. TC 00-xxxx, Draft 2.0, 1995

[19] Siemens-Nixdorf: WorkParty Documentation, 1996

[20] Wodke, D.;et al..: The Mentor Project: Steps Towards Enterprise Wide Workflow Management. Int. Conf. On Data Engineering, New Orleans, March 1996

A Communication Computation Model for developing Computer Supported Cooperative Work Systems

Masahiro Hiji† Hiroshi Nunokawa‡ Masatoshi Miyazaki†

†:Graduate School of Information Sciences,
Tohoku University
Katahira Aoba, Sendai 980-77, Japan
E-mail : {hiji, miyazaki} @dais.is.tohoku.ac.jp

‡:Institute for Science Education,
Miyagi University of Education
Aoba aza Aramaki Aoba, Sendai 980, Japan
nunokawa@ipc.miyakyo–u.ac.jp

Abstract

Cooperative work is done not only by workers but with some software used by each worker. They do various communications including interaction between worker and software in cooperative works. To develop computer supported cooperative work systems, it is important to support various communications among them. A computational model that can represent communicative objects such as workers and software and various communications among them are necessary for that purpose.

We propose this communication computation model as such a model. A worker and software are represented as the AutonomousObject, and a group is represented as the GroupField in this model. Various communications are represented by communicative functions that correspond to each communication form. We show that our model is effective for modeling various communications in cooperative work.

Keywords: Human Communication, Model, CSCW, Autonomous Object

1 Introduction

Due to the spread of computer networks, a decentralized system is used for individual work and for support of a communication between people and cooperative work on the basis of it. All kinds of researches on the system that supports cooperative works are being done, and many systems such as an electronic meeting/conference system [2,3],a sharing window system [4,5,6] and a work flow system [7,8] are being developed in these researches [1]. There are many systems enclosed as individual application programs in these systems. Therefore, these systems can see a specific side with cooperative work, and can only support the part of cooperative work from this side and cannot support the whole cooperative work with various sides.

Only person who does cooperative work does not do it. Software such as the application program and the database that are used for the work of an each person become necessary in the cooperative work, too. Persons share a processing data and an operation on data with other persons in going on with cooperative work, and do various communications. This communication has various forms of which a one to one real-time conversation, sending documents, meeting, etc. A person uses these forms properly according to the purpose of communication and the state of partners. Cooperative work consists of all kinds of objects such as persons and application programs that they use. These objects exchange information by various forms. In order to develop system to support cooperative work, it is necessary to establish computation model that can uniformly represent these objects and exchanges of information of various forms that are done among them. By modeling various objects concerning with cooperative work and exchanges of information among them, we can develop system for supporting cooperative work easily, and a developed system can support various communications in cooperative work.

We propose the communication computation model as a computational model for that purpose, then show an example of modeling a cooperative work based on this model. This model is based on the viewpoint saying that we consider exchanging information among all kinds of these objects as a communication that is similar to the human communication. Because these software receive structured messages, execute received messages and send processed result. These messages and data are a kind of information, exchanging this information is communication. This model represents various primary objects such as a person who does cooperative work and the application programs that he uses in cooperative work as AutonomousObject, and represents a group who does cooperative work as GroupField. This model represents communication that is done by various forms among these objects as communicative functions corresponding to each form. By using our model that we propose, communication among users, interaction between a user and software and sending request message and receiving processing data among software is able to represent as exchanging information among AutonomousObject and GroupField. As a result, we can model the software that constitutes a system and user that use a system uniformly. The system on the basis of this model can support cooperative work that is being done by users and software.

2 Human Communication on a Decentralized System

In order to clearly define a computational model that will be the basis for the development of computer supported cooperative work systems on a decentralized system, our meaning of communication is as follows [9,10]. A communication can be considered as a sender choosing a from selection of communicative partners, and exchanging information to or among them through some communication form. That is to say, communication is defined by triple terms – communicators, communication form and information –.

Communicator means a set of objects which exchange information and interprets the information. Communicator includes not only the primary object but also a group of primary objects. Also, we call a primary object belonging to a group a member object. A member object is not always the same, members of a group change other objects and number of member objects belonging to a group varies. As the result, it is necessary that communicators exchange information with all or some member objects at that time when they communicate with the group as communicator.

The communication form means the mode of exchanging information among communicators. This form is defined as a triplet of – the number of communicators (number), the direction of information flow (direction) and the time of information exchange (time). That is to say, communication form = <number, direction, time>. In these parameters, the number of communicators means the combination of number of senders and receivers. It is represented in the form of "number of senders TO number of receivers", and is any one of these for types of "1 TO 1", "1 TO n", "n TO 1" and "n TO m ", where n and m can be any number greater than 1. The direction of information flow means the direction of information flow between senders and receivers. This direction of information flow consists of "One-sided direction" and "mutual direction". "One-sided direction" is the case where only sender can send information to receiver but the opposite is not true. "Mutual direction" is the case when sender and receiver exchange information mutually. The time of information exchange means the time necessary in exchanging information between senders and receivers. This time of exchanging of information consists of "real-time" and "nonreal-time". The "real-time" means that information will not only be sent to the receiver in real time but also be read by the receiver at once. The "nonreal-time" means that not assuring neither a real time sending nor an immediate reading of information.

The information means the specific contents sent to partner communicators through a communication form.

The type of the number of communicators is decided when the communicators are selected. The type of both direction and time is decided by the purpose of communication and conditions of communicators. A communication form is decided by these.

To describe above, a communication consists of a set of communicators and communication form defined purpose of communication and condition of each communicator and information exchanged among these communicators. A communication process is defined in terms of a sequence of this communication [11].

This communication process is dynamic [10]. It means that a communication is not only, some specific communicators, for instance an exchange of information based on specific communication form among communicators, but also multimodal communication according to communication process. In this communication process, communicators and communication form vary with time. Based upon above communication, this variation is thought of changes of elements of communication as follows.

This variation of communicator not only means the case when the partner communicator is changed, but also means when the member or members of the group that is the communicator partner are changed. A variation in the number of communicators includes changing from one-to-one communication into one-to-n communication and so on. A variation of the time of communication flow means changing from real-time communication into nonreal-time communication and so on. A variation of the direction of communication flow means changing from one-sided communication into mutual communication or changing from mutual communication into one-sided communication.

3 Communication Computation Model

3.1 Modeling a Communication

We propose a computation model for modeling a communication based on the communication analysis in chapter 2. Our model uses communicators and an exchanging of information among communicators in modeling. In this model, contents of information exchanged in communications and interpretations of information are not taken into consideration. These are implemented by programming language based on our model.

A communicator is a set of objects that exchanges information through some communication form, creates and interprets it. Software can be regarded as a communicator. Because software receives a command from other software and users, executes it, and sends a processing result to this software and users. That is to say, software is a communicator in these meanings. Therefore a communicator is a set of computational objects abstracting an object (person, group and software, etc.) which exchanges and interprets information. Communication computation model represents these communicators as AutonomousObject and GroupField. An individual communicator within these communicators is represented as an AutonomousObject. A group within these communicators is represented as a GroupField.

An AutonomousObject consists of interior conditions, a script that described behavior itself and interpreter that executes script. An AutonomousObject can change interior conditions and script by executing the script in message received from other AutonomousObjects. The interpreter in AutonomousObject has a high-order function that interprets data as script. An AutonomousObject has a CommunicationScope as a conceptual scope that recognizes partner being possible to communicate in real-time. A real-time communication is then possible among AutonomousObjects with the same CommunicationScope in AutonomousObjects.

A GroupField is a kind of an AutonomousObject. A GroupField is a boundary that distinguishes each group of AutonomousObject from each other. And it is the medium for sending message received from some AutonomousObject to AutonomousObjects within its GroupField. A GroupField has a CommunicationScope the same as an AutonomousObject.

A number of communicators in triplet of communication form is decided at the time of selecting communicators. Therefore AutonomousObject and GroupField can represent some communication form by having communicative function corresponding to each type of the direction of information flow and the time of information exchange in triplet of communication form. Communication computation model represents "real-time" as CooperativeSend and "nonreal-time" as InformSend. The time of information exchange is represented as these two kinds of communicative functions. The CooperativeSend is a mode that assures that information will not only be transmitted to a receiver in real time but also be read out by the receiver at once. The InformSend is a mode that differs from CooperativeSend in sense that it does not assure neither a real time sending nor an immediate reading out of information. It leaves the reading out as the responsibility of the receiver. Communication computation model represents "one-sided direction" as Send and "mutual direction" as ContractSend. The direction of information flow is represented as these two kinds of communicative functions. The Send is a mode that transmits information from sender to receiver. The ContractSend is a mode that always knows other communicators and transmits information mutually. The time of information exchange and the direction of information flow in communication form are represented as combinations of individual communicative functions above. That is to say, these are InformSend, InformContractSend, CooperativeSend and CooperativeContractSend. Information is described as script. The script is a program executed by an interpreter that AutonomousObject has. A script is exchanged as a structured message by using above communicative function.

That is to say, communication computation model represents a communicator as AutonomousObject and GroupField. This model represents exchanging information by some communication form as exchanging message having script that described information by using communicative function corresponding to the time of information exchange and the direction of information flow in triplet communication form. A communication process is represented as a sequence of these communications.

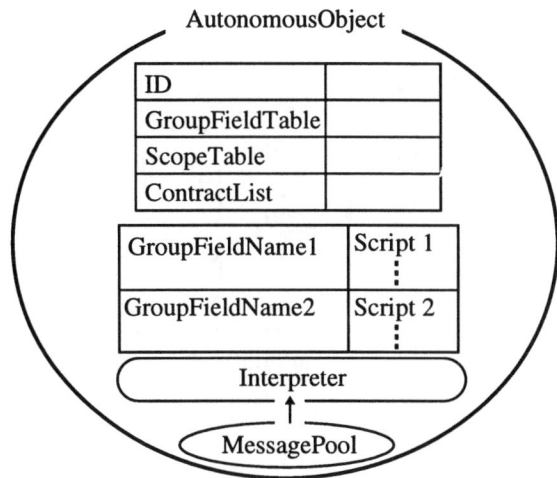

Figure 1: Structure of AutonomousObject

3.2 AutonomousObject Structure

An AutonomousObject structure is shown in Figure1. An ID specifies individual AutonomousObject. A GroupFieldTable is a list of ID of GroupField to which AutonomousObject belongs. This parameter manages that AutonomousObject belongs to what GroupField currently. A ScopeTable is a list of names of CommunicationScope. This parameter manages which CommunicationScope has been opened so that AutonomousObject shows that a communication can be done in real time. A ContractList is a list of ID of AutonomousObject or GroupField of partner at the time of exchanging message by using ContractSend. This parameter manages a partner communicator exchanging message by using ContractSend. An AutonomousObject set this parameter on ID of partner at the time of starting ContractSend and deletes the ID of partner from the ContractList at the time of ending ContractSend. The script is program that is executed by the interpreter of AutonomousObject. The script is described for every GroupField to which AutonomousObject belongs. A script has a case that it is executed by a direct interpreter and other case that a script defined in an AutonomousObject is executed. A MessagePool temporarily maintains a message that other AutonomousObject and GroupField send. An AutonomousObject read out a message maintained in a MessagePool when it is able to read.

An AutonomousObject consists of InformSend, InformContractSend, CooperativeSend and CooperativeContractSend in section 3.1. An AutonomousObject represents various communication forms as communicative function corresponding to each form. An AutonomousObject can use these communicative functions properly corresponding to the purpose of a communication and the condition of a partner communicator.

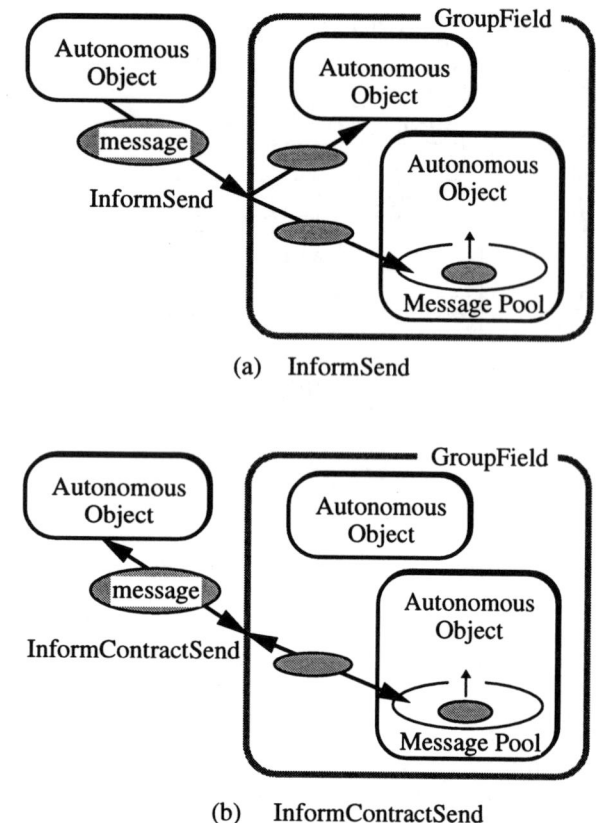

(a) InformSend

(b) InformContractSend

Figure 2: Exchanging message by InformSend and InformContractSend

3.3 Communication Among AutonomousObjects and GroupField

A script is exchanged as a structured message as follows.

Message format: (<Name>, <To>, <From>, <script>)

In this format, <Name> parameter is a name of communicative function sending this message, is any one of InformSend, InformContractSend, CooperativeSend and CooperativeContractSend. <To> parameter is name of receiver, <From> parameter is a name of sender, and <script> parameter is contents to send. An AutonomousObject translates script to message, sends this formatted message by using a suitable communicative function.

In case that <To> parameter in a message is AutonomousObject, the message is maintained to MessagePool in its AutonomousObject. AutonomousObject reads out the message whose <Name> parameter is ContractSend prior to other message from MessagePool, and interpret <script> in its message. In case that <To> parameter in a message is GroupField, GroupField sends the receiving message to an AutonomousObject that belongs to its GroupField.

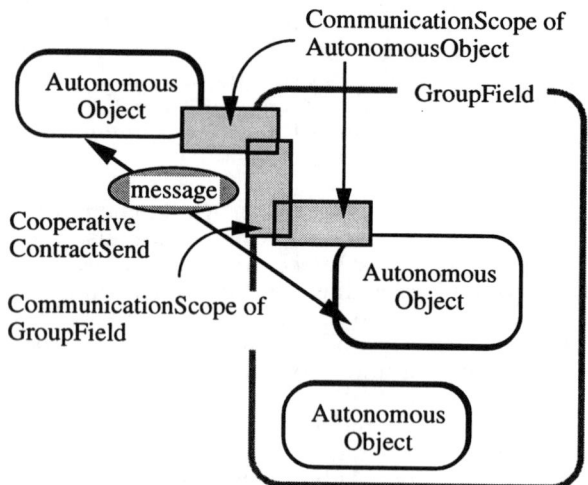

Figure 3: Exchanging by CooperativeContractSend

Then the way to send the receiving message differs by the name of the communicative function specified <Name> parameter. In case that <Name> parameter is InformSend, GroupField sends the receiving message to all AutonomousObjects that belong to its GroupField at that time (Figure 2(a)). In case that <Name> parameter is InformContractSend, GroupField sends the receiving message to certain AutonomousObject that belongs to its GroupField (Figure 2(b)). In case that <Name> parameter is CooperativeSend, GroupField sends immediately the receiving message to all AutonomousObjects that belong to its GroupField at that time. In case that <Name> parameter is CooperativeContractSend, GroupField sends immediately the receiving message to all AutonomousObjects that belong to its GroupField at that time and open the same CommunicationScope that GroupField opens (Figure 3).

The partner that is possible to exchange a message in real time changes with time. If it changes from a condition of possibility of exchanging information of which a partner AutonomousObject is real-time to an impossible condition in a certain point of time, AutonomousObject can know this change by CommunicationScope. AutonomousObject is changed from a CooperativeContractSend used for it to a InformContractSend at that time. An AutonomousObject can select a communicative function corresponding to a condition of partner AutonomousObject this way.

An AutonomousObject sends to another AutonomousObject the message having the script that changes its interior conditions and scripts. Another AutonomousObject executes the script in receiving message, changes its own interior conditions and scripts.

We design the programming language based on communication computation model, and make out its implementation on a decentralized system. This pro-

216

Table 1: Functions for describing communicator

Programming Functions	Outline of Functions
(CreateAutonomousObject <Name><Behavior>)	create an AutonomousObject
(CreateGroupField <Name><Condition>)	create a GroupField
(DeleteGroupField <GroupName>)	delete a GroupField
(InGroupField <GroupName><Obj.Name>)	an AutonomousObject participate a GroupField
(OutGroupField <GroupName><Obj.Name>)	an AutonomousObject leaves a GroupField
(OpenScope <Scope><Obj.Name>)	open a CommunicationScope
(ShutScope <Scope><Obj.Name>)	close a CommunicationScope

Table 2: Functions for describing communication figure

Programming Functions	Outline of Functions
(InformSend <To><From><Message>)	send <Message> to <To>
(InformContractSend <To><From><Message>)	exchange <Message> among <To> and <From>
(CooperativeSend <To><From><Message>)	send <Message> to <To> in real time
(CooperativeContractSend <To><From> <Message>)	exchange <Message> among <To> and <From> in real time
(Contract <To><From><Condition>)	request using ConstractSend
(ContractAccept <To><From><Yes/No>)	inform acceptance or not of request using ContractSend
(ContractEnd <To><From>)	inform the end of CooperativeSend
(WhoInTheScope <ScopeName>)	query for possible real time partners

gramming language adds programming functions that describe AutonomousObject, GroupField and communicative functions to DeLis (Decentralized Lisp Interpreters)[12]. The DeLis is a programming language that has programming functions for describing data communication and graphical user interface. We show these programming functions in Table 1 and Table 2.

4 Application to systems to support cooperative work

4.1 System to support workflow including exceptional cases

The work process that describes as workflow proceeds through exchanging a standard form among specific workers and departments and doing each work based on its form. An exceptional work breaks out in process of proceeding work because of the deficiency in an entry in a standard form. It is necessary for workflow management system to support the work process including exceptional cases. We afford an instance of such a work process. The work processes of application for connecting to the computer network proceed by the workflow of following.

(1). An applicant sends to an administrator an application form for connecting to the computer network.

(2). The administrator checks the entry in the received application form.

(3). If there are no problems, the administrator sends to a network operation department the application form.

(4). The network operation department receives the application form and transfers it to some operator in this department.

(5). The operator receives the application form and works for connecting to the computer network based on entry in the application form.

(6). After the operator completes this work, he registers the necessary entry on a database.

(7). After the operator completes the register of the entry, he sends its result to the administrator and the applicant.

In case that it is a problem in the entry in (3) of the above process, the work for solving the problem is an exceptional work. In this case, for example, the administrator and the applicant have an exchange of information as following.

217

(3-1). The administrator inquires of the applicant the problem in real time, and he solves the problem by exchanging the problem points and its reply with the applicant in real time.

(3-2). If the applicant cannot accept the inquiries in real time when the administrator inquires of the applicant the problem in real time, the administrator sends to the applicant the application form.

(3-3). If the applicant cannot accept the inquiries in real time when the administrator inquires of the applicant the problem in real time, the administrator looks for a person who is inquiring in real time, and he solves the problem by exchanging the problem points and its reply with this person in real time.

The above-mentioned work process proceeds through the applicant, administrator, network operation department and operator who belongs it exchange information that is expressed as the application form in non-real time. And in case that it is a problem in this work process, the administrator tries to exchange information for solving the problem in real time. After all, the administrator changes communication form in case of an exceptional case.

That is to say, a work process including exceptional works can model as Figure 4 in terms of the communication computation model. The communication computation model represents an applicant, an administrator, an operator and database software as each applicant A-Object, administrator A-Object, operator A-Object and database A-Object that is AutonomousObject. And this computation model represents network operation department as network operation GroupField that is GroupField. The administrator A-Object has the script that described procedure for checking the application form. And the operator A-Object has the script that described the entry to need to register on a database and procedure to register this entry. The network operation GroupField has the script that described how the network operation GroupField selects an operator A-Object to transfer a received application form. We consider remittance of an application form and an inquiry about problem to be communication. This computation model represents sending an application form in terms of InformContractSend. The applicant sends to the administrator the application form is represented as that the applicant A-Object send to the administrator A-Object a message having the script that described the application form by using InformContractSend. And this script is description of items and its value that the application form consists of and method for displaying these.

In case of the above-mentioned (3-1), to exchange inquiry and its reply in real time is represented in terms of CooperativeContractSend. The applicant cannot accept the inquiries in real time in (3-2) is represented as that the applicant A-Object shut its CommunicationScope. And the administrator A-Object

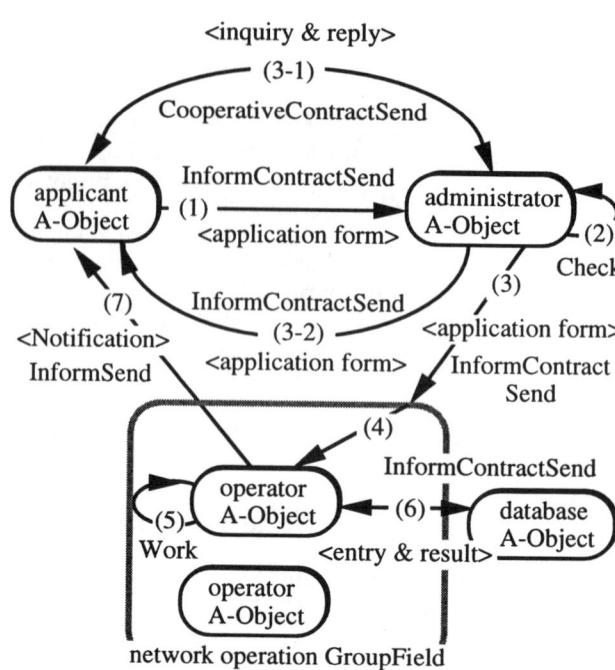

Figure 4: Modeling Workflow including an Exceptional case

recognizes this condition by closing its CommunicationScope, and send to the applicant A-Object a message having the script that described checked application form by using InformSend. The administrator looks for a person who is inquiring in real time like (3-3) is represented as follows. The administrator A-Object obtains an AutonomousObject who opens CommunicationScope that means to reply its problem in real time, and exchanges problem points and reply with obtained AutonomousObject by using CooperativeContractSend . The operator registers the necessary entry on a database means to send to a database the necessary entry and to receive its result. This is represented as that the operator A-Object and the database A-Object exchange the message having the script that described the necessary entry and its result by using InformContractSend.

Communication computation model can uniformly represent software that users use as well as users as AutonomousObject. And this model can uniformly represent various exchanges of information among users and software in terms of communicative functions. As the result, a workflow management system that is developed based on this model can manage the progress of the work process and support the real time communication in case an exception case occurs.

4.2 System to support accommodative work

Accommodative work is a kind of cooperative work, and needs to make workers having a different estima-

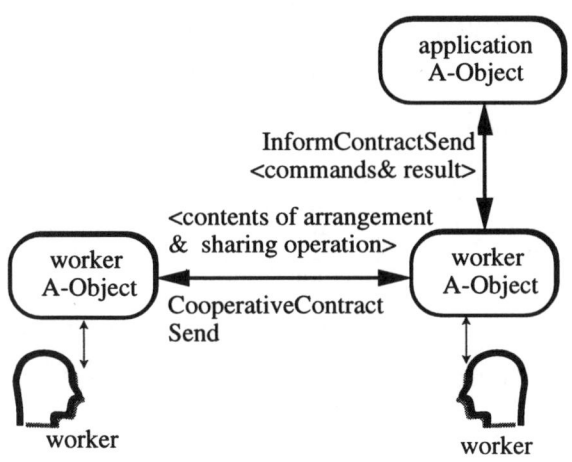

Figure 5: Modeling a Accommodative Work

tive standard come to an understanding based upon each's estimative standards. In order to progress accommodative work, it is necessary that they exchange a changed data and each's estimate, and make the best suitable plan in all. Necessary data, a method for displaying data and an application program using evaluation differs from every worker in accommodative work because they estimate plan from each's best method. But they share a part of data and operations to sharing data. And operation to sharing data influences estimation of other workers sharing its data. Therefore, workers display data by form suiting with an estimative standard of each worker, share operation to a part of displaying data by different form, make meeting and arrangements among workers and go on with an accommodative work.

We consider this accommodative work to be a work which workers and application programs exchange information such as data, operation of a sharing data and contents of meeting and do ones work based on this information. That is to say, the workers and application programs are communicator exchanging information and interpreting it. Therefore communication computation model can represent accommodate work as Figure 5. The communication computation model represents the works and application program which workers use as each worker A-Object and application A-Object that is AutonomousObject. Each worker A-Object has a user interface for displaying data which the application A-Object processed. A method to display data is described as script which each worker A-Object has. A difficult user interface realizes by describing each worker A-Object a different script. For example, a worker A-Object having Gant chart script shows processing data as Gant chart, but a worker A-Object having spread sheet script shows the same data as spread sheet. By giving the best script which worker describes his worker A-Object, worker is possible to see processing data by best form.

A worker operates application program and receives processing data of application program. This is represented in terns of InformContractSend. That is to say, worker A-Object sends to application A-Object the message having the script that described commands for operating an application program by using InformContractSend. And the application A-Object executes the script in received message, and sendings to the worker A-Object the message having the script that described processing data by using InformContractSend. And worker A-Object receives the message which the application A-Object sends.

A operation of sharing data on a different user interface is represented in terms of CooperativeContractSend. A worker A-Object manages with a sharing data, translates the operation for data displayed on a user interface into what operation for which data, sends to the worker A-Object sharing data the message having the script that described the translated operation by using CooperativeContractSend. Other worker A-Object interprets the script in the received message by interpreter and changes sharing data on user interface. An arrangement among workers is represented in terms of CooperativeContractSend. Worker A-Objects exchanges the message having the script that described contents of arrangement by using CooperativeContractSend.

As above, communication computation model can uniformly represent software as well as a worker as AutonomousObject. And this model can uniformly represent various exchanges of information among workers and software in terms of communicative functions. A worker can exchange information with software like other worker only changing communicative partner. Through each worker A-Object, a worker can select the best communication methods corresponding to a situation.

5 Related Work

MCM (Milan Conversation Model) [13] is system to support communication among users. This system consists of various communicative methods and user interfaces called message switches for selecting these methods. These methods are one-to-one or one-to-n real time communication and nonreal-time communication. We consider this system as communication computation model from the point of consisting any communicative function corresponding to communication form and selecting these functions. But a user is necessary to select communicative function through a user interface themselves. This system does not have a framework for describing how to change communicative methods like an AutonomousObject.

RT-Michele[14] represents various elements concerning cooperative work – user and document – as an agent. And RT-Michele represents two kinds of communication form of real time and nonreal-time by introducing a special field of which real-time message sending is called meeting environment. An agent usually sends to other agent message by message passing. And if an agent needs to send message in real time, agents move the meeting environment and send to other agents a message. Other agents in the meeting environment receive this message in real time. That is to say, RT-Michele can represent the time of information exchange in the parameter that determines a communication form and changes of these. But it can not represent the various communication forms that include the number of communicators and the direction of information flow and a change of such communication form. And, RT-Michele cannot describe a change of a communication form as it is because it is necessary to describe the action that does not relate directly to the change of communication form by introducing a special field called meeting environment. In case agents communicate in real time, it is necessary that these agents enter in the meeting environment. This action is not relate to a real-time communication. And, as for meeting environment, agents in this environment are only possible to communicate in real-time, and agent cannot know which agent is possible to communicate with in real-time.

In the groupware field, how to share data and operation on a similar window among users is being studied [2,4,5]. These systems force users to see the sharing data by the same form. As a result, these systems can not support cooperative work like sharing operation of data by different form. Our model represents all kinds of objects concerning cooperative work (e.g., worker, application program, etc.) as an AutonomousObject. By each AutonomousObject having different script for displaying data, it is possible to share operation on different displaying data. And a system to support cooperative work based on our model can support such cooperative work.

6 Conclusion

In this paper, we propose the communication computation model that is able to model all kinds of objects concerning cooperative work and communication among them. This model represents workers, software and group as an AutonomousObject and a Group-Field. And this model represents exchanging information based on various communication forms as using a communicative function corresponding to each communication form. In addition to this, we design a programming language based on our model. We model all kinds of objects concerning cooperative work based on our model, and can develop system to support cooperative work using this programming language. A system to support cooperative work that is developed based on this model can support seamlessly communication among users and software.

In future work, we will develop more system to support cooperative work and evaluate our computation model through it. We will also consider establishing how to estimate a developed system based on our model.

References

[1] C.A.Ellis, S.J. Gibbs and G.L.Rein, " Groupware : Some Issues and Experiences", *Comm. ACM*, Vol.34, No.1, pp.38-58, 1991.

[2] K. Watabe, S. Sakata, K. Maeno, H. Fukuoka and Y. Ohmori, "Distributed Multiparty Desktop Conference System : MERMAID," *Proc. of the Conf. on Computer-Supported Cooperative Work* , CSCW'90, pp.27-38, LosAngeles, 1990.

[3] A. Wiliam and S. Buxton, "Telepresence: Integrating Shared Task and Person Space," *Proc.of Graphic Interface 92*, pp.123-129, 1992.

[4] H. Ishii, M. Kobayashi and J. Grudin, "Integration of Interpersonal Space and Shared Workspace : ClearBoard Design and Experiments", *ACM Trans. Information Systems*, Vol.11, No.4, pp.349-375 1993.

[5] T. Crowley, P. Milazzo, E. Baker, H. Forsdick and R. Tomlinson, "MMConf: An Infrastructure for Building Shared Multimedia Applications", *Proc. of the Conf. on Computer-Supported Cooperative Work*, CSCW'90, pp.329-342, LosAngeles, 1990.

[6] M. Stefik,G. Foster, D.G. Bobrow, K. Kahn, S. Lanning and L. Suchman, "Beyond The Chalkboard : Computer Support for Collaboration and Problem Solving in Meetings", *Comm. ACM*, Vol.30, No.1, pp.32-47, 1987.

[7] R. Medina-Mora, T. Winograd, R. Flores and F. Flores, "The Action Workflow Approach to Workflow Management Technology", *Proc. of the Conf. on Computer Supported Cooperative Work*, CSCW'92 , pp.281-288, Toronto, 1992.

[8] H. Ishii, "Message-driven Groupware Design based on Office Procedure Model" *OM-1 J. Inf. Process*, Vol.14, No.2, pp.184-191, 1991.

[9] M. Hiji, H. Nunokawa and M. Miyazaki, "Computation Model for Human Communication", *Proc. of the 6th Int. Conf. on Human-Computer Interaction*, Vol.2, pp.521-526, 1995.

[10] M. Hiji, H. Nunokawa and M. Miyazaki, "Communication Computation Model for Modeling Dynamic Human Communication", *Trans. of IEICE A*, Vol.J79-A, No.2, pp.197-206, 1996.

[11] M. Hiji, H. Nunokawa and M. Miyazaki, "Representation of Communication Protocol in Communication Computation Model", *Trans. of IEICE B-I*, Vol.J79-B-I, No.5, pp.321-328, 1996.

[12] T. Mitsuishi, H. Nunokawa, M. Miyazaki and S. Noguchi, "Programming Language Delis for Decentralized Environment", *IPSJ SIG Notes*, 10-PRG-8, pp. 57-64, 1993.

[13] G.D. Michelis and M.A.Grasso, "Situating Conversations within the Language/Action Perspective : The Milan Conversation Model ", *Proc. of the Conf. on Computer-Supported Cooperative Work* , CSCW'94, pp.89-100, Chapel Hill, 1994.

[14] Y.Nakauchi, E.Miyoshi, T.Okada and Y.Anzai, "Computer Supported Cooperative Work Environments Based on Mulit-Agent Model", *IEICE Trans. Inf. & Syst.*, Vol.J75-D-II, No.11, pp.1874-1883, Nov. 1992.

An Architecture to Support Distributed Trade Documents

H. Wing and R. M. Colomb

CRC for Distributed Systems Technology*
Department of Computer Science
The University of Queensland
Brisbane, Qld 4072, Australia
Email: wing@cs.uq.edu.au

Abstract

By integrating the many different technologies and approaches involved in electronic education and commerce, we can effectively support advanced applications in these areas. This paper describes an architecture which can be used to facilitate the specification, execution, and distribution of incremental trade documents among heterogeneous interchanges. We argue that the conventional, top-down, centralized approach to workflow, in its current state, is not suitable to support the undetermined nature of an advanced trade application. What this approach lacks is a unified object-oriented model which can be used to design interoperable trading systems with a single set of well-defined abstractions. This paper, in an attempt to overcome the limitations imposed by the conventional approach, presents an overview of principles relating to object interoperability, semantic heterogeneity, and the document-centric paradigm associated with trade documents. A view object model, and its relevant components are also discussed.

1 Introduction

Worldwide co-operation coupled with increasing competition in every aspect of education and commerce has forced both institutions and business to be more flexible and efficient. Consequently, while still trying to minimize operating cost, education and business enterprises need effective ways to mass-market their products and extend their operations to the open global markets. This must be achieved without losing the quality expected in conventional education and commerce.

The demand then, is for a just-in-time, on-demand, team-based, networked, geographically dispersed, and automatic approach in providing 'on-line' education and commerce in the immediate future. In providing solutions towards this demand, this paper presents a generic framework to support the distribution, execution, sharing and management of trade documents associated with education (or business) brokerage on the internet [14]. In particular, we provide an overview of the essential principles related to object interoperability, semantic heterogeneity, and the document-centric paradigm. Also examined is the relevant object model and essential components.

*The work reported in this paper has been funded in part by the Co-operative Research Centre Program through the Department of the Prime Minister and Cabinet of the Commonwealth Government of Australia.

This paper is organized as follows: Section 2 presents related work and the motivation for this work. Section 3 introduces the architecture and its formal definitions. Section 4 examines the practical utility and effectiveness of the architecture's components. Section 5 presents concluding remarks.

2 Motivation and Related Work

An example of a trade is an education brokerage on the internet. In this environment, a course customer, provider, and education broker may not know each other's 'ways of doing business' (eg. trade procedures which can always be systematically processed) prior to commencing a trade such as taking or providing a course. Each participant therefore views each other's internal processes as 'black-boxes'. An integrated trade procedure can be formulated, however, after a series of trading steps and negotiations. This is an example of a 'bottom-up' approach to workflow since the relevant business process can only be formed on the fly, step-by-step. This is in contrast with the top-down approach in which an overall workflow must be known in advance and available for partitioning, monitoring and execution purposes. Associated with these advanced applications are the trade documents which represent the various structured information records necessary to instantiate a trade.

For instance, within the electronic commerce area, a *Home-Buying* dossier may contain the loan application, buyer particulars, credit information, and escrow instruction, etc. With respect to the electronic education area, a *Course-Taking* folder may be used to store the student's education needs, background, test information, and course templates, etc. In advanced trade applications, these information dossiers need to be captured, represented, collaborated, and distributed among the various trading interchanges. Based on a person X's organizational role and computing capability, X's workspace (eg. workstation, PC, or mobile computer) may store and process several such information dossiers. The idea reflected here is that each of the dossiers contains sufficient knowledge (eg. workflow's meta data) about its part of the business process in order to correctly coordinate, execute, monitor, and manage the various trade messages and documents.

In the context of the above example of the education brokerage on the internet, we observe the following char-

221

acteristics: 1) *There exist many different types of trade documents* (eg. digital cash, test templates, course credits); 2) *Trade documents are compose-able* (eg. credit information, buyer particular, and property information, etc., when combined, form a loan application); 3) *Trade documents are cross-platformed* (eg. a test template may be originally designed and formulated in a Unix-based machine in which a student who may use another totally different platform (a PC, workstation, or mobile computer for instance) can still operate on this test template); 4) *Trade documents must be EDI-processable*, if they are to be systematically interpreted and processed.

The different documents' characteristics require different management techniques and treatments. In order to support the management and distribution of these documents across heterogeneous platforms and environments, we have to integrate and extend the many different technologies and approaches.

To support advanced applications in electronic education or commerce, the following technologies are essential:

• The Distributed Object Computing (DOC) framework allows computational objects to be widely distributed across different platforms and native environments [25, 7, 19, 26]. A system such as CORBA[19, 26] supports this type of distribution. It provides a mechanism in which objects can be distributed among different network operating systems, programming languages, database management systems, and object-oriented user interfaces.

• To support groupware application, Microsoft's OLE [6] or Apple's Open/Doc [27] allow objects to be linked and embedded as a collection of loosely-coupled entities known as compound objects.

• To deal with the semantic heterogeneity problems (or systematically process and exchange trade messages) [10, 25] we need EDI's third party standard messages (eg. UN/EDIFACT or X12 [17]), closed trade procedures [5, 28, 8], and/or open trade procedures [16].

• To support the specification, coordination, execution, and management of the trade activities and data, we need the technologies provided by the Workflow Management Systems (WFMSs) [21, 15, 28, 13].

• Finally, technologies provided in the Transactional Workflow systems (TWFs) [23, 24, 1, 3] help to incorporate the well-defined failure semantics and sophisticated recovery features of the advanced transactional models (ATMs) [9, 29] into the workflow context.

The successful emergence of these new technologies indicates that the various inter-organizations' compound trade document and procedures will become the primary paradigm for capturing trade commodities common in electronic commerce. This is because these technologies, when combined, will become the overall framework for managing non-record oriented trade information.

Problems arise when these technologies treat work entities (eg. trade documents) as a collection of encapsulated - stand alone - computational objects. The objects' collaborated behaviours and infrastructures are often vaguely represented and somewhat ignored. The difficult task of specifying and coordinating the various work entities is left to the application, thus leaving the specification and coordination unguided to the individual trade designer. This leads to more, larger applications and added maintenance

cost.

Building and supporting a simple trade application may not be cost effective due to the tremendous programming and support requirements associated with the different object models. To overcome the complexity problem, we allow trade documents to be specified, executed, monitored, and managed under one unified object model instead of having to deal with the many different sets of localized programming abstractions and supporting environments. In doing this we need to integrate the absolutely essential components of the application while eliminating the unnecessary features associated with the underlying supporting models.

Recent works [3, 1] suggest that failure handling, high availability, replication and navigational flexibility are the essential components in large scale workflow systems. In these systems, business processes are defined in advance, so it is not possible to evaluate all possible exceptions and error conditions beforehand. we must therefore provide means to cope with such situations. Recovery techniques can be integrated into the workflow model to support the transactional capabilities [23, 3, 24]. In particular, ATM's backward recovery feature of Sagas [12] can be implemented in the proposed object model to compensate and undo failed activities, while the forward recovery feature of Flex Transaction [11, 4] guarantees a process which makes progress in the presence of execution failures based on the navigational capability of the different execution paths.

There are other problems which arise from the top-down approach. These include communication bottle-neck, low availability, an inflexible system configuration, and lack of resilience to system failure and arise when the overall knowledge and definitions of a business process are centralized in one node [1, 2]. These limitations emphasise the need for a bottom-up approach to workflow; one which can support trade interoperability based on the semantics and availability of the traded documents and does not rely on overall business processes.

In recent work, a document-centric type of application [21] will allow a structure workflow to be implemented and executed on the document-based Lotus Notes' native workflow concepts. Combined with the distributed advantage of CORBA, [18] provides the different workflow run-time architectures by briefly extending the CORBA IDL. While these works are useful in addressing the interoperability and document-centric aspects of the trade applications, they still lack a simple, unified object model which can support the essential features of the underlying supporting technologies.

3 The VOM Architecture

In the context of distributed object computing, CORBA 'turns everything into nails and it gives everyone a hammer'[20]. View Object Model (VOM) architecture goes one step further; it gives everyone a choice of a prefabricated design kit with instructions on how to assemble it. In a sense, VOM model is designed to allow designers to specify and model fully interoperable trade systems.

Imagine, a house-buying process, involving tasks like choosing the house, participating in the negotiation, applying for the home loan, finalizing the payment, obtaining the title etc. which can be finalized within a matter of hours - not days or weeks. This would require a compre-

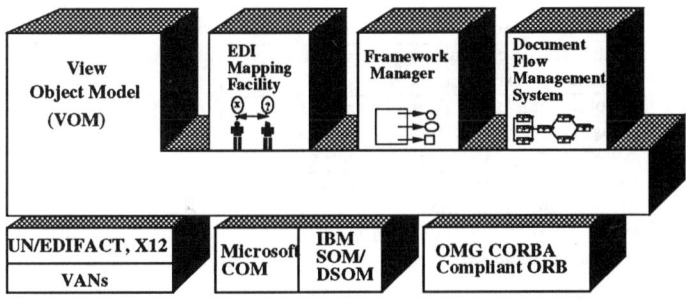

Figure 1: An architecture for global trade documents

The ABC of Communication Theory	Interoperability Type	Object Type
[A]. Ability to transmit symbols (transmision)	Connectivity	Element
[B]. Ability to understand a group of symbols (deals with content)	Interoperability (concern with semantic)	Part
[C]. ability to understand and put contents to good uses in appropriate situation (context)	Co-operation (join together to solve a task)	Dossier or Document

Table 1: VOM First Principle - The ABC of Behaviour Sharing

hensive set of supporting technologies such as CORBA, Lotus Notes, IBM/FlowMark, IBM/Message Queues, and UN/EDIFACT etc. The VOM architecture aims to unify these technologies by alleviating their shortcomings, while removing the unnecessary features of the underlying systems.

The top part of Figure 1 depicts the VOM architecture's three components: EDI Mapping Facility, Framework Manager, and Document Flow Management System (DFMS). The interface layer shows the VOM object model used as an interface between the above three application supporting tools and the following supporting technologies: CORBA, SOM, COM object models.

The next subsection introduces the two basic principles associated with the VOM object model.

3.1 VOM's Two Principles

The primary function of specifying the object model is to capture the various behaviours and properties of the underlying objects, thus enabling the specification of a more generic language for building co-operative systems. Closely related to the VOM model are two principles: the first is concerned with interoperability and the second deals with the component software methodology.

3.1.1 First Principle - The ABC of Behaviour Sharing

To examine this principle, it is useful to note a passage from one of the earliest papers on communication theory, by Warren Weaver [22]:

> Relative to the broad subject of communication, there seem to be problems at three levels. Thus it seems reasonable to ask, serially:
> Level A. How accurately can the symbols of communication be transmitted? (the technical problem.)
> Level B. How precisely do the transmitted symbols convey the desired meaning? (the semantic problem.)
> Level C. How effectively does the received meaning affect conduct in the desired way? (The effectiveness problem.)

Similarly, [7] distinguishes three different levels of interoperability with respect to inter-operable systems: A. connectivity, B. interoperability, and C. co-operation.

Combining the basic concepts about communication, interoperability and the notion of object granularity, we

form the basic hypothesis about behaviour sharing associated with object interoperability. We refer to this combined notion as 'the ABC of behaviour sharing'.

Table 1 shows the relationships between communication, interoperability and VOM objects. This simple but powerful combination forms the of this research. To illustrate the practical utility of this principle, we next describe sample solutions to the trade problems associated with each of the interoperability levels (eg. A,B,C).

With reference to the first principle shown in table 1, level A is concerned with the trade problems associated with the sending and receiving of trade information. This problem can be effectively solved by the technical solutions provided by the maturity of network technologies such as ONC RPC, DCE RPC, or TCP/IP.

Level B relates to the ability of the application to systematically understand trade documents. This problem has been partially solved by the technical solutions provided by the third party standards such as ANSI ASC X12, UN/EDIFACT, and Open/EDI.

Level C is concerned with how trade documents can co-operate to solve a task. This problem has not been fully addressed by the abovementioned technologies. However, these technologies provide piecemeal solutions to the problem. In our research, we address the issue as a whole by combining these technologies to solve the problem.

Imagine a scenario where A wants to send a message to B, but for some reason it arrives at the wrong addressee, C. If C and A speak the same language, then C can decode - but without the context to understand - A's message. However, if C and A do not speak the same language, then A's message is seen by C as a collection of symbols with no apparent syntax or construct. Alternatively, if the message arrives at the right address, B being the intended receiver, should have the background and context to decode and understand the message.

In a sense, the C level of interoperability allows trade documents to co-operate based on the context of the interchanged documents. In designing the VOM architecture, we aim to facilitate both the B and C levels of interoperability.

3.1.2 VOM's Second Principle - The Document Centric paradigm

To see the benefits of the second principle, it is useful to distinguish between 'document-centric' versus 'application-

Characteristic	Document-centric	Application-centric
Management of control	control and diverse information is contained in the document definitions	control and diverse information is contained in the programming codes
Incremental development	is possible since new behaviours can be added and removed without disturbing other processes associated with the document	is not possible, since a complete application design must be coded and compiled before a working application can be installed
Flexible route	the paths which a document travels can be dynamically updated	the flow of the application is fixed by the program's algorithms
Adaptable to role	document transforms its behaviours to suit the actor roles; here the actor's roles may change at any time	the program's functions and who can use it must be previously determined
Responsive to rule	at each interchange, a document may react to a set of local rules to carry out the local semantic requirements; here semantics which govern the application may also change at any time	it is slow to react to the changes in the application's rules and semantics since both rules and applications must be updated and re-compiled
Flexible EDI messages	semantic heterogeneity can be masked out by either EDI standards or type hierarchies hence ad-hoc and new business messages can be dynamically added to	semantic heterogeneity can only be masked out by EDI standards which are too time consuming to adjust to new business messages due to the required elaborate negotiations

Table 2: VOM Second Principle: to promote document-centric model

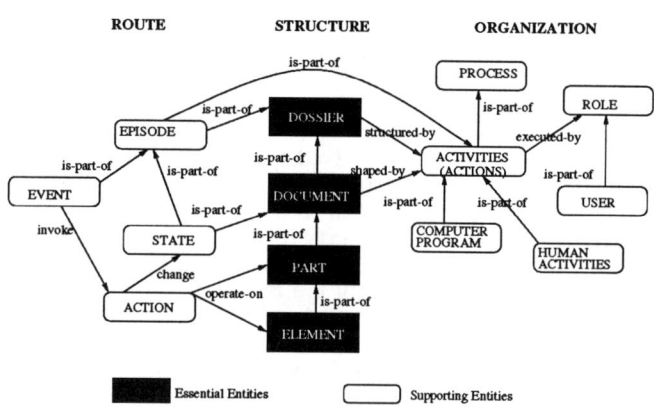

Figure 2: Elements of the document-centric model

centric' paradigms.

One common property in the programming languages, C, C++, Java, SQL, etc. is that they are designed to promote what we consider as an 'application-centric' type of programming, since all of the program controls and other diverse type of information are captured, designed and embedded into the applications' codes. In contrast, the 'document-centric' type of programming shifts the controls from the application programs to the documents themselves. A document contains all of the necessary information about itself and the useful relationships associated with other documents. Moreover, it allows its properties and behaviours to adapt to each of the associated roles and actions.

Table 2 compares the different characteristics associated with the above two paradigms. Flexibility and responsive characteristics of the document-centric application suggest that shifting from application-centric to document-centric requires a specialized object model.

The next subsection describes the information flow associated with the VOM object model.

3.2 Information flow

As we mentioned previously, the core of a document-oriented workflow application is the dossier and its constituent parts. A collection of documents form a dossier (eg. a *Course-Taking* dossier may contain the various university related documents). A document may be comprised of many different parts (eg. a university application may consist of a student's personal details, educational background, and needs etc). Similarly, each part may be comprised of many different primitive elements (eg. a personal details may consist of name, address, date of birth, etc.). As shown in Figure 2, associated with a dossier is a collection of episodes and documents. An episode defines the complete semantic unit for triggering a particular set of events in which various actions can be applied onto a set of documents. Sometimes it is necessary for actions to be carried out on behalf of the document elements. In a sense, the episodes can be used to capture the dynamic aspects of a business process. Work described in [5] uses these episodes to fabricate the so called Documentary Petri-Net (DPN).

The episode specifies the vital EDI messages which are necessary to carry out the collaborated activities. Each complete semantic unit of the episode is either executed completely or not at all. In the VOM model, it is used to support the commit, abort and rollback notions associated with trade documents.

In order to support the organizational aspects of a workflow, the process, activities, role, and user attributes are part of the dossier's supporting entities. Their relationships with the dossier and its constituent parts are illustrated in Figure 2.

We have described the essential and supporting entities of the document-centric model. The next subsection details a special property which is designed to correctly execute and manage the related trade objects.

3.3 'ATOMICS' Property

It is well recognized that most WFMSs lack the ability to ensure the correctness and reliability of the workflow execution in the presence of concurrency and failure [13]. As [3] points out, TWFs provide the useful workflow context based on the traditional transaction concepts and database oriented aspects. However, this solution is only a fraction of the overall picture, much in the same way that transaction management is only one of the many parts of data systems. Tying the VOM model to a particular transaction property will result in major restrictions that will limit its applicability and usefulness as an advanced trade application supporting tool. For this reason, the following 'ATOMICS' property is designed to ensure the correctness of the distributed trade objects, in the same way that the ACID property is does in the classic transaction models.

- **Atomicity:** is a stand alone, useful trade artifact which represents an 'all or nothing' proposition similar to the A function of the ACID property.
- **Transmittency:** artifact can be sent and received regardless of the underlying platforms and network protocols (eg. A level of interoperability).
- **Organizational-oriented:** artifact can only be understood and interoperated under the scope of a particular organization (eg. C level of interoperability).
- **Modularity:** related trade messages are modularized into a series of episodes so they can be effectively managed and reused.
- **Interoperable:** In addition to the isolation function of the ACID property, all artifacts should be identifiable by a global identification scheme similar to OIDs in ODBMS, and be able to be operated across heterogeneous platforms (eg. B level of interoperability).
- **Composable:** artifacts can be combined with others to form another unique and identifiable part.
- **Semantically-complete:** all of the information needed (by the receiver from the sender) must be in the information unit. This is a mechanism to ensure that no business commitment can be expected of a recipient until all of the necessary information is received.

ACID vs. ATOMICS property: The ACID property is used to reflect the current state of the data in real time. In contrast, the workflow model is good at reflecting the changing states of information over time, but does not do very well when it comes to reflecting the current state of the data in real time. This is because the workflow model is not transaction oriented in the ACID sense, and therefore does not use the two-phases commits to synchronizing distributed changes across heterogeneous environments. Thus, supporting a new class of groupware applications (eg. trade monitor), which are neither transactional nor workflow-oriented in a pure sense - but a combination of both - requires a new set of properties. For this reason, the ATOMICS property is designed to reflect the creation, flow, and tracking of trade information which is to be distributed across heterogeneous interchanges over time, in addition to reflecting the current state of trade information in real time.

The new ATOMICS property integrates the ACID and workflow properties in order to support the overall interoperability and semantic heterogeneity aspects that are lacking in either the transactional or workflow models above. More specifically, the ATOMICS property is designed to specify the behavioural properties of the trade information. It is used to indicate whether or not a trade object is reusable, compose-able, exchangeable (eg. EDI-processable)

and interoperable among the different frameworks. In short, the ACID property specifies the behavioural notions associated with transactions, whereas the ATOMICS property specifies the behavioural notions associated with trade information.

3.4 VOM's Run-time architecture

In the document-centric model, the monitoring and execution of a workflow does not need to be decomposed and distributed as it does in the top-down approach. Each dossier contains the necessary routing definitions to direct the documents to the desired places, and the necessary semantics to invoke the required actions to be employed in its constituent parts. By definition, each dossier by itself, is a distributed component of a workflow. Each dossier is thus in charge of its own part of the trading process. To exhibit the one-to-many relationship, it acts as the node manager by directing the control and data flow associated with other dossiers.

For a particular trading, a specific type of dossier is requested and downloaded from a global, standardized repository (built and maintained by a particular trade organization) onto a node. A node may contain one or more dossiers. During an execution, each dossier creates a dossier table which contains the static information describing the various relationships between that dossier and other related dossiers (resident in other nodes). A dossier table is created to describe the static information (metadata) about that dossier's infrastructure, route and organizational aspects.

Associated with each dossier table is a dossier thread. The dossier thread is used to manage the execution of a dossier's instances. The dossier thread uses the information stored in the dossier table to communicate with other dossiers. During the execution of the dossier thread, an instance table is created to keep track of the run-time information associated with each of the trading steps. In particular, the trading steps describe the 'When-Which-What' notion associated with a dossier event. The When notion is used to tell when a particular transaction is started or stopped, the Which notion is used to tell which of the dossier's constituent parts may be affected, and the What notion is used to tell what activities can be carried out on behalf of which parts.

Communication between the dossier instances can be facilitated by the PUT/GET calls of the message queue which is stored at the dossier's node. In this way, a node failure relating to any particular instance can be recovered by the persistent feature of the message queue. By using the transactional PUT and GET calls of the message queue we can introduce the transactional capability into the VOM model. Moreover, the persistent message queue facilitates asynchronous communication between the applications that run at different points in time.

Due to the space restriction, we cannot provide a detailed example of this execution model. Instead, we wish to demonstrate that the execution of a dossier does not require the existence of an overall business process. The flow of the documents can be formulated based on the information available from the trading steps stored in the instance table.

In this subsection, we have described the basic architectural components associated with the execution of the VOM model. In particular, we have briefly introduced the components which are necessary to support the so-called incremental approach to distributed workflow. In this approach, crash failures can be isolated and recovered due to the distributed and persistent nature of the dossier and its message queue. Furthermore, replication can be used to support critical applications. Finally, in the presense of node failures, transactional recovery can also be supported.

3.5 Supporting a Trade Application

Figure 3 is used to demonstrate how a trade application can be supported. The relevant supporting steps are described below.

Step 1: Document design Depending on the characteristics of the business process, an advanced trade application can be built based on the following two approaches: top-down or bottom-up. In the top-down approach, a user may use C++ or Smalltalk combined with the appropriate APIs to define a workflow and its constituent parts. The workflow definitions can then be compiled to generate runtime objects such as CORBA, OLE, and FlowMark etc. A trade designer may alternatively use the VOM to specify the document's infrastructure, flow, EDI class hierarchies, and interface definitions associated with a bottom-up approach. The document infrastructure defines how the document is formed. The document flow specifies the coordination of the document's constituent parts and elements. The EDI class hierarchies act as the dictionaries to systematically interpret the document's messages. Finally, the use of well-defined interface definitions facilitates the object's distribution, interoperability and semantic heterogeneity.

Step 2: Document execution The VOM document's class definitions (eg. Java applets) are the formal representations of the Lotus Notes, OLE , or FlowMark objects. These Java-like mini-applications can be compiled to generate a collection of Java-based VOM bytecodes. In this way, run-time document objects can be executed and monitored on any hardware and software platforms due to the high-level, machine-independent codes which can interpret and run on any machine on which the Java interpreter and run-time system have been implemented. If necessary, Java bytecodes can be dynamically translated into native machine code if required by performance demand. As the documents migrate from one dossier to another, they take on different shapes depending on the different dossier's guidelines. In governing these guidelines, the dossier thread and instance thread communicate with the dossier table and instance table for the necessary static and run-time information.

Step 3: Document usage End users can interact with the trade documents by activating the various document management tools. By using the bottom up approach, a user can build an advanced trade application without the complications and limitations inherent in the conventional top-down approaches. The following section describes the management tools associated with the model.

4 Document Management Tools

To illustrate the utility of the proposed model, this section describes the essential facilities which are designed to support the interpretation, creation, and coordination of the trade documents.

4.1 EDI Mapping Facility

Hoping to overcome the semantic heterogeneity associated with the trade objects, this component is designed to

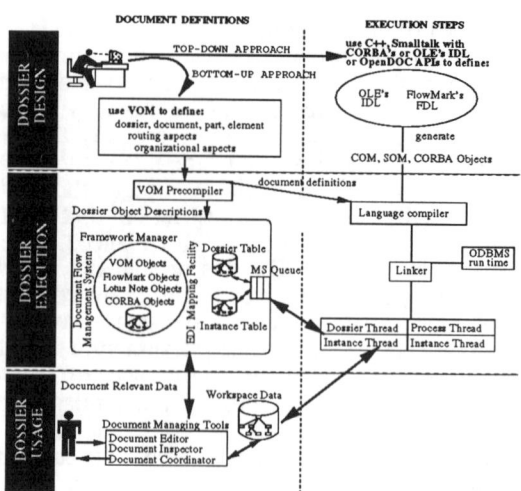

Figure 3: Supporting an advanced trade application

support the different types of trades: 1st trade, established trade, and trade with revision.

Embedded in this component is a key information structure - the Electronic Documentary Dossier - a compound framework containing computer processable information such as trade procedures, different types of trades, and the EDI message protocols etc. This information structure supports the interoperability of information parcels in an open trade environment. In a sense, the dossier construct is essentially a documentary folder that is capable of storing and managing the various types of trades and associated commodities (eg. purchase order, letter of credit, invoice, etc.) which can both be formally represented and interoperated across heterogeneous platforms.

In the context of DFMS, the EDI mapping facility aims to support the following problems:

- How the standardized trade messages and procedures can best be coordinated and managed;
- What the various information properties which can be embedded in these trade messages and procedures are;
- How these entities can be shared and distributed;
- And most importantly, how these trade entities can be considered as 'EDI processable' (eg. systematically interpret and process the trade information) and 'system interoperable' (eg. populate and migrate trade information from one platform to another);

In doing so, the VOM architecture formally captures and represents the trade commodities such as documentary petri nets (eg. to represent 'way of doing business' agreed between the buyer and seller) and episodes (eg. to represent the interoperable trade messages). These entities are encapsulated in a so-called electronic document dossier.

A dossier is a 4-tuple (D,A,R,E) where D represents a finite set of documents (eg. Home-buying, Grad-school-application, Paper-submission etc.) belonging to the workspace. The internal structures of a dossier can be described as: a document in D comprising of a finite set of acts A; an act A comprising of a finite set of roles R; and each role R, formally described by a finite set of episodes E.

Each episode is also a 4-tuple(EV,IS,AC,FS) to model

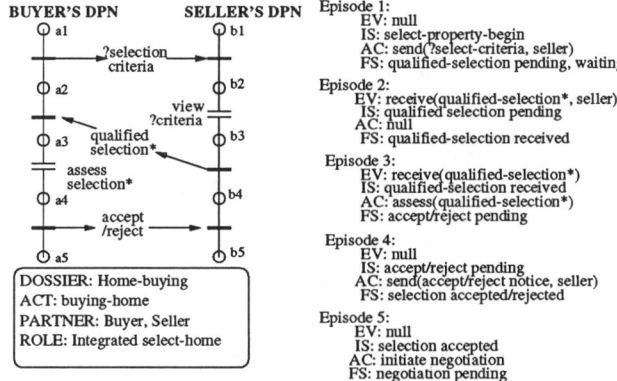

BUYER'S DPN   SELLER'S DPN

```
Episode 1:
    EV: null
    IS: select-property-begin
    AC: send(?select-criteria, seller)
    FS: qualified-selection pending, waiting
Episode 2:
    EV: receive(qualified-selection*, seller)
    IS: qualified-selection pending
    AC: null
    FS: qualified-selection received
Episode 3:
    EV: receive(qualified-selection*)
    IS: qualified-selection received
    AC: assess(qualified-selection*)
    FS: accept/reject pending
Episode 4:
    EV: null
    IS: accept/reject pending
    AC: send(accept/reject notice, seller)
    FS: selection accepted/rejected
Episode 5:
    EV: null
    IS: selection accepted
    AC: initiate negotiation
    FS: negotiation pending
```

DOSSIER: Home-buying
ACT: buying-home
PARTNER: Buyer, Seller
ROLE: Integrated select-home

Figure 4: Documentary Petri-Net and its episodes

the Complete Semantic Unit (CSU) of a role R. In a sense, the CSUs specify the vital EDI messages which are necessary to carry out the collaborated activities required by a particular role. Each CSU is either executed completely or not at all. This feature is similar to the 'all or nothing' proposition (eg. Atomicity) of the ACID property. In our work, it is designed to support the commit, abort and rollback notions associated with trades.

A CSU comprises event (EV), initial state (IS), final state (FS), and action (AC). There are at least two types of events (others are being investigated): timer-event (eg. if a camera-ready version of the paper is not received within 10 days then send a reminder notice) and obligatory-event (eg. if the buyer's bank receives a confirmation about 'title transferred' then it is obliged to release the payment to the seller's bank). The structure of the dossier construct is illustrated in Figure 2.

As an example, Figure 4 illustrates the documentary petri-net and its relevant episodes which show the trade relationships between the seller and buyer participating in a particular role.

4.2 Framework Manager

This component of the VOM architecture allows applications to create and maintain work items associated with inter-organizations' trade documents.

These documents are the integrated containers of different frameworks, composite parts and atomic elements. Document objects can be edited or activated in place. In a framework document, parts and atomic elements can be inserted. Parts are the fundamental building blocks of the VOM framework. Similarly, atomic elements are the elementary building block of the VOM part. A part may contain different types of data eg. spreadsheet cells with formula, images, fax, graphics, or video clips etc. A part may contain other parts. The framework handlers are the independent Java-based source codes (applets) that manipulate and display the different framework objects. There are three types of framework handlers: editors, inspectors, and auditors.

The Framework Editor is used to display a framework's contents and provide a user interface for modifying those contents. The Framework Inspector allows a user to carefully examine the framework's content without editing them. The Framework Coordinator allows the user to

place constraints which enforce certain behaviours onto the framework's content. The part and element handlers have similar functionalities to the framework handlers. However, the part and element handlers are designed to work with finner grained objects such as documents' parts and atomic elements.

For example, in relation to the home-buying scenario, a buyer can use the Element Editor to create a loan request message and send it to the bank; the loan officer can use the Element Inspector to examine the content of this message; the loan officer can use the Framework Editor to put together a loan application document (unfilled). The other users such as the bank manager, buyer, seller, and title company can use the Part Coordinator to control and regulate the mortgage contract part to enforce business rules such that if the buyer has not received the title within 10 days after signing the contract, the payment to the seller will be re-credited back to the buyer's bank account.

4.3 Document Flow Management System (DFMS)

The DFMS, a component of the VOM architecture, helps to bring the information specifically to the people who can act on it. The DFMS has a similar functionality to the conventional WFMS. However, their approaches to workflow are different because DFMS is designed to support the document-centric type of application whereas the conventional WFMS is designed to support the process-oriented type of application.

In the context of the top-down approach, submitting a home mortgage application and having it go through in a matter of hours would require a good workflow model which coordinates the different parts of the work, and tracks these parts to make sure the work gets done by the right people. Thus, this 'global workflow' provides the necessary semantics to direct and monitor the flow of the trade information. However, there are times when this global workflow may not exist. In this case, the knowledge and semantics required for the coordination and management of the trade information can only be derived from the information stored in the different dossiers. This enables the DFMS to reason, make plans, and correctly distribute these objects.

An existing workflow by itself cannot do too much, but in combination with Email, EDI services, persistent object stores, desktop productivity tools and the embedded behaviours and properties of the trade documents, it can create some dynamite combinations. This is one of the important aims of this component.

The reason is that the behaviours and properties of every trading object can be identified and managed due to the atomic properties associated with these objects. As pointed out earlier, each of the dossier contains sufficient knowledge about a business process in order for it to correctly coordinate, execute, monitor, and manage the various trade messages and documents.

Open trading often leads to obscure knowledge about the trade information, since definitions and the semantics of the workflow may not be available up-front. As an example, a student may submit his or her personal profile, education needs and background to an education placement service without really knowing what kind of advice (or contingencies) he/she may get in return. This is due to the fact that different institutions may have different placement policies and internal factors (eg. lack of courses) which may affect the placement outcomes. Often, these policies and factors are 'black-boxes' to the applicants. The point is that in the document-oriented model, it is natural to model the exceptions related to trades. For example, contingencies to a particular university application can be easily expressed as additional contingencies. Where the rules to represent the contingencies may not be known in advance, the top-down approach is not very useful. The bottom-up approach is more suitable for supporting incremental trade applications.

5 Conclusion

This paper introduces an intergrated architecture which is designed for the purpose of specifying, executing, and managing trade documents. A view object model representing a collection of programming abstractions can be used to specify, share, route and re-use the different trade documents and procedures. This model can be used to design interoperable trading systems with a single set of well-defined abstractions instead of having to deal with the many different sets of localized properties and behaviours of the individual supporting technologies.

References

[1] G. Alonso, D. Agrawal, A. Abbadi, K. Kamath, R. Günthör, and C. Mohan. Failure Handling in Large Scale Workflow Management Systems. IBM Research Report RJ9913, IBM Almaden Research Center, Nov. 1994. available in http://www.almaden.ibm.com /cs/exotica.

[2] G. Alonso, D. Agrawal, A. Abbadi, K. Kamath, R. Günthör, and C. Mohan. Exotica/FMQM: A Persistent Message-Based Architecture for Distributed Workflow Management. In *Proc. IFIP Working Conference on Information System for Decentralized Organizations*, 1995.

[3] G. Alonso, D. Agrawal, A. Abbadi, K. Kamath, R. Günthör, and C. Mohan. Advanced transaction models in workflow contexts. In *Proc. IEEE 12th International Conference on Data Engineering*, 1996.

[4] M. Ansari, L. Ness, M. Rusinkiewicz, and A. Sheth. Using Flexible Transactions to Support Multi-System Telecommunication Applications. In *VLDB Conference*, 1992.

[5] R. Bons, R. Lee, R. Wagenaar, and C. Wrigley. Modelling Inter-organizational Trade Procedures Using Documentary Petri Nets. In *The Hawaii International Conference on System Sciences*, 1994.

[6] K. Brockschmidt. *Inside OLE, 2nd Edition*. Microsoft Press, Redmond, WA, 1995.

[7] M. L. Brodie. The promise of distributed computing and the challenges of legacy information systems. In D. Hsiao, E. J. Neuhold, and R. Sacks-Davis, editors, *Proc. IFIP TC2/WG2.6 Conference on Semantics of Interoperable Database Systems*, Amsterdam, 1993.

[8] J. L. Dietz. Business modelling for business redesign. In *Proceedings of 27th Hawaii International Conference on System Sciences (HICSS)*, 1994.

[9] A. Elmagarmid, editor. *Database Transaction Models For Advanced Applications*. Morgan Kaufmann, 1992.

[10] A. Elmagarmid, J. Chen, and O. Bukhres. Remote System Interfaces: An approach to overcoming the heterogeneity barrier and retaining local autonomy in the integration of

heterogeneous systems. *International Journal of Intelligent and Cooperative Information Systems*, 2(1):1–22, 1993.

[11] A. K. Elmagarmid, Y. Leu, W. Litwin, and M. Rusinkiewicz. A multidatabase transaction model for interbase. In *Proc. of the 16th VLDB Conference*, 1990.

[12] H. Garcia-Molina and K. Salem. Sagas. In *Proc. 1987 SIGMOD International Conference on Management of Data*, 1987.

[13] D. Georgakopoulos, M. Hornick, and A. Sheth. An Overview of Workflow Management: From Process Modeling to Workflow Automation Infrastructure. *Distributed and Parallel Databases*, 3(2):119–153, 1995.

[14] M. Hämäläinen, A. B. Whinston, and S. Vishik. Electronic Markets for Learning: Education Brokerages on the Internet. *Communications of the ACM*, 39(6), June 1996.

[15] IBM. *FlowMark - Modeling Workflow, Version 2.1*. IBM Document No. SH19-8241-00, Mar 1995.

[16] ISO/IEC/JTC1/WG3. The Open-EDI Reference Model, Working Draft Document N255, 1994.

[17] P. Kimberley. *Electronic Data Interchange*. McGraw-Hill, Inc., New York, New York, 1991.

[18] J. Miller, A. Sheth, K. Kochut, and X. Wang. CORBA-Based Run-Time Architectures for Workflow Management Systems. *Journal of Database Management, Special issue on Multidatabases*, 7(1), 1996.

[19] Object Management Group and X Open, Document Number 91.12.1 Revision 1.1. *The Common Object Request Broker: Architecture and Specification, Revision 2.0*, Jul 1995.

[20] R. Orfali, D. Harkey, and J. Edwards. *Essential Client/Server Survival Guide*. John Wiley & Sons, 1994.

[21] B. Reinwald and C. Mohan. Structured Workflow Management with Lotus Notes Release 4. In *Proc. 4th IEEE Computer Society International Conference (CompCon), digest of papers*, 1996. available in http://www.almaden.ibm.com/cs/exotica.

[22] C. E. Shannon and W. Weaver. *The mathematical Theory of Communication*. The University of Illinois Press, Urbana, Illinois, 1949.

[23] A. Sheth and M. Rusinkiewicz. On transactional workflows. *IEEE Data Engineering Bulletin*, 16(2), 1993.

[24] A. Sheth and M. Rusinkiewicz. Specification and execution of transactional workflows. In W. Kim, editor, *Modern Database Systems: The Object Model, Interoperability, and Beyond*. AMC Press, 1995.

[25] A. P. Sheth and J. A. Larson. Federated Database Systems for Managing Distributed, Heterogeneous, and Autonomous Databases. *ACM Computing Surveys*, 22(3), Sept. 1990.

[26] S. Vinoski. Distributed Object Computing with CORBA. *C++ Report Magazine*, July 1993.

[27] G. Williams. OpenDoc Is Cross-Platform, 1994. http://www.cilabs.org /xplaform.html, Appeared in the November 1994 Edition of Apple Directions.

[28] Workflow Management Coalition. *Workflow Management Coalition Workflow Reference Model Specification*, Jan 1995. http://www/aia.ed.ac.uk.wfmc.

[29] A. Zhang, M. Nodine, B. Bhargava, and O. Bukhres. Ensuring Relaxed Atomicity for Flexible Transactions in Multidatabase Systems. In *Proc. 1994 SIGMOD International Conference on Management of Data*, 1994.

Session 3C

Applications: Factory Automation

Chair

Hermann Kopetz

A Robot with a Decentralized Consensus-making Mechanism Based on the Immune System

Akio ISHIGURO, Yuji WATANABE, Toshiyuki KONDO,
Yasuhiro SHIRAI and Yoshiki UCHIKAWA

Dept. of Information Electronics, Graduate School of Eng.,
Nagoya University Furo-cho, Chikusa-ku, Nagoya, 464-01, Japan
Phone: +81-52-789-3167, Fax: +81-52-789-3166
Email: {ishiguro/yuji/kon/shirai/uchikawa}@bioele.nuee.nagoya-u.ac.jp

Abstract

In recent years much attention has been focused on behavior-based artificial intelligence(AI), which has already demonstrated its robustness and flexibility against dynamically changing world. However, in this approach, the followings have not yet been resolved: how do we construct an appropriate arbitration mechanism, and how do we prepare appropriate competence modules. In this paper, to overcome these problems, we propose a new decentralized consensus-making system inspired by the biological immune system. And we apply our proposed method to behavior arbitration for an autonomous mobile robot, namely garbage collecting problem that takes into account of the concept of self-sufficiency. To verify the feasibility of our method, we carry out some simulations. In addition, we investigate two types of adaptation mechanisms, and try to evolve the proposed artificial immune network using reinforcement signals.

1 Introduction

In recent years much attention has been focused on *behavior-based AI*, which has already demonstrated its robustness and flexibility against dynamically changing world. In this approach, intelligence is expected to result from both mutual interactions among *competence modules* (*i.e.* simple behavior/action) and interaction between the robot and environment. However, there are still open questions: 1) how do we construct a mechanism that realizes appropriate arbitration among multiple competence modules, and 2) how do we prepare appropriate competence modules.

Brooks has showed a solution to the former problem with the use of *subsumption architecture* [17, 18]. Although this method demonstrates highly robustness, it should be noted that this architecture arbitrates the prepared competence modules on a fixed priority basis. It would be quite natural to vary the priorities of the prepared competence modules according to the situation. *Maes* proposed an another flexible mechanism called *behavior network system* [15, 16]. In this method, agents (*i.e.* competence modules) form the network using cause-effect relationship, and an agent

suitable for the current situation and the given goals emerges as the result of activation propagation among agents. This method, however, is difficult to apply to a problem where it is hard to find the cause-effect relationship among agents.

On the other hand, the immune system has various interesting features such as immunological memory, immunological tolerance, pattern recognition, and so on viewed from the engineering standpoint. Recent studies on immunology have clarified that the immune system does not just detect and eliminate the non-self materials called *antigen* such as virus, cancer cells and so on, rather plays important roles to maintain its own system against dynamically changing environments through the interaction among *lymphocytes/ antibodies*. Therefore, the immune system would be expected to provide a new methodology suitable for dynamic problems dealing with unknown/hostile environments rather than static problems.

Based on the above facts, we have been trying to engineer methods inspired by the biological immune system and there application to robotics [1, 2, 3]. We expect that there would be an interesting AI technique suitable for dynamically changing environments by imitating the immune system in living organisms. In this paper, we propose a new decentralized consensus-making system inspired by the biological immune system. We then apply our proposed method to behavior arbitration for an autonomous mobile robot, namely *garbage collecting problem* that takes into account of the concept of *self-sufficiency*. In order to verify the validity of our method, we perform some simulations. In addition, we try to evolve the proposed artificial immune network using reinforcement signals. Finally, we show an another adaptation mechanism from the selectionist standpoint.

2 Overview of the biological immune system

The basic components of the biological immune system are *macrophages*, *antibodies* and *lymphocytes* that are mainly classified into two types, namely *B-lymphocytes* and *T-lymphocytes*. B-lymphocytes are the cells maturing in *bone marrow*. Roughly 10^7 distinct types of B-lymphocytes are contained in a human body, each of which has distinct molecular structure and produces "Y" shaped antibodies from its surfaces. The antibody recognizes specific *antigens*, which are the foreign substances that invade living creature, such as virus, cancer cells and so on. This reaction is often likened to *a lock and key relationship* (see Fig.1). To cope with continuously changing environment, living systems possess enormous repertoire of antibodies in advance. On the other hand, T-lymphocytes are the cells maturing in *thymus*, and they generally perform to kill infected cells and regulate the production of antibodies from B-lymphocytes as outside circuits of B-lymphocyte network (idiotypic network) discussed later.

For the sake of convenience in the following explanation, we introduce several terms from immunology. The portion on the antigen recognized by the antibody is called an *epitope* (antigen determinant), and the one on the corresponding antibody that recognizes the antigen determinant is called a *paratope*. Recent studies in immunology have clarified that each type of antibody also has its specific antigen determinant called an *idiotope* (see Fig.1).

Based on this fact, *Jerne* proposed a remarkable hypothesis which he has called the *"idiotypic network hypothesis"*, sometimes called *"immune network hypothesis"* [8, 9, 12, 13, 14]. This network hypothesis is the concept that antibodies/lymphocytes are not just isolated, namely they are communicating to each other among different species of antibodies/lymphocytes. This idea of *Jerne's* is schematically shown in Fig.2. The idiotope Id1 of antibody 1 (Ab1) stimulates the B-lymphocyte 2, which attaches the antibody 2 (Ab2) to its surface, through the paratope P2. Viewed from the standpoint of Ab2, the idiotope Id1 of Ab1 works simultaneously as an antigen. As a result, the B-lymphocytes 1 with Ab1 are suppressed by Ab2. On the other hand, antibody 3 (Ab3) stimulates Ab1 since the idiotope Id3 of Ab3 works as an antigen in view of Ab1. In this way, the stimulation and suppression chains among antibodies form a large-scaled network and works as a self and not-self recognizer. Therefore, the immune system is expected to provide a new parallel distributed processing.

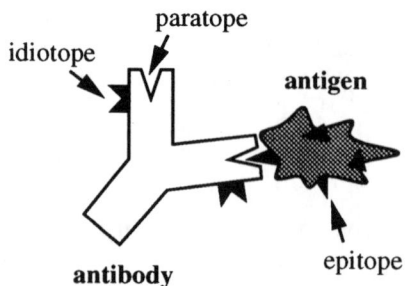

Fig. 1: Structure of an antigen and an antibody.

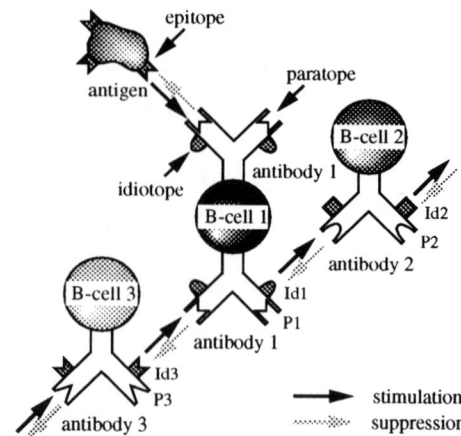

Fig. 2: Jerne's idiotypic networks hypothesis.

3 Proposed consensus-making network based on the biological immune system

3.1 Action selection problem and the immune system

As described earlier, in the behavior-based AI, how to construct a mechanism that realizes appropriate arbitration among the prepared competence modules must be solved. We approach to this problem from the immunological standpoint, more concretely with use of immune network architecture. Fig.3 schematically shows the action selection system for an autonomous mobile robot and the immune network architecture. As shown in this figure, current situations, (*e.g.* distance and direction to the obstacle, *etc.*) detected by installed sensors work as multiple antigens, and a prepared competence module (*i.e.* simple behavior) can be regarded as an antibody (or B-lymphocyte), while the interaction between modules is represented by stimulation and suppression between antibodies. The basic concept of our method is that the immune system equipped with the autonomous mobile robot selects a competence module (antibody) suitable for the detected current situation (antigens) in a bottom-up manner.

3.2 Problem

For the ease of the following explanation, we firstly describe the problem used to confirm the ability of an autonomous mobile robot with our proposed immune network-based action selection mechanism (for convenience, we dub the robot *"immunoid"*). To make the immunoid really autonomous, as *Pfeifer et al.* advocated, it must not only accomplish the given task, but also be *self-sufficient*[19, 5]. Inspired by their works, we adopt the following *garbage collecting problem* that takes into account of the concept of self-sufficiency. Fig.4 shows the environment. As can be seen in the figure, this environment, surrounded by walls, has a lot of garbage to be collected. And there exist a garbage

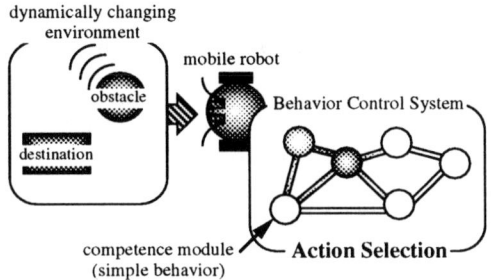

(a) An autonomous mobile robot with an action selection mechanism.

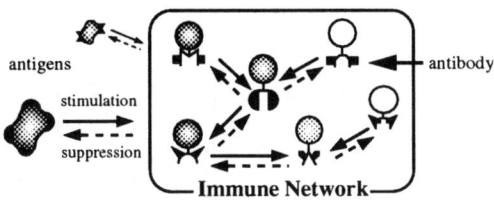

(b) Immune network architecture.

Fig. 3: Basic concept of our proposed method.

can and a battery charger in a home base. The task of the immunoid is to collect the garbage, and put it into the garbage can without running out of its energy (*i.e.* battery level). Note that the immunoid consumes some energy as it moves around the environment. This can be similar to the *metabolism* in the biological system.

In this study, we assume that prespecified quantity of initial energy is given to the immunoid, and the current energy level can be detected by the simulated internal sensor installed in the immunoid. For quantitative evaluation, we also use the following assumptions:

1. The immunoid consumes energy E_m with every step.

2. The immunoid loses additional energy E'_m when it carries garbage.

3. If the immunoid collides with obstacles (*i.e.* walls), it loses some energy E_c.

4. If the immunoid reaches the home base, it instantaneously obtains full energy.

5. If the energy level of the immunoid is high, go_to_home_base behavior might not emerge to avoid over-charging.

Based on the above assumptions, we calculate current energy level as:

$$E(t) = E(t-1) - E_m - E'_m - E_c, \qquad (1)$$

where $E(t)$ denotes the energy level at time t.

For ease of understanding, we explain why this problem is suitable for the behavior arbitration problem in detail using the following situations. Assume that the immunoid is in the far distance from the home base, and its energy level is low. In this situation, if the immunoid carries out the garbage, it will run out of its energy due to the term E'_m in equation (1). Therefore, the immunoid should select the go_to_home_base behavior to fulfill its energy. In other word, the priority of the go_to_home_base behavior should be higher than that of the garbage_collecting behavior. On the other hand, if the immunoid is in the near distance from the home base. In this situation, unlike the above situation, it would be preferable to select the garbage_collecting behavior. From these examples, it is understood that the immunoid should select an appropriate competence module by flexibly varying the priorities of the prepared competence modules according to the internal/external situations.

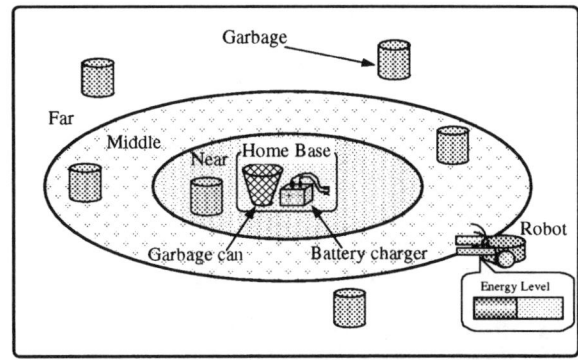

Fig. 4: Environment.

3.3 Definition of the antigens and antibodies

As described earlier, the detected current internal/external situation and the prepared simple behavior work as an antigen and an antibody, respectively. In this study, each antigen informs the existence of garbage (direction), obstacle (direction) and home base (direction and distance), and also current internal energy level. For simplicity, we categorize direction and distance of the detected objects and the detected internal energy level as:

- direction $\rightarrow$ *front, right, left, back*
- distance $\rightarrow$ *far, middle, near*
- energy level $\rightarrow$ *high, low.*

Next, we explain how we describe an antibody in detail. To make the immunoid select a suitable antibody against the current antigen, we must look carefully into the definition of the antibodies. Moreover, we should notice that our immunological arbitration mechanism selects an antibody in a bottom-up manner through interacting among antibodies. To realize the above requirements, we defined the description of antibodies as follows. As mentioned in the previous section, the identity of each antibody is generally determined by the structure (*e.g.* molecular shape) of its

233

paratope and idiotope. Fig.5 depicts our proposed definition of antibodies. As depicted in the figure, we assign a pair of precondition and action to the paratope, and the ID-number of the stimulating antibody and the degree of stimuli to the idiotope, respectively. The structure of the precondition is the same as the antigen described above. And we prepare seven kinds of actions for the immunoid: *move forward, turn right, turn left, turn back, explore, catch garbage, search for home*.

In addition, for the appropriate selection of antibodies, we assign one state variable called *concentration* to each antibody.

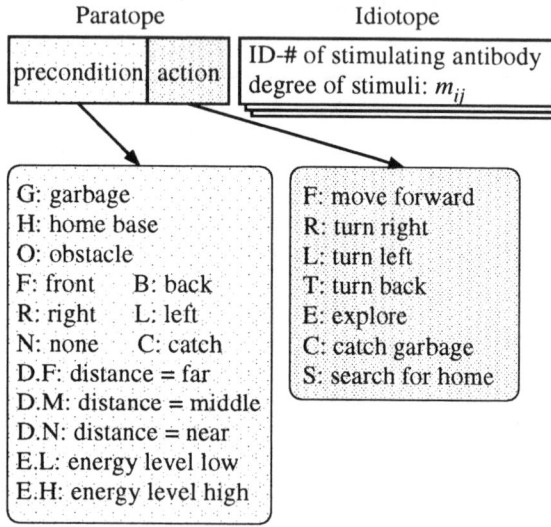

Fig. 5: Definition of antibody.

3.4 Interaction between antibodies

Next, we explain the interaction among antibodies ,that is, the basic principle of our immunological consensus-making networks in detail. For the ease of understanding, we assume that the immunoid is placed in the situation shown in Fig.6 as an example. In this situation, three antigens listed in the figure possibly invade the immunoids interior. Suppose that the listed four antibodies are prepared in advance that respond to these antigens. For example, antibody 1 means that if the immunoid detects the home base in the right direction, this antibody can be activated and would cause `turn_right` action. However, if the current energy level is high, this antibody would give way to other antibodies represented by its idiotope (in this case, antibody 4) to prevent over-charging.

Now assume that the immunoid has enough energy, in this case antibodies 1, 2 and 4 are stimulated by the antigens. As a result, the concentration of these antibodies increases. However, due to the interactions indicated by arrows among the antibodies through their paratopes and idiotopes, the concentration of each antibody varies. Finally, antibody 2 will have the highest

concentration, and then is allowed to be selected. This means that the immunoid finally catch the garbage.

In the case where the immunoid has not enough energy, antibody 1 tends to be selected in the same way. This means that the immunoid ignores the garbage and tries to recharge its battery. As observed in this example, the interactions among the antibodies work as a priority adjustment mechanism.

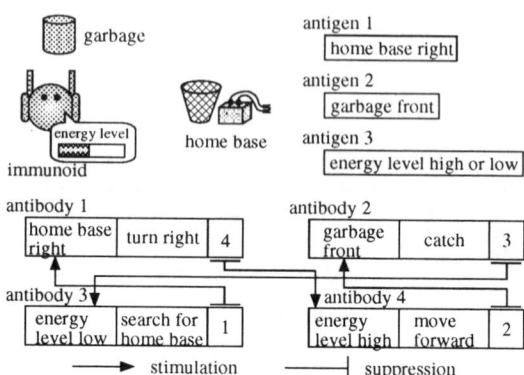

Fig. 6. An example of consensus-making network by interacting among antibodies.

3.5 Dynamics

The concentration of i-th antibody, which is denoted by a_i, is calculated as follows:

$$\frac{dA_i(t)}{dt} = \left(\sum_{j=1}^{N} m_{ji}a_j(t) - \sum_{k=1}^{N} m_{ik}a_k(t) + m_i - k_i \right) a_i(t) \qquad (2)$$

$$a_i(t+1) = \frac{1}{1 + \exp(0.5 - A_i(t))} \quad , \qquad (3)$$

where, in equation (2), N is the number of antibodies. m_{ji} and m_i denote affinities between antibody j and antibody i (i.e. the degree of stimuli), and antibody i and the detected antigen, respectively. The first and second terms of the right hand side denote the stimulation and the suppression from other antibodies, respectively. The third term represents the stimulation from the antigen, and the forth term the dissipation factor (i.e. *natural death*) [9]. Equation (3) is a squashing function to ensure the stability of the concentration. In this study, selection of antibodies is simply carried out on a *roulette-wheel manner* basis according to the magnitude of concentrations of the antibodies. Note that only one antibody is allowed to be selected and act its corresponding action to the world.

3.6 Results

To verify the feasibility of our proposed method, we carried out some simulations. In this study, we prepared 22 antibodies of which paratope and idiotope are described *a priori* (Fig.7). As a rudimentary stage of investigation, we determined the degree of stimuli of each antibody heuristically. In the figure, note that the degrees are omitted in each idiotope for lack of space.

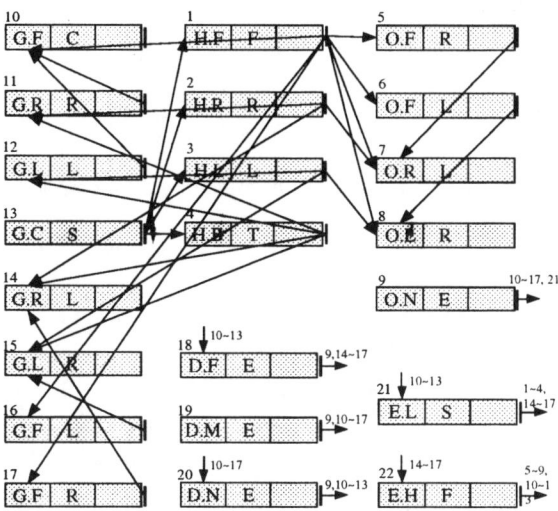

Fig. 7: Prepared immune network.

At the beginning of the simulations, we equipped the immunoid with the maximum energy level (*i.e.* 1000). Typical results obtained in the simulations are as follows: while the energy level is enough, the immunoid tries to collect garbage and carry to the home base. As the remaining energy runs out, the immunoid tends to select an antibodies concerned with `go_to_home_base` and/or `search_for_home_base` behaviors. After successful reaching the home base, the immunoid starts to explore again. Such a regular behavior could be frequently observed in the simulations.

In order to evaluate the ability of our proposed arbitration mechanism, we furthermore carried out simple simulations by varying the initial energy level. Fig.8(a) and (b) are the resultant trajectories of the immunoid in the case where the initial energy level is set to 1000 (maximum) and 300, respectively. In case 1, due to the enough energy level, the immunoid collects the garbage B and successfully reach the home base. On the other hand, in case 2, due to the critical energy level the immunoid ignores the garbage B and then collects the garbage A. From these figures, it is understood that the immunoid selects an antibodies suitable for the current situations by flexibly changing the priorities of the antibodies.

To make our proposed method convincing, demonstration in real environments is highly necessary. We applied our proposed method to a real experimental mobile robot and could observe above typical results. Fig.9 illustrates the experimental robot (Khepera^{TM}).

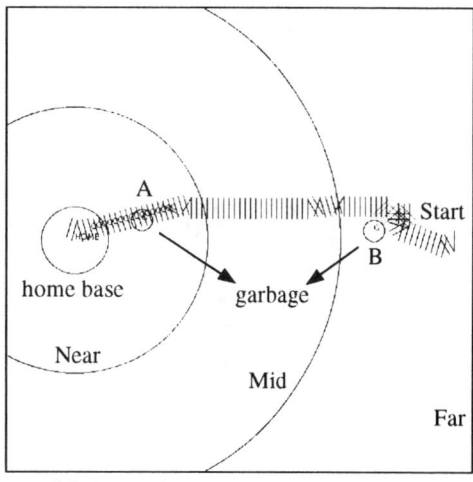

(a) Case 1 (initial energy level = 1000).

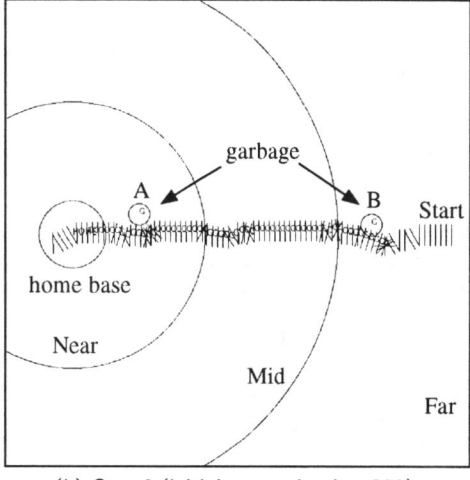

(b) Case 2 (initial energy level = 300).

Fig. 8: Resultant trajectories.

4 Adaptation mechanisms

For more usefulness, as some researchers have been pointed out, the introduction of some adaptation mechanisms is highly indispensable. Adaptation mechanism is usually classified into two types: *adjustment* and *innovation*[4, 10]. In the followings, we propose an adjustment mechanism suitable for the proposed consensus-making system, and show a possible/promising innovation mechanism.

4.1 Adjustment mechanism

For an appropriate consensus-making, it is necessary to appropriately determine the ID-number of the stimulating antibody and its degree of stimuli m_{ij}, *i.e.* priorities among antibodies. To realize this aim, we propose an on-line adaptation mechanism using the advantages of the prementioned description of antibodies. Additionally, it is desired that this mechanism can even work under the situation where the idiotopes

Fig. 9: Real experimental robot (Khepera^{TM}).

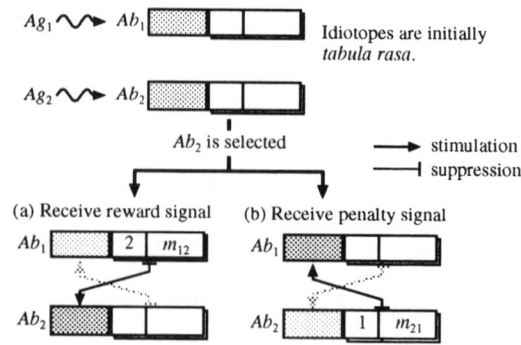

Fig. 10: Proposed adjustment mechanism.

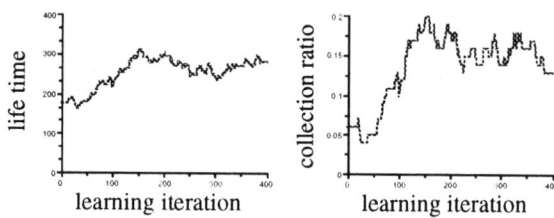

Fig. 11: Transition of life time and collection ratio.

of the prepared antibodies are initially *tabula rasa* (*i.e.* blank).

For the ease of the following explanation, we assume that antigen 1 and 2 invade the immunoids interior (see Fig.10). In this example, antibody 1 (Ab_1) and 2 (Ab_2) are simultaneously stimulated by each antigen. Consequently, the concentration of each antibody increases. However, since the priority between Ab_1 and Ab_2 is unknown (because idiotopes are initially *tabula rasa*, there are no stimulation/suppression chain), in this case, either of them can be selected randomly.

Now, assuming that the immunoid randomly selects Ab_2 and then receives a positive reinforcement signals as a reward. To make the immunoid tend to select Ab_2 under the same or similar antigens(situation), we record the ID-number of Ab_2 (*i.e.* 2) in the idiotope of Ab_1 and increase a degree of stimuli m_{12}. In this study, we simply modify the degree of stimuli as:

$$m_{12} = \frac{T_p^{Ab_1} + T_r^{Ab_2}}{T_{Ab_2}^{Ab_1}} \quad (4)$$

$$m_{21} = \frac{T_r^{Ab_1} + T_p^{Ab_2}}{T_{Ab_2}^{Ab_1}} \quad (5)$$

where $T_p^{Ab_1}$ and $T_r^{Ab_1}$ represents the number of times of receiving penalty and reward signal when Ab_1 is selected. $T_{Ab_2}^{Ab_1}$ denotes the number of times when both Ab_1 and Ab_2 are reacting to their specific antigens.

We should notice that this procedure works to raise the relative priority of Ab_2 over Ab_1. In the case where the immunoid receives a negative reinforcement signal, we record the ID-number of Ab_1 (*i.e.* 1) in the idiotope of Ab_2 and modify m_{21} in the same way. This works to decrease the relative priority of Ab_2 over Ab_1.

To confirm the validity of this adjustment mechanism, we carried out some simulations. Fig.11 denotes transition of life time and collection ratio. From these results, it is understood that both are improved gradually as iterated. We are currently implementing this mechanism into the real experimental system.

4.2 Innovation mechanism

In the above adjustment mechanism, we should notice that we must still describe the paratope of each antibody in a top-down manner. One obvious and promising candidate to avoid such difficulties is to combine an innovation mechanism with the proposed adjustment mechanism. In the biological immune system, the *metadynamics function* can be instantiated as an innovation mechanism [6, 11, 7]. The metadynamics function works to maintain appropriate repertoire of antibodies by incorporating new types (these are generally generated as *quasi-species* through the proliferation process of the activated antibodies) and removing useless ones. Fig.12 schematically shows the concept of the metadynamics function. Incorporating this mechanism is currently under investigation.

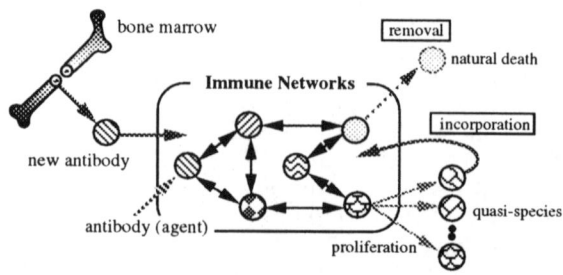

Fig. 12: Metadynamics function.

5 Conclusions and Further work

In this paper, we proposed a new decentralized consensus-making mechanism based on the biological immune system and confirmed the validity of our proposed system by applying to an behavior arbitration for an autonomous mobile robot. And we showed an adaptation mechanism for an appropriate arbitration using reinforcement signals.

Since this study is still in a rudimentary stage of investigation, we designed antibodies a priori in a top-down manner. For more usefulness, we must clarify how to combine the proposed adjustment and innovation mechanisms. This is currently undertaking.

Acknowledgments

This research was supported in part by a Grant-in-Aid for Scientific Research on Priority Areas from the Ministry of Education, Science, Sports and Culture, Japan (No. 07243208).

References

[1] A. Ishiguro, S. Ichikawa and Y. Uchikawa, "A Gait Acquisition of 6-Legged Walking Robot Using Immune Networks", *Journal of Robotics Society of Japan*, Vol.13, No.3, pp.125-128, 1995 (in Japanese), also *in Proc. of IROS '94*, Vol.2, pp.1034-1041, 1994

[2] A. Ishiguro, Y. Watanabe and Y. Uchikawa, "An Immunological Approach to Dynamic Behavior Control for Autonomous Mobile Robots", *in Proc. of IROS '95*, Vol.1, pp.495-500, 1995

[3] A. Ishiguro, T.Kondo, Y. Watanabe and Y. Uchikawa, "Dynamic Behavior Arbitration of Autonomous Mobile Robots Using Immune Networks", *in Proc. of ICEC'95*, Vol.2, pp. 722-727, 1995

[4] B.Manderick, "The importance of selectionist systems for cognition", *Computing with Biological Metaphors*, Ed. R.Paton, Chapman & Hall, 1994

[5] D.Lambrinos and C.Scheier, "Extended Braitenberg Architecture" *Technical Report, AI Lab*, No. IFIAI95.10, Computer Science Department, University of Zurich, 1995

[6] F.J.Valera, A. Coutinho, B.Dupire and N.N.Vaz., "Cognitive Networks: Immune, Neural, and Otherwise", *Theoretical Immunology*, Vol.2, pp.359-375, 1988

[7] H.Bersini and F.J.Valera, "The Immune Learning Mechanisms: Reinforcement, Recruitment and their Applications", *Computing with Biological Metaphors*, Ed. R.Paton, Chapman & Hall, pp.166-192, 1994

[8] H.Fujita and K.Aihara, "A distributed surveillance and protection system in living organisms", *Trans. on IEE Japan*, Vol. 107-C, No.11, pp.1042-1048, 1987 (in Japanese)

[9] J.D.Farmer, N.H.Packard and A.S.Perelson, "The immune system, adaptation, and machine learning", *Physica 22D*, pp.187-204, 1986

[10] J.D.Farmer, S.A.Kauffman, N.H.Packard and A.S.Perelson, "Adaptive Dynamic Networks as Models for the Immune System and Autocatalytic Sets", *Technical Report LA-UR-86-3287*, Los Alamos National Laboratory, Los Alamos, NM, 1986

[11] J.Stewart, "The Immune System: Emergent Self-Assertion in an Autonomous Network", *in Proceedings of ECAL-93*, pp.1012-1018, 1993

[12] N.K.Jerne, "The immune system", *Scientific American*, Vol.229, No.1, pp.52-60, 1973

[13] N.K.Jerne, "The generative grammar of the immune system", *EMBO Journal*, Vol.4, No.4, 1985

[14] N.K.Jerne, "Idiotypic networks and other preconceived ideas", *Immunological Rev.*, Vol.79, pp.5-24, 1984

[15] P.Maes, "The dynamic action selection", *Proc. of IJCAI-89*, pp.991-997, 1989

[16] P.Maes, "Situated agent can have goals", *Designing Autonomous Agents*, pp.49-70, MIT Press, 1991

[17] R.Brooks, "A Robust Layered Control System for a Mobile Robot", *IEEE Journal of R&A*, Vol.2, No.1, pp.14-23, 1986

[18] R.Brooks, "Intelligence without reason", *Proc. of IJCAI-91*, pp.569-595, 1991

[19] R.Pfeifer, "The Fungus Eater Approach to Emotion –A View from Artificial Intelligence", *Technical Report, AI Lab*, No. IFIAI95.04, Computer Science Department, University of Zurich, 1995

Design of Local Communication for Cooperation in Distributed Mobile Robot Systems

Tamio ARAI

Dept. of Precision Machinery Engineering,
The University of Tokyo
7-3-1 Hongo, Bunkyo-ku, Tokyo 113 Japan
arai@prince.pe.u-tokyo.ac.jp

Eiichi YOSHIDA

Mechanical Engineering Laboratory,
AIST, MITI
1-2 Namiki, Tsukuba-shi, Ibaraki 305 Japan
eiichi@mel.go.jp

Abstract

This paper presents a novel design methodology of local communication system for cooperation in distributed mobile robot systems. Our goal is to design a local communication system so as to transmit task information to necessary robots in minimum time without excessive propagation. In this paper, we propose a layered methodology, i.e. design from spatial and temporal aspects based on analysis of information diffusion by local communication between robots. The spatial design gives the optimal communication area minimizing transmission time for various cooperative tasks. In the temporal design, we derive the information announcing time to prevent excessive information diffusion. Finally, the simulations and experiments demonstrate that the design methodology is effective in constructing an efficient local communication system.

1 Introduction

Distributed mobile robot systems are currently expected to accomplish complicated tasks by intelligent cooperation. One of the major essential issues for cooperation in such distributed systems is communication between robots.

Communication in such distributed systems can be classified into the following two types shown in **Fig. 1**:

1. Communication for notification of a task to the number of robots required by the task.
   <u>Content of information</u>: attributes of multiple tasks (e.g. the place and type of task)
2. Communication for task execution.
   <u>Content of information</u>: status of task execution (e.g. the map being constructed in the case of cooperative map generation task)

The above procedure can be generally applied to most of many-robot cooperation.

Global communication has been often utilized in previous studies for robotic systems consisting of only a few robots [1]~[3]. Distributed robotic systems, however, require much greater number of robots for the sake of flexibility and robustness of cooperative task execution as shown in [4] [5].

Global communication has the following problems when applied to distributed many-robot system:

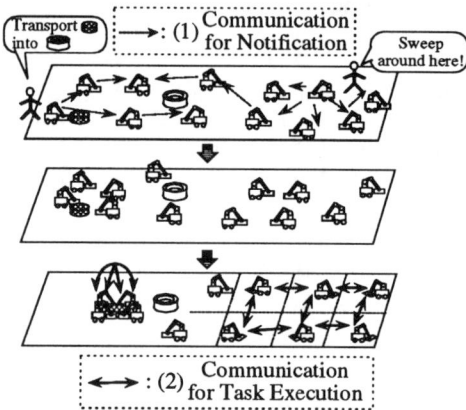

Fig. 1: Two types of communication for cooperation

- The efficiency of information transmission becomes low when a single communication medium is assigned to many robots, like radio network.
- If a central station manages the communication in the system, increasing load may cause communication bottleneck and insufficient fault-tolerance.

For these reasons, local communication has been frequently brought into use in recent research [5] [6]. Authors have been working on local communication shown in **Fig. 2** for many mobile robots [7] [8].

In this simple communication model, a robot sends out information in the form of packet within a limited area. This model has following advantages:

- The communication can be easily implemented using infrared device [9] or camera image [10].
- Information transmission takes place in distributed and concurrent manner, which reduces excessive information processing.
- The overall system becomes robust against addition, removal, or breakdown of robots.

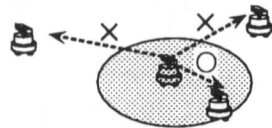

Fig. 2: Local communication between mobile robots

The information contained in a packet is diffused between robots by repeated local transmission like hearsay, along robots' movement. In both types of communication (1) and (2) mentioned above, the information needs transmitting to the number of robots necessary for cooperation *efficiently*, namely in minimum time without excessive information diffusion. So it is important to know how information is diffused so as to design an efficient local communication system.

Despite many studies conducted recently on application of local communication to distributed mobile robotic systems[5] [6], its design has hardly been discussed based on mathematical analysis, but on only computer simulations. This simulation-based try-and-error design method requires whole simulation process from scratch even for slightest change of system.

Related studies in the field of communication theory [11] did not take account of various cooperative tasks or transmission to limited number of robots.

From these backgrounds, this paper aims to give analytical guidelines on design of local communication in distributed mobile robotic systems. The analysis will be proceeded in two steps:

(A) Spatial design Maximization of the efficiency of spatial transmission of information between two robots to minimize overall transmission time.
Design Parameter: Local communication area

(B) Temporal design Improvement of communication efficiency to transmit the information to necessary robots without excessive diffusion by repeated spatial transmission.
Design Parameter: Information announcing time

The above designs will be made based on the "equation of information diffusion," derived in chapter 2, first Spatial and next temporal design in chapters 3 and 4 respectively.

2 Formulating Information Diffusion

We will formulate information diffusion by local communication as a fundamental analysis the following chapters will be based on. The "equation of information diffusion" will be derived after explaining our local communication model.

2.1 Local Communication Model

We will employ a simplified model of local communication as shown in **Fig. 3**. Principal parameters of the model are listed together:

ρ: Density of robot population
R_c, ϕ, A: Radius, visible angle and area of output range of information ($A = 0.5R_c^2\phi$)
x: Average number of robots in output range ($= \rho A$)
p_e: Probability of information output from a robot
c: Information acquisition capacity
$r(t)$: Ratio of informed robots at time t
m: Total number of robots in the system
n_e: Desired number of robots the information is transmitted to
$\mathcal{M}$: Set of parameters that decide movement of robots (velocity etc.)

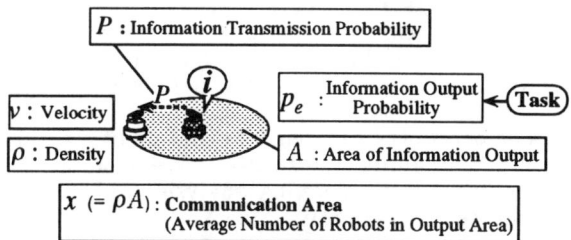

Fig. 3: Local communication model

In this model, communication takes place as described below:

(i) Each robot outputs information in the form of a "packet" within a limited area A, with certain probability p_e usually determined by task.

(ii) There is an upper limit in number of robots from which each robot can obtain information.

(iii) Each robot executes information reception process at every time unit which is long enough for acquisition. If there are any reachable information, the robot receives it.

This simple model, as stated in chapter 1, has such advantages as load distribution and easy implementation. The parameter x, given as the product of ρ and A, represents the average number of robots in output area A. We refer to x by the term "communication area" since it is clearer than using the value A directly.

We define the upper limit in (ii) as "information acquisition capacity" c. If a robot finds more than c robots that output information, two cases are possible:

(a) The robot cannot receive information from any robots. [*interfering communication*]

(b) The robot can receive information from c robots. [*non-interfering communication*]

2.2 Equation of Information Diffusion

Information used in cooperative tasks is diffused among robots by repetition of local communication as shown in **Fig. 4**. As explained in chapter 1, a packet contains such information as set of attributes (type, place,...) of several tasks in (1) communication for notification; and status of task execution updated by each robot in (2) communication for task execution. Then repeated transmission of a packet leads to diffusion of information about multiple tasks or status updated by multiple robots. We will therefore proceed the analysis assuming that the information about each task or the status change made by each robot is diffused independently.

Paying attention to a specified content of information $\mathcal{I}$, we define "I-Robots" as those robots that received the content $\mathcal{I}$, and N-Robots not received. The

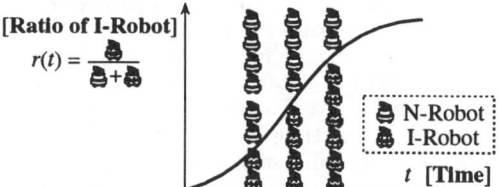

Fig. 4: Information diffusion among robots

ratio of I-Robots at time t is represented by $r(t)$.

The transmission time can be defined as the number of time units described in (iii) before the information is received by the number of robots determined by the task.

Let us derive the differential equation of $r(t)$ which describes the diffusion process. We just show an overview here, more detailed analysis is given in [7]. The increase of $r(t)$ per time Δt, $\Delta r(t)$, corresponds to the percentage of newly generated I-Robots at time t. We define the "*Information Transmission Probability*" P as the probability that a robot can successfully obtain information from others at time t, which is a function of c, p_e, x and t. The increment $\Delta r(t)$ is in proportion to ratio of N-robots $1 - r(t)$ and $P(c, p_e, x, t)$. Then the diffusion process is modeled by the following differential equation:

$$\frac{dr(t)}{dt} = \beta(\mathcal{M}, x)\, P(c, p_e, x, t)\, \{1 - r(t)\} \quad (1)$$

where $\beta(\mathcal{M}, x)$ stands for the effect of robot motion. We call this equation (1) as the "*equation of information diffusion*".

The analyses and design in the following chapters are conducted based on this equation. The spatial design of local communication, as mentioned in chapter 1, contributes to information transmission to necessary robots in minimum time. This will be realized by maximizing the Information Transmission Probability P in Eq.(1), which increases "diffusion velocity" $dr(t)/dt$. We should analyze the property of P and then carry out spatial design before temporal design, since the overall behavior of Eq. (1) cannot be known without the analysis of P.

We will thus deal with the spatial design in chapter 3, and next, the temporal design in chapter 4.

3 Spatial Design

Figure 5 illustrates the spatial design of local communication in distributed mobile robotic systems.

The evaluation function and related parameters in the spatial design are as follows:

- **Evaluation function:**
  Information transmission time W
  1. to an arbitrary robot
  2. to multiple robots
- **Design parameter:**
  Communication area x
- **Input parameters:**
  ρ: Robot density (determined from environment)
  p_e: Information output probability (determined from task)
  c: Information acquisition capacity (determined from communication capacity of robots)

We will utilize communication area x as the design parameter, since others concerning spatial information transmission are determined by the environment, task or robot capacity. The evaluation function, information transmission time W should be minimized for efficient local communication. We define the *optimal communication area* x_{opt} as the value of x which gives the minimum value of W.

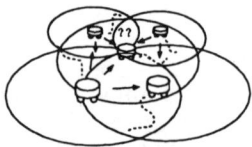 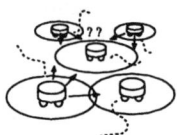

(a) Output range too large (b) Output range too small
Fig. 5: Control of Local Communication Range

We will model some typical cooperative tasks before this optimization. The optimal communication area x_{opt} will be derived first for transmission to an arbitrary robot and next to multiple robots by extending the first method.

3.1 Models of the Cooperative Tasks

We model the three typical cooperative tasks as follows (Fig. 6):

(I) **Random search of the area** [12][13] :
Robots move randomly to search the environment.

(II) **Cooperative transfer** [14][15] :
Robots transport a circular object by griping the edge (Fig. 6(II)).

(III) **Search in assigned area** [16][17] :
Each robot searches an assigned area (**Fig. 6(III)**).

The radius of object in (II) and the length of the edge of assigned area in (III) are normalized as 1 unit length. As mentioned before, the parameter to be designed is x for (I) and (III). Task (II) is a particular case in which R_c is used as design parameter, but the basic analytical method does not change.

3.2 Information Transmission Probability

In this section, we will derive information transmission probability, which will be computed from spatial distribution of robots, and will be used to derive information transmission time.

3.2.1 Modeling Distribution of Robots

The spatial distribution of robots is modeled for each task in 3.1 by the following probability:

$$\Pr[i \mid i \subset \mathcal{S}(x)] \equiv \Pr[i \text{ robots exist in area } x] \quad (2)$$

(I) Random Search of the Area When objects are disposed randomly on a plane, it is generally known that the number of objects in a certain area is Poisson distributed. This characteristics can be also applied to the case where robots move individually. The distribution of robots is expressed as follows:

$$\Pr[i \mid i \subset \mathcal{S}(x)] = \frac{\{\rho A\}^i}{i!} e^{-\rho A} = \frac{x^i}{i!} e^{-x} \quad (3)$$

(II) Cooperative transfer When the number of robots is m, the average number of robots

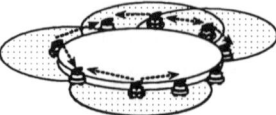 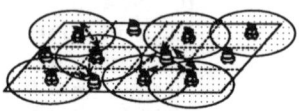

(II) Cooperative Transfer (III) Search in Assigned Area
Fig. 6: Model of cooperative tasks

x in an area is expressed as product of the density $(m-1)/(2\pi)$ and length of an arc of the object that is included in the area. Therefore, $\Pr[i|i \subset \mathcal{S}(x)]$ is expressed using binomial distribution as follows:

$$\Pr[i|i \subset \mathcal{S}(x)] = {}_{m-1}C_i(\frac{x}{m-1})^i(1-\frac{x}{m-1})^{m-1-i}$$

$$\text{where} \quad x = \frac{2(m-1)}{\pi}\sin^{-1}\frac{R_c}{2} \qquad (4)$$

(III) Search in Assigned Area For this task, it is difficult to obtain an analytical model of spatial distribution of robots applicable to all over x. We approximate the spatial distribution using normal distribution with average μ and variance V given by computer simulation as follows:

$$\mu = \sqrt[3]{x^3+1} - 1, \quad V = 0.6\sqrt{\mu} \qquad (5)$$

3.2.2 Information Transmission Probability

We will derive the information transmission probability P in this section and use it to calculate the information transmission time as evaluation function.

First, the P will be computed in the case of transmission between arbitrary two robots. Next, the methodology is extended to the case that the information diffuses to multiple robots.

We distinguish P_I and P_N for interfering and noninterfering communication respectively.

To an Arbitrary Robot: We define the probability Q_{ij} that there are i robots in communication area and j out of i robots are sending out information. As the product of $\Pr[i|i \subset \mathcal{S}(x)]$ and binomial distribution, Q_{ij} is expressed as follows:

$$Q_{ij}(p_e, x) = \Pr[i|i \subset \mathcal{S}(x)] \, {}_iC_j p_e^j(1-p_e)^{i-j} \qquad (6)$$

Using Q_{ij}, we derive P, namely the probability that a robot can successfully receive information sent out by another robot:

$$P_I(c, p_e, x) = \sum_{i=1}^{c}\sum_{j=1}^{i} Q_{ij} + \sum_{i=c+1}^{\infty}\sum_{j=1}^{c} Q_{ij} \qquad (7)$$

$$P_N(c, p_e, x) = P_I(c, p_e, x) + \sum_{i=c+1}^{\infty}\sum_{j=c+1}^{i}\frac{c}{j}Q_{ij} \qquad (8)$$

To Multiple Robots: The probability $\bar{r}_j(t)$ that there is at least one I-Robot for $\mathcal{I}$ among j robots is expressed as $\bar{r}_j(t) = 1-(1-r(t))^j$. The information transmission probability $P(c, p_e, x, t)$ is derived from (7) and (8) as follows:

$$P_I(c, p_e, x, t) = \sum_{i=1}^{c}\sum_{j=1}^{i} Q_{ij}\bar{r}_j(t) + \sum_{i=c+1}^{\infty}\sum_{j=1}^{c} Q_{ij}\bar{r}_j(t) \quad (9)$$

$$P_N(c, p_e, x, t) = P_I(c, p_e, x, t) + \sum_{i=c+1}^{\infty}\sum_{j=c+1}^{i} Q_{ij}\bar{r}_c(t) \quad (10)$$

3.3 Optimal Communication Area

First, we will consider information transmission time, and then, we will derive the optimal communication area x_{opt} that will minimize the time.

3.3.1 Information Transmission Time

To an Arbitrary Robot: A robot sends out the information a number of times because it might not be received immediately due to interference. So we define the information transmission time W as the average time it takes for the output information to be successfully received.

The probability that the information sent out for the first time can be received is expressed as P. Then the probability that it takes i time units to succeed in transmission is expressed as geometric distribution $P(1-P)^{i-1}$. Thus, W is expressed as $W = 1/P$[18].

Therefore, the optimal communication area x_{opt} that maximizes P also minimizes W.

To Multiple Robots: In this case, the optimal communication area x_{opt} is the area that minimizes information diffusion time to multiple robots. The equation of information diffusion (1) is rewritten using $P(c, p_e, x, t)$ of (9) or (10) as follows:

$$\frac{dr(t)}{dt} = \beta(\mathcal{M}, x) \, P(c, p_e, x, t) \, \{1-r(t)\} \qquad (11)$$

In the equation, $\beta(\mathcal{M}, x)$ is coefficient that stands for the effect of robot motion [7]. The detail of this coefficient will be shown in chapter 4 where the information diffusion is analyzed.

The area x_{opt} is obtained as x maximizing the part $\beta(\mathcal{M}, x)P(c, p_e, x, t)$ in the right-hand side of (11).

3.3.2 Optimal Communication Area

To an Arbitrary Robot: We will derive the optimal communication areas x_{opt} that maximize information transmission probability P. First, we express P as the function of (p_e, x) when information acquisition capacity c is given, as 3-dimensional graph shown in Fig. 7 (as an example, random search in interfering communication, $c=1$). Next, according to the probability of information output p_e determined from the task, the optimal communication area x_{opt} is obtained as x maximizing P indicated by "Max of P_I" in Fig. 7.

The values of x_{opt} are plotted in terms of p_e in Fig. 8. It was projection to (p_e, x)-plane of the curve

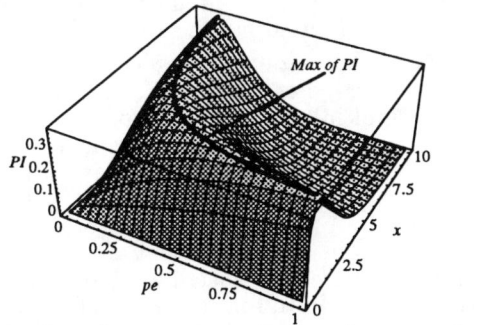

Fig. 7: P plotted versus (p_e, x) (random search, $c=1$)

241

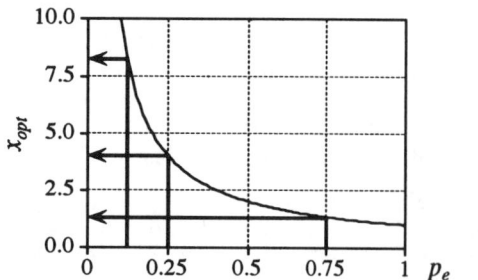

Fig. 8: x_{opt} plotted versus p_e (random search, $c=1$)

represented "Max of P_I" in Fig. 7. We can easily obtain the x_{opt} for various p_e with this graph.

In the case of random search for interfering communication, $P_I(c, p_e, x)$ is arranged from (3) and (7):

$$P_I(c, p_e, x) = e^{-p_e x}(\sum_{k=0}^{c} \frac{(p_e x)^k}{k!} - 1) \qquad (12)$$

By solving $\frac{d}{dx}P(c, p_e, x) = 0$, we can derive x_{opt} that maximizes P in a simple formula:

$$x_{opt} = \sqrt[c]{\frac{c!}{p_e^c}} = \frac{\sqrt[c]{c!}}{p_e} \qquad (13)$$

In (13), x_{opt} is in inverse proportion to p_e, the probability of information output from each robot.

This means that communication area should be small (large) when information transmission is frequently (sparsely). And x_{opt} increases when information acquisition capacity c increases. These characteristics agree with our senses. However, we must pay attention to the fact that x_{opt} is not simply proportional to c.

We can also obtain the optimal radius $R_{c\,opt}$ of communication area for cooperative transfer. By using x of (4), P_I is expressed as follows:

$$P_I(c, p_e, x) = \sum_{i=1}^{c} {}_{m-1}C_i(\frac{p_e x}{m-1})^i(1 - \frac{p_e x}{m-1})^{m-1-i} \qquad (14)$$

We can obtain x_{opt} that maximizes P_I in (14) as inversely proportional function of given p_e. It is easy to calculate $R_{c\,opt}$ from x_{opt} using (4).

For non-interfering communication or in other tasks, x_{opt} is computed in the same way.

To Multiple Robots: The optimal communication area can be derived for information transmission to multiple robots by maximizing the part $\beta(\mathcal{M}, x)P(c, p_e, x, t)$ in the right-hand side of (11). Since this is obtained in a similar way with that for an arbitrary robot, we are not going into details.

The analyses so far make clear the relationship between the optimal communication area in spatial design and parameters of distributed robotic system. This is of great help to spatial design of local communication system in distributed mobile robotic systems.

4 Temporal Design

Temporal design is to determine how long to let the information be diffused for the information to be transmitted to desired number of robots.

The evaluation function and parameters involved in the design are listed below:

- **Evaluation function:**
  Diffusion rate (Ratio of I-Robots) $r(t)$
- **Design parameter:**
  Task announcement time T_{ann}
- **Input parameters:**
  n_e: Desired number of robots the information is transmitted to (determined from task)
  x: Communication area (designed in chapter 3)
  $\mathcal{M}$: Set of parameters that decide robot motion (determined from task, motion capacity of robots)

When a robot sends out information in communication for notification (**Fig. 1 (1)**) or for task execution (**Fig. 1 (2)**), the particular information, we call $\mathcal{I}$, is diffused to multiple robots by repeated local communication and robot motion as already shown in **Fig. 4**. The information $\mathcal{I}$ should be obtained by n_e robots to start or to continue the task in both communication types.

The *temporal design* of local communication is, based on this diffusion process, to determine the information announcing time T_{ann} so that the information $\mathcal{I}$ is transmitted to n_e robots without excessive diffusion. T_{ann} can be controlled by setting a expiration time to the information since it is generated.

We assume the information about each task in communication (1), or the status change made by each robot in communication (2), is diffused independently as stated in section 2.2.

4.1 Analysis on Effect of Robot Motion

Here we will mainly make clear the effect of robot motion on the information diffusion. The derived equation of information diffusion was:

$$\frac{dr(t)}{dt} = \beta(\mathcal{M}, x) P(c, p_e, x, t) \{1 - r(t)\} \qquad (11)$$

We make an assumption that, in each task, robot motion can be modeled as a random motion that each robot moves straight with velocity v and changes its direction randomly within the range θ at every τ time units; $\mathcal{M} = \{v, \theta, \tau\}$. The coefficient $\beta(\mathcal{M}, x)$ in (1) is given using the area $S(\mathcal{M})$ each robot sweeps per time unit as:

$$\beta(v, x) = \frac{1 - e^{-\rho S(\mathcal{M})}}{1 - e^{-x}} \quad \text{where}$$
$$S(\mathcal{M}) = 2R_c v \frac{1}{\tau}\{(\tau - 1) + \frac{1}{\theta}\int_0^\theta \cos\alpha d\alpha\} \qquad (15)$$

The diffusion process $r(t)$ can be obtained by numerically solving the differential equation (11) using (15). Since the information transmission probability $P(c, p_e, x, t)$, derived as eqs. (9) or (10), is generally nonlinear, it is difficult to understand the characteristics of diffusion process $r(t)$ from numerical solutions.

The next section will introduce a linear approximation to clarify the relationship between transmission time and other parameters.

4.2 Calculation of Diffusion Time using Logistic Equation

Let us consider the simplest case of information diffusion to see its fundamental property, namely interfering communication with information acquisition capacity (c) equals to 1. This corresponds to ordinary case where robots' communication capacity is rather limited. Calculating (9) with $c = 1$, (11) can be arranged as:

$$\frac{dr(t)}{dt} = ar(t)\,\{1 - r(t)\}$$
$$\text{where} \quad a = \beta(\mathcal{M}, x)e^{-p_e x}p_e x \tag{16}$$

This is a logistic equation, then $r(t)$ is derived as:

$$r(t) = \frac{r(0)e^{at}}{r(0)\{e^{at} - 1\} + 1} \tag{17}$$

Figure 9 shows calculated $r(t)$ for communication area $x = 0.4$, 0.85, 1.6. Other parameters are $\rho = 0.125$, $\phi = 360[\deg]$, and parameter sets of robot motion (v, θ, τ) are $(0.2, 60[\deg], 3)$. Initial value of $r(0)$ is given as $0.02 = 1/50$, assuming that one robot of 50 initially has the information. Here, the optimal communication area x_{opt} is computed as 0.85 using the optimization methodology in section 3.3. As seen in Fig. 9, the information is diffused the most rapidly with derived $x_{opt} = 0.85$.

The diffusion time $T(n_e)$ required so that the information is diffused to n_e out of m robots can be easily calculated from (17) as:

$$T(n_e) = -\frac{1}{a}\log\left\{\frac{m - n_e}{(m - 1)n_e}\right\} \quad \text{where}$$
$$a = \beta\, e^{-p_e x}p_e x = \beta\, P_I(1, x, p_e) \tag{18}$$

Equation (18) describes the characteristics of diffusion time for given parameters. The diffusion time $T(n_e)$ is inversely proportional to β and P_I (to an arbitrary robot). Since β represents the swept area per time unit as shown in 4.1, (18) implies that diffusion becomes faster as robots move with greater velocity.

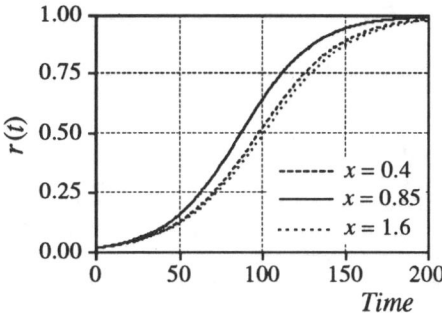

Fig. 9: Information diffusion $r(t)$ (c=1, p_e=1.0)

Information announcing time T_{ann} can be determined taking account of some margin such as 3σ of transmission time $T(n_e)$ [7].

For further analysis of information diffusion, we apply a linear approximation to equation of information diffusion (11). One of the situations where the analysis of information diffusion has great significance is, when tasks are announced in an environment where the robot density is low. The estimation of time required to transmit the task information to desired robots should be correct, otherwise a slight error in estimation makes a large difference of task execution time. On the other hand, the effect of this difference is relatively small when robot density is high, which means the diffusion velocity is fast.

If the robot density is low, then there is little possibility that there are two or more robots in its possible communication area. The information transmission probability is, therefore, approximated by $\Pr[i \geq 1 | i \subset \mathcal{S}(x)]$, the probability that there are at least one robot in its communication area, regardless of interference and information acquisition capacity. If robots are randomly distributed in the plane, $\Pr[i \geq 1 | i \subset \mathcal{S}(x)]$ is approximated with low robot density as:

$$\Pr[i \geq 1 | i \subset \mathcal{S}(x)] = 1 - e^{-p_e x} \simeq p_e x \tag{19}$$

The error of resultant logistic function is about 10% even if $p_e x$ is as much as 0.5, which is rather condensed case. The approximation simplifies (18) into the following form.

$$T(n_e) = -\frac{1}{\beta(\mathcal{M}, x)p_e x}\log\left\{\frac{m - n_e}{(m - 1)n_e}\right\} \tag{20}$$

Equation (20) allows us to estimate the diffusion time in a simple manner, and clearly explain the diffusion time $T(n_e)$ and such parameters as x, n_e.

The analytical results obtained here are of great help in temporal design of local communication.

5 Simulations and Experiments

Some simulations and experiments have been undertaken to demonstrate the validity of design of local communication addressed so far in spatial and temporal design respectively.

5.1 Simulations of Information Diffusion

In the following simulations, multiple mobile robots which perform the tasks modeled in section 3.1 are implemented on computer, and information transmission is simulated for transmission either to an arbitrary robot and to multiple robots. The simulation results will be compared to analytically predicted values. We calculate for the cases interfering communication here, but non-interfering cases can also be verified likewise.

5.1.1 Verification of Spatial Design

Figure 10 shows the results of the simulations. The information acquisition capacity $c = 1$, probability of information output $p_e = 0.3, 0.5, 0.8$, and simulation time is 500 time units for each tasks. The number of robots is 25 for (I) and (III), 10 for (II).

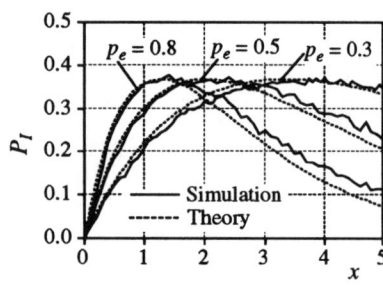
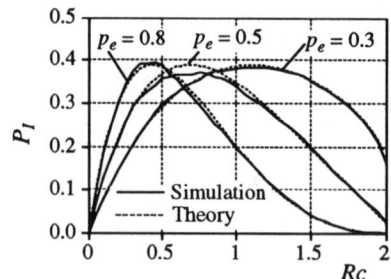
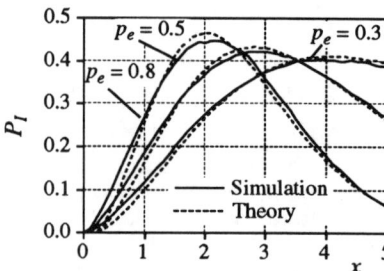

(I) Random search　　　(II) Cooperative transfer　　　(III) Search in assigned area

Fig. 10: Information transmission probability P_I versus p_e

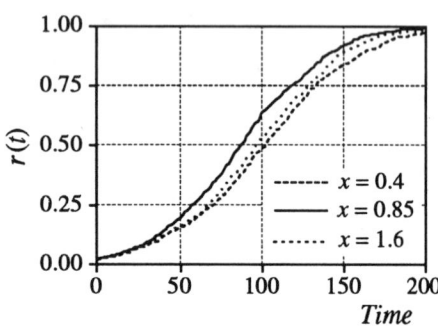

Fig. 11: Simulation of diffusion $r(t)$ ($c=1$, $p_e=1.0$)

The values of analyses matches well with those of simulations and P in simulation takes the maximum value at x_{opt} or R_{copt} which give the maximum of P in the analysis; the accuracy of the analyses is shown.

5.1.2 Verification of Information Diffusion

We implement an environment on computer, in which 50 robots search randomly in the workspace 20×20.

Figure 11 illustrates the simulation result of information diffusion processes with the same parameters of calculation of (17) in Fig. 9.

Since the calculated diffusion process in Fig. 9 models well the simulation results, the effectiveness of information diffusion model has been verified.

Furthermore, as predicted in analysis in Fig. 9, information diffusion is the most rapid with calculated $x_{opt} = 0.85$. in Fig. 11. This demonstrates the effectiveness of the optimal communication area in transmission to multiple robots.

It was shown that the optimal communication area derived by the analyses is effective from simulations.

5.1.3 Verification of Temporal Design

Tasks performed cooperatively by mobile robots are simulated to verify the temporal design of information announcing time.

The specifications of the cooperative task simulation are as follows:

1. Tasks are consecutively announced from "task signboards" during the period T_{ann}.
2. Task signboards show the locations of task execution and the remaining announcing duration to robots.

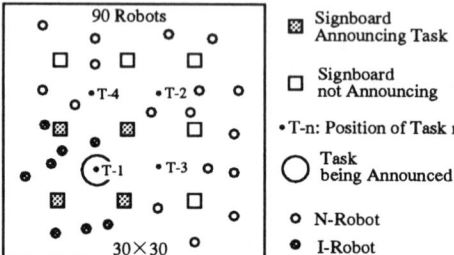

Fig. 12: Environment of Cooperative Tasks Simulation

3. I-robots continue random walking until the remaining announcing duration expires.
4. After the termination of the task announcement, robots start moving straight to the task point.
5. Tasks start as soon as the required number of robots gather to the task point.

The temporal design is utilized to determine the period during which tasks are announced.

We consider "task signboards" in the simulation environment which are disposed with the constant density ρ_{sign} as in Fig. 12 where different periods of task announcement T_{ann} are tested. The desired number of robots n_e to start the task is 10 out of all 90 robots.

As shown in Fig. 12, the environment is 30×30, and density of robots is 0.1, with parameter of random motion (v, θ, τ) are $(0.1, 60[deg], 3)$.

We will analyze *task execution rate* and *total information transmission* in this simulation. The task execution rate is the ratio of executed tasks to all the announced tasks. The total information transmission stands for how many times information is passed among robots per task. The relationships between these two indices and the announcing period T_{ann} are shown in Figs. 13 and 14.

Using the formula (20), we can derive diffusion time $T(n_e)$ as 74.5. To ensure the information transmission T_{ann} is calculated as 142.6 by adding three times of the standard deviation of $T(n_e)$.

From Fig. 13, it is observed that all the tasks are executed if T_{ann} is longer than 142.6, which shows the derived T_{ann} is effective to transmit the information to necessary robots.

It could be concluded that the announcement time T_{ann} must be long enough to make tasks performed surely by robots from the analysis of task execution rate. However, the longer we take T_{ann} for reliable ex-

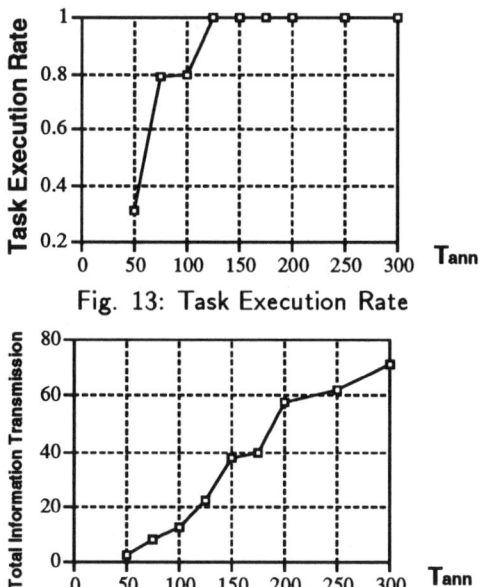

Fig. 13: Task Execution Rate

Fig. 14: Total Information Transmission per Task

ecution of tasks, the more total information transmission becomes as Fig. 14 indicates. This leads to higher cost of communication because of unnecessary information processing. Task announcement time T_{ann} obtained using our temporal design has effectiveness also from this point.

5.2 Experiment of Transmission

We have carried out some simple experiments on information transmission to verify derived principles of spatial design. Local communication is realized using infrared device so that communication area can be limited and adjusted to the desired distance as Fig. 15.

5.2.1 System configuration

The configuration of infrared local communication system is shown in Fig. 16. We use infrared LEDs and sensors available as remote-controller of TV set. Modulation is FSK (Frequency Shift Key) and the transmission rate is 2400[bps]. Communication distance can be adjusted by changing the current to LED from PC via D/A board, from 1 to 4[m].

Infrared LEDs are arranged circularly together with four sensors in the center to have 360[deg] of communication range. Fig. 17 shows the communication device mounted on a mobile robots equipped with relative

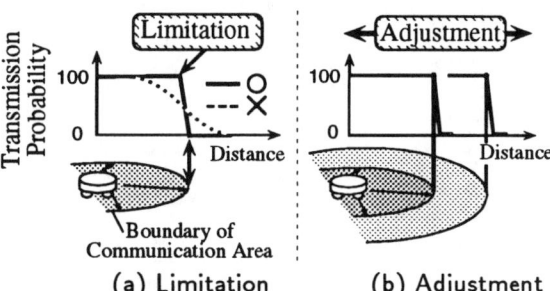

(a) Limitation   (b) Adjustment

Fig. 15: Control of communication area

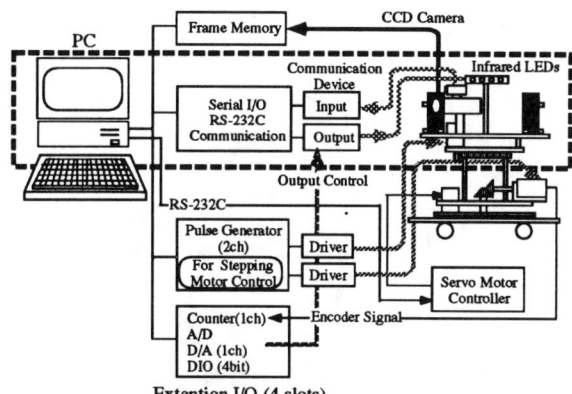

Fig. 16: System configuration of communication system

Fig. 17: Photo of infrared communication device

position/orientation sensing system using LED signboards and video camera [10].

5.2.2 Verification of Spatial Design

We have chosen random search task as basic one. We measured the information transmission probability P of this *receiver* robot from transmitter robots randomly distributed within 2.5 [m] of distance as shown in Fig. 18. This is interfering communication, with the density of robots $\rho = 0.092$, the information output probability $p_e = 1.0$.

Transmitter robots sends out 200 [bytes] of data changing the communication distance of $1.0 \sim 2.5$ [m].

Figure 19 shows the experimental result of information transmission probability P compared to theoretical value calculated using the methodology shown

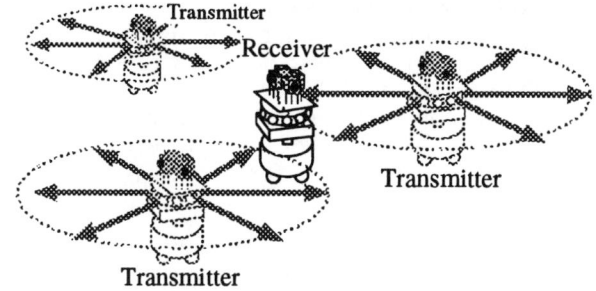

Fig. 18: Verification of optimal communication area

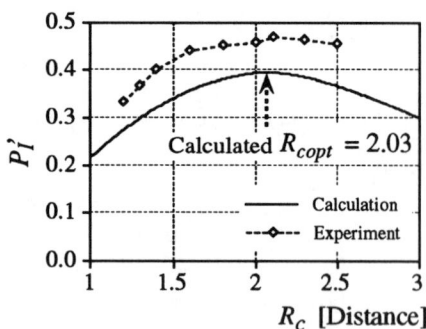

Fig. 19: Experimental results of P

in 3.3.2. It is first observed in Fig. 19 that P takes the maximum in experiment at the analytically derived optimal communication area R_{copt}. This demonstrates the effectiveness of spatial design of local communication.

We can also see that there is an offset between experimental and analytical value. One of the reason is because the device does not not realize interfering communication completely, i.e. when there are two robot on a line from the receiver robot, the far robot is shaded by the near one so that interference may not occur. This is to be studied in future development of this work.

Although we showed only the case of random search task here, the effectiveness of design for other tasks, of course, can be verified in the same way.

6   Conclusion

In this paper, we have presented a design methodology of local communication for cooperation in distributed mobile robotic systems. Efficient cooperation requires communication system in which task information is transmitted to necessary robots in minimum time, without excessive diffusion. We proposed to divide the design into two phases, spatial and temporal design based on analysis of information diffusion. This methodology allows us to construct efficient local communication system in a plain and systematic fashion.

We examined the property of equation of information diffusion and showed that the design can be classified into spatial and temporal one. The evaluation function and design parameters in both designs were clearly specified to give the good perspective.

Spatial design optimizes the communication area which leads to minimum transmission time. On the other hand in temporal design, given optimal communication area, information announcing time is derived to transmit the information to necessary robots without causing superfluous diffusion.

It has been shown that, from some simulations and experiments, proposed design method is helpful to build efficient cooperative system.

References

[1] S. Yuta, et al.: "State Information Panel for Inter-processor Communication in an Autonomous Mobile Robot-Controller," *Proc. IEEE Int. Workshop on Intelligent Robots and Systems* , 1990.

[2] H. Asama, et al.: "Functional Distribution among Multiple Mobile Robots in an Autonomous and Decentralized Robot System," *Proc. IEEE Int. Conf. on Robotics and Automation*, 1921-1926, 1991.

[3] F. R. Noreils: "An Architecture for Cooperative and Autonomous Mobile Robots," *Proc. IEEE Int. Conf. on Robotics and Automation*, 2703-2709, 1992.

[4] M. J. Mataric: "Designing Emergent Behaviors: From Local Interactions to Collective Intelligence," *Proc. 2nd Int. Conf. on Simulation of Adaptive Behavior: From Animals to Animats 2*, 432-441, 1993

[5] S. Ichikawa, et al.: "An Experimental Realization of Cooperative Behavior of Multi-Robot System," *Distributed Autonomous Robotic Systems*, 224-234, 1994, Springer-Verlag.

[6] J. Wang: "On Sign-board Based Inter-Robot Communication in Distributed Robotic Systems," *Proc. IEEE Int. Conf. on Robotics and Automation*, 1045-1050, 1994.

[7] T. Arai, et al.: "Information Diffusion by Local Communication of Multiple Mobile Robots," *Proc. IEEE Int. Conf. on Systems, Man and Cybernetics*, Vol. 4, 535-540, 1993.

[8] E. Yoshida, et al.: "A Design Method of Local Communication Range in Multiple Mobile Robot System," *Proc. IEEE/RSJ Int. Conf. on Intelligent Robots and Systems*, Vol.2, 274-279, 1995.

[9] S. Suzuki, et al.: "An Infra-Red Sensory System with Local Communication for Cooperative Multiple Mobile Robots," *Proc. IEEE/RSJ Int. Conf. on Intelligent Robots and Systems* , Vol.1, 220-225, 1995.

[10] T. Arai, et al.: "Real-time measuring system of relative position on mobile robot system," *Proc. Int. Symp. on Industrial Robots*, 931-938, 1993.

[11] H. Takagi, et al.: "Optimal Transmission Ranges for Randomly Distributed Packet Radio Terminals," *IEEE Transactions on Communication*, Vol.COM-32, No.3, 246-257, 1984.

[12] L. Steels: "Cooperation between Distributed Agents through Self-Organization," *Decentralized AI*, 175-196, 1990, North Holland.

[13] K. Singh, et al.: "Map Making by Cooperating Mobile Robots," *Proc. IEEE Int. Conf. on Robotics and Automation*, Vol.2, 254-259, 1993.

[14] D. J. Stilwell, et al: "Optimal Control for Cooperative Mobile Robots Bearing a Common Load," *Proc. IEEE Int. Conf. on Robotics and Automation*, 58-63, 1994.

[15] J. Ota, et al.: "Transferring and Regrasping a Large Object by Cooperation of Multiple Mobile Robots," *Proc. IEEE/RSJ Int. Conf. on Intelligent Robots and Systems* , Vol.3, 543-548, 1995.

[16] R. Beckers, et al.: "From Local Actions to Global Tasks: Stigmergy and Collective Robotics," "Artificial Live IV", R. A. Brooks and P. Maes eds., 181-189, 1994, MIT Press.

[17] D. Kurabayashi, et al.: "An Algorithm of Dividing a Work Area to Multiple Mobile Robots," *Proc. IEEE/RSJ Int. Conf. on Intelligent Robots and Systems* , Vol.2, 286-291, 1995.

[18] R. M. Metcalfe et al.: "Ethernet: Distributed Packet Switching for Local Computer Networks," *Communications of the ACM*, Vol.19, No.7, 395-404, 1976.

Proposal and Development of an Autonomous Decentralized System for Newspaper Production

Hiroshi ITO, Shinji FUJISAWA and Youichi HATTORI
Production Control Department, Asahi Shimbun Tokyo Head Office,
5-3-2 Tsukiji, Chuo, Tokyo 104-11, Japan

Abstract

Over the past sixteen years, newspaper PCSs (Production Control Systems) have been developed to support newspaper production, so that newspaper publishers can offer the most up to date information to the greatest possible number of readers in a timely manner while keeping production costs low. Up to now, the chief goal has been to produce a predetermined number of copies of a stylized newspaper (using a set layout and set number of pages) as efficiently and accurately as possible. However, in today's information-oriented society, readers and advertising clients have come to expect low cost newspapers that reflect their various interests and demands. To satisfy these, a PCS should aim at providing a reliable and flexible production system that allows immediate changes in format and circulation volume according to article contents, as well as coordinated production (printing and delivery) between locally distributed printing plants.

In this paper, a new newspaper PCS architecture and data-processing and networking techniques are discussed based on an autonomous decentralized system (ADS). The architecture provides a system that allows flexible change of the production schedule, while ensuring reliability, because it has autonomous filtering functions that distinguish between different kinds of data, namely control data and man-machine data. The Asahi Shimbun newspaper company currently uses this highly effective system in its Setagaya printing plant.

1. Introduction

The Asahi Shimbun newspaper company was founded in 1879. By actively adopting the latest available technology since the 1970s, Asahi Shimbun has been able to computerize its entire pre-press production process, such as editing. From early in its history, the company has automated its post-press processes, namely printing and distribution.

Generally speaking, a newspaper's mission is to offer correct and timely information to as many readers as possible at low cost. The Asahi Shimbun distributes 8.3 million copies of its morning edition and 4.4 million copies of its evening edition nationwide from locally distributed printing plants. To make it possible to print so many papers

with up-to-date news in a short time and distribute them to delivery agents, the production management in each plant has been automated being undertaken by computers.

This type of system technology allows newspapers to be swiftly delivered to readers. However, reflecting the information orientation of today's society, readers and advertising clients are demanding not only an improvement in the quality and timeliness of articles, but flexible content services as well.

On the other hand, concepts of Autonomous Decentralized Systems (ADS) have been discussed to realize reliable and flexible computer systems [1]-[6].

In this paper, a Production Control System (PCS) that offers the kind of flexible production described above is discussed. This PCS is based on an ADS. User and system requirements will be discussed in Secs. 2 and 4, respectively. Section 5 will propose an ADS-style architecture for a newspaper PCS, after looking at existing systems. Section 6 describes the system currently in operation at Asahi Shimbun's Setagaya printing plant.

2. User Requirements

This section discusses reader requirements.

2.1 The State of Newspaper Production in Japan

In Japan, the seven largest newspaper publishers each supply over 1 million copies daily. Asahi Shimbun prints 8.3 million copies of its morning-edition national newspaper and 4.4 million copies of its evening-edition national newspaper (April, 1996). This places Asahi Shimbun firmly in the ranks of the world's leading newspapers in terms of total circulation. To supply such a large number of copies nationwide, each printing plant is equipped with high-speed web offset presses and folders that pile, fold and cut the papers. Each press can print a maximum at 72,000 copies per hour and 8 pages at a time. Asahi Shimbun holds 191 of those presses and 47 folders in 19 printing plants nationwide.

With Tokyo Head Office, there are four Asahi Shimbun printing plants that have 95 presses and 23 folders, and six contract printing plants. These papers are delivered to local delivery agents by truck. The transportation area covered by truck has a radius of several hundred kilometers [7][8]. (Fig. 2.1)

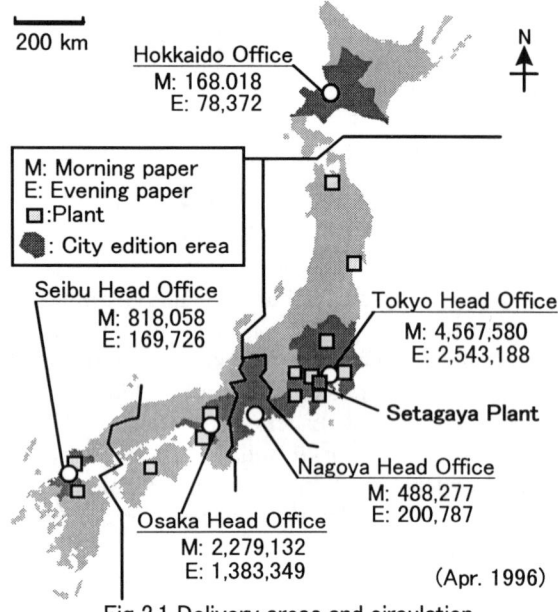

Fig 2.1 Delivery areas and circulation

Over 90% of the newspapers sold by Asahi Shimbun are directly delivered to the doorsteps of subscribers by door-to-door delivery network. These papers are delivered to readers from local delivery agents by moped and bicycle. Asahi Shimbun organizes 4,000 delivery agents spread throughout Japan, and employs about 90,000 workers. Each plant dispatches a large number of trucks almost every minute to supply papers to these agents.

To be able to deliver papers to all readers by the determined deadline, papers must be printed and distributed according to an order based on the distance from a local printing plant to the customer. During the intervals between printings, changes can be made to the editions.

The Asahi paper usually has five major editions and 140 local editions nationwide. Major editions are revised to offer the most up-to-date information. Local editions have articles and advertisements for each local area.

A morning paper usually has 24 to 48 pages. Local papers consist of pages from the major edition and local edition. Tokyo Head Office publishes 53 types of morning papers and 5 types of evening paper. At voting time, the number of major editions increases to 6 or 7, and the number of local editions also increases.

2.2 Reader Requirements

Generally, what readers want from newspapers are reliability and flexibility.

(1) Reliability

In terms of delivery, it is important that newspapers be delivered at a regular time (usually 6:00) every day. Sometimes the delivery could be delayed because of rotary press accidents and so on. However, as matters now stand, readers expect papers to arrive on time and newspapers companies firmly believe that timely delivery is a very important part of their mission.

(2) Flexibility

Readers have come to desire more finely defined services from newspapers than before. For example, when an important topic arises, readers want to be able to get detailed coverage of the topic within an appropriate number of pages. The number of pages could be also extended to include advertisements when many clients want to have their ads on a specific day.

3. Outline of Newspaper Production

This section outlines newspaper production.

3.1 The Newspaper Production Processes

Broadly speaking, newspaper production can be broken down into four processes: plate-making; printing; addressing; and delivery. Brief descriptions of these processes are as follows[9]. (Fig 3.1)

1) Plate-making Process

The printing plates to be attached to the web offset press are created in this process.

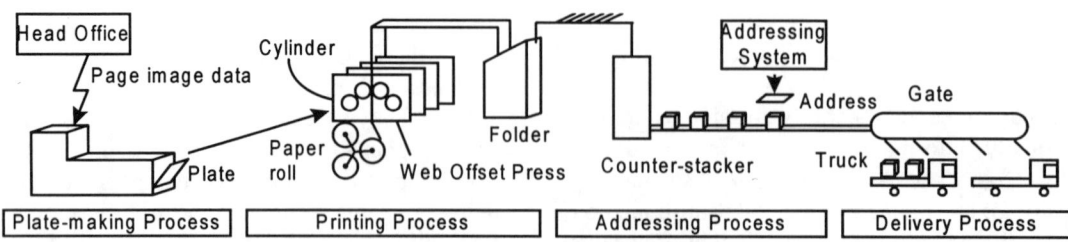

Fig 3.1 Newspaper Production Processes

248

First, when pagination is completed, the page image is sent by facsimile from the pre-press computers to the plate-making machine. The machine then reads the page information and creates the corresponding aluminum offset plates. The number of plates created will change according to the page's target area and the number of web offset presses required to print the newspaper.

2) Printing Process

The newspaper is printed during this process.

The plant operators take the printing plates created by the plate-making machine and install them on the cylinder of the web offset press. The positioning of these plates changes depending upon the number of pages in the paper, the positioning of color pages and so forth. The pages of the newspaper are printed as the cylinders of the web offset press revolve. The papers are then cut and folded to take the shape of a newspaper as we know it.

3) Addressing Process

In this process, the number of papers ordered by each delivery agent are gathered and labeled for delivery.

The newspapers printed, cut, and folded in the previous process are sent by conveyor belt to the counter-stacker (CS). This CS counts out the number of papers ordered by each delivery agent. Then an online addressing system prints the agent's address and places this on top of the stack of newspapers.

4) Delivery Process

In this process, the newspapers are loaded onto trucks that have arrived at the gate.

The addressed stacks of papers are sent to the gate via carts or trays. The truck driver registers his/her truck upon arrival by inserting an ID card into the loading system. When this is done, the stacked papers progress down the conveyor belt and onto the appropriate truck

Web Offset Presses

3.2 Daily Operations

Fig. 3.2 shows the daily operations in Setagaya plant.

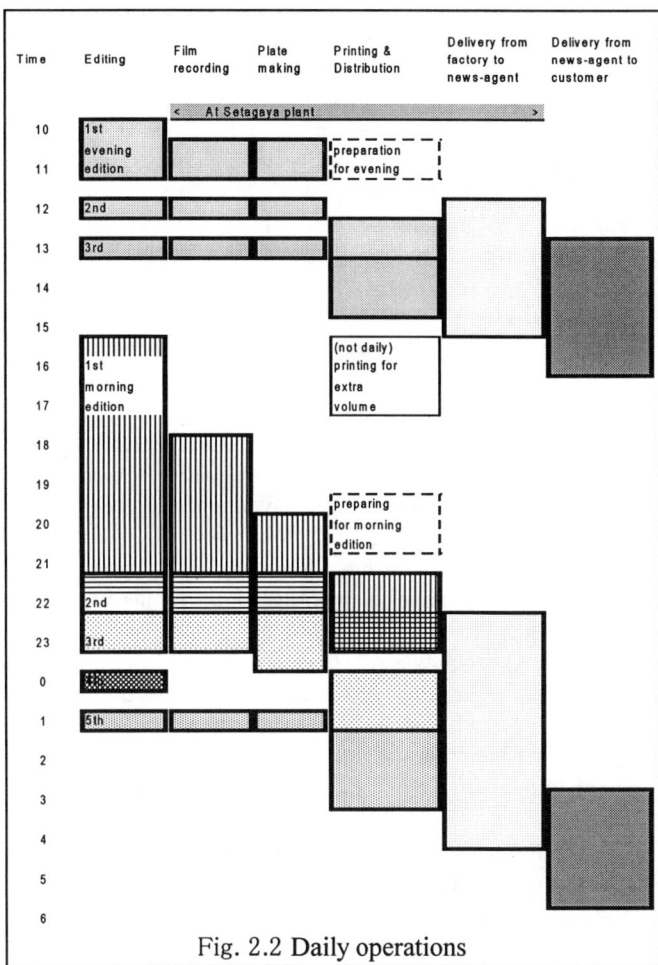

Fig. 2.2 Daily operations

4. System Requirements

In order to meet the user requirements discussed in Sec. 2, the requirements for the newspaper PCS will be discussed.

4.1 Operation Requirements

(1) Reliability

To satisfy the reader needs shown in 2.2(1), reliability at a plant and between plants is necessary.

<Reliability at a plant>

Four main processes occur at each printing plant. When this series of events occurs without fault, the papers can finally be distributed. These processes were originally done by people and each process could be an autonomous unit. This occurred until the present when centralized PCSs were developed with a preference for efficiency. However, from now on it is also necessary for processes to continue even if other process do not.

<Reliability between plants>

Even if reliability within plants is improved, there will be

249

times, for whatever reason, when a plant cannot print newspapers. When this occurs, that plant's work will be done at a different plant. The rescheduled data for the layout of the pages etc. would then be sent from the PCS at the head office to each plant. However, many distribution trucks are not integrated into the system that creates this change. From now on there will be a demand for newspaper production process management systems to swiftly cause this change as well. News paper production reliability will increase when this is seen as newspaper production work, because of cooperation between plants.

(2) Flexibility

Moreover, the following flexibility is needed to satisfy the requirements of 2.2(2).

<Format changes>

In order to satisfied the aforementioned user requirements, it is important to be able to change the number of copies printed, the layout, and the newspaper format, even during printing.

For example, when a big news story hits a newspaper, it is necessary to replace the printing plates on the web offset press for the new story immediately, in order to publish the latest news changes. Therefore, PCS should give information about the replacement, such as the number of plates to be newly made, to operators in a few seconds. An useful user interface of terminals is also necessary to change the positioning or the layout of color pages easily.

<Changes in circulation>

With the exception of the changing numbers of papers sold on the street, there is daily contact with stores for circulation changes, but there is a deadline for these changes. Operators must input revisions occurring after the deadline by hand.

In this, the information age, as readers directly subscribe to newspapers online, it is important to have a flexible system that can change circulation, even during printing, so there will be no excess or deficiency in the number of copies printed and distributed.

4.2 Development Requirements

There are development requirements as well as operation requirements as follows;

(1) Step-by-step expandability

Newspaper companies operate continuously over 350 days a year. Moreover, an extra issue may have to be published at any time with important news. As a result of this, there is demand for a system that can be constructed incrementally without affecting the existing system. Also, if the system can be expanded incrementally, there is the possibility of decreasing the initial investment for system construction.

(2) Openness

Recently, accompanying the increase in PC performance,

the improvements in software environments, and satisfaction with application software on PC, users have had the desire to develop man-machine interfaces by themselves to reduce the cost and period needed for system development. So far, there has been no standard for interfaces to connect computers and facilities. Therefore, there is a demand for openness in a variety of situations from the hardware level to the software level.

4.3 Maintenance Requirements

There is also a requirement for easy modification of the system when facilities need to be replaced and parameters need to be changed. This is because, facilities can only be replaced on the newspaper holidays, and their parameters must also be modified on the same day. Other data modification, such as changing local edition types, occurs almost everyday. Like the development requirement, maintenance is also required to be carried out without suspending system operation.

5. PCS based on ADS

In this section, an architecture and techniques for a PCS based on an ADS is proposed, after discussing the limits of a centralized PCS.

5.1 The Limits of a Centralized PCS

Prior to this, existing PCSs had been developed as centralized systems to secure real-time-ability in controlling low intelligent facilities and to use computers effectively.

Figure 5.1 shows the centralized PCS architecture. Head

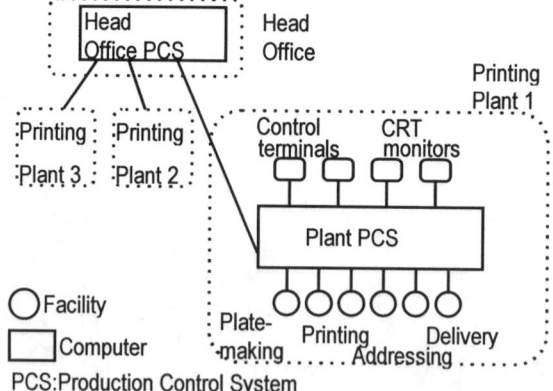

Fig 5.1 Centralized PCS Architecture

offices and printing plants are geographically dispersed and each has its own centralized PCS. The head office PCS receives page information from the pre-press process system connected to it. Then the PCS prepares a printing plan for each plant and sends the production plan information and page information for each plant (page information is several hundred bytes) to each plant. The head office PCS also receives information from the plants, every two minutes, during the printing progress [10].

The centralized PCS has been shown to have the following limitations with regard to meeting the new system requirements. Since the plans were rescheduled by the head office PCS, response is limited. Furthermore when a PCS is down, most of the facilities cannot function effectively. Moreover, when adding or maintaining system functions, the facilities, etc. it is necessary to stop the entire system.

5.2 Basic Architecture

The basic idea of the ADS-based PCS is to construct the system as an integration of autonomous units corresponding to each operation, to improve reliability and flexibility.

An 'autonomous unit' is a unit which can execute its own function independently and which is not directly controlled by other units. The unit should have the data required to execute the function locally, although data that must be shared is open to the other units. A data field (DF) is a logical field where all of the data is shared by all units. The various kinds of data circulate logically in the DF. The data circulating in the DF has no destination address, but each data has an ID which represents the meaning of the data. Each unit has the ability to autonomously select the necessary data according to the ID and receive it. To execute this process, a platform is installed in every autonomous unit [11]-[13].

Figure 5.2 depicts the proposed PCS architecture.

This system consists of autonomous units which correspond to the printing plants and the head office, and the DFs that connect the units, for the purpose of interplant cooperation in newspaper production. This structure ensures that new printing plants can easily be added.

In each printing plant, the system structure consists of three layers of autonomous units: the man-machine layer; the process-management layer; and the facility layer. These units are connected to the information DF and the control DF to satisfy the reliability and flexibility described in Sec. 4.

The man-machine layer consists of monitor terminals and control terminals for the operators at the facilities and in the control room. The monitor terminal displays the current status of the printing process. The control terminal outputs orders to the facilities and the process-management layer. Both types of terminals are located at the facilities and are connected to the information DF. The information DF is also connected to the DF between printing plants. Therefore, required production data held in the head office on printing plants can be sent to the DF.

Next, there is the process-management layer that is connected to the information DF. This layer consists of autonomous units that manage each process. These units send the current printing status to the information DF, and

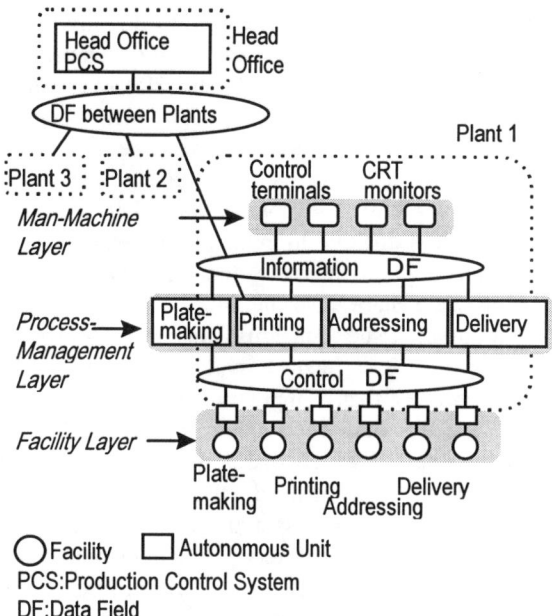

Fig 5.2 PCS Architecture based on

receive requests from operators and the required production data from the information DF. These process-management units can exchange data among themselves via the DF to coordinate their processes. On the other hand, these units are also connected to the control DF, to which the autonomous units that manage facilities are also connected. The process-management unit controls the facilities and receives information on their status.

The facility layer consists of autonomous units that manage the facilities. These units receive control data from the control DF and report the status of the facilities to the DF.

This architecture ensures on-line expandability and maintainability for development and maintenance requirements. This system structure can also provide both flexibility and reliability of operation as follows.

<Reliability>

Since each layer autonomously shares data through the DF, the reliability mentioned in Sec. 4 can be realized. Reliability means the ability of the rest of the system to continue functioning autonomously, even if a part of the system stops. For example, when the system is printing, if the autonomous unit of the printing management process stops, each facility can still continue its function by utilizing its own data.

<Flexibility>

In a conventional system, the head office's PCS reschedules the production process in response to requests to change the number of pages and their layout. It is quicker, however, to reschedule at the local printing plant.

To make it possible to change a newspaper's format flexibly, operators should be able to directly input various requests at a terminal and the facilities must be able to respond to the request immediately. However, there are two issues concerning data sharing of the man-machine request data and the control data within the same DF.

The first issue concerns performance. Each facility sends and receives real-time control data that is mostly small, however, it is difficult to maintain the real-time property of the control data when the control data and bulky man-machine data share the same DF.

The other issue concerns system reliability when an operator makes a mistake when inputting a request, or inputs a request that is impractical in the current situation.

Judging from the above, the proposed architecture does not have one DF which shares the man-machine layer and the facility layer. Instead, these layers are indirectly connected via the process management layer. This process management layer has filtering functions that can send and receive the data through both of the different quality DFs and verifies the requests from the operators.

5.3 Filtering Function

A filtering function has been installed in the autonomous units in the process management layer at an application level and system level in order to connect the information DF, through which large amounts of data flow but not in real-time, and the control DF, through which a small amount of information flows in real-time.

For the application level filtering function, autonomous units in the process management layer verify whether the change is suitable in its applications.

On the other hand, for the system level filtering function, the process management layer has the following functions.

<The Filtering Functions between DFs>
There are the following three functions. (Fig 5.3)
(1) Segmenting/Blocking
The data segmenting function receives data from a DF, breaks the data down into smaller parts, and sends the results to another DF. The data blocking function

collects many pieces of data, integrates the data into one data package, and then sends the results to another DF.
(2) Synchronizing/Ordering
This function causes all data to be sent into the DF at a given time in a certain order. In the timing part of the function, data can be sent cyclically or in response to changes in the condition of the equipment or when something happens. In the ordering part of the function, data is sent based on equipment priority.
(3) Transforming
The transforming function receives data from one DF, changes it into a form recognizable by the other DF, then sends the data to the other DF.

E.g.. 1) As an example of the functions working from the information DF to the control DF, the sequence of functions is explained with regard to the web offset press and the addressing system when it necessary to make a change in a edition.

First, the autonomous unit where printing process control occurs receives the data for the necessary change from the information DF. (1) At this time, the data is broken down into smaller pieces by the segmenting function. Then the separated data is sent using the synchronizing and ordering functions. The autonomous unit where printing process control occurs sends the data indicating a stoppage to the control DF. (2) (3) However, the web offset press and addressing system do not stop immediately upon receipt of this message. Within the addressing system, the newspapers are bundled into groups of predetermined size ordered by each delivery agent; if work was to be stopped midway through the bundle, all the newspapers already printed for that bundle would become unnecessary surplus to be thrown away. As a result of this procedure, the addressing system, via the control DF, sends the data regarding the remainder of the copies to be printed to the web offset press. Since this data needs to be sent quickly, the autonomous units where printing process control occurs, sends large amount of data for the following edition (5), just after receiving the data from the addressing system about the stop in the work (4).

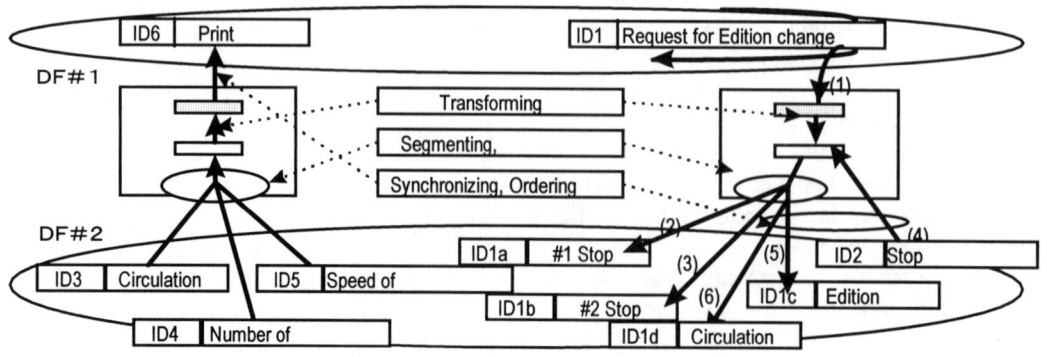

Fig 5.3 Filtering Function between different kinds of

E.g. 2) As an example of the functions working from the control DF to the information DF, the creation of display data for monitor terminals from the printing results is discussed. In the control DF, the bit code data on printing conditions flows from each piece of facility. This data is integrated by the blocking function and the contents are changed into byte code by the transforming function, both of which are located in the autonomous units where printing process control occurs. This data is then sent in a cyclical manner to the information DF by the synchronizing function.

<Filtering Process Management>

The filtering process for the previous three functions occurs in response to two things: a specific event that occurred; or on a predetermined cycle. In terms of an event, the process starts its execution after receiving the necessary data from DF. Otherwise, the filtering function occurs according to a timer.

6. Application Example

In this section, the system currently being used at Asahi Shimbun's Setagaya printing plant and the effect of the system's introduction to the plant are explained.

6.1 The Setagaya Printing Plant PCS

The Setagaya printing plant is one of four printing plants owned by Asahi Shimbun's Tokyo Head Office. At the Setagaya plant, including color presses, there are 22 web offset presses and 4 folding machines. The autonomous decentralized PCS proposed in this paper at the Setagaya printing plant is developed. Development have been broken down into two phases. During the first phase, the plate-making and printing processes are developed. The addressing and delivery processes will be developed in

phase 2. The first phase system is already operational.

Figure 6.1 shows the system structure of the autonomous decentralized PCS currently in use at the Setagaya printing plant.

Since the head office PCS and the equipment at the Setagaya printing plant were already in place and operational, these were used as is. Everything else was newly developed as an autonomous decentralized PCS. Each data field (DF) corresponds physically to each Ethernet. This network of DF between printing plants is located in the head office and is connected to the information DF located in the printing plant by a router. Currently, only the information server of the PCS's front end computer and the Setagaya printing plant facilities are connected on the network. However, there are plans to connect the other printing plants' routers to the network in an incremental manner.

In the network's information DF, 28 PCs (Personal Computer) are connected as monitors and control terminals. Each autonomous unit of the man-machine layer corresponds to these PCs. The autonomous units that control the plate-making and printing processes in the process management layer also physically correspond to control computers. In the network's control DF, three PCs are connected to control all the various plate-making facilities. The web offset presses are also connected to the network by a PLC (Programmable Logic Controller). There are graphic panel displays showing the printing conditions of the web offset presses via the PLC. From these graphic panel displays, even if one of the computers in the process management layer has a problem, the web offset press printing conditions will still be supervised and controlled.

This type of autonomous software platform is installed in every computer and PLC. This platform has high openness and can be installed in a variety of computers and

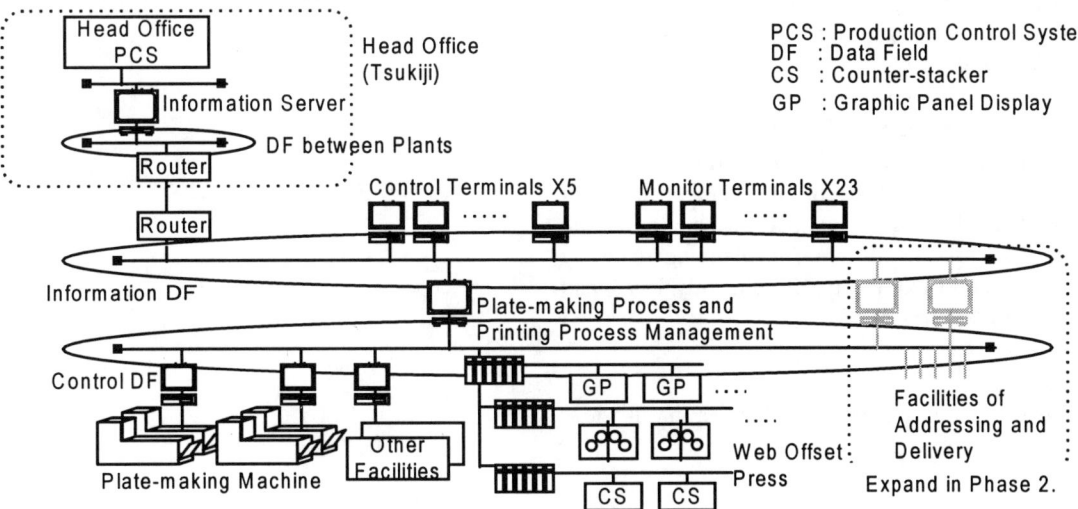

Fig 6.1 Setagaya Printing Plant's PCS

253

controllers. As a result of the installation of this software, without concern to which type of component, data can be shared equally between the autonomous units. In reality, UDP/IP protocol message broadcasting is used. Data can be shared between market applications and the platform on the PC.

6.2 Effect

1. Operation

As a result of the introduction of this system, operational reliability and flexibility has improved. More concretely, compared to previous production, the number of print procedures has fallen by 5% and plate-making procedures by 20%. Also, the percentage of unused pages printed and the percentage of unused printing plates made have decreased by 3% each.

2. Development

With this system, since the autonomous units use a common platform for the transmission of applications, compared to the previous system, the personnel needed to develop the system has been halved and the time to construct the system has been reduced by 2/3.

3. Maintenance

None of the facilities and functions in the system have been changed. However, since all facilities and applications are more independent of one another than the centralized system, system change ix expected to cost less when compared to the previous system.

7. Conclusion

This paper proposed the architecture for an autonomous decentralized PCS for newspaper production, and described the practical application of this system. This system was constructed by using architecture with three DFs: the DF between various printing plants, the information DF, and the control DF. Computers were arranged to control the filtering function between the DFs to satisfy user requirements for improved page flexibility, while still maintaining high reliability.

The proposed technology is currently being used at Asahi Shimbun's Setagaya printing plant and it has proven effective in terms of the plate-making and printing process. Compared to previous production, the number of print procedures required has fallen by 5% and plate-making procedures by 20%. Asahi Shimbun is planning to apply this technology to the Osaka Head Office in July 1997, and the new Kawasaki plant in December 1997.

Acknowledgment

We would like to thank Hitachi, Ltd. for helping us to set up the PCS at the Setagaya plant.

References

[1] R.Popescu-Zeletin, et al... "Y-Architecture for the Integration of Autonomous Components", Proc. Int. Symposium on Autonomous Decentralized Systems (ISADS), April 1993, pp. 21-27

[2] H. Kopetz, et al.. "Real-Time System Development: The Programming Model of MARS", Proc. ISADS, April 1993, pp. 290-299

[3] S. Chand, "Decentralized Monitoring and Diagnosis of Manufacturing Processes", Proc. ISADS, April 1993, pp. 384-389

[4] A.S. Lim, "A Uniform Software Architecture for Cooperation, Reliability and Reconfiguration of Autonomous Decentralized Systems", Proc. ISADS, April 1995, pp. 33-39

[5] J. Agre et al.. "Autoconfigurable Distributed Control Systems", Proc. ISADS, April 1995, pp. 162-168

[6] H. Yamamoto, A. Yoshizawa, et al.. "On-line Software Test Technique based on Autonomous Decentralized System", Proc. FTDCS, 1993

[7] "Introduction to Asahi Shimbun", Asahi Shimbun, 1995

[8] "Asahi Shimbun (Brochure of Asahi Shimbun)", Asahi Shimbun, 1995/1996

[9] T. Nagai, "Concepts and Trends of Production Control Systems", Newspaper Technology, Vol. 2, No. 148, 1994, pp. 22-25 (In Japanese)

[10] M. Funayama, "Asahi Shimbun Tokyo Head Office's new PCS (Production Control System)" Newspaper Technology, Vol.2, No.148, 1994, pp. 12-17 (In Japanese)

[11] H. Ihara and K. Mori, "Autonomous Decentralized Computer Control Systems", IEEE Computer, vol.17, no.8, Aug.1984, pp. 57-66

[12] K. Mori, "Autonomous Decentralized Systems: Concept, Data Field Architecture and Future Trends" Proc. ISADS, April 1993, pp. 28-34

[13] H. Wataya, K. Kawano, and K. Hayashi, "The Cooperating Autonomous Decentralized System Architecture", Proc. ISADS, April 1995, pp. 40-47

Session 4A

Architecture II

Chair

Luciano Lenzini

Composing Distributed Objects in CORBA

Jeff Magee, Andrew Tseng and Jeff Kramer
Department Of Computing
Imperial College of Science, Technology and Medicine
180 Queen's Gate, London SW7 2BZ, United Kingdom
E-mail:{jnm,jk}@doc.ic.ac.uk

Abstract

The paper addresses the problem of structuring and managing large distributed systems constructed from many distributed objects. Specifically, the paper proposes a component model which can be used to compose objects into manageable entities. Components are specified using Darwin, an architecture description language developed by the authors. A mapping of distributed objects into Darwin components is described together with an outline of how Darwin and its associated tools are implemented in a CORBA compliant environment.

Keywords: distributed software architecture, components, object management

1. Introduction

A recent study in software maintenance for distributed systems [12] has indicated that the move to distribution has contributed to the simplification of primitive software components in a distributed architecture. However, this benefit is often overwhelmed by the increased complexity of the overall distributed system and its architecture. The architectural description language (ADL) Darwin[7] is designed specifically to help manage this problem. Darwin is a declarative language, intended as a general purpose notation for specifying the structure of systems composed from diverse components using diverse interaction mechanisms. It deliberately separates the description of structure from that of computation and interaction in order to provide a clear separation of concerns. Darwin supports the specification of both static structures fixed during system initialisation and dynamic structures which evolve as execution progresses [8].

In this paper, we describe the use of Darwin to structure distributed systems implemented using CORBA[1] distributed objects. The approach is to define a a mapping from the CORBA object model into the Darwin component model. Essentially, the concerns of structuring a large system from a set of components are taken to be orthogonal to the concerns of programming its functionality as a set of interacting objects. Object interaction and the associated

issue of interface compatibility are the concern of the Object Request Broker(ORB) and the Interface Definition Language (IDL), while the concerns of organising a set of distributed objects into a manageable architecture, we argue, are best dealt with using an ADL such as Darwin and its associated tools. We take the view articulated by Nierstrasz and Tsichritzis [9] that components are complimentary to objects and that whereas current object oriented languages emphasize programming, components are concerned with composition.

In the following, using a simple example, we motivate the use of a component model to structure a system of distributed objects. Section 3. then outlines how the Darwin component model and the ability to compose components into complex systems can be superimposed on the basic CORBA object model without requiring modifications to either the model or the associated services and facilities. Details of a prototype implementation are presented together with some examples of the use of Darwin to structure CORBA systems.

2. Objects & Components

To illustrate the role of objects and components in constructing distributed systems, we will use the simple example depicted in Figure 1 of a pipeline of filter objects which is fed a stream of integer values by a producer object at one end and connected to a consumer object at the other. Each filter object is instantiated with its own integer value and removes multiples of that number from the stream. For example, filter(2) removes all even numbers. Each object may potentially be located on a different processor.

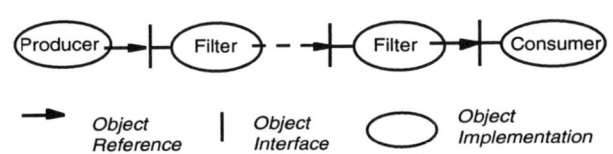

Figure 1 - Filter pipeline example

The Common Object Request Broker Architecture (CORBA)[1] is primarily concerned with facilitating transparent access to remote objects. It clearly separates the specification of the interface of an object from the implementation of that object. Interfaces in CORBA are specified in an Interface Definition Language (IDL). This interface specifies exactly those operations which can be invoked by a client and it is used to generate proxy objects in an implementation language which hides distribution from the client. CORBA specifies mappings into different implementation languages such as C and C++. In the example, it is clear that the interface to the filter objects and those to the consumer object should be of the same type. A suitable IDL specification of an interface which provides a single operation for communicating an integer value is given below:

```
interface IntStream {
    void put(in long x);
};
```

This interface must be implemented by the filter and consumer objects. However, we are left with the problem of how the consumer object finds out which filter object it should send values to and how each filter object knows which is the next filter in the pipeline. To do this, we must extend the interface of each object to include an operation which can be used to inform it of the reference of the next object in the pipeline as shown below:

```
interface Producer {
    void next(in IntStream f);
};

interface Filter : IntStream {
    void next(in IntStream f);
};

interface Consumer : IntStream {
};
```

The Filter interface now extends the IntStream interface with the operation next which can be used to send the reference of either another filter object or the reference to the consumer object since the Consumer interface is also derived from IntStream. The remaining design problems are how do we instantiate the structure of Figure 1 and how do give each filter object the value it is intended to filter out of the integer stream. CORBA itself does not specify a standard way of creating objects, however, the associated Common Object Services Specification (COSS)[2] outlines a scheme using factory objects. The interface for a simple factory object which can create filter objects is given below:

```
interface FilterFactory {
    Filter createFilter(in long x);
};
```

where x is the value which would be passed to the constructor of the new filter object. To create the pipeline, another "builder" object is required which constructs the system of Figure 1 by creating objects using the factories and passing references to the newly created objects using the their next operations.

In summary, to create the simple system of Figure 1, in addition to programming the implementation of the Producer, Filter and Consumer objects, the programmer must provide factory and builder objects to construct the system. The IDL specifications do not give a clear picture of the structure of the system. They do not clearly indicate that each Filter component has a constructor parameter. The programmer has been forced to program a next operation which has little to do with the function of the objects. In a more realistic example, an object might require references to many other objects and the programmer is required to ensure that these bindings are made correctly. The structure of the system is actually embedded in the implementation of the builder object. In a larger more realistic system, this would lead to maintenance problems. These structuring and object interconnection problems are precisely those addressed by architectural description languages such as Darwin[7].

2.1 Components

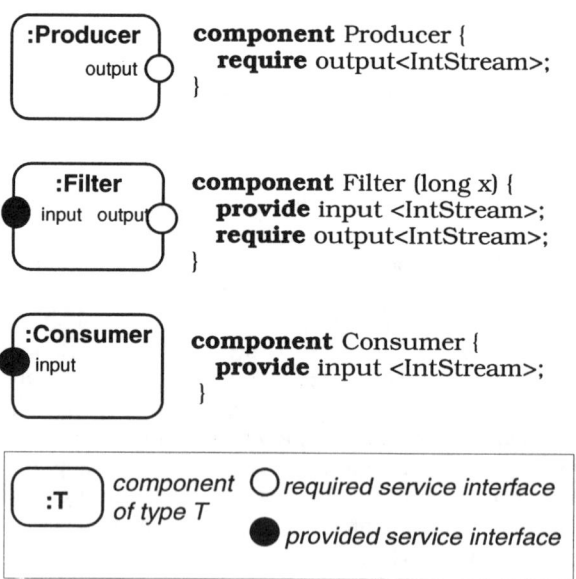

Figure 2 - Darwin component types

A provided service interface refers to a service implemented within a component whereas a required service interface refers to an interface provided outside the component. In Figure 2, the type of each service interface whether provided or required is IntStream. IntStream is

258

the interface specified using IDL as before. Components thus interact using CORBA remote object invocation protocols as before. The component specification can also include initialisation parameters which are required to instantiate the component as in the Filter component example. In contrast to CORBA objects, a component type in Darwin specifies the complete context for the implementation of that component. This context describes both provided and required service interfaces and initialisation parameters. In general, components may provide zero or more services and require zero or more services.

Systems are specified in Darwin by describing the set of component instances and the set of bindings between required and provided services as shown in Figure 3 for a simple filter system consisting of a Producer, a Consumer and a Filter component. The specification declares instances of the three component types with their actual parameters where appropriate. Component composition is accomplished by declaring bindings between required and provided services as in P.output -- F.input. The Darwin compiler checks that bindings are made between requirements and provisions of compatible interface types. To do this it must have access to the IDL interface descriptions. A Darwin specification such as that of Figure 3 is intended to be used directly to construct the desired system at runtime. The implementation effect of a binding is to place a reference to a provided service in the component requirement bound to that service. In general, many requirements may be bound to a single provided interface in exactly the same way that many CORBA clients may access a service provided by a server object.

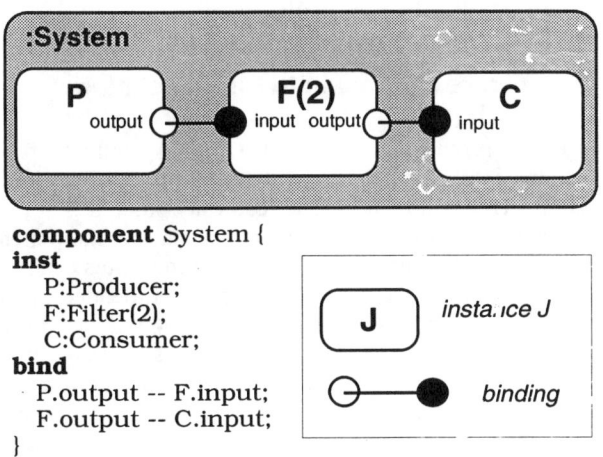

```
component System {
inst
    P:Producer;
    F:Filter(2);
    C:Consumer;
bind
    P.output -- F.input;
    F.output -- C.input;
}
```

Figure 3 - Simple Filter system

2.2 Composite Components

The Producer, Filter and Consumer components are examples of primitive components which have a computational behaviour defined in a programming language such as C++. We will see in the next section that primitive components can be implemented in a CORBA environment in a way which is consistent with the implementation of CORBA interfaces. To manage the structural complexity of large systems, Darwin has the capability to allow composite components to be constructed from more primitive component types. These composite components have no computational behaviour other than the composite behaviour of their constituent components.

The example of Figure 4, shows a composite component which encapsulates a pipeline of filter components. The number of filter instances is determined by the parameter n substituted when an instance of the Pipe component is elaborated at runtime. The Pipe component type is implemented by an array of filter instances dimensioned by the array declaration. The replicator construct forall range declares the actual instances and their bindings. Each Filter instance is declared explicitly since they have different parameter values determined by the function fvalue. This function is defined in the implementation language and can either compute filter values or read them from a file. The guard construct when expression, only includes associated bindings and instances in an elaborated system if expression evaluates to true.

The example of figure 4 locates each filter instance F[k] on a different host computer by means of the annotation @k+1. The integer machine identifiers are mapped to real machine addresses by the distributed component factory service. This level of indirection in mapping permits portable specifications. In general, instances are located at the machine on which the enclosing component is elaborated unless they are annotated.

The Pipe component can be substituted for the Filter component in the simple filter system of Figure 3 to give a system with multiple filters.

Further details of Darwin and its associated toolset may be found in [3,5,6,7,8]. We have attempted to demonstrate in this section that while the CORBA approach deals elegantly with helping the programmer deal with distribution by clearly separating the specification of interfaces from object implementations and by providing transparent access to remote services, it does not support an architectural view of distributed applications. The architecture of a CORBA distributed application is embedded in the various factory and builder objects which construct the system. The use of an ADL such as Darwin which describes a system as a set of interconnected components and which thus provides an explicit structural specification of the system augments CORBA with such an architectural view. The importance of software architecture in the design, construction and maintenance of large complex systems has recently been realized by a number of researchers and industrialists[4,13].

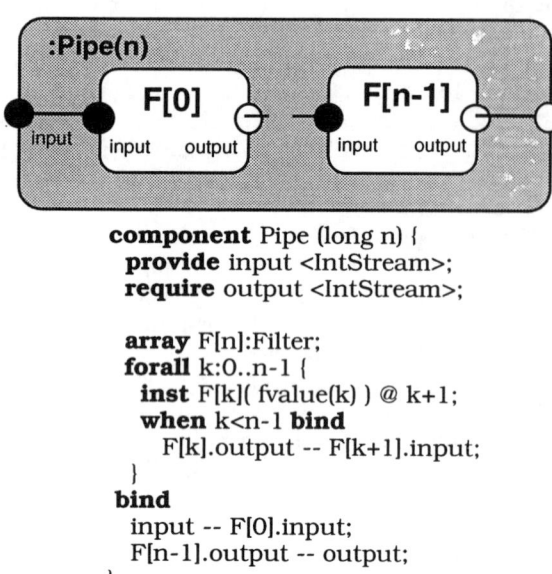

```
component Pipe (long n) {
  provide input <IntStream>;
  require output <IntStream>;

  array F[n]:Filter;
  forall k:0..n-1 {
    inst F[k]( fvalue(k) ) @ k+1;
    when k<n-1 bind
      F[k].output -- F[k+1].input;
  }
  bind
    input -- F[0].input;
    F[n-1].output -- output;
}
```

Figure 4 - Composite Component Pipe

3. Components in CORBA

The Darwin descriptions of the previous section specify the structure of the filter application while embedding its computational behaviour in primitive components. The behaviour is determined by object implementations and these object implementations interact via interfaces specified in IDL using the ORB in the normal manner. Primitive components encapsulate objects and specify their instantiation, their required interfaces and provided interfaces. As depicted in Figure 5, a primitive component may embed one or more objects.

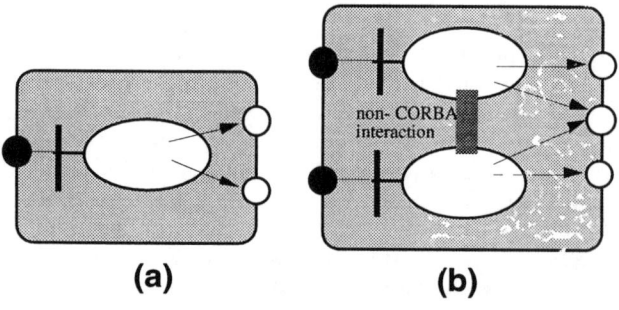

(a)   **(b)**

Figure 5 - Embedding objects in components

A primitive component may contain more than one object implementation if the objects require to be collocated. Typically, collocation is required because the object implementations interact directly, perhaps for efficiency reasons passing pointers to shared resources,

and consequently do not use CORBA interaction and do not have IDL interfaces specified for these interactions. However, the usual case is that each primitive component contains a single object and thus has a single provision as depicted in Figure 4(a).

3.1 Primitive Component Implementation

To implement components within the CORBA environment, they must be given a representation which can be manipulated and managed by CORBA services. Consequently, each component is implemented as an object which has an interface specified in IDL. Both the interface specification and the object implementation (in C++) are generated by the Darwin compiler from the component description. For example, the IDL generated from the Filter component Darwin description of Figure 2 is given in Figure 6.

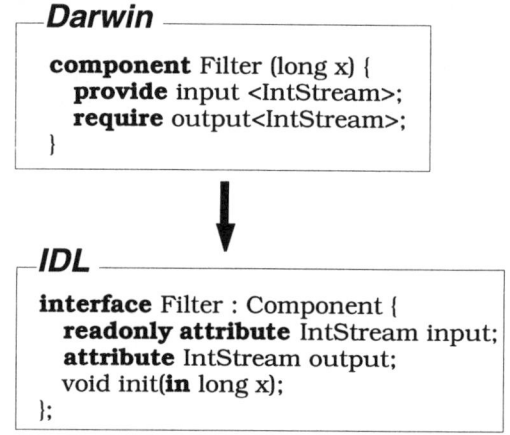

Figure 6 - Darwin to IDL translation

Each provision in the Darwin specification is translated into a read only attribute of the correct object reference type. The attribute is read only since it is set by the implementation of a component when the component is instantiated and must not be modified from outside since it is a reference to the application object embedded in the component. Each requirement is similarly mapped to an attribute which is not read only since it is set externally to reflect the binding of the component instance. Component parameters become parameters to the operation init which when invoked creates the application object. IDL interfaces whether primitive or composite are derived from the IDL Component interface type. This permits component objects to be treated uniformly by the distributed Darwin elaboration algorithm which constructs the system at runtime.

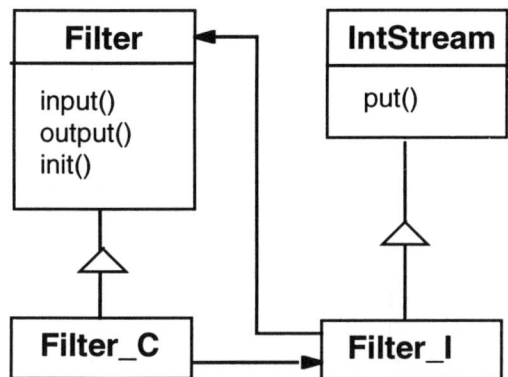

Figure 7 - Filter Component Class diagram

Figure 7 is an OMT[11] class diagram depicting the relationships between the various classes generated by the IDL[1] compiler and by the Darwin compiler for a primitive component. Filter and IntStream are the classes generated by the IDL compiler from the component and service interfaces respectively. Filter_C is the component implementation generated by the Darwin compiler. It implements the init operation which creates a new instance of the filter implementation class Filter_I. The Filter_C class maintains a reference to the Filter_I object since this is the object reference returned by the input attribute. The Filter_I class is written by the programmer who implements the functionality of the filter component. It differs from normal C++ implementations only in the way external interfaces are accessed. To access the required interface output, the reference is obtained by accessing the Filter_C object. A reference to Filter_C is passed by init when the filter implementation object is created.

The schema depicted in Figure 7 was adopted for two reasons. Firstly, it permits the component management object to control more than one application object and, secondly, the alternative schema considered which used only a single object to represent a component was not strictly CORBA compliant since it meant that a single object had two interfaces. While Orbix[10], the CORBA implementation used in prototyping, permits an object to have more than one interface (using TIE objects), this is not standard.

3.2 Composite Component Implementation

Composite components are implemented in exactly the same way as primitive components in that the Darwin

---

1. In the interests of simplifying the exposition, we have omitted the details of the classes necessary for interworking with the Basic Object Adapter (BOA) and of the CORBA::Object class from which all interface classes must be derived.

compiler generates an IDL description of the component interface and a C++ implementation for that component. For example, the IDL interface generated for the Pipe composite component is depicted in Figure 8.

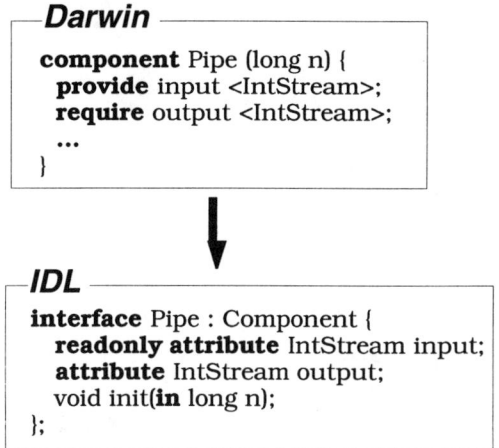

Figure 8 - Pipe composite component IDL

Composite components have no application computational behaviour, consequently, there is no application object. The composite component implementation object is compiled directly from the Darwin specification for that component. These objects compiled from Darwin descriptions are builders for the application structure. An application constructed from a Darwin specification starts by instantiating the top-level composite component. This instantiates its constituent components until the leaf components instantiate the application computational components they embed. When instantiation completes, the builder objects representing each composite component cooperate to bind requirements to provisions as specified by the Darwin structural description. Binding is accomplished by obtaining provided references from the attributes of primitive components and setting the required interface attributes to these values. A detailed specification for the Darwin elaboration may be found in [7]. This was first implemented in the Regis system[6] and adapted for the current CORBA prototype. The algorithm results in primitive components being bound directly to primitive components as depicted in Figure 9. There is no indirection overhead and object invocations are not routed through intermediaries. A CORBA system structured using Darwin incurs no extra overhead on distributed invocations between objects.

In contrast to the Regis system where the composite component builder entities terminate after instantiation is complete, in the CORBA system, we have chosen to keep builder objects so that they may be interrogated by management tools to display and modify the structure of the system[3].

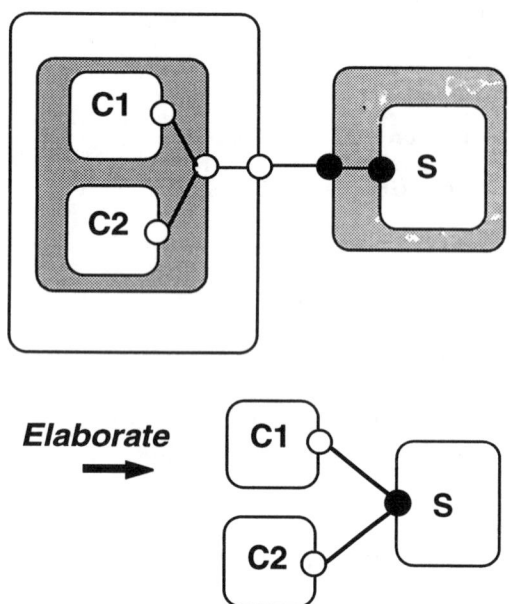

Elaborate

Figure 9 - Elaborating composite components

In this section, we have outlined how Darwin components, both primitive and composite, may be implemented in a CORBA environment as objects with interfaces specified in IDL. This means that builder objects may communicate using the standard object invocation protocols and that these objects representing components may be created and accessed using standard services. An application constructed from a Darwin specification results in exactly the same configuration of computational objects as if the builders and factories were programmed explicitly. The levels of indirection used in the specification to structure a large system into a hierarchical composition of components has no runtime overhead since Darwin elaboration results in direct binding between primitive components.

4. Conclusion

In this paper we have attempted to show that there is a need for techniques and tools to help structure systems of distributed objects, even for simple examples. Environments such as CORBA help in dealing with distribution by clearly separating the specification of interfaces from object implementations and by providing transparent access to remote services, but do not support an architectural view of distributed applications. The architecture of a CORBA distributed application is embedded in the various factory and builder objects which construct the system. Furthermore, they lack the ability to compose objects so as to build more complex structures.

In order to overcome this problem, we advocate the use of an architectural description language such as Darwin which views a system as a set of interconnected components, which provides for hierarchical component composition and which provides an explicit structural specification of the system. In addition, these structures can be parameterized and can include component instance and binding replication, conditionals and even recursion. We have shown how Darwin can be used to augment CORBA with an architectural view and how to map distributed objects into Darwin components. The Darwin specification is used directly to construct the required system. The Darwin description of a distributed application thus relieves designers and programmers of the non-trivial task of explicitly providing code to construct and initialise the distributed application. As we have seen from the previous section, the component, instantiation and binding abstractions used in Darwin can be mapped to existing CORBA abstractions and services. The use of Darwin does not impose any extra application object interaction overheads.

The ideas outlined in this paper have been prototyped using Orbix 2.0[10] for Solaris. This is a multi-threaded CORBA implementation. We have not discussed the synchronisation issues raised by the approach. For example, at what point during elaboration when do required interfaces become valid for use by an application object. The prototype has adopted the approach to this problem documented in [7] which specifies an elaboration algorithm is both concurrent and distributed.

We have concentrated in this paper on describing how CORBA objects may be composed using the Darwin component model. However, applications constructed using Darwin may freely communicate with other distributed CORBA applications which do not use Darwin. Darwin includes the ability to export services into a name space and import services from a name space. This facility can use the CORBA name service to access external services and provide services which can be accessed by others. This facility is not currently implemented in the prototype but is planned for the near future. In addition, the current prototype uses an ad-hoc implementation for the factory service used to create components. We plan to revise this in line with the CORBA Common Object Services specification for life cycle management.

Finally, we think that a major contribution of this work has been to demonstrate that it is possible to impose a component model on CORBA without requiring extensions to the CORBA IDL. We believe that the IDL is an elegant solution to specifying interfaces. It is not appropriate to extended it for tasks for which it was not intended - namely architecture specification.

Acknowledgments

The authors would like to acknowledge discussions with our colleagues in the Distributed Software Engineering Section Group during the formulation of these ideas - in particular, Naranker Dulay.

References

[1] The Common Object Request Broker Architecture and Specification, *Object Management Group*, Revision 2.0. 1995

[2] CORBAservices: Common Object Services Specification, *Object Management Group*, Revision Edition March 31, 1995.

[3] S. Crane, N. Dulay, H. Fosså, J. Kramer, J. Magee, M. Sloman, K. Twidle, Configuration Management for Distributed Systems, *Proc. of the IFIP/IEEE International Symposium on Integrated Network Management (ISINM 95)*, Santa Barbara, May 1995, Chapman and Hall 1995.

[4]]D. Garlan, D.E. Perry, Introduction to the Special Issue on Software Architecture, *IEEE Transactions on Software Engineering*, 21 (4), April 1995, pp 269-274.

[5] Keng Ng, Jeff Kramer, Jeff Magee and Naranker Dulay, The Software Architect's Assistant - A Visual Environment for Distributed Programming, *Proceedings of Hawaii International Conference on System Sciences (HICSS-28)*, January 1995.

[6] J.Magee, N. Dulay, J. Kramer, Regis: A Constructive Development Environment for Distributed Programs, *Distributed Systems Engineering Journal*, Vol. 1, No. 5., pp 304-312.

[7] J. Magee, N. Dulay, S. Eisenbach and J. Kramer, Specifying Distributed Software Architectures, *Proc. of 5th European Software Engineering Conference, ESEC '95*, Barcelona, September 1995.

[8] J. Magee, J. Kramer, Dynamic Structure in software Architectures, *Proc. 4th ACM SIGSOFT Symposium on the Foundations of Software Engineering*, San Francisco, October 1996.

[9] O. Nierstrasz, D Tsichritzis, Object Oriented Software Composition, *Prentice-Hall: The Object-Oriented Series*, 1995.

[10] Orbix -2, Progarmming and Reference Guides, *IONA Technologies Ltd.*, Release 2.0, Novv. 1995.

[11] J. Rumbagh, M. Blaha, W. Premerlani, F. Eddy, W. Lorenson, Object-Oriented Modelling and Design, *Prentice-Hall Internation Editions*, 1991

[12] S. Schneberger, Software Maintenance in Distributed Computer Environments: System Complexity versus Component Simplicity, *Proc. of IEEE International Conference on Software Maintenance ICSM 95*, 1995, pp304-313.

[13] D. Soni, R. Nord, C. Hofmeister, Software Architecture in Industrial Applications, *Proc. 17h IEEE International Conference on Software Engineering (ICSE17)*, Seattle, Italy, 1995, pp196-210.

Guidelines for Computational Modelling
in CORBA Environments

Harrold Korte and Richard Westerga
KPN Research, The Netherlands

Abstract

Transferring the behaviour of a computational design in a CORBA implementation is not always a trivial job, especially when the design incorporates the special feature of objects having multiple interfaces, such as those abandently used in TINA specifications. This paper shows this is possible, and give guidelines prescribing the structure of C++ objects that represents a computational object in the implementation. An example application, that includes the seven special features for which the guidelines can provide solutions, is used throughout the paper.

1. Introduction

Is a CORBA system considered to be ODP conformant? *Yes*, is the initial and intuitive answer of most ODP and TINA DPE experts. This paper focus on the implications of such a statement from one particular point of view, that is, the one of the application programmer.

In ODP, the computational model [1] offers powerful design concepts with objects having multiple interfaces, interfaces that can be dynamically created and destroyed during the lifetime of an object, together with configuration aspects in the form of capsule management. This paper provides guidelines how these can be reflected in CORBA implementations using the OMG-IDL/C++ language binding in the ORBIX platform of IONA[1] [2].

The guidelines are intended for designers and application programmers prototyping TINA systems using a CORBA platform. Applying the guidelines should help in structuring the code,

thereby enabling high-level designs in terms of the computational model.

The guidelines are presented using small parts of C++ code. Furthermore, the guidelines heavily use specific structuring mechanisms in C++ [3], e.g. inheritance. To visualize the structures, we use a graphical representation based on OMT.

This paper is organised as follows: Section 2 provides the rationale of multiple interfaces applied in telecommunications. Section 3 introduces the example that will be used throughout the paper to illustrate the guidelines. Requirements on the guidelines, i.e. which concepts they should incorporate, are listed in Section 4. The guidelines themselves are presented in Section 5. Conclusions are given in Section 6.

2. CORBA in Telecommunications

The Telecommunications Information Networking Architecture Consortium (TINA-C) aims at the provisioning of an architecture that combines the best of both telecommunications and information technology [4]. Key concept of TINA is the Distributed Processing Environment (DPE). All aspects of telecommunication services and systems are modelled as computational objects running on top of the DPE, e.g. service control and management, connection management, accounting, etc.

Where the DPE was originally based on ODP principles, all TINA services has been modelled using the computational model of ODP. In particular the concept of an object having multiple interfaces has been considered to be relevant in a telecommunications context. For example, an object offering its services on one interface, and offering another interface for management purposes (Figure 1).

---

[1] This paper uses the OMG-IDL/C++ binding of IONA's Orbix 1.3. This binding has been adopted by OMG with minor modifications, and the standardized OMG-IDL/C++ is implemented in IONA's Orbix 2.0.

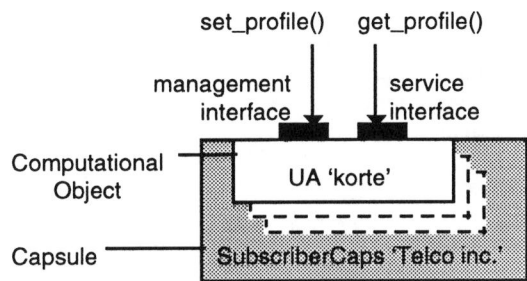

Figure 1: Multiple interfaces and configuration concepts

The figure also illustrates the concepts available in a computational design, together with configuration considerations of the engineering viewpoint: A capsule is configured with a number of COs, where each CO support a number of interfaces that can be dynamically instantiated during the lifetime of the CO. In Figure 1 the capsule consists of a number of User Agents (UAs), each of them representing a user in the network.

Where traditionally managed resources are usually associated with hardware systems, e.g. switches, the influence of information technology on TINA resulted in an increased number of *software* resources. Multiple interfaces is a desirable feature for those resources, for example to implement different access rights on the service and management interface. Not only for this; there are numerous examples in TINA, where different interfaces model the different services of one object.

With the momentum of OMG's CORBA, the DPE technology of TINA drifted away from ODP towards CORBA. However, the concept of multiple interfaces is still considered to be relevant. As such, a Request for Proposals has been issued by OMG to gather solutions for this problem domain [5].

3. The Banking Example

The example in this paper is a variation of the often used banking example. The design of the application is shown in Figure 2.

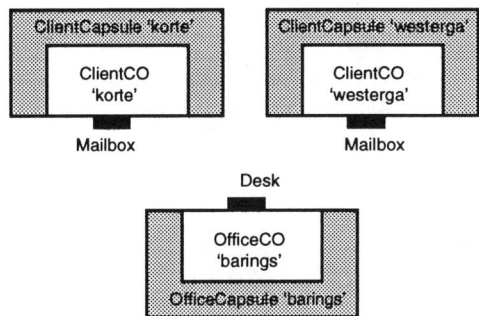

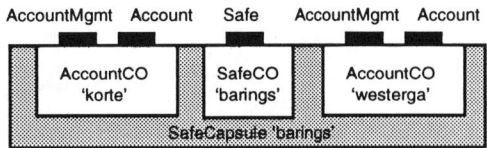

Figure 2: Capsules, objects and interfaces in the banking example

The application is only a simulation of the activities carried out by a bank, and is not a real-world system; its purpose is to demonstrate the guidelines, where it incorporates the basic concepts of the Computational Model. The three capsules represent the three different players in the example, i.e. a client that contacts an office of the bank to open account. As a result, an account will instantiated for him in the safe. After this, the client will use the account to make deposits and withdrawals. The account will eventually be closed.

For illustration purposes, the SafeCapsule is the most interesting: At instantiation of the capsule, a SafeCO will also be instantiated. During the lifetime of the capsule, a number of AccountCO are instantiated and destroyed, when clients open, respectively close an account. Furthermore, the AccountCO contains two interfaces that are created during the lifetime of the object.

A typical scenario of a client opening, using and closing an account is illustrated in Figure 3. The figure is a message sequence chart, where:

- Vertical lines denote Computational Object (CO) instances during their lifetime.
- A horizontal black arrow is an invocation between COs; If the invocation is synchronous, a grey arrow is the return, where the client blocks during the invocation. This in contrast with an asynchronous invocation, when there is no return and the client does not block.
- Dashed arrows are requests for a binding to a CO, or the request for instantiation of a CO.

- The numbered bullets are examples of basic actions that are required by the application programmer, as will be described in Section 4 and Section 5.5.

❶ Get/set data encapsulated in the CO. This is a basic assumption of all object models, i.e. the behaviour can set and retrieve the encapsulated data of the object.

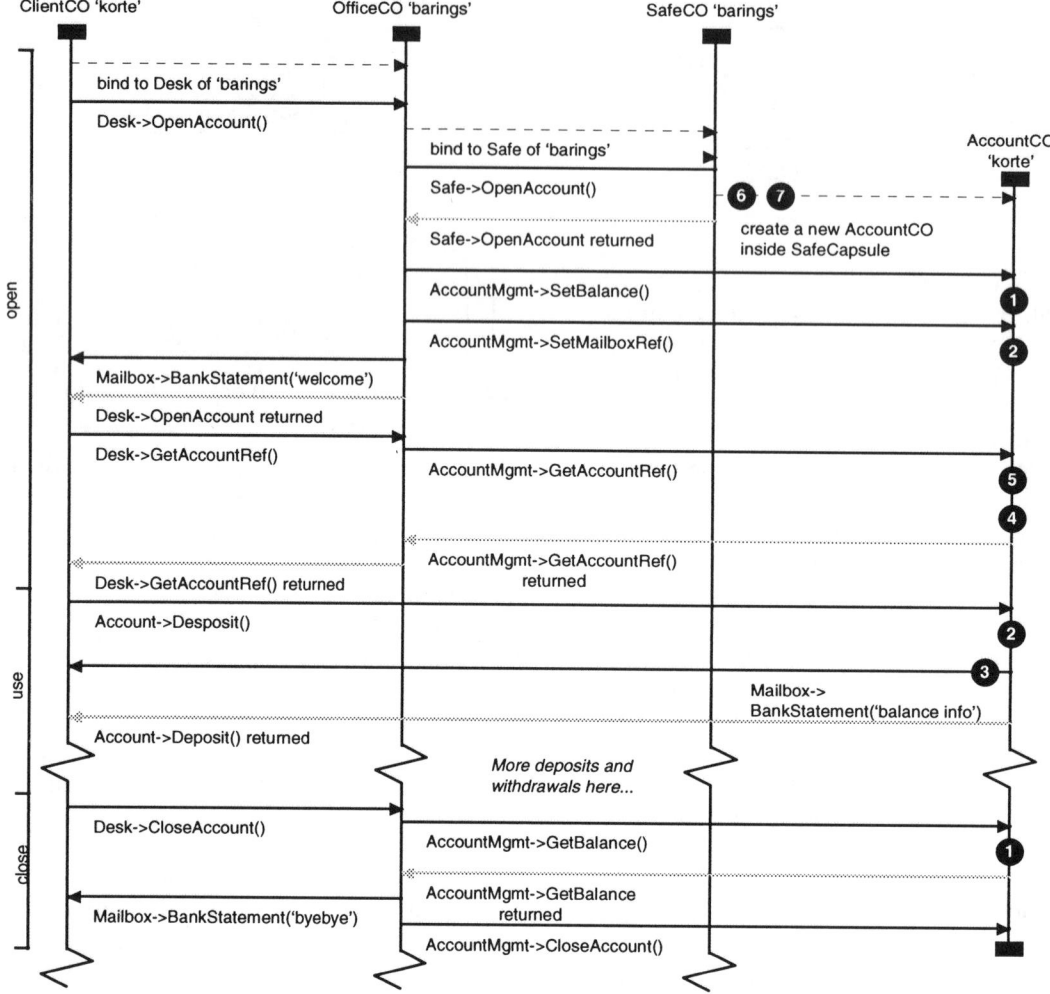

Figure 3: Banking example scenario

4. Requirements on the Guidelines

The guidelines mainly structure the support of features that are available to the application programmer, and that he needs in order to implement the application which has been designed on the computational level. In a CORBA environment, the application specific part of the behaviour of a Computational Object is located in the specification of the operations of its interfaces. In programming this behaviour, it should be possible to perform the following list of basic actions (examples are shown in Figure 3, indicated by corresponding numbered bullets):

❷ Get/set the reference of an interface that is used in the CO in the role of a client. References to these interfaces are treated in the same way as encapsulated data of the object. Retrieval of the interface reference is usually carried just before invocation of an operation on that interface.

❸ Invoke an operation on an interface using the reference, where the server interface can be in the same CO, the same capsule or in another capsule. This uses the basic remote procedure call support of the platform. In the invocations, data (and interface references) can be passed either as argument or result.

❹ Get the reference of an interface instance within the same CO. This is required when the

266

result of an invocation is a reference to another interface in the same CO.

❺ Create/destroy an interface instance within the same CO. During an invocation, a new interface in the same CO might be instantiated. Usually, the reference to this newly created interface is the result of the invocation.

❻ Create/destroy a CO within the same capsule. Operations that request the creation and destruction of COs within the same capsule, are actual capsule configuration management operations. Their behaviour is not purely computational, as it is configuration specific.

❼ Request creation/destruction of an interface of a different CO within the same capsule. This is closely related to the previous item, i.e. capsule management operations usually instantiate a CO, the CO instantiates one of its server interfaces and the reference is passed as the result of the capsule management operation. Again, this is not purely computational, as it is configuration specific.

5. Guidelines

5.1 Introduction

When implementing computational designs in a structured way, the programmer is confronted with the general framework of the C++ specification, i.e. which structure of C++ objects implements a computational object with multiple interfaces and their dynamics? The guidelines presented in this section proposes a specific structure, that has been developed after careful consideration of various alternatives, keeping the requirements presented in the previous section in mind.

Note that the reasoning about different possibilities that arise when developing the guidelines are only partially motivated. Most of the time, they are just posed as a rule, where only the most interesting trade-offs are explained a bit further.

The most important guideline concerns the structuring of a CO. A CO consists of:

- *Interface* objects (Section 5.2),
- a *container* and a *core* object (Section 5.3).

There is also the *capsule* object (Section 5.4) that deals with the creation and deletion of CO structures inside a capsule. Objects in this section means C++ objects, instead of objects in terms of the computational object model in the previous sections.

The guidelines are supported by a library package called ODP. The library consists of a single class ODP, which is a subclass of the CORBA class of Orbix. The ODP class contains a number of nested class definitions:

```
class ODP: public CORBA
{
public:
  typedef CORBA::Object Interface;
  typedef CORBA::ObjectRef IfRef;

  class Container
  {
  public:
    // Operations are declared pure
    // virtual. They are defined in a
    // superclass.

    virtual ODP::IfRef createIf(
      String _ifName
    ) = 0;

    virtual void destroyIf(
      String _ifName
    ) = 0;
  };

  class Core: virtual public Container
  {...};

  class Capsule
  {...};
};
```

5.2 Interface Object

C++ objects that implement an interface come into two flavours. They either used by clients as a proxy for the remote interface, or they implement the interface operations at the server side. It is common practice in platform technology, that the interface objects are (partially) generated from the interface definition:

- Client proxy objects are completely generated so that they can be used by clients without any further notice.

- The server interface objects are partially generated: The generated part schedules incoming invocations to the part that implements the operations. The latter part is, of course, hand-coded.

The structure of C++ objects that implement a client proxy or a server interface object is prescribed in the CORBA specification, where the structuring mechanism that is used is inheritance, if the BOA approach is used, that is defined in CORBA 2.0. In CORBA, interface objects are called *CORBA objects* and they are (partially) generated from an OMG-IDL interface specification.

In the guidelines, there are three rules for the hand-coded part of a server interface object:

1. It is a subclass from the generated part (prescribed by CORBA).
2. It contains the operations of the interface (prescribed by CORBA).
3. It contains a pointer to a container object that is related to the Computational Object it is part of (described in Section 5.3). The pointer to the container object is the first argument of the constructor of the server interface object.

For the server interface object of the Account interface, this results in the signature listed below. The OMT structure behind the generated part is shown in Figure 6, at the end of this paper.

```
// AccountImpl is the hand-coded server
// interface object; AccountBOAImpl is
// generated from the OMG-IDL interface
// specification.

class AccountImpl: public AccountBOAImpl
{
private:
   // pointer to an AccountContainer
   AccountContainer *container_;
public:
   inline AccountImpl(
     AccountContainer _container
   ): container_(_container) {};

  long Deposit(
    long _amount, CORBA::Environment &
    );
  long Withdraw(
    long _amount, CORBA::Environment &
    );
};
```

Client proxy objects with corresponding interface references are used as ordinary CORBA client objects. The redefinition of CORBA Objects in the ODP package described in the previous section rename them to type ODP::Interface.

As already pointed out in Section 4, the application specific part of the behaviour of a CO is specified in the operations of its interfaces, i.e. in the server interface objects. How this behaviour can be specified using the guidelines is presented in Section 5.5, after the introduction of supporting structuring principles.

5.3 Container and Core Objects

The basic concept of all object models is the encapsulation of data in the object. This is incorporated in the guidelines by the introduction of a *container* object. In the structured collection of C++ objects that model a CO, the container object holds the data that is encapsulated by the CO, which can be accessed by the other objects of the CO, e.g. the server interface objects. There are three categories of data in the container object:

- Data specific to the application, e.g. the balance in the AccountCO.
- A list of interface references that are used by this CO in the client role, e.g. a reference to the Mailbox in the AccountCO.
- A list of interface references the CO supports, e.g. the Account and AccountMgmt interfaces in the case of the AccountCO. If, for example, the result of an operation is a reference to another interface in the same CO (requirement ❹ in Section 4), it can be obtained via the container object.

There are three rules in the guidelines for container objects:
1. It is a virtual subclass from ODP::Container. A virtual subclass is a C++ concept; see [3].
2. The hand-coded part of the server interface objects are declared to be *friend* classes of the container, so that they can access the data encapsulated in the container object. Furthermore, the encapsulated information in the container is only protected, so that it can be used by a subclass, that is, the Core object described later on.
3. The type of the references stored, both the ones used in the client role and those of server interface objects, is ODP::InterfaceRef, regardless of the interface type. An alternative option is to use the advanced type features of CORBA, and to store references to specific interface types instead (e.g. MailboxRef, AccountMgmtRef, etc.). This has not been applied in order to minimise the dependencies of the container object to specific interface types.

The resulting signature for the container object of the AccountCO is listed below:

```
class AccountContainer
 : virtual public ODP::Container
{
// Server interface objects are
// declared as friend.
friend class AccountMgmtImpl;
friend class AccountImpl;
protected:
   // Application specific data.
   unsigned long balance_;

   // List of interface references used
   // in the client role;
   // type is always ODP::IfRef.
   ODP::IfRef mbox_;

   // List of references to interfaces
   // the CO supports;
   // type is always ODP::IfRef.
   ODP::IfRef accountMgmt_, account_;
};
```

In the computational model, COs can offer more than one interface, and interfaces are dynamically instantiated and destroyed during the lifetime of the CO. There are basically two approaches to reflect this in CORBA implementations, called here the *inheritance approach* and the *composition approach*. Both approaches use a *core* object, that is the glue in the structured collection of C++ objects that from a CO.

The *inheritance approach* is probably the most intuitive one and also the one that is currently most used. In this approach, the core object is a subclass of all its interfaces; it inherits all the server object classes the CO supports. However, there are a number of conceptual drawbacks:

- Interfaces are created/destroyed simultaneously with the core object that represents the CO, as a result of the inheritance. This is not strictly in line with the semantics of the Computational Model.
- The operation names in the different interfaces of a CO may not be overlapping, since the core object is a subclass of al its interfaces, and name clashes would occur.
- A CO cannot offer more than one instance of the *same* interface, as a result of the inheritance. There are definitely computational designs where it is desirable to have this feature.

Because of these limitations, the composition approach is proposed here. A core object is subclass of the container object, and is a composition of its interfaces. Furthermore, the server interface objects have generic pointer relations with the container, in order to access the encapsulated data of the CO. The composition of the core and its interfaces can be precisely one (interface is instantiated when the CO is instantiated), zero or one (interface is instantiated during the lifetime of the CO), zero or more (more than one instance of one interface type), etc.

Figure 4 gives a visual example of the two approaches.

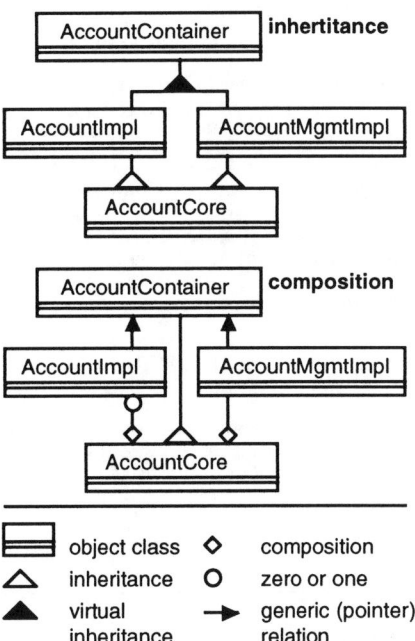

Figure 4: Alternative structures for a CO

The identified limitations of the inheritance approach are solved in the composition approach. On the other hand, the number of C++ object instances that correspond to a CO instance increases (three for AccountCore).

There are three rules in the guidelines for core objects:

1. It is a subclass of ODP::Core and the corresponding container object.
2. It contains pointers to its interfaces reflecting the applied composition structures (see Appendix 1) of the core object and its interfaces .
3. It redefines the createInterface and destroyInterface functions, which are declared as pure virtual in ODP::Container (see Section 5.1).

For the core object of the AccountCO, the result is the following object signature:

```
class AccountCore
:public ODP::Core,
 public AccountContainer
{
private:
   // server interfaces
   AccountImpl *account_;
   AccountMgmtImpl *accountMgmt_
public:
   // Pure virtuals of ODP::Container
   ODP::IfRef createIf(String _ifName);
   void destroyIf(String _ifName);
};

ODP::InterfaceRef
AccountCore::createInterface(
   const String &_interfaceName
```

```
){
  if (_ifName == "AccountMgmt") {
    accountMgmt_
      = new AccountMgmtImpl(this);
    AccountContainer::accountMgmt_
      = (ODP::InterfaceRef)accountMgmt_;

    return
      AccountContainer::accountMgmt_;
  }
  else if (_ifName == "Account") {
    // similar code as above if clause
  }
  return NULL;
}

void AccountCore::destroyInterface(
  const String &_ifName
){
  // Release is prescribed by CORBA to
  // destroy server interface objects.
  if (_ifName == "AccountMgmt")
    accountMgmt_->_release();
  else if (_ifName == "Account")
    account_->_release();
}
```

5.4 Capsule Object

The final structuring object is the capsule object, that configures a number of COs in a capsule. It has already been mentioned that its use is not a purely computational one, however it is needed to complete an executable specification. The structuring mechanism for a capsule is straightforward; it is composed of one or more core objects. In the banking example, the SafeCapsule contains precisely one SafeCore, and zero or more AccountCore's, as shown below. The complete object of the SafeCapsule is shown at Figure 6, at the end of this paper.

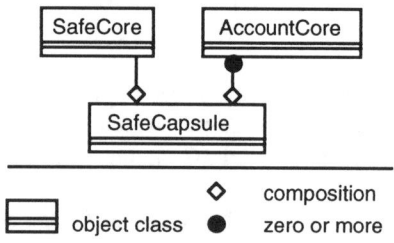

Figure 5: Core objects in a capsule

There are three rules in the guidelines for capsule objects:
1. It is a subclass of ODP::Capsule.
2. It contains pointers to the core objects it contains (reflecting the composition structure).
3. It contains operations to create/destroy supported core objects. The create/destroy operations are declared as static members. Thereby, they are globally accessible inside the capsule. The result is a pointer to the newly

created core object (required to implement action ❼, see Section 5.5)

The object signature of the SafeCapsule object is shown below (together with example usage in the main function of the capsule):

```
class SafeCapsule: public ODP::Capsule
{
private:
  // array of AccountCore objects
  static AccountCore **accounts_;

  // SafeCore object
  static SafeCore *safe_;
public:
  static ODP::Core *createAccountCore();
  static ODP::Core *createSafeCore();
};

main()
{
  SafeCapsule::createSafeCore();
}
```

5.5 Putting it together

After presenting the structure of a CO, imposed by the guidelines, this section will illustrate their benefits, assessing the requirements listed in Section 4. The examples are the same as depicted in Figure 3.

❶ Get/set encapsulated data. The balance is obtained and set via the SetBalance operation on the AccountMgmt interface of AccountCO. The balance is retrieved in the GetBalance operation on the Account interface of AccountCO.

```
void AccountMgmtImpl::SetBalance(
  long _balance, CORBA::Environment &
){
  container_->balance_ = _balance;
}

long AccountImpl::GetBalance(
  CORBA::Environment &
){
  return container_->balance_;
}
```

❷ Get/set, and ❸ use the reference of an interface that is used in the CO in the client role. The reference is obtained and set via the SetMailbox operation on the AccountMgmt interface of AccountCO. The reference is retrieved just before using it in the Deposit operation on the Account interface of AccountCO.

```
void AccountMgmtImpl::SetMailbox(
  ODP::InterfaceRef _mbox,
  CORBA::Environment &
){
  container_->mbox_ = _mbox;
```

```
  container_->mbox_->_duplicate();
}

long AccountImpl::Deposit(
  long _ammount, CORBA::Environment &
){
  core_->balance_ += _amount;

  Mailbox *mbox
  = Mailbox::_narrow(container_->mbox_);
  mbox->BankStatement("deposit conf");
  mbox->_release();

  return container_->balance_;
}
```

❹ Create a server interface in the same CO, and ❺ retrieve the reference as the result of an operation. The reference is requested in the GetAccountRef operation on the AccountMgmt interface of the AccountCO. During the operation, the Account interface is instantiated.

```
ODP::IfRef
AccountMgmtImpl::GetAccountRef(
  CORBA::Environment &
){
  // The container_ attribute actually
  // points to an AccountCore instance.
  // Therefore, the redefined createIf
  // of AccountCore is invoked
  container_->createIf("Account");
  container_->account_->_duplicate();

  return container_->account_;
}
```

❻ Create a CO in the same capsule, and ❼ request the instantiation of an interface in that CO. **Warning:** configuration specific code! In the banking example, the reference is requested in the OpenAccount operation on the Safe interface of the SafeCO. The SafeCO implements the SafeCapsule configuration management aspects.

```
ODP::IfRef SafeImpl::OpenAccount(
  CORBA::Environment &
){
  // Operation is static.
  ODP::Core *core
  = SafeCapsule::createAccountCore();

  ODP::Ifref *accountMgmt
  = core->createIf("AccountMgmt");

  return accountMgmt->_duplicate();
}
```

6. Conclusions and Future Work

In the previous sections, guidelines for engineering computational designs in CORBA environments have been presented. The guidelines are mainly implementation driven, based on the following approach:

- Identification of basic actions in a design according to the semantics of the computational model. Seven actions are identified so far (from which two not pure computational).
- For each of the seven actions, corresponding C++ structures are identified that reflect the semantics in the implementation.
- No tool support is required, though this could be helpful.

In the few exercises we had so far, the advantages are twofold: First, it becomes more easy, and therefore faster, to implement the top-level design, as this is prescribed by the guidelines. Note that the detailed design can still be arbitrary complex. Secondly, a structured implementation is easier to understand by other people than the programmer; which is beneficial in a development team.

For future work, the next step should be, of course, application in various projects. Experiences might lead to different actions and implementation structures, as those proposed in this paper. What follows, is tool support, that translates an ODL like specification into a CORBA implementation based on the guidelines. An example is the Platypus project [6]. At the same time, it is also needed that the behaviour of computational designs can be expressed in a more high-level language. Currently, only scenarios using message sequence charts are attached to the design. An example of high-level language tool support can be found in the ACE environment [7]. Furthermore, in the ACTS Project DOLMEN, guidelines for computational modelling in SDL-92 environments have been developed, on the basis of the ideas presented in this paper [8]. If all of this is further developed, the envisaged service creation methodology for TINA systems enables high-level design specifications, supported with various tools to derive the implementation.

Acknowledgements

The ideas presented in this paper are the result of work that has been carried out in the past two years. The guidelines were applied and refined in various projects, both external and internal to our company. It originated as a spin-off in the Targeting Activity of the RACE II Project SCORE during the past two years. This activity has been one of the early initiatives that explicitly addressed service creation in an ODP

271

and TINA context, from 1993 to 1995. However, the presented ideas do not necessarily reflect the view of the SCORE project. We kindly thank Aart van Halteren and Dirk Los for their valuable feedback in developing and applying the guidelines.

References

[1] N. Natarajan et. al., *Computational Modelling Concepts*, TINA Baseline document, December 1994.

[2] *ORBIX distributed object technology Programmer's Guide - Release 1.3.1*, IONA Technologies Ltd., February 1995.

[3] Margaret A. Ellis, Bjarne Stroustrup, *The Annotated C++ Reference Manual*, AT&T Bell Laboratories, New Jersey, December 1990.

[4] Martin Chapman, Stefano Montesi, *Overall Concepts and Principles of TINA*, TB_MDC.018_1.0_94, February 1995.

[5] Andrew Watson, OMG, ed., *Revised Multiple Interfaces and Composition RFP*, orb/96-01-04, available in PostScript format at http://www.omg.org/public-doclist.html

[6] Nigel Hooke, Barry Kitson, *A CORBA-based ODL/C++ Language Mapping*, Proceedings of the TINA '96 Conference, Heidelberg, Germany, September 1996.

[7] TINA ACE Home Page, http://andromeda.cselt.stet.it/ace/ACE.html

[8] ACTS AC036 DOLMEN, *Component Modelling and Design Guidelines*, Document No. MC-DEL02, December 1996.

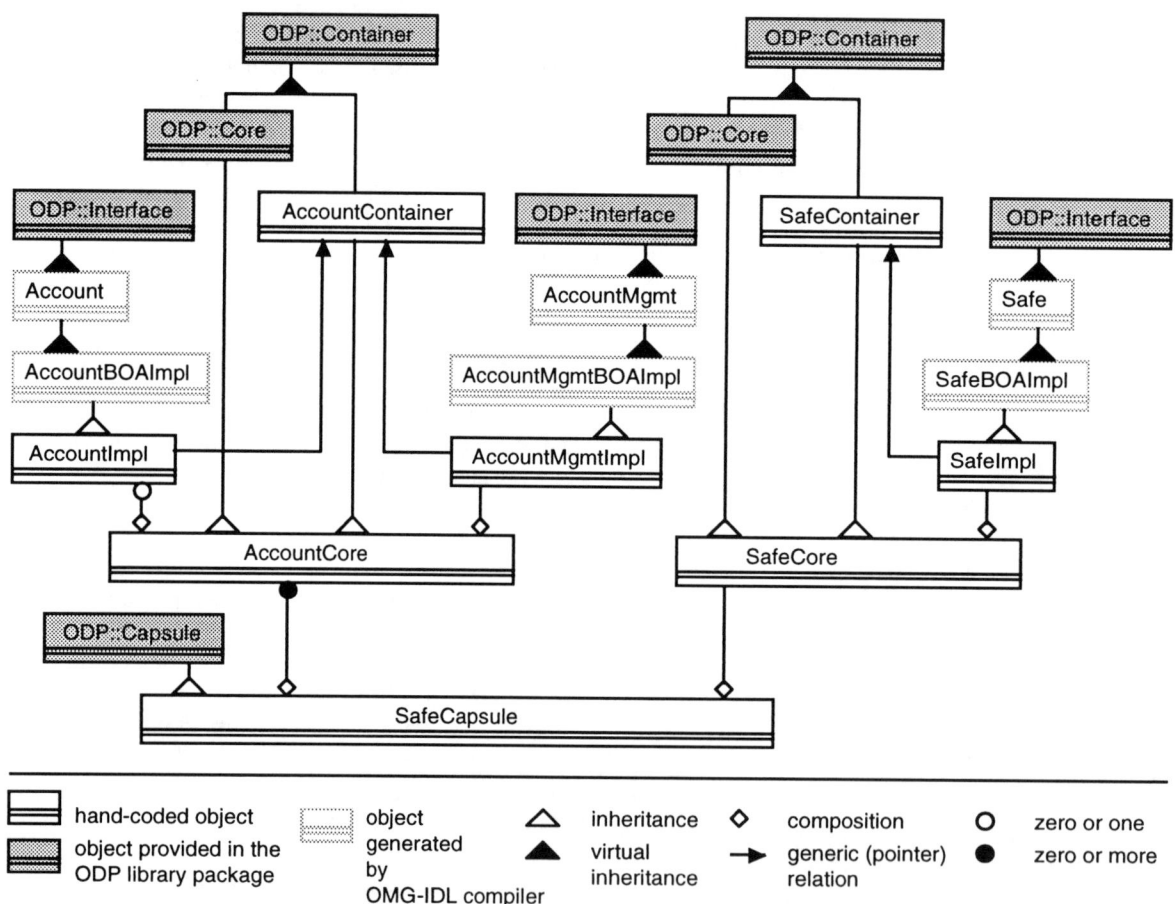

Figure 6: Complete design of the SafeCapsule

Implementing Replicated Services in Open Systems Using a Reflective Approach

Joni Fraga, Carlos Maziero, Lau C. Lung, Orlando G. Loques Filho[*]

Laboratório de Controle e Microinformática
Departamento de Engenharia Elétrica - UFSC
88.049-900 Florianópolis SC - BRAZIL
e-mail: {*fraga,maziero,lau*} *@lcmi.ufsc.br*

[*]Pós-Graduação em Computação Aplicada e Automação - UFF
24210-240 Niterói RJ - BRAZIL
e-mail: *loques@caa.uff.br*

Abstract

In this paper we evaluate the use of an object-oriented open platform based on the CORBA standard [15] for the implementation of replicated services. To improve the flexibility of the implementation, we use a reflective approach [13], which allows for separation of aspects related to the replication model from those related exclusively to the service being replicated. This separation makes it possible to modify the replication protocol according to the fault tolerance level desired, without any implications for the application code.

Keywords: fault tolerance, object groups, CORBA, computational reflection.

1. Introduction

Distributed systems have been recently characterized by their increase in dimensions and their heterogeneity. These systems have adopted the idea of open architecture, obtaining the interoperability of their components by the homogeneity of their corresponding interfaces. An effort made in terms of open programming is the CORBA standard (Common Object Request Broker Architecture), the result of the work of various companies which are part of the Object Management Group [15], whose aim is the integration of different programming systems based on objects. The use of CORBA standards, therefore, permits the interaction of objects distributed in the system, regardless of their coding languages, machine architecture or operational systems.

The concept of group processing has been introduced in distributed programming models with the aim of providing support for cooperative work (groupware), making possible an increasing availability of shared resources, or in replicated processing, due to fault tolerance. The use of CORBA standards has evolved in recent years in terms of incorporating group processing services. The Group Server abstractions are object of specification for inclusion in the OMA architecture [2]. The purpose of Group Server is similar to the approach used in ANSA (Advanced Networks Systems Architecture) [1], presenting a concentrator element in group communications, which is a handicap in the performance and reliability of a system.

Furthermore, various prototypes and even products of CORBA platforms have been developed, offering support to group processing. Specifically, we may mention the ORBs (Object Request Brokers) RDO/C++ [11], Orbix+Isis [10] and Electra [14]. These platforms make use of tools such as Isis [5] and Horus [18] that provide group communication based on the reliable broadcast concept. The tools cited above offer more reliable bases than the solutions sought in the specifications of Group Server in OMG.

In this article, we have set out to present our work on the integration of replication techniques into an open system, according to the patterns of the CORBA proposal, in order to make available mechanisms of fault tolerance to the applications distributed on that platform. The implementation of replication techniques is backed by the use of ORBs presenting a support for group processing.

With the aim of minimizing the replication reflexes on the programming of applications, a programming model was adopted, based on the *computational reflection* [13]. This paradigm permits the complete separation

273

of the coordination mechanisms among the replicas from the application in itself. This separation, besides simplifying the programming of the replicated application, introduces a great flexibility into the system by allowing the alteration of the replication protocols, without interfering with the application functionality, or even involving changes on the level of execution support, which would be difficult, considering the nature of open systems.

The programming model presented was used successfully in the integration of different replication techniques. As an implementation support, use was made of Electra, an ORB with support to process groups. In the present article, we will present only the active competitive replication technique described in [17], to illustrate the advantages offered by this model in the environment under consideration.

The article is structured as follows: in section 2 we present the active competitive replication model; in section 3 we introduce the concepts of computational reflection and set out to structure the model according to this approach; in section 4 we describe the CORBA standard and the ORB Electra with its extensions for group support; finally, in section 5, we present in detail the integration of the reflective model proposed with the CORBA platform utilized and the results obtained in its implementation.

2. Software component replicated models

Replication techniques are an alternative that enables services to continue in distributed systems, even when failed nodes are present. The unit of replication is a software component (objects, processes, etc.), encapsulating data identified as *replica state*. The replicas are distributed among different sites in the network. The coordination of the replication defines the way the different replicas must interfere in the processing, in terms of maintaining the consistency and transparency of the whole.

The techniques vary according to the degree of synchronism and the types of replicas involved. In the literature, passive, active and semi-active replication models are identified [17]. In the passive replications, a privileged replica executes the processing referring to the input data, while the others have their states updated by the privileged one, using *checkpointing* (state transfer mechanisms). The *coordinator-cohort* model, presented in [4] is an example of this type of replication.

In active replication models, all the components receive the input data, process them simultaneously and produce the same outputs. In these models, identified as *State Machine*, the consistency of the replica state neces-

sarily implies *determinism of replica*, which can be obtained by consensus about the input data and its order [19]. Some authors identify semi-active replications, in which, although all the replicas work in competition, only one produces the output. The order of the inputs is imposed by a privileged replica. The *leader-followers* technique described in [17] is an example of semi-active replication.

In [12] exhaustive studies are carried out on replication techniques and their implementation aspects. In this text we limit ourselves to the active competitive replication model described in [17].

2.1 Active competitive replication

In the competitive replication model all the replicas are active but only one responds to a given input data request. The main characteristic of this model is the competition among the replicas: only the fastest replies to the request. The coordination of the technique is distributed: each replica has an associated controller, responsible for receiving, broadcasting and comparing messages, with the corresponding replica dedicated to request processings. To guarantee replica consistency, all the messages among them are transmitted by means of atomic broadcasts. The competitive replication model can be devised so as to tolerate two sets of faults [17]: *timing faults*, involving semantics of *crash*, omission and timing errors; *arbitrary faults*, that take in the whole spectrum of failure semantics. For clarity and economy of space, we shall limit ourselves, in this text, to the first set of faults.

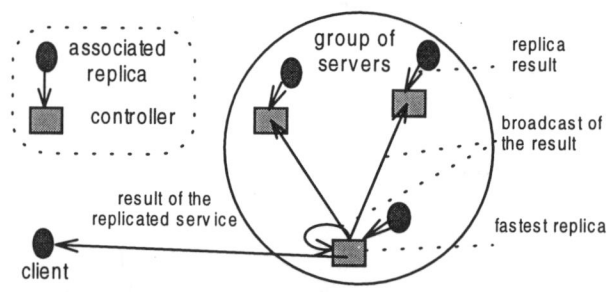

figure 1: Competitive replication model for temporization faults.

Figure 1 illustrates competitive replication, in a simplified manner, under the assumption of timing faults. In this case, considering the client/server model using a replicated server, a client request broadcast in the server group is received by the controllers, which send it to the corresponding replicas. On receiving the result of a processing of its associated replica, each controller verifies whether it has already received the message with the result of the same processing from another controller of the group. In the absence of these messages, the controller concatenates an identifier to the result, and broadcasts the resulting message to the controllers group. If the controller receives its own message first, it finds out that its replica is the fastest and therefore is the one responsible for sending the reply to the client; otherwise its message is discarded. This algorithm guarantees that only one replica answers to the client request, because all the messages broadcast in the group are observed by each member in the same relative order (a total order imposed by the use of a atomic broadcast protocol).

Finally, the broadcast of a message *end_of_processing*, after sending the results to the client, by the controller of the fastest replica, closes the processing cycle in terms of the client request. This message makes it possible to work out strategies to detect faults in the fastest replica controller and its substitution by another controller for sending results to the client.

The protocol shown above covers up errors due to timing faults. Concerning treatment of failed elements, two detection procedures are foreseen in the original literature. [17]:

- A weak coupling is perceived between the controller and its replica. In this case, *time-out* mechanisms are maintained in the controller to detect the lack of an associated replica ;
- Competitive replication gives a privilege to the fastest replica and, consequently, can lead to considerable asynchronism in the set of replicas. This asynchronism is dealt with, by periodically having a *rendezvous,* in which all the controllers broadcast the results of their replicas among themselves and the last to broadcast is the one which sends the results to the client. This *rendezvous* is limited in time, so as to allow for the detection of failed controllers.

3. Reflective structure for the competitive model

The *computational reflection* paradigm allows a system to execute processing on itself, in order to modify its behavior. In [13], the reflective paradigm is introduced into the object oriented programming using the *meta-objects* approach. Here, the functional and non-functional aspects are separated through the use of *base-objects* and *meta-objects*. A meta-object is associated with each base-object. Through their methods, the base-objects express the application functionalities, while the associated meta-objects carry out control procedures that determine the behavior of their corresponding base-objects. The calls to the base-object methods are trapped, so as to activate the meta-methods that make it possible

to modify base-objects behavior or add functionalities to their methods.

In this study, computational reflection is used to develop an integration model for replication techniques in open systems. The reflective paradigm allows us to assign to the base-object the functionalities of a replicated application, while meta-objects execute replica coordination protocols. This model allows the use of different replication techniques while the base-objects maintain their characteristics; to this end, all that is needed is to change the associated meta-objects.

The structure proposed for incorporating active replication concepts into the reflective processing model is presented in figures 2 and 3. Each replica was mapped under the form of an base-object, with which a meta-object, assuming the functionality of controller, is associated. The competitive replication that we use follows a failure semantic of crash. Since we accept a strong coupling between the controller and the associated replica, the errors generated will be attributed to both; in the *crash* failure, the controller and associated replica will cease their execution.

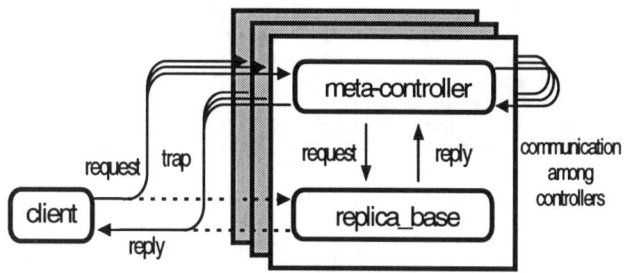

figure 2: Reflective structure for the active competitive replication model.

A request multicast by the client into a group of replicas is trapped to the respective controllers, that, in turn, have to interact in order to implement the coordination protocols of the replication scheme used. The actions of a controller are succinctly described in the code of figure 3. Each base-object method is associated with a meta-method in the controller, responsible for its activation (*method base_1 and meta_method_1,* in the figure cited).

The *meta_control* method implements the coordination protocol among the replicas described in the preceding section. The basic behavior of the algorithm consists of iterating between the choice of a replica for the reply to the client (`first`) and the closing procedure, until there is a confirmation that the reply has actually been sent (`concluded` condition of the `while` loop). It is simple to verify the termination of the request processing: if, after multicasting the method `closing`, the fastest replica (`first`) is still alive (into the *membership*), then the reply was actually sent. Otherwise, a new replica is chosen and the process is repeated. This procedure eliminates the need to multicast a message about the end of the processing. In the algorithm, the activations of the methods `multicast_id` and `closing` are transmitted to all the replicas of the group, in a totally ordered manner.

```
class meta_controller {
   // declaration of variables

   method meta_method_1 (parameters) {
      method_base_1 (parameters);
      meta_control   (parameters);
   };

   ... // declaration of further meta-methods

   // implementation of the meta-control method
   method meta_control (parameters) {
      first := null ;
      concluded := false ;
      my_id := get_system_id () ;
      while not concluded do
         if (first = null) then
            group.multicast_id (my_id);
         end ;
         if (first = my_id) then
            // first replica to reply
            return ; // return reply to the client
         else
            if not concluded then
               group.closing () ;
            end ;
         end ;
      end ;
   }

   method multicast_id (int id) {
      if (first = null) then
         first := id ; // id of the fastest replica
      end ;
   }

   method closing () {
      if (first ∈ membership) then
         concluded := true ;
      end ;
      first := null ;
   }
}
```

figure 3: Competitive replication meta-controller.

Both the competitive replication model and the support utilized give a privilege to the fastest replica, what may cause a lack of synchronism in the slower replicas. The periodical execution of the global *rendezvous* technique, proposed in [17] is not used here, due to its cost implications in the system performance. The solution adopted is based on the property of *virtual synchronism* [5], maintained by the lower layers of the support used in this implementation. In this way, as long as the replica belongs to the *membership* of the group, it will have the same messages in the same order as the others. When the

input buffer in the communication support associated with the slowest coordinator/replica pair, reaches the limit of its capacity, the support withdraws the replica from the *membership*. A replica can detect its exclusion and reintegrate itself to the group, through the `view change` *(view)* method, defined in the interface BOA of Electra and activated automatically by the support for each change in the *membership*. The activation of this method is not preemptive, occurring only after processing the current method. The body of the method `view change` is defined according to the application characteristics. In this way, in our implementation we carried out a membership test in the body of this method: if the replica has been excluded (`view.number=1`), the BOA function `join` *(group)* is activated, thereby effecting its reintegration into the group.

Regarding the crashes that may occasionally occur in the evolution of the system, our implementation provides procedures for recuperating the degree of replication. If the number of active replicas in the group falls below a preestablished limit, the oldest replica takes the initiative of producing new replicas, in this way, reestablishing the ideal number of replicas. The code referring to these recuperation procedures is based on a *membership* test (`view.number < quorum_minimum`), and it is inserted into the `view_change` method cited above.

Our replica state recovery approach differs from that proposed in [7], in which the recovery occur through meta-methods making updates in public attributes of their associated replicas, with the use of appropriated coordination protocols. In our approach, we utilize more support-provided primitives and fewer coordination protocols, what simplifies the state recoveries. The state recovery, in our system, is based on the primitive *join*, offered by the support, and activated through the `view change` method.

In object-oriented languages, each meta-object is an instance of a class on the meta-level that defines its structure and behavior. In this article we limit ourselves to talking only about meta-objects because we are interested in emphasizing the aspects of execution time of the meta-objects approach. In [8], these aspects added to other referents to the use of the same approach in real-time applications, are approached within the structure of a language that is being developed.

4. The CORBA support utilized

The implementation of the replication model presented in section 3 presupposes the existence of an run-time support that offers facilities for programming distributed objects. A platform conceived based on the concepts of the CORBA (*Common Object Request Broker*

276

Architecture) standard is to provide the necessary support for distributed object-oriented applications. In this section we briefly describe the Electra system, a CORBA platform utilized in our implementations.

4.1 CORBA architecture

CORBA specifications form a set of standards and concepts proposed for open systems by OMG (Object Management Group) [15]. CORBA architecture is composed of an ORB (Object Request Broker) kernel, that implements communication abstractions among distributed objects and an interface management structure that contains static and dynamic invocation interfaces, object adapters, interface and implementation repositories (figure 4).

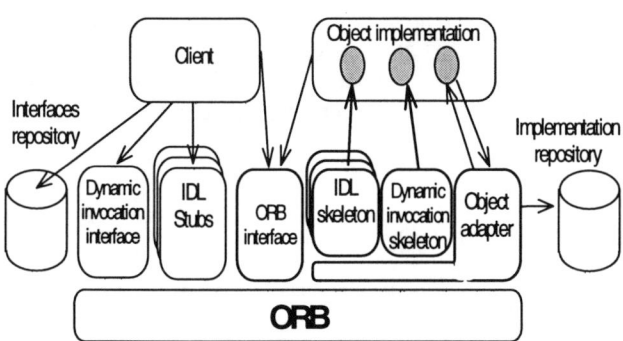

figure 4: The CORBA architecture.

In a CORBA environment, each object has its interface specified through an Interface Description Language (IDL), a declarative language with syntax and predefined types based on the language C++. The interactions follow the client/server model. The CORBA client, in a service request, utilizes stubs generated in the compilation of the IDL specification of the server object, or builds this request, using the dynamic invocation interface DII. To allow for dynamic invocations, object interfaces must be stored in the interface repository. The client's request is transmitted over the network, using the ORB, that transfers the control to the object adapter to activate the operation in the implementation of the server object, by means of the IDL skeletons.

The original CORBA proposal does not provide for adequate support mechanisms to groups of objects. To fill in this gap, some extensions to the CORBA standard have been proposed in terms of incorporating this concept. Electra [14] is a product of these efforts.

4.2 Electra

Electra [14] is an *Object Request Broker* (ORB), compatible with the CORBA standard [15], presenting support to object groups. This platform combines the benefits of the CORBA standard with the power of lower level tools for group processing, such as Horus [18], Isis [5], and others. Interactions in Electra can occur as reliable multicast or point-to-point communications. Ordering mechanisms (total, causal and fifo) are offered to guarantee consistency among members of the object group. The client makes use of a given method invocation model, regardless of whether the server is a single object or a group. These invocations may be synchronous, asynchronous (*one-way*) or semi-synchronous (*deferred-synchronously*), through static or dynamic interfaces. On multicasting a method call, through a CORBA static or dynamic invocation interface, the programmer has at his disposal two modes of group communication in Electra:

- Transparent: the group is seen as a simple and completely available object, and the client only receives the final result furnished by the group;
- Non-transparent: permits access, in an invocation, to the results of each individual member of the group of objects.

In the interface BOA (*Basic Object Adapter*) of Electra, services referring to group management are added, such as *creating* a group of objects, *including* objects in the group or *excluding* them from it, *selecting* a protocol of *multicast, membership* and transfer of state, and so on. These services are provided by the lower level tools used, such as Horus or Isis.

5. The integration model of replication techniques in open systems

In the previous section, we could see that the CORBA/Electra platform offers adequate support for group processing. In this section, we describe the integration of the reflective model proposed in section 3 over the Electra system.

5.1 The integration model in CORBA platforms

Figure 5 explicits the integration model of replication techniques within the CORBA context. The access to the support provided by a CORBA platform is available both to the server and to the client by entities represented as *meta-objects (client and server)* and identified generically as *meta-communication*. These entities are actually nothing more than the set of *stubs* for the client and server, *stubs* for the communication among the replicas and the BOA interfaces providing the group management. All these *stubs* are generated by the translation of the IDL [16] declaration of a server object. The use of the term "abstract object" given to meta-communication on the model follows some authors [9] and has the sense of a simple separation for greater clarity. In reality, these interfaces are generated in Electra as a set of methods that will be composed of multiple inheritance in the client and controller meta-objects (section 5.2).

In the model the client introduces itself within a *client-base* structure, that represents the application behavior, and a *meta-client*, that does not present an active function in our implementation, but that could be used in managing the replicated client, or to implement mechanisms for handling exceptions in the client. The structure of each server replica is similar to that of the client: a *replica-base* object, carrying out the replicated service, and a *meta-controller*, responsible for executing the coordination protocol of the replication, like one described in figure 3.

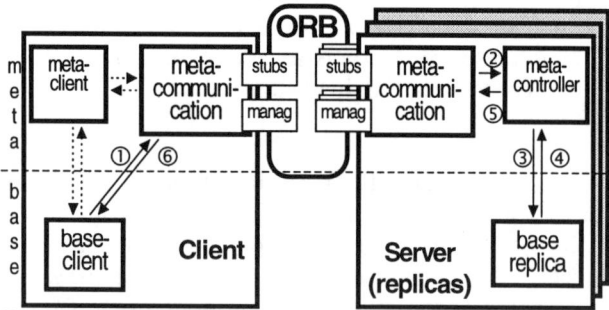

figure 5: Structure of the model on a CORBA support.

The numbered arrows in figure 5 indicate the normal way of a client request: The request made by the client base (1) is then broadcast using a stub appropriated in the client meta-communication. In each replica, the meta-communication, by means of a local stub, receives the request and transfers it to the meta-controller (2), which then activates the local replica (3). On receiving the reply (4), the meta-server executes the coordination protocol, by means of the meta-communication so as to interact with other replicas. The processing and interactions on the level of the meta-controllers are conditioned at this time by the replication model utilized. Later, the reply is then sent back to the client (5 and 6).

This model can be used in other replication techniques, the differences centering mainly in server replicated meta-controllers. In some techniques the meta-communication entities may gain functionality, besides that of concentrating methods of access to CORBA sup-

porting services. For example, in the use of active replicas with voter and adjuster mechanisms, the implementation of voting or adjustment can be programmed on the client's side in a more simplified form. Transparency could be achieved in this case, implementing these mechanisms in the client meta-communication entity, which, with the addition of this functionality, takes on the characteristics of a real object.

5.2 Building replicated services following the integration model

The first step in the building of a system on a CORBA/Electra platform is a description in IDL of the meta-controller interface, following the specification of the replicated service provided by the server to the client. Besides this interface, due to some limitations imposed by the Electra, it is necessary to declare a second one, for implementing the replica coordination, composed by methods that provide communications among replicas.

```
// IDL

interface meta_controller_1
{
    // Description of the data types employed

    // Description of the server methods
    boolean meta_method_1 (parameters);
    ...
    boolean meta_method_n (parameters);
};

interface meta_controlller_2
{
    // Description of the meta_controller methods
    boolean broadcast_id (in int id);
    boolean closing ();
};
```

figure 6: IDL interface of the replicated server.

Figure 6 presents both IDL declarations of a replicated server in according to the specifications described in figure 3. The interface *meta_controller_1* allows clients access to the services offered, while the interface *meta_controller_2* declares the methods necessary for intra-replica interactions. It should be pointed out that both interfaces are actually two facets of the same server (or, in our case, of the same group of objects).

In compile-time, the Electra/IDL compiler automatically generates the whole support for communication (*stubs*) among the entities involved, including, as well, the functionalities for group management of the BOA (in Electra, every object is an instance of a sub-class of the

BOA class). The compiler also generates files containing structures (declarations of variables and methods) for including the client and server codes. The programmer, then, is responsible for the implementation of the replicas (base-objects) and the replica coordination suitable (meta-objects), by filling the bodies of the methods declared in the interfaces. With this implementation scheme, illustrated in figure 7, the client and server base-objects are kept devoid of all activities that are not related to the application itself. All aspects related to the replica coordination and the interactions in the CORBA context, are concentrated at the meta-level.

Our implementation was carried on a UNIX platform, where each associated pair base-object/meta-object was intended to share the same process, making the interactions between them local, without the need for ORB. The needs for concurrence between base-objects and meta-objects within a process are satisfied by the use of a threads library offered by the Electra support. However, the current version, (1.0) of this platform does not support pre-emptive threads, which limits the degree of concurrence in dealing with client requests. As a result of this restriction, it becomes difficult to implement replication techniques, which has forced us to seek alternative implementation solutions. The solution adopted consists of separating the functionalities of the meta-controller into two UNIX processes.

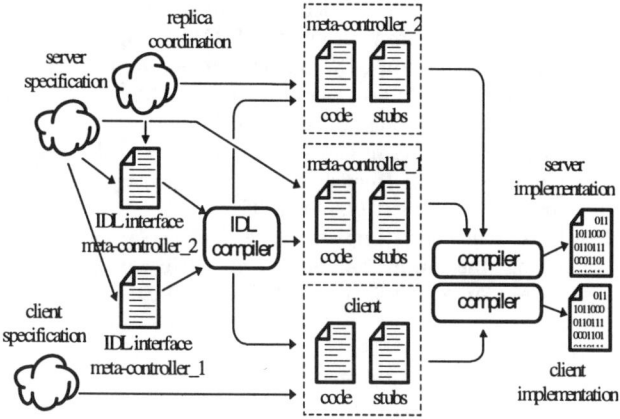

figure 7: Application building process.

Due to the fact that the language used (C++) has no specific constructions to support reflection. The reflection is implemented artificially, through the direct activation to the meta-method, in the client code. The use of a language supporting reflection, as is the case with *Open C++* [6], might eliminate this problem, but in this case, the IDL compiler of the CORBA environment used should support this language.

6. Considerations about the results

Redundancies and fault tolerance implementations can follow several approaches [7]. The implementation of fault tolerance techniques by means of runtime support offers on the application level some degree of transparency concerning the coordination aspects of the technique used. The disadvantage is that once the fault assumptions and the replication technique are chosen in configuration time, we will have defined a specific execution support. The approaches of library and languages for the implementation bring aspects of coordination to the programmer level, without, however, separating them from functional application aspects.

Computational reflection permits independence of the replica codes in relation to the coordination protocols, leading to a greater flexibility in the system: changing the technique or altering it by meeting desired degrees of fault tolerance, may simply result in switching the coordination protocols on the meta level, involving no alteration in application algorithms or in the run-time support what is suitable in open systems. The use of the reflective computing for implementing fault tolerance techniques is proposed in [3] [7], and the separation between the coordination and the replicas has already been recommended in [17].

The model presented was used for implementing the active competitive replication protocol, described in section 2. The implementation carried out makes intensive use of the Electra support facilities, which makes easier the coordination needs of the technique implemented. Furthermore, the uses of a CORBA platform has allowed the implementation of our application on an heterogeneous system (local network of machines running SunOS 4.X and Solaris), facilitating aspects of interoperability.

The integration structure proposed has proved to be quite flexible, other replication techniques can be easily implemented. Up to now, we have implemented the *primary/secondary* , *leader/followers* and *cyclic redundancy* techniques, using the same integration model. The necessary changes for the substitution of replication techniques in the integration model are limited to the IDL meta-controller interface and their codes that implement the coordination protocols.

These replication models were applied in the implementation of a multimedia application (animation viewer accepting the *MPEG* format). Simulations of *crashes* were carried out, utilizing these implementations. In all these replication techniques utilized, the continuity of service was obtained in case of failures, since the premises of each technique were respected. At present, we are working out detailed measurements on the performance of the replication techniques cited using the Electra platform. We are also porting our work to the Orbix+Isis platform [10].

7. Conclusion

An integration model for replication techniques in open distributed systems was presented in this article. The use of computational reflection concepts makes it possible to obtain the necessary flexibility for developing and implementing different replication models for fault tolerance in these environments.

Within this context, the work presented in this article continues at present in various directions. The validation of the model proposed through application in real situations and the incorporation of language constructions in terms of facilitating the programming of the reflective model are some of the current activities involving this work.

The programming model presented in this article is part of a cooperative research project, sponsored by the brazilian state agency CNPq (PROTEM-CC project), and has as its aim to build an object-oriented environment that supports distributed applications with requirements of real-time and fault tolerance.

Acknowledgment

We would thank S. Maffeis, author of the Electra system [14], for his kind attention helping us to solve our main difficulties in the beginning of this work.

References

[1] E. Oskiewicz, N. Edwards, **"A Model for Interface Groups"**, ANSA Phase III technical report APM.1002.01, Cambridge-UK, may 1994.

[2] R. M. Adler, **"Group-Oriented Coordination Extensions to OMG´s OMA/CORBA"**, OMG Presentation, San Jose - CA, June 1995.

[3] G. Agha, S. Frolund, R. Panwar, D. Sturman, **"A Linguistic Framework for Dynamic Composition of Dependability Protocols"**, Proceedings of the DCCA-3, 1993.

[4] K. Birman, T. Joseph, F. Schmuck, **"ISIS - A Distributed Programming Users Guide and Reference Manual"**, The ISIS Project, Department of Computer Science, Cornell University, Ithaca - NY, march 1988.

[5] K. P. Birman, **"The Process Group Approach to Reliable Distributed Computing"**, Technical Report TR 91-1216, Cornell University Computer Science Department, Ithaca, N.Y., July 1991.

[6] S. Chiba, **"Open C++ Programmer's Guide"**, Technical Report 93-3, Department of Information Science, University of Tokio, 1993.

[7] J. Fabre, V. Nicomette, T. Pérennou, R. J. Stroud and Z. Wu, **"Implementing Fault Tolerant Applications using Reflective Object-Oriented Programming"**, Proceedings of the 25th IEEE International Symposium on Fault-Tolerant Computing, Pasadena (CA), June 1995.

[8] J. Fraga, J.-M. Farines, O. Furtado, F. Siqueira, **"A programming model for real-time applications in open distributed systems"**. Proc. of the 2nd IEEE Workshop on Future Trends in Distributed Computing Systems, august 1995.

[9] O. Hagsand, H. Herzog, K.P. Birman, R. Cooper, **"Object-Oriented Reliable Distributed Computing"**, 2nd IEEE International Workshop on Object-Orientation in Operational Systems, 1992.

[10] Isis Distributed Systems Inc., IONA Technologies, Ltd. **"Orbix+Isis Programmer's Guide"**, 1995. Document D070-00.

[11] Isis Distributed Systems Inc., **"RDO/C++ Tutorial Release 1.0.3"**, Apr. 1994.

[12] M. C. Little, **"Object Replication in a Distributed System"**, PhD. Thesis, University of Newcastle upon Tyne Computing Laboratory, September 1991.

[13] P. Maes, **"Concepts and Experiments in Computational Reflection"**, OOPSLA 87 Proceedings, pp. 147-156, October 1987.

[14] S. Maffeis, **"Adding Group Communication and Fault-Tolerance to CORBA"**, In Proceedings of the 1995 USENIX Conference on Object-Oriented Technologies, Monterey - CA, June 1995.

[15] Object Management Group, **"The Common Object Request Broker: Architecture and Specification"**, Revision 1.2, OMG Document, December 1993.

[16] Object Management Group, **"IDL C++ Language Mapping Specification"**, OMG Document 94-9-14, 1994.

[17] D. Powell, **"Delta-4 Architecture Guide"**, Esprit II P2252, Delta-4 Phase 3, August 1991.

[18] Robbert V. Renesse and Kenneth P. Birman, **"Protocol Composition in Horus"** Dept. of Computer Science of the Cornell University, Mar 1995.

[19] F. B. Schneider, **"Implementing Fault-Tolerant Service Using the State Machine Approach: A Tutorial"**, ACM Computing Survey, 22(4):299-319, Dec 1990.

Session 4B

World Wide Web Information Management

Chair

Ralf Steinmetz

Design of a Database and Cache Management Strategy for a Global Information Infrastructure

Paul Francis   Shin-ya Sato

NTT Software Laboratories
{francis,sato}@slab.ntt.co.jp

Abstract

NTT Software Labs is producing a distributed, self-configuring information navigation infrastructure designed to scale to global proportions. For reasons of large scale, unreliability (of the Internet, its connected computers, and the implementations), and the complete autonomy of the participants, a number of difficult database and cache consistency problems arise that are not solved by techniques commonly used either for the Internet (i.e. DNS), or for existing distributed database systems. This paper describes a set of strategies designed to solve these problems. In particular, it focuses on the use of third-party detection and notification of database and cache inconsistency.

1 Introduction

A team of researchers in the Software Labs of NTT is undertaking an ambitious program to design and implement a fully distributed information discovery infrastructure that can scale to global proportions and operate in the current Internet. We call this information discovery infrastructure Ingrid <http://www.ingrid.org/>.

The basic idea behind Ingrid is as follows. Each site that currently has a web server (and wishes the contents of that web server to be globally, or locally, discoverable) installs Ingrid server software. For each document on the web server, this software automatically searches for and creates bi-directional links with certain documents that have similar content (as determined by commonality of keywords), and that are on web servers that are already running Ingrid server software. The result is a distributed topology, parallel to and in some respects similar to the current HTML topology, which we call the Ingrid topology.

Unlike the HTML topology, however, the Ingrid topology can be efficiently (we hope) keyword searched. And, unlike large web-wide search services such as Alta Vista, this search is distributed. That is, at search time, the local Ingrid discovery client software executes a series of queries to Ingrid servers. Each Ingrid server refers the Ingrid client to other Ingrid servers that have better knowledge about the keywords being searched. These Ingrid servers are in turn queried until the Ingrid servers with the best matching documents are found.

We believe (without here giving any particular jus-

tifications) that this approach may in the long run have several advantages over the current centralized search engine approach:

- It will scale better by volume (that is, will allow a larger number of documents to be searched).

- It will scale better by document insertion/deletion latency (that is, new or modified documents will be searchable sooner, and deleted documents will be unsearchable sooner).

- It will give information providers more control over what documents are searched and how they are searched.

At the same time, we believe that our distributed approach has several disadvantages. One of these is the fact that information providers will have to take an active role in making their documents discoverable, by installing and configuring the Ingrid software. Another disadvantage is the added complexity of maintaining the correctness of what amounts to a distributed database among fully autonomous and non-coordinating participants. The success of Ingrid hinges on the satisfactory resolution of these disadvantages.

In this paper, we focus on the second issue — that of maintaining the correctness of the global distributed database and its supporting caches. Because of the intended scale of Ingrid information infrastructure (virtually all publicly available information on the globe), the dynamics of the infrastructure (some information may come and go in a matter of minutes), the complete autonomy of the participants (anybody attached to the Internet), and the general unreliability of the systems involved, we believe the problems we address here are unique. We believe the individual solutions we propose are also, if not truly unique, at least not well-known. Taken together, however, our solutions comprise a unique approach to database and cache consistency.

1.1 Paper Outline

Section 2 outlines the characteristics of the Internet and of information discovery salient to the issues that we address. Section 3 gives a brief overview of Ingrid itself as it pertains to the problem of database consistency. Section 4 describes a number of current technologies and how they meet or fail to meet our needs. Sections 5 through 7 each describe the solution to one aspect of Ingrid's database consistency problems.

2 Characteristics of the Current Internet

The Internet is a world-wide computer network consisting of a huge number of subnetworks which are managed by different organizations. Once an organization gets a class of IP address and domain names associated its subnet, it can use the network without particular coordination with other networks. So, the Internet is a global composition of many autonomous subsystems. This architecture leads to rapid, bottom-up evolution of the Internet.

This architecture also leads to the flaky characteristics of the Internet. Computers in an organization may be down or overloaded from time to time. The Internet infrastructure itself, though improving, still suffers outages and frequent minor glitches.

We now have various network services on the Internet, such as e-mail, NetNews, WWW, etc. Most of these inherit the "bottom-up" characteristics of the Internet (or we may be able to say that those services can be widely used because they are congenial with the style of the Internet). For instance, WWW allows users to publish their information without any tight coordination with other people. The web has been, and still is growing rapidly, and this also accelerates the speed of information update. For example, indexes generated by a web robot[1] sometimes point to resources that are no longer available.

These Internet characteristics suggest the following requirements for a global information infrastructure:

- The system (infrastructure) should follow the style of the Internet, i.e., it should be distributed, autonomous, and scalable.

- We should not assume that every element of the (distributed) system is working properly. The system should be flexible so that it can cope with partial disorder.

It is within this framework that the distributed database and cache management of Ingrid must operate.

3 Overview of Ingrid Architecture

In this section, a brief and partial description of Ingrid is given. In particular, only that which is required to understand the context of the database consistency problem is described. For a more complete description, please see [6], or the Ingrid web page <http://www.ingrid.org/>.

Ingrid can be viewed as a distributed system consisting of a distributed infrastructure accessed by applications (see Figure 1). *Sending applications* insert units of information into the infrastructure in the form of *Resource Profiles* (RPs). An RP typically describes an Internet accessible information resource, though is

not constrained to this. An RP contains, at a minimum, information on how to retrieve the resource, and some terms that describe the resource and can be used by other applications to find the resource:

$$\{resource\ location;\ term_1, term_2, ..., term_N\}$$

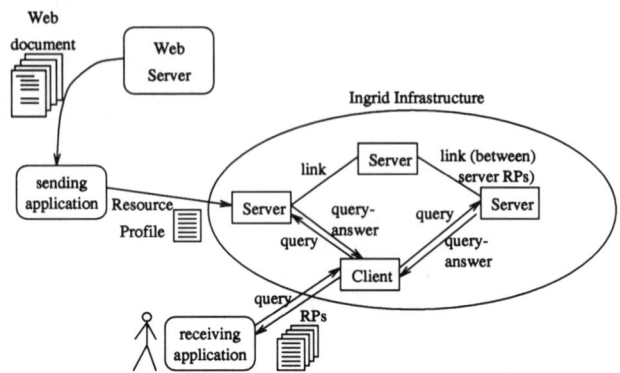

Figure 1: Ingrid Architecture

Receiving applications retrieve RPs from the infrastructure by submitting a query that consists of one or more terms. The infrastructure then returns RPs that contain matching terms to the receiving applications.

3.1 Ingrid Resource Profiles (RPs) and their Links

For the purposes of this paper, we can assume that there are two kinds of systems in the infrastructure, *Ingrid Servers* and *Ingrid Clients* (or just servers and clients for short). When referring equally to both types of systems, we use the term *Ingrid System* (or just system). Each sending application sends its RPs to the server it is associated with. When a server receives an RP, it searches the infrastructure for RPs with similar terms, and installs *Ingrid Links* (or just links) between its RP and the similar RPs (one link for each other RP), potentially on other servers. (The RPs chosen to be neighbors are selected according to a specific organizational structure whose properties are that 1) it can be efficiently searched, and 2) it does not require an excessive number of links. It is outside the scope of this paper to describe that structure. Again, please see [6].

To install the link, the installing server I sends a link install request packet to the server N containing the neighbor RP. The packet contains the identity of both RPs, and the terms common to the two RPs. (Note also that all packets contain the DNS name of the system that sent the packet and the system that received the packet. Note too that all packets between servers and clients are connectionless — they run over UDP[9].) The neighbor server stores the information and sends back an acknowledgment. As a result, the two servers each contain a *half-link* which together comprises the link:

Server I's half-link:

$\{system\ name_N; rp_{N_1}; term_1, term_2, ..., term_j\}$

Server N's half-link:

$\{system\ name_I; rp_{I_1}; term_1, term_2, ..., term_j\}$

3.2 Queries and Query-Answers

Because of Ingrid's caching strategy (described below), both servers and clients can be the originators or recipients of queries. (The authors acknowledge that this is a bit confusing. The reader should think of a system as being a server by virtue of its having RPs, and of a system as being a client by virtue of its looking for RPs.)

A *query* consists of the set of terms that are being searched:

query: $\{term_1, term_2, ..., term_q\}$

The resulting *query-answer* consists of the list of servers and clients that have information relevant to the query. Each entry in the list indicates what type of information it is (*persistent, answer-cache,* or *query-cache*), and which of the terms in the query apply:

query-answer:

$\{system\ name_1; info\text{-}type_1; term_{11}, term_{12}, ..., term_{1q}\}$
$\{system\ name_2; info\text{-}type_2; term_{21}, term_{22}, ..., term_{2q}\}$
$\{system\ name_3; info\text{-}type_3; term_{31}, term_{32}, ..., term_{3q}\}$
...
$\{system\ name_k; info\text{-}type_k; term_{k1}, term_{k2}, ..., term_{kq}\}$

The info-type of persistent can only be returned by a server. It refers to either 1) information about one of the server's own RPs, or 2) information about one of the server's links (that is, information about a neighbor server's RPs). The two "-cache" info-types are derived from two different types of cached information, and are explained below.

3.3 Caching in the Ingrid Infrastructure

Clients and servers each have a cache. When a server gets a query from a client, it caches the terms of the query and the name of the client. When it answers a query, entries from this cache are of type query-cache. When a client gets a query-answer, it caches the entries of information type persistent. When the client answers a query, entries from this cache are of type answer-cache (see Figure 2).

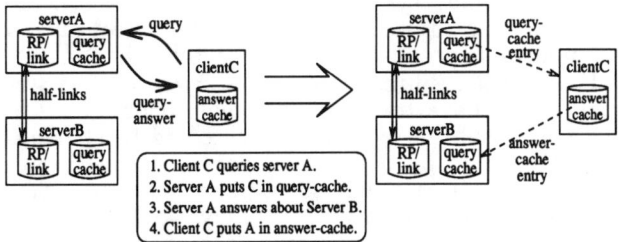

Figure 2: Generating Query-Cache and Answer-Cache Entries

We use this caching structure in order to be able to take advantage of the work done in previous searches

by clients. It is not enough for only clients to cache the answers they receive (the answer-cache). Servers must remember which clients have done which searches, so that they can refer other searching clients to those clients (thus the query-cache).

The reader should note that servers and clients are fairly selective about what they cache. First, information is quickly downgraded as it is spread around. Received persistent information is sent out as answer-cache information, and received answer-cache information is not sent out at all (nor is received query-cache information). This prevents chains of referals, where system A knows that system B knows something, and system B knows that system C knows something, and so on.

Second, systems tend not to keep multiple entries that have the same set of terms, because such information is generally redundant. (This is because of the nature of the organizational structure of the links, which isn't explained here.) Specifically, among cache entries that have the same terms:

- answer-cache entries overwrite query-cache entries (and the cache entry is refreshed),

- persistent entries overwrite answer-cache entries (or, more accurately, the answer-cache entry is deleted when the persistent entry is placed in the RP/link database),

- newer cache entries of the same type replace older ones (and the cache entry is refreshed).

Finally, systems strive to delete cached information as soon as it gets out-of-date. Specifically, as soon as a client has deleted an answer-cache entry, the query-cache entries that refer to that client (for that set of terms) should be deleted. Likewise, as soon as persistent (RP or link) information is deleted from a server, the answer-cache entries that refer to that server should be deleted.

How to delete these entries as soon as possible but with a minimum of overhead is one of the major topics of this paper, and is covered in section 5. The difficulty here is the highly dynamic nature of the cache, and the large number of other systems the cache may refer to.

The second major topic of this paper, covered in section 6, is how to maintain consistency between the servers on either end of a link. The difficulty here has primarily to do with the size of the database containing the RPs and links. Servers with 100,000's of RPs and 1,000,000's of links are expected to be common.

Section 7 briefly covers the remaining database consistency problem — that between the applications and servers.

4 Existing techniques

In this section, we list some important existing techniques in distributed data management([2], [3]) and discuss how they apply or don't apply to Ingrid.

System architecture Currently, the most dominant architecture for distributed systems is client-server. Distributed database systems (or, distributed persistent storage systems) are normally organized as client-server systems. As this architecture does not scale so well, systems aiming for high scalability adopt other architectures, such as peer-to-peer (e.g., SHORE[4]). Ingrid is also based on a peer-to-peer architecture.

Replication Replication of data is a common method to improve efficiency of data access, and to guarantee fault-tolerancy. It entails, however, some cost for managing consistency. Ingrid cuts down this functionality and lightens operations to gain scalability (at the expense of availability). As a result, replication mechanisms commonly used with database management systems do not apply to Ingrid.

Transaction and concurrency control Ingrid does not require transactions nor concurrency control, which is an important basis for claiming Ingrid can be scalable. This is because the information organizational structure of Ingrid minimizes and localizes operations for data update. Adding a resource to the infrastructure is achieved by establishing a link between the resource and an existing member of the infrastructure, involving at most two servers. Furthermore, adding/deleting a link to/from a server does not come into conflict with others. Again, as a result, transaction and concurrency control commonly used with database management systems do not apply to Ingrid.

Caching Ingrid makes heavy use of cached information to aid in the efficiency of infrastructure searches. Good examples of autonomous caching systems in the Internet include DNS[5] and proxy caching of web documents[8]. DNS caching requirements are different from Ingrid in that DNS objects are relatively persistent (usually measured by days). Thus, the time-to-live mechanism of DNS is not appropriate for Ingrid.

The persistence of cached web documents is much more variable (though still not as variable as in the Ingrid case). The typical web proxy mechanism for discovering out-of-date information is to go to the document source and check for changes using the if-modified-since mechanism of HTTP. This mechanism is too heavy-weight for Ingrid. Ingrid objects are small compared to web documents, and the cost of checking the validity of an entry is similar to the cost of obtaining the entry in the first place. Therefore, the caching strategies used in proxy servers are not appropriate for Ingrid.

Database consistency While Ingrid does not have the heavy-weight replication and concurrency requirements of common distributed database systems, it none-the-less does require that peers in the linked structure maintain consistent information. We find that routing protocols in the Internet[7] share many of the requirements of Ingrid — database consistency in the face of faulty

systems and autonomous operations. While to some extent we mimic the timeout/refresh style of database consistency of Internet routing protocols, we find, particularly because of Ingrid's severe scaling requirements, that we must enhance these with techniques of our own (Section 6).

5 Consistency of Cached Information

The simplest way and most efficient way to operate a cache (in general — not just for Ingrid) is to simply let older information in the cache be replaced by newer information. This is appropriate if the lifetime of information in a cache is much shorter than the lifetime of the information itself. The next simplest way is to flush information from the cache after it has been there for a certain amount of time. This is appropriate for the case where the approximate lifetime (or lower-bound thereof) of the information is known, and is used by DNS[5].

Neither of these cases work for Ingrid caches. The information in an Ingrid cache has widely varying lifetimes. Ingrid is a general-purpose infrastructure for information discovery of all kinds. We cannot predict in advance what kind of information will be used with it, but can expect a wide variability (from permanently archived documents on one hand to information about the current topic of a chat session on the other). In addition, we cannot expect the application to always, or even often, know the lifetime of the information it hands to the Ingrid infrastructure. Nor do we want to complicate applications by requiring them to determine the lifetime of the information even in the cases where they can.

In any event, even if we did learn lifetime information from applications, the difficulty of taking advantage of it is compounded by by the fact that Ingrid caches not only keep *first-order* cached information (answer-cache), but also *second-order* cached information (query-cache). In fact, query-cache entries strictly speaking do not represent what *is in* other caches, but rather what is *expected to be* in other caches.

Since we cannot use a purely local mechanism (overwrite or timeout) to flush old information with Ingrid, we must rely on some kind of external trigger. One potential such mechanism would be to have the system whose information is being cached (the *cached system*) send a message to the *caching system* when the information is known to be invalid. We reject such a mechanism on account of its high overhead.

Its overhead is high in two ways. First, cached systems would have to retain information about all of the related caching systems, which could be substantial. (It would have to remember every query and query-answer it sent over some period of time.) Second, the deletion of a cached entry would result in packet overhead, either because the cached systems informs the caching system that some information is no longer valid, or because the caching system informs the cached system that it is no longer caching. In terms of number of packets sent, the cost of deleting the cached information would be similar to the cost of

286

the original query or query-answer that generated the cached information in the first place.

5.1 Ingrid's Third-party Cache Consistency Mechanism

Instead, we use a third party as the outside trigger. In addition to the cached and caching systems defined above, we define a *querying system* to be a system that queries a caching system, and learns about the cached system in the query-answer. The querying system is the third party. (Note that here we are not concerned with the cache in the querying system, or the cache entry in the caching system that results from the querying system's query. Our only concern is the cache entry in the caching system that points to the cached system. It doesn't matter to this discussion if this entry is answer-cache or query-cache.)

When a querying system gets a query-answer from a caching system that it just queried, it of course saves the answer, and uses it and other answers to determine which system to query next. In our mechanism, in addition to saving the answer, it saves the name of the caching system from which it received the answer.

When the querying system subsequently queries the cached system, it checks to see if the information it received from the cached system matches the information it expected to receive (based on the query-answer previously received from the caching system). If it does not, the querying system informs the caching system that the cached information no longer exists at the cached system and should be flushed.

For example, assume that (querying) system Q queries (caching) system A for the terms 'cat', 'dog', 'mouse'. System A answers that (cached) system B is known to have recently made a query about the terms 'cat', 'mouse', and so is expected to have more information. System Q queries system B, but learns that in fact system B has no such information. System Q then sends a message to system A to flush the cached information about system B (see Figure 3).

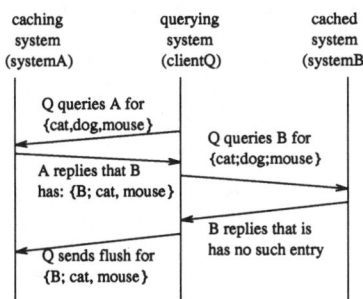

Figure 3: Third-Party Flush Notification

Two related open issues are 1) when should the caching system refresh entries in the cache, and 2) whether the caching system should flush only when the cached system explicitly indicates that the cached information no longer exists (*flush-if-nack*), or also if no query-answer was received from the cached system (*flush-if-unreachable*), either because it is down or just unreachable via the network.

Regarding the first issue, cache entry refresh, there are two possible strategies. First, the entry can be refreshed only when the original information is re-learned (or, as described in Section 3.3, is replaced by better or newer information with equivalent terms). We call this the *first-order refresh strategy*. Second, the entry can be refreshed any time that it is included in a query-answer. This is the *second-order refresh strategy*. This can be done on the assumption that, if it is in fact no longer good, then the third-party flush notification will take care of it.

The reverse logic (and, the actual logic that suggests the second-order refresh strategy) is: if no third-party flush notification is heard, then it can safely be refreshed. This logic holds in all cases except when the cached system is unreachable *and* the original information no longer exists (including the case where the cached system is permanently down). In this case, the entry in the caching system will incorrectly persist if 1) it is queried often enough to keep it refreshed, and 2) the cached system remains unreachable for a long time.

Therefore, if we use the second-order refresh strategy, then we must use a flush-if-unreachable policy as well. Unfortunately, while the second-order refresh strategy results in better information in the caches, the flush-if-unreachable policy results in cases where a cache entry is flushed even though the cached system is only down for a very short time (or only temporarily unreachable because of a network glitch). Our compromise solution to this is to include a bit in the flush packet that indicates whether the flush is because of an explicit negative indication or because of system unreachability. This gives the caching system the option of not flushing the cached information, but rather temporarily not including it in query-answers.

5.2 Efficiency of the Third-party Mechanism

This mechanism is efficient in terms of memory in two ways. First, the caching and cached systems do not need to maintain any extra information. This is particularly important when these systems are servers, which already have a heavy memory load (RPs, links, and cache). Second, the systems that do maintain extra information (the querying systems) only need to keep one extra piece of information (the name of the caching system) tagged to a piece of information they already need to keep as part of their normal infrastructure search (the query-answer). The big-O order of scaling remains unchanged — memory is increased only by a constant factor.

This mechanism is also efficient in terms of packets sent. First, no packets are required to flush entries that are unused. They simply eventually get overwritten by a newer entry. The only time an explicit flush packet is required is when 1) a cache entry exists in the caching system, 2) the corresponding cache entry or RP does not exist in the cached system or the cached system is unreachable, and 3) a querying system queries for the cached entry.

Consider the case where a cache entry in the caching system is query-cache. In general, if the cache entry is refreshed in one of the systems, it will also be refreshed in the other, since the querying system will tend to query the cached system after having queried the caching system, resulting in a refresh in both caches. Thus, for this case, the explicit flush event described above will be relatively rare (relative to the number of times a cache is used).

Note that, as an optimization, the flush packet is not acknowledged (nor, therefore, ever re-transmitted). The idea here is that, if the flush packet is dropped, it is not a big deal, since cached information is not depended on to be reliable anyway. The next access of the cached information will result in another flush packet, and the entry can (probably) be flushed then.

5.3 Problems with the Third-party Mechanism

The main concern with the third-party mechanism is that of bogus flush information. A potential security attack on the infrastructure would be for a system to transmit many bogus flush packets, thus causing good information in caches to be removed. The solution to this threat is of course for the caching system to query the cached system itself to verify that the cached information is in fact not there. This solution, however, incurs extra overhead.

Our inclination at this time is to not require the verification for the case of cache flush packets, though we still consider this an open problem. There are several reasons. First, the correctness of Ingrid does not rely on the cached information. Caching only improves efficiency. Thus, the premature deletion of cached information is not as bad as it might otherwise be.

Second, a flush packet can only flush one entry at a time. Further, the flush packet can only flush an entry that actually exists. Therefore, the attacker has to know what entries already exist in the attacked system, which requires that the attacked system be queried first by the attacker. Therefore, such an attack is relatively costly to the attacker, particularly if the attack is made to a large number of other systems.

If the attacker attacks only a single server, however, it is more problematic. Possibly the server can detect the excessive number of flushes, but since the attacker could masquerade as a different system each time, it would be impossible to distinguish the attacker's flushes from valid flushes. Upon detecting excessive flushes, the attacked system could switch to a policy of checking the validity of flush messages, with the additional overhead that this implies.

6 Link Consistency

As described in Section 3.1, links in Ingrid always consist of a pair of entries, one at the head-end of the link, and one at the tail-end. Each entry refers to the neighbor server and the RP of the neighbor server, and is essentially a mirror image of its partner. When either RP is deleted, then the link should also be deleted (by deleting the half-link on each end).

The basic mechanism for deleting a link is an explicit delete message. Unlike the case with caches, the explicit delete message is here efficient because the servers at either end of the link need no extra information to support the use of explicit delete. By virtue of the (bi-directional) link information itself, they know what other server contains information about their own RP. (The authors recognize that this is obvious enough as to hardly be worth mentioning.)

If the servers in Ingrid were 100other mechanism to maintain link consistency. This is however not the case, and the possibility exists for the information for one half-link to become inconsistent with the other (that is, one end deletes the entry while the other end does not).

One such reason is because one of the servers is down for an extended period while the other server is trying to delete the link. In our current implementation, a server deleting a link will try to contact the other server for a period of two weeks, but failing that it will simply consider the link to be deleted and stop trying. If the other server recovers after that time, it will (incorrectly) consider the link to be still valid. Other reasons include database corruption, for instance because of faulty software, faulty hardware (i.e., a disk crash), or faulty administration (i.e., 'rm *').

6.1 Periodic Refresh of Half-Links

Without an additional mechanism, such an inconsistency would remain as long as the RP of the still-active half-link remained, which in many cases is essentially forever. The most straight-forward mechanism for maintaining link consistency is a periodic refresh mechanism. In this case, each half-link has a timeout period. If the other end does not send a packet confirming the continued existence of the link before the timeout expires, the half-link is (unilaterally) deleted. This sort of approach is common with Internetwork routing protocols [7].

The problem here is that a server can contain potentially hundreds of thousands of RPs, and millions of links. Further, the majority of the RPs may be rarely accessed. A simple refresh mechanism may account for a significant percentage of Ingrid activity of a given server. For instance, assume that a server has 100,000 RPs and 1,000,000 *external* (to other servers) links. Even a relatively slow refresh period of one-per-month results in an average of 1 refresh every 2.5 seconds.

Whether or not this is prohibitively excessive depends on a number of factors, and cannot be answered universally here. These factors include the bandwidth and processing capacity at the server, and on the actual percentage of bad links relative to the percentage that can be tolerated. The longer the refresh period, the higher the percentage of bad links. The higher the percentage of bad links, the higher the frequency of useless queries (because clients will be referred to servers that in fact no longer have the RP referred to). Thus, there is some optimal refresh rate for each server. If the approximate optimal refresh rate (assuming that it can be found) is such that the bandwidth and processing capacity of the server is sufficient, then the simple refresh mechanism is an acceptable solution.

6.2 Supplementing Periodic Refreshes with Third-Party Flushing

Because of the uncertainty and general high overhead of the simple refresh mechanism, we reduce (but do not eliminate) the need for an explicit refresh by using the third-party flush and the second-order refresh mechanisms described in the previous section. Unlike the previous case, however, a flush notification is *not* sent in the case where the "cached" server is unreachable. (The term 'cached' is here a misnomer, because we are referring to a link rather than a cached entry per se. We continue to use the term here, but in quotes.) The reason for this is given shortly. In addition, a half-link is never deleted unless the absence of the neighbor RP is directly verified.

In order to take advantage of third-party flush to reduce the frequency of explicit refreshes, we must change the initiator of the explicit refresh from that of what it normally is. Normally the refresh is initiated by the controller of the actual information. That is, the semantics of a refresh is "please don't delete your entry about me because it is still valid". For our case, we change it so that the initiator is the other end. That is, the semantics becomes "I'm about to delete my entry about you, so if it is still valid, please tell me so". In essence, we go from a refresh mechanism to a *refresh-request* mechanism.

Specifically, both servers keep a refresh-request timer for each half-link. When the timer expires, the server sends a refresh-request to the other end of the link. If the server gets no response (after a suitable number of attempts spread over a very long period of time), the half-link is unilaterally deleted. Otherwise, the server either deletes the link or resets the refresh-request timer according to the contents of the response.

Servers reset the refresh-request timer for a link whenever it includes that link's information in a query-answer. As with the analogous case of refreshing a cache entry, this is done on the assumption that, if the neighboring RP no longer exists, the third-party flush will notify the server. And, as with the cache case, the assumption is incorrect in the case where the caching system is unreachable and the cached information no longer exists.

However, there are several differences between refreshing a link's refresh-request timer and refreshing a cache entry that requires us to handle links differently. First, link information is more important that cache information. It is important that links remain active when the neighbor RP still exists. Therefore, it is certainly inappropriate to delete a half-link simply because the neighbor server is unreachable. It is also inappropriate to temporarily disable the half-link while the neighbor server is unreachable, because of the cost of determining precisely when the neighbor becomes reachable again. (A link about a non-existent RP is much less damaging than no link for an existing RP.)

Second, if the neighbor server is permanently down, then the half-link will persist more easily than in the caching case. This is because the time period for the refresh-request timer is generally much longer than the lifetime of a cache entry.

For these reasons, querying systems do not send explicit flush notifications in the case where "cached" servers are unreachable. In addition, servers monitor the reachability of neighbor servers over long periods. If no packets are received from a neighbor server for some period of time, then the neighbor server is pinged even if no refresh-request timers have expired. If the neighbor server is eventually determined to be permanently down, all half-links for that server are unilaterally deleted.

The time period of per-server monitoring can be on the order of one month. The time period of the refresh-request timer can be set longer, perhaps on the order of one year. It basically acts as a long-term garbage collection mechanism for those few link-halves that 1) weren't deleted properly to start with, and 2) weren't detected by the third-party flush.

7 RP to Actual Resource Consistency

RPs are added to and deleted from Ingrid Servers by applications via an open protocol mechanism. Typically, an RP describes a real-world resource of some sort, and consistency between the RP and the resource must be maintained. It is the responsibility of the application to keep the RP up to date, and to inform the server about any changes of the state of the RP. What we are concerned with here is the problem of maintaining consistency between the application's version of the RP and the server's version of the RP.

We would like to be able to use a third-party flush mechanism here as well. Doing so, however, is problematic. One problem is that interactions between application and the Ingrid infrastructure are always initiated by the application. Ingrid applications do not listen on sockets. Thus, the receiving application (the one discovering the inconsistency) would have to inform the server, which would then have to wait until accessed again by the original sending application (by which time the inconsistency may have been discovered by the original application itself).

Another problem is that receiving applications don't have an opportunity in their normal operation to verify the consistency of an application's RP with the server's RP. A receiving application normally goes directly from the RP learned from the server to the actual resource. Thus, the receiving application needs to deduce that the RP is out-of-date based on its interaction with the original resource. This may sometimes not be possible.

As a result, we use only the following common mechanisms:

1. The primary mechanism is an explicit delete from the sending application to the server.

2. A standard refresh mechanism is used for each RP, but with a very long timeout (of order many months). The refresh is initiated by the application. The server deletes any RPs not refreshed.

3. A somewhat shorter refresh mechanism (a few weeks) is used to refresh the existence of the appli-

289

cation at all. If the server doesn't hear anything from the application for this time, it deletes all RPs learned from the application.

8 Conclusions and Future Work

This paper presents a design for dealing with the problems of a global, distributed, autonomous information information infrastructure. It does not, however, present conclusions about the performance of the design. We believe that the design presented here has some interesting and useful characteristics. However, until it is implemented and tested in a real environment, we cannot make any definite conclusions.

We are currently testing many aspects of Ingrid with real users. The performance of the cache is not among the most interesting or serious problems we have to address, so it is not yet clear when we will get around to doing a careful analysis of the performance of the cache. It is, however, also an interesting and important problem, so we will do it eventually, and probably sooner rather than later. The interested reader should consult the Ingrid web page <http://www.ingrid.org/> for up-to-date information.

References

[1] M. Koster, "World Wide Web Robots, Wanderers, and Spiders,"
<http://info.webcrawler.com/mak/projects/robots/robots.html>

[2] D. Bell and J. Grimson, "Distributed Database Systems," Addison-Wesley, 1992.

[3] M. J. Feeley "Methods and Models for Management of Distributed and Persistent Data,"
<http://www.cs.washington.edu/homes/feeley/generals/DataManagement/DataManagement.html>

[4] N. Hall, "SHORE Project Home Page,"
<http://www.cs.wisc.edu/shore/>

[5] P. Albitz and C. Liu, "DNS & BIND," O'Reilly & Associates, Inc. 1992.

[6] P. Francis et. al., "Ingrid: A Self-Configuring Information Navigation Infrastructure," Proceedings Fourth International World Wide Web Conference, Boston, Dec. 1995, pp. 519 – 537.

[7] C. Huitema, "Routing in the Internet," Prentice Hall PRR, 1995.

[8] M. Abrams et. al., "Caching Proxies: Limitations and Potentials," Proceedings Fourth International World Wide Web Conference, Boston, Dec. 1995, pp. 119 – 133.

[9] W. Stevens, "Unix Network Programming," Prentice Hall, 1994.

Management of a Secure WWW-Based Document Store

Michael Tschichholz, Michael Höft, Mehran Roshandel

GMD FOKUS, Hardenbergplatz 2, D-10623 Berlin, Germany

{tschichholz | hoeft | roshandel}@fokus.gmd.de

Michael Gehrke

DeTeBerkom, Goslaer Ufer 35, D-10589 Berlin, Germany

micky@deteberkom.de

Abstract

This paper describes the design of a manageable WWW based secure multimedia document store. This store consist of two main components, one offering the core service functionality and the other providing the management. Only the seamless integration of these components in conjunction with the store's security features into a comprehensive multimedia teleservice will offer the opportunity to use it in the currently evolving pan-European "Open Service Market". The design approach discussed in this document is based o ODP concepts, covering an enterprise model, an computational model and a technology model.

The focus of this paper will be on the service session model, the CORBA and Java based management system and the security architecture.

1. Introduction

This paper discusses the design of a managed WWW-based secure multimedia document store, called *WebStore*, which provides the necessary features to become a commercial service in the evolving pan-European "Open Service Market" (OSM) environment [1][2]. The service will allow customers to store and retrieve arbitrary documents, ranging from plain text to compound documents, using the well-known interface of the WWW. To overcome today's shortcomings of Internet based services, i.e. no guaranteed quality of service (QoS) for multimedia services, insufficient security, service monitoring and control, customer care, etc., required security and management functions are being developed based on ODP and TINA concepts.

The *WebStore* could be offered as a stand-alone service to provide, for example, a visionary "electronic mall", or it could be part of more complex composed services which are value-added combinations of a number of other services. The use of a teleservice in a commercial environment makes the availability of appropriate

Service management and security will become essential requirements in an OSM environment, as outlined in [3]. Both requirements are not sufficiently solved with today's Internet technology.

The service provider, on the one hand, must be able to obtain the necessary information to bill customers as well as to offer the service with a meaningful performance. On the other hand, the customers must be able to adapt the service parameters to satisfy their requirements as well as to supervise the costs that service usage is producing. To fulfil those expectations, *WebStore* will provide a rich set of management functionality, including subscription, accounting, customer administration, service provider administration and service configuration. The integration of service management and core service functionality will be transparent to the end users, thus emphasising the equal importance of both areas.

Furthermore, a bilateral authentication between customer and service provider as well as the provisioning of access control and encryption mechanisms is an essential requirement for a teleservice product in the envisioned OSM environment.

To prove the suitability of our design, a first prototype of the *WebStore* service is being developed as part of the ACTS project PROSPECT [4], where it will be an integral component of a tele-educational service environment in two pan-European trials (see Figure 1). The focus is on the design and implementation of solutions for composed services and their management systems for competing and co-operating service providers of pan-European teleservices [5].

This paper is structures as follows: Section two

291

presents the enterprise model for the *WebStore* service as part of a more complex tele-educational environment, identifying the involved domains, roles, and resources, as well as the major requirements on management and security.

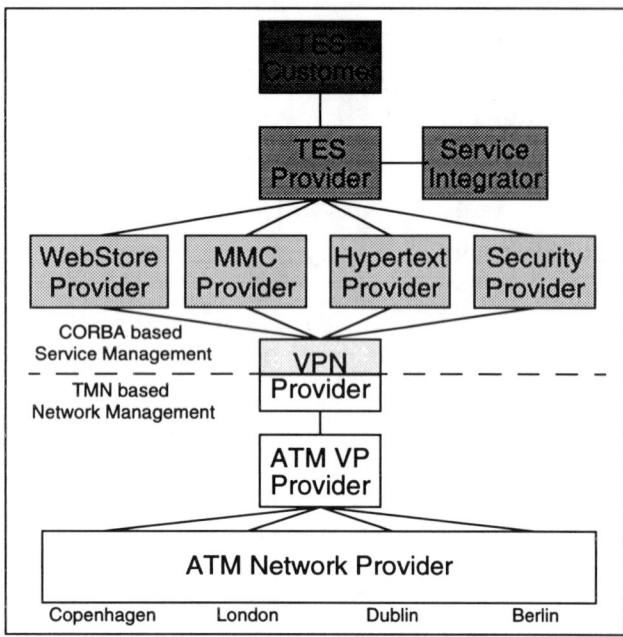

Figure 1: The PROSPECT tele-educational environment (PTEE)

The third section introduces the computational model with the basic computational objects and associated interfaces, followed by the technology model identifying the technology being used. The features of the *WebStore* management services, its role specific management interfaces and the characteristics of the management agent (server) are outlined in section 4. Available security services and their interworking with a certification authority are described in section 5.

2. The PROSPECT tele-educational environment

2.1. The PROSPECT trial 1 enterprise model

PROSPECT's first pan-European trial envisages a broadband network infrastructure consisting of an international ATM backbone and access networks run by the national hosts. Various Integrated Broadband Communication (IBC) services, such as Virtual Private Network (VPN), multimedia mail and conferencing, interactive hypermedia systems, secure multimedia *WebStore*, etc. are being provided on this network infrastructure by different service providers (see Figure 1). In the trial, these IBC services are composed to establish the PROSPECT tele-

educational environment. The management of the tele-educational services and the general purpose tele-services which support them, form the basis of the end-user service management provided by PROSPECT.

The *WebStore* is being used in this trial to store tutorial material, to support group work in conjunction with e-mail, and to enable students to study on their own. A contract is established with the tele-educational service (TES) provider which grants to all students the right to access the related course material as well as the course specific storage area (allowing them to upload their own documents).

The service provider domain represents the organisation that offers the *WebStore* service to potential customers (see Figure 2). The customer domain represents an organisation (e.g. a TES provider) that purchases storage capacity from the *WebStore* service provider to store arbitrary documents.

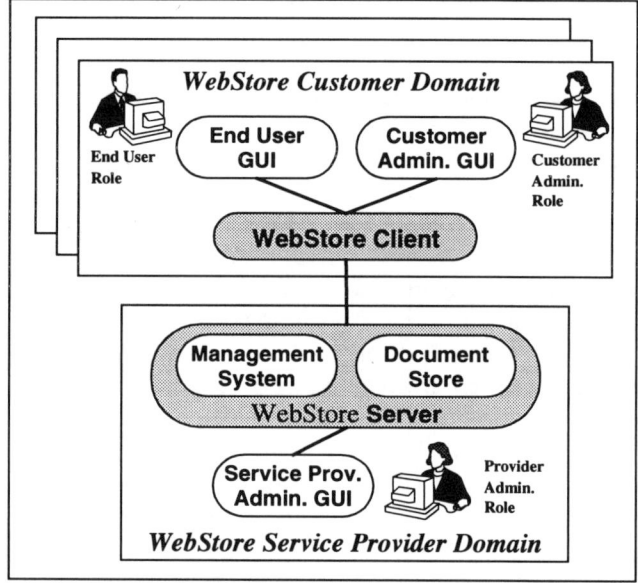

Figure 2: WebStore enterprise model

There exist a number of roles that are relevant in the *WebStore* service scenario. Within the customer domain these roles are: *end user* and *service administrator*.

- The end users provide and/or consume the documents represented in the document store.
- The customer service administrator is concerned with subscription, accounting, billing, and user administration.

Within the service provider domain there is only one role: the *service administrator*.

- The service provider administrator operates the *WebStore*, handles the customer administration, and is concerned with contracts and accounting.

The composed teleservices with integrated management services like subscription and accounting will be provided to the relevant roles in the customer and service provider domains via appropriate interfaces.

2.2. The need for a WebStore session concept

To support the need of service or content providers for individual usage metering and accountability in conjunction with end-user needs for "one-stop-shopping" and "user session authentication", a *WebStore* session concept is required to enable a coupling of associated interactions with the *WebStore*.

In the context of PROSPECT, the *WebStore* is being used to support tele-educational sessions. This session is defined as the time when a student (or teacher) takes part in a tutorial, in group work, or studies alone. Within this session he/she may use material stored on the *WebStore*. The *WebStore* session exists from the first access until the student or teacher leaves the tele-educational session. Within this timeframe no further authentication is required and all interactions with the *WebStore* are associated with common usage metering logs being stored in "user specific service session MIBs".

As seen from the *WebStore* service provider's perspective, a session will start if access is granted to a user (content provider or end user) and terminates when this user leaves the session explicitly (logout) or implicitly when an individually defined time period finishes (user profile).

This general approach will also be valid if the *WebStore* is used to implement an "electronic mall": the *WebStore* session starts when a user enters the mall. It will be associated to a specific service or content provider if he/she enters a specific shop within this mall, might be associated to other service providers in the case of "shopping sessions", and will terminate when this user leaves the mall. This concept of the "user session" is intended to support a variety of customised accounting or security models.

2.3. WebStore management requirements

The functional requirements can be associated with the three roles that were identified in section 2.1. Besides the core service functionality that the end user role requires, such as authentication and the storage/retrieval of documents, there is also a need for management functionality, such as obtaining information about stored documents (size, access rights, etc.), retrieving quotas (used space, allowed space, etc.) and accounting information per user as well as threshold controls. It is also desirable that the core and management functionality can be granted individually to each end user.

The customer administrator role requires management specific functionality only. The obvious requirements are user administration and the handling of subscriptions (such as add/remove customer sites or cancel subscriptions), but it is also necessary to obtain information on document as well as quota and accounting information, the latter for individual users or accumulated for all users of a customer. The accounting information could be based on a basic charge, the used disk space and the number of read and write accesses. The manipulation of user specific access rights requires an Access Control List (ACL) editor.

The Service Provider Administrator role requires management functionality in a similar manner to the Customer Administrator role. The main functionality needed will be handling of subscriptions, customer administration, accounting (including initialisation/termination of billing mechanism, collection of usage information, calculation of bills, etc.). Furthermore, there is the need for document handling capabilities, for example to remove expired documents. Another desirable functionality is the support for service publishing via trading or directory services.

2.4. WebStore security requirements

The PROSPECT environment in which the *WebStore* is developed has three important characteristics which have to be taken into account when considering security requirements for the *WebStore*. First, it is a commercially oriented trial where information with significant value will be communicated over an open network. This was the most essential reason for covering security issues in the project. Second, the project is focused on the management of tele-services, not on the services themselves. Therefore, we had to cover security as well as security management aspects. Finally, because the Open Service Environment is assumed to consist of a network of competing and cooperating service providers and customers we had to provide security mechanisms supporting a multi-domain environment.

A requirements analysis performed at the beginning of the project obtain the following results regarding information security:

"Half of all end-users, teachers and managers are expected to find almost all these aspects of security to be of very high importance." The aspects mentioned can be summarized as:

- Authenticity of participating humans
- Authenticity and integrity of content (data)
- Confidentiality of contents

Additionally, all agreed that "third parties should not be able to get information on the interactions that have taken place during a course". This was taken as a confidentiality requirement on the exchange of usage data between management functions.

There were also some exceptions to the rule stated above: "Most teachers don't see any point in constraining interactions between course participants to take place only when required by the course schedule. Similarly, many teachers, end-users and managers don't mind if course participants can gain access to other participants' session".

In section 5 we focus on the implementation of the three security requirements listed above.

3. The WebStore models

3.1. Computational model

This section discusses the initial computational architecture of the *WebStore* service taking technology constraints into account (see Figure 3). The service provider domain contains the following major computational objects (COs):

- *WWW Server*: This CO provides the core service functionality, i.e. the storage and retrieval of the multimedia documents. It does not offer the storage capacity itself but handles and synchronises access to the document store.

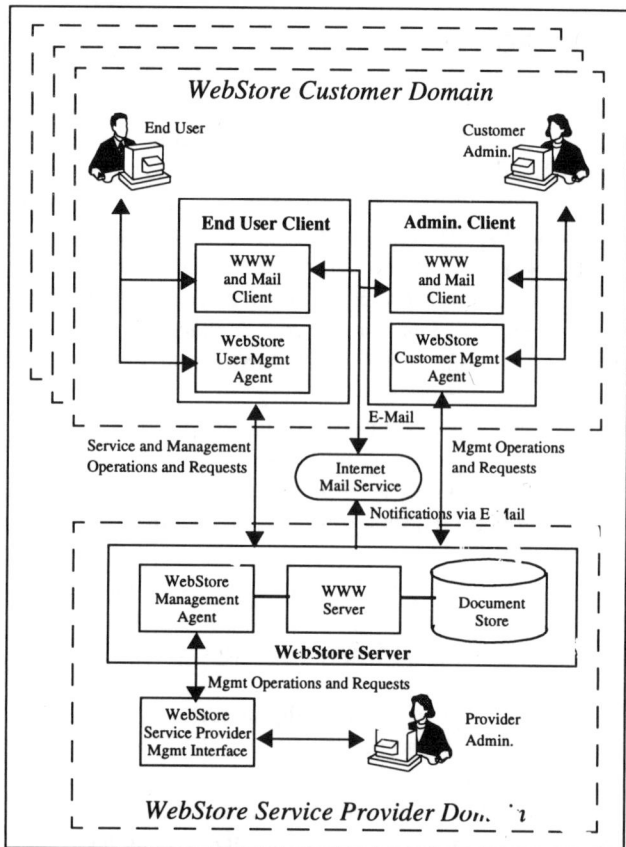

Figure 3: WebStore computational model

- *Document Store*: The multimedia documents are physically stored within this CO. For performance reasons it will be normally located on the same end

system as the WWW server, but this is not an architectural requirement.

- *WebStore Management Agent*: This CO handles all the management requests from the different components of the *WebStore* service. It is directly connected to the WWW server and offers its functionality via a standardised interface which makes the distribution and the extension of the *WebStore* services much easier.
- *WebStore Service Provider Management Interface*: This CO contains the user interface for the service provider administrator. It accesses the management functionality of the service via the interface of the *WebStore* Management Agent described above.

The *WebStore* customer domains contain:

- *WWW and Mail Client*: The core service functionality on the client side will be provided by this CO. The approach to accessing the service via the well known WWW will facilitate the usage of the service.
- *WebStore User Management Agent*: This CO presents the user interface to the management functionality for the end users in the *WebStore* customer domain. The management functionality is accessed via the interface of the *WebStore* management agent.
- *WebStore Customer Management Agent*: This CO represents the user interface to the management functionality for the customer administrators in the *WebStore* customer domain. Analogous to the CO described above, the management functionality is accessed via the interface of the *WebStore* management agent.

3.2. Technology model

There are three areas concerning the realisation of W.*Store* and its management system where the underlying technology has to be selected: the implementation language, the technology for the core service functionality, and the technology for the management functionality. The technology has been selected based on the business requirements of the *WebStore* provider stakeholder: strict use of object-oriented technology, fast service development/provisioning, availability on the most common OS platforms, compatible integration of management and security with the standardised WWW service and use of HTTP 1.1 [6] to enable uploading of documents. Furthermore, the WWW server to be selected must provide a management interface or must at least support management and security related enhancements.

The implementation language that has been chosen for the *WebStore* service is Java. The main advantages are:

easy integration of functionality into WWW pages, availability on most common OS platforms which provides portability of the *WebStore* management agent and of the graphical user interfaces, the fast implementation cycles, the consistent object-oriented approach, and last but not least, the powerful support of the thread concept. One drawback is that there are currently no experiences with the performance of complex Java systems, but it is expected that the use of Java just-in-time compilers will overcome possible problems.

Unfortunately there are currently no WWW servers available that support HTTP 1.1 or provide management interfaces. There are already several proposals for a management interface to HTTP servers [8][9]. In addition, there exist some proposals for protocol independent MIBs developed by the University of Twente [10][11]. But due to the early stage in development they are quite unstable and not supported by any server.

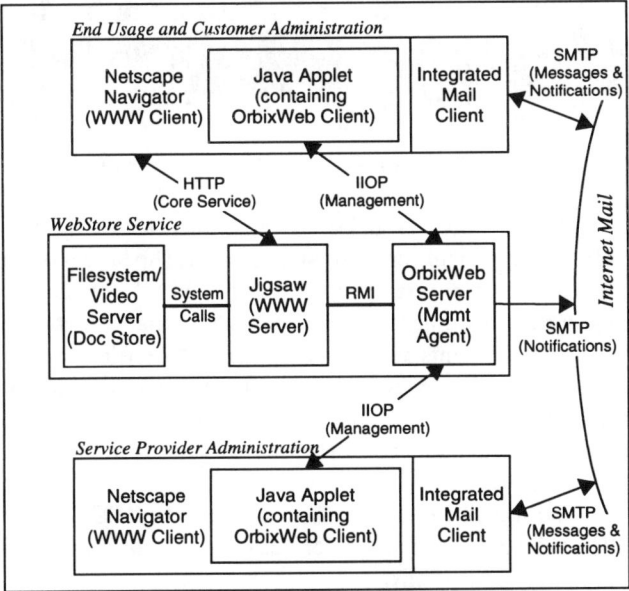

Figure 4: WebStore technology model

To be able to implement the necessary management functionality, some available WWW server implementations were reviewed with respect to their implemented functionality and extensibility (based on initial information from [12][13]). As a result of the review the *Jigsaw* server [14] was chosen to be used on the server side. Jigsaw is a new development of the World-Wide Web Consortium (W3C). It is an implementation in Java and due to its object-oriented design very easy to extend. Since the whole *WebStore* service will be implemented in Java it fits perfectly in the whole system.

On the WWW client side the choice is much easier. However, although there are currently numerous client implementations available, there are a few requirements that have to be satisfied by the client of choice. The *WebStore* approach of using Java applets on the *WebStore* client side makes the support of Java an essential requirement. The support of most operating systems and the availability of extensions, enabling for example on-line audio and video playback, is desirable. For these reasons the *Netscape Navigator 3.x* was chosen as the reference WWW client for the *WebStore* service.

The necessary communication functionality for the management integration is based on CORBA 2.x [15][16]. There are several other possibilities, like OSI/CMIS/P or SNMP based approaches, but CORBA has established itself recently as a de facto standard for distributed applications. Furthermore, the use of Java enables easy on-line downloading of a complete communication environment via applets, avoiding any pre-installations and the related software management. This technique allows a very flexible integration of communication abilities into the customer system that is, moreover, totally transparent for the user.

The CORBA implementation that will be used for the *WebStore* development is *OrbixWeb* from *IONA*, a Java based CORBA 2.x system. It offers client and server capabilities and provides a very lightweight implementation which is important when used as part of a downloadable Java applet.

The selection process discussed above leads to the following technology for the *WebStore* prototype.

- Java as the programming language
- *Jigsaw* as a Java based WWW server
- *Netscape Navigator 3.x* as a WWW client (also covering an e-mail client)
- *IONA's OrbixWeb* as a Java based ORB (client and server side)

Consequently, the technology model depicted in Figure 4 can be derived from the computational architecture.

4. WebStore management

4.1. Management services

This section describes the management specific functionality that will be realised within the *WebStore* service, separated by the three roles that were identified in Section 2.1.

User management functionality

- document handling (store, retrieve, delete)
- sending an e-mail to selected addresses containing the URL of a newly stored document (manually via an integrated mail agent, or automated by the *WebStore* service)

- retrieving document information (size, access rights, expiration date, etc.)
- obtaining quota information per end user (used space, allowed space, limits, etc.)
- threshold control per end user (including the automatic generation of warnings via e-mail)
- getting usage data per end user
- searching for specific documents via keywords
- subscription to mailing lists allowing users to receive information (e-mail) about specified subjects
- handling of virtual private document stores (i.e. create, delete, modify and share folders)
- security management (i.e. changing password, set/unset security level, choice of security policy)

Customer management functionality

- user administration (i.e. add, delete, modify users)
- retrieving usage data for each individual end user as well as accumulated for all end users of the customer (based on a basic charge, the used disk space, and the number of read and write accesses)
- subscription handling (authorise/bar end users, add/remove customer sites, cancel subscriptions)
- getting document information (size, access rights, expiration date, etc., including sorting and filtering abilities)
- obtaining quota information for each single end user or accumulated for all end users of the customer (used space, allowed space, limits, etc.)
- threshold control, accumulated for all end users of the customer (including the automatic generation of warnings via e-mail)
- ACL (Access Control List) editor to define and manipulate document access rights per user

Provider Management functionality

- customer administration (i.e. add, delete, modify)
- subscription handling (authorise/bar end users, add/remove customer sites, create/cancel subscriptions)
- accounting (initialise/terminate billing mechanism, collect usage information, calculate bills, send bills)
- store administration (changing available capacities, disk checks, back up, etc.)
- document handling (e.g. delete expired documents)
- service publishing (via trader or directory service)

4.2 Role specific management interfaces

The management functionality for the different roles is presented via graphical user interfaces (GUI) within a WWW browser. All of these GUIs are realised as Java applets. To start a session the user has to run a stand-alone Java program providing the authentication interface which enables the access to the *WebStore* service. After success-

ful authentication, the Netscape Navigator will be started with a user specific WWW page containing the Java applet that provides access to the core functionality as well as to the management functions.

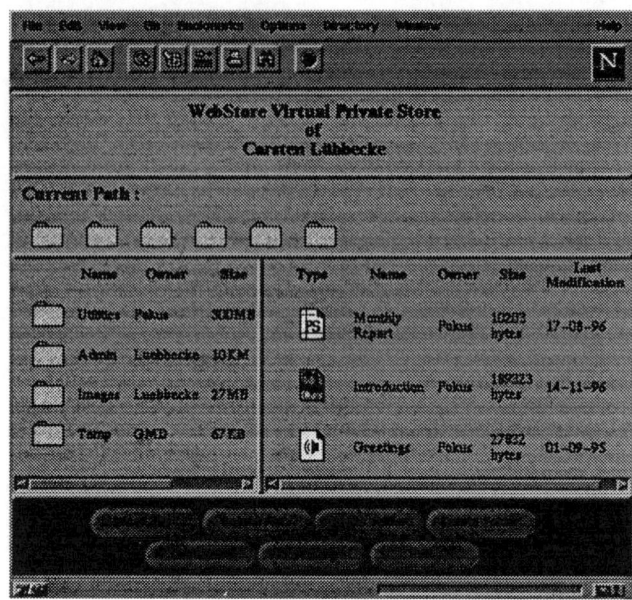

Figure 5: WebStore end user GUI

Figure 5 depicts an example of such a role specific WWW page (in this case for an end user). The layout of the page can be configured by the user. It allows the creation of new virtual document stores (i.e. folders) that can contain specific documents such as video files, audio files or exercises. The access to those documents can be granted to certain individual users or user groups. The GUIs for the other roles supported by the *WebStore* service have a similar look-and-feel.

The use of Java applets instead of pure HTML code allows the design of user interfaces that are similar to conventional applications.

4.3 WebStore management agent

The *WebStore* management agent implements the management related computational objects (COs) which contain the necessary information for the management of the Jigsaw WWW server (see Figure 3).

The subscription COs are generic, i.e. service independent COs that handle all the subscription information needed for subscription oriented tasks and secure authentication.

The accounting COs are also service independent COs that contain tariff information and are responsible for the collection of service usage logs. Furthermore, they support notifications and provide interfaces for the control of service usage, the setting of tariffs and the creation of bills.

The service specific COs contain information about the service configuration (e.g. store capacity), the performance (e.g. available capacity), the folders (e.g. owners and access rights), and the documents (e.g. size, owners and access rights). They offer interfaces for the manipulation of this information (like setting the maximum available capacity, modifying thresholds, or handling folders and documents). They also contain the Java RMI interfaces to the Jigsaw server needed to access and manipulate the folders and documents on the document store.

5. WebStore security

In general, the security concept of the *WebStore* and that of many other services in the Open Service Market will consist of two complementary levels of security:

- On the network level security functions will be applied for packet filtering (firewall).
- On the application level additional functions for features such as authentication and access control will be applied.

It is obvious that for a commercial service both levels of security should be implemented and harmonized. In this paper we deal only with the application security level.

5.1. Security services

User Authentication

User authentication provides for the unambiguous verification of a user's identity. In PROSPECT the authenticity of the *WebStore* is also considered a requirement. Therefore, an authentication protocol for mutual authentication is applied between *WebStore* and its clients.

It was decided that public key cryptography provides the appropriate support for the requirements of multidomain security. Because authentication protocols based purely on asymmetric cryptography have been shown to have weak performance, a hybrid 2-way protocol will be applied which combines the advantages of asymmetric cryptography (high security) with that of symmetric algorithms (high performance).

Access Control

The access control service allows the control of who accesses which documents under what conditions on the *WebStore*. The access control function also supports the administration of access rights.

The access control functions for the *WebStore* should at least support the definition of access rights dependent on a number of context variables including user identity, role (user, admin), document class and accounting status.

Certificate Management

The certificate management service supports the generation of private/public key pairs, the issuing of public key certificates, the distribution of public key certificates via a public key directory, and the issuing of smartcards containing the private key of a user [17].

5.2. Security architecture

The security architecture depicted in Figure 6 describes the application of the security services to the *WebStore* technology model (cf. Figure 4).

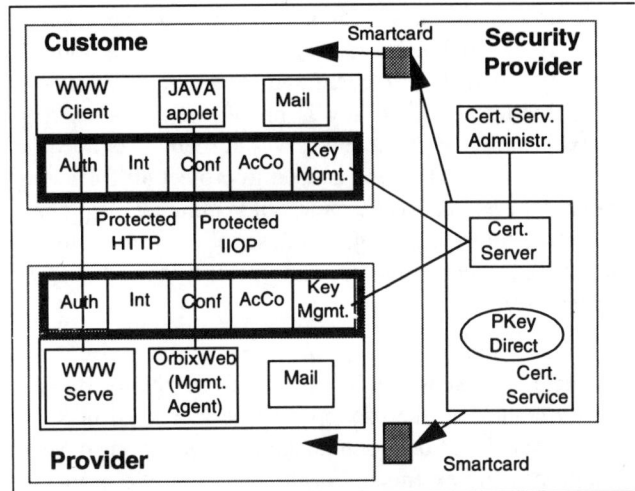

Figure 6: *WebStore* security architecture

The communication protocols are protected on the application level by using the different security services, shown with a thick grey border in Figure 6. To retrieve keys of communication partners an on-line access to public key certificates is provided. In the PROSPECT enterprise model this service is provided by the security service provider. In the security service provider domain a certificate service administration application is used which provides a user interface to the certificate service. This application allows an administrator to carry out several management operations, including administration of a queue of certification requests, revocation lists, notifications of users and viewing of log files [17].

Another main task of the security service provider is the generation of asymmetric key pairs and the issuing of smartcards. The key generation process is already executed by the smartcard.

5.3. Issues in Security Integration

"Security integration" is understood as the process of designing and implementing the use of security technology to realise security functions in an application or system. Concerning the *WebStore*, security integration has the objective to maximise usability of the security services for *WebStore* users. In particular, this means that security features are integrated transparently to the user during a "user session".

The *WebStore* session concept discussed in section 2.2 is exploited to implement "user session authentication". With user session authentication a user interaction with the *WebStore* client is only necessary during the session access phase. For the lifetime of the session the authenticity of the communication link is automatically maintained by the *WebStore* components.

Therefore, in the session access phase the authentication service, which is based on public key cryptography, is used between *WebStore* client and server. A security context including a session key is negotiated. This security context is stored at client and server in their "user specific service session MIBs" and is used to protect all following session communication by means of symmetric cryptography. In the following we give a description of the integration of the secure session concept into the *WebStore* architecture.

IIOP

Security tokens are generated from the respective security context in the user specific service session MIB. The IDL specifications for management operations have to be appropriately extended. For the purposes of confidentiality and integrity we shall use function hooks provided by OrbixWeb.

HTTP

Security in the WWW is currently a widely discussed topic. Several approaches have been suggested (SHTTP, SHTML, SSL, etc.), but unfortunately there does not appear to be a single standard solution to this problem. We shall not repeat the pros and cons of the various approaches here but instead summarize our own proprietary one:

Security tokens will be exchanged by using the standard HTTP header fields "WWW-authenticate" and "Authorization". Commercial WWW products are not necessarily prepared for integrating tokens generated by a smartcard into the header fields. Therefore, we shall use specialised secure clients which are placed between WWW client and server and which intercept the HTTP traffic and perform security operations on that traffic.

6. Conclusion

In the not too distant future the Internet may have decided on common management interfaces and a single standard for security. An open problem will remain, namely, whether companies outside the US will place sufficient

trust in products equipped with security technology originating from the US. We believe we have taken an approach whereby security of Internet services can be implemented independent of a particular technology.

The implementation of the *WebStore* can be seen as one example of an enhanced Internet based service which forms the basis for a commercial teleservice in an OSM environment. By using the *WebStore* in the PROSPECT tele-educational trials the prototype versions will be evaluated and its functionality and characteristics will be improved over the coming months.

7. References

[1] *Europe and the global information society. Recommendations to the European Council* (Bangemann Report), European Commission, Brussels, May 1994.

[2] European Telecommunications Standards Institute (ETSI), *Report of the Sixth Strategic Review Committee on European Information Infrastructure, Part A: Summary and Recommendations*, Sophia Antipolis, June 1995.

[3] E.K. Adams and K.J. Willetts, *The Lean Communications Provider*, McGraw-Hill, New York, 1996.

[4] PROSPECT Consortium, ACTS Project AC052, *A Prospect of Multi-Domain Management in the Expected Open Service Market, Technical Annex*, September 1995.

[5] Tschichholz, M. (ed.), PROSPECT Consortium, *The Initial Prospect System Model*, February 1996.

[6] Fielding, R., Gettys, J., Mogul, J. C., Frystyk, H., and Berners-Lee, T., *Hypertext Transfer Protocol – HTTP/1.1*, Internet Draft, IETF, June 1996, work in progress.

[7] Schulzrinne, H., *World Wide Web: Whence, Whither, What Next?*, in IEEE Network, March/April 1996.

[8] Kalbfleisch, C. W., *Applicability of Standards Track MIBs to Management of World Wide Web Servers*, Internet Draft, IETF, June 1996, work in progress.

[9] Hatzewinkel, H., van Hengstum, E., and Pras, A., *Definition of Managed Objects for HTTP*, Internet Draft, IETF, April 1996, work in progress.

[10] Hatzewinkel, H., van Hengstum, E., and Pras, A., *Definition of Managed Objects for an Information Store*, Internet Draft, IETF, April 1996, work in progress.

[11] Hatzewinkel, H., van Hengstum, E., and Pras, A., *Definition of Managed Objects for an Information Retrieval Service*, Internet Draft, IETF, April 1996, work in progress.

[12] WebCompare, *Server Features Comparison*, URL: http://www.webcompare.com/server-main.html, June 1996.

[13] WebCompare, *Browser Features Comparison*, URL: http://www.webcompare.com/browser-main.html, May 1996.

[14] World Wide Web Consortium, *Jigsaw Overview*, URL: http://www.w3.org/pub/WWW/Jigsaw, October 1996.

[15] OMG, ed. Soley, R. M., *Object Management Architecture Guide, Revision 3.0*, Third Edition, John Wiley & Sons, Inc., June 1995.

[16] OMG, *The Common Object Request Broker: Architecture and Specification, Revision 2.0*, July 1995.

[17] Gehrke, M., Hetschold, T., "Management of a public key certification infrastructure - Experiences from the DeTe-Berkom Project BMSec", Proceedings JENC7, Budapest, May 1996.

A Client-Oriented Distribution Architecture
for Web Search Agents

Hyunsuk Seung

Multimedia Laboratory
Corporate Technical Operations
Samsung Electronics Co., Ltd.
Suwon, Korea
hseung@samsung.co.kr

Doo-Hwan Bae

Department of Computer Science
Korea Advanced Institute of
Science and Technology
Taejon, Korea
bae@selab.kaist.ac.kr

Abstract

Among the agents deployed on the Web, search agents which can query multiple search engines simultaneously are gaining popularity among end-users. As the number of users and search engines increase, however, establishing multiple connections to remote search engines has the potential to saturate the network and increase the load on the server running the search agent. To resolve these issues, we propose a client-oriented distribution architecture for Web search agents. Our architecture allows having multiply replicated agents distributed throughout the network initiate the queries while localizing the network traffic to the client's machine. Special attention has been made on the design and implementation of a search agent based on our proposed architecture that limits its use of network resources and prevents skewing the logs on the remote sites. We also present the evaluation results of our Web search agent.

1 Introduction

The role of agents in the field of information retrieval has become increasingly important as the amount of information resources available over the World Wide Web[1] has increased at an alarming rate in recent years. Among the agents deployed on the net, *search agents* which can query multiple Web search engines, such as AltaVista, Excite, Infoseek, and Lycos, are gaining popularity among end-users[2] who would rather delegate the task of searching to competent agents instead of searching on their own. Often termed as simultaneous or multi-threaded search service, search agents

are a form of an information gathering agent, providing an integrated interface to accessing multiple search engines.

Currently, search agents speed up their search process by submitting the queries to the remote search engines in parallel. As the number of users and search engines increase, however, Koster [8] points out that establishing multiple connections to remote search engines has the potential to saturate the network and increase the load on the server running the search agent. Also, since it is the search agent that queries the search engines, the access logs on the search engines will be skewed. Furthermore, the location of the search agent may induce inefficient routing of query responses over the network.

Our study focuses on resolving these issues by having multiply replicated agents distributed throughout the network initiate the queries while localizing the network traffic to the client's machine. Special attention has been made on the design and implementation of a search agent that limits its use of network resources, prevents skewing the logs on the remote sites, and acts in accord with the guidelines suggested for robot writers[10].

The organization of this paper is as follows. In section 2, we provide a brief overview of various Web search agents in operation. In section 3, we present our client-oriented distribution architecture for operating Web search agents along with its implementation details. In section 4, we evaluate the performance of our client-oriented Web search agent. Section 5 states our conclusions.

2 Web Search Agents

In this section, we provide an overview of Web search agents currently under research and the potential

---

[1] Herein referred to as "Web."
[2] "End-user" and "user" should be read interchangeably throughout this paper.

299

problems they are known to exhibit. Web search agents can be divided into two: *server-side search agents* that operate from a single server for multiple users, and *client-side search agents* that operate from the client machine for the benefit of a single user.

2.1 Server-Side Search Agents

Server-side search agents, as the name suggests, operate from a single centralized system for the benefit of multiple users. Users initially connect to the agent's home page and retrieve the query form written in HTML. When the form is filled out by the user and submitted to the agent, the agent maps the query to a form suitable for each search engine it plans to contact. The agent, then submits the translated queries to the search engines and retrieves the results, often performing the tasks in parallel by multithreading the network connections. Finally, the results sent from the search engines are gathered, collated, and sent back to the user who initially requested them. Some of the well-known server-side search agents include SavvySearch [4], ProFusion [6], and MetaCrawler [13].

At the benefit of providing abstraction and integration of many search services tied into a single uniform tool, server-side search agents pose several critical problems [8]. The most noticeable ones are:

1. The network and server load on the server running the agent increases.

2. The distances the queries and their responses would have to traverse increases.

3. Log skewing problem on remote search engines.

To avoid making unnecessary remote queries, SavvySearch [4] selects the search engines to contact based on the keywords, domain information submitted by the user, and past knowledge about which search engines had returned the most useful results. Also, it goes about to reduce the amount of parallelism to be exploited by predicting the expected network and system load as well as other data collected from user feedbacks. Similar to Savvy Search, ProFusion [6] automatically identifies the topic of the queries submitted by the user, thereby reducing the number of search engines to contact. MetaCrawler [13], on the other hand, always contacts all the search engines on its list. It does not go about reducing the amount of parallelism. The reason for this, the authors argue, is based on the observation that users follow references reported by a variety of different search engines. Another way to reduce the possibility of network saturation is to submit the queries in a sequential fashion just as how a human

user would perform when accessing multiple search engines, but this would sacrifice the speed, often making it less appealing for its users since the added benefit is minimal compared to other more comprehensive agents.

2.2 Client-Side Search Agents

Client-side search agents have advantages over the server-side search agents in that network usage, retrieval, and filtering process are performed on the client's machine. The log skewing problem disappears since searches are initiated directly from the client's machine. Examples of client-side search agents are Fish Search [2], TkWWW robot [14], and WebCompass [12].

Fish Search was one of the first client-side search agents to appear on the Web. Embedded in a modified version of Mosaic Web browser, Fish Search sequentially retrieves matched documents from a user-specified starting page and displays them along with their relevance score. TkWWW robot is dispatched from the TkWWW browser to aid the user in intelligent browsing. Users could program the TkWWW robot via TkWWW Tcl extensions to retrieve documents matching user-supplied keywords, construct Web index, or collect specific data types. WebCompass, on the other hand, is a client application that searches and indexes topics specified by the user by accessing multiple search engines. WebCompass agent keeps the customized indices up-to-date by querying the search engines periodically in the background.

Unfortunately, there are several drawbacks to client-side search agents [9].

1. Client-side searches could cause considerable overhead on several servers in a far shorter time than a manual retrieval by the user, even servers that do not provide the specific information in question.

2. Problems will arise when the search engines change their query syntax, program location, or even their URL addresses, making the queries sent by the agents obsolete, and not all users will realize this misbehavior.

3. Since the agent is serving a single user, indices or cache generated from the retrieval process cannot be shared with others.

4. A carelessly programmed filtering module by the user could create havoc on the network and increase load on the servers by failing to recognize infinite loops created from dynamic documents

or by revisiting the same document over and over again.

Problems exhibited by both the server and client-side search agents have prompted Eichmann [5] to write up a code of "Agent Ethics," for authors of various search agents to adhere. The guidelines suggested for server-side agents are identity, openness, moderation, respect to other servers, and authority. For client-side agents, they are identity, moderation, appropriateness, and vigilance.

3 A Client-Oriented Distribution Architecture

In this section, we propose a new architecture, called the *client-oriented distribution architecture*, for operating search agents on the Web. As we will see in the following sections, the client-oriented distribution architecture as applied to Web search agent systems has significant advantages over the traditional methods of operating a single server-side search agent.

The client-oriented distribution architecture provides a framework for building agent systems that allows dynamic distribution of search requests submitted by the clients among multiply replicated agents dispersed over the Web. Distribution of search requests minimizes network consumption and reduces the system load as compared with a single centralized agent system. There are two types of agents that operate in the architecture: *broker agent* and *task agent*. The broker agent acts as an intermediary between clients and task agents. The task agent is an agent that provides some useful service to the client. It is the service of the task agent the client is interested in. The job of the broker agent is then to introduce the most appropriate task agent to each connecting client.

3.1 The Method

Below is a high-level procedure on how a client establishes a service transaction with a Web agent employing the client-oriented distribution architecture:

1. In order to initiate a service transaction, the client must first request a query form from the broker agent. This is the same as how other Web-based transactions initially start off.

2. The broker agent responds with a dynamically created query form that lists the URL address of a specific task agent. The task agent is selected according to factors such as the client's location, the network and server load of each participating task agent, and so forth.

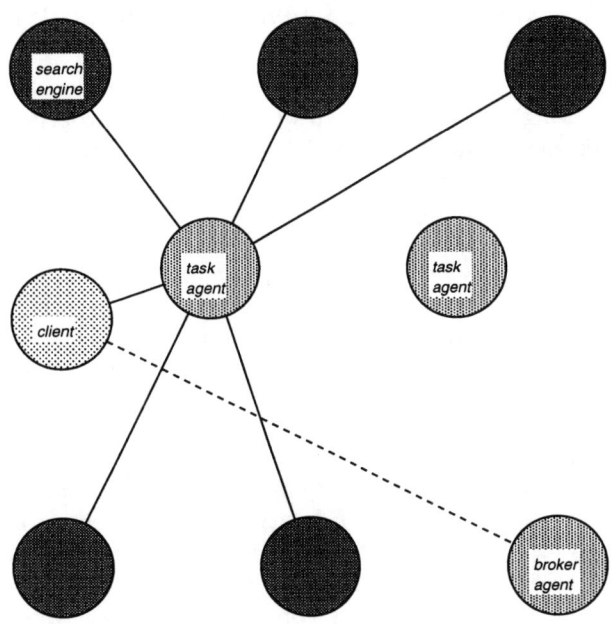

Figure 1: Client requests a query form (dotted line) and submits the filled out form to the nearest task agent, which in turn submits it to multiple search engines.

3. Upon receiving the form, the client fills it out with keywords and other appropriate parameters and submits it. Unlike conventional server-side agent systems, however, this query is not submitted back to the server (in this case, the broker agent) that provided the query form. It is instead submitted to the task agent as specified on the form. Figure 1 shows a client submitting to the nearest task agent as chosen by the broker agent.

4. The task agent receives the filled-out query form from the client and processes it. In the case of a search agent, it contacts a number of remote databases.

5. When the task agent is completed with its job, the result is sent back to the client.

6. If the result is satisfactory, the service transaction is completed. If not, the client could change the keywords or parameters and resubmit the form for further refinement of the query, and the process repeats from step 4. Note that in such cases, successive transactions take place between the client and the assigned task agent, since the broker is out of the picture after step 2.

301

3.2 Advantages of Client-Oriented Distribution Architecture

By providing an architecture for intelligently distributing the clients' queries to the most appropriate task agent, the Web search agent system employing the client-oriented distribution architecture has the following advantages:

- *Network consumption in the immediate proximity of the server running the agent is reduced.* This is achieved by two ways: through distribution of client's request and through localization of network traffic.

- *Separation of the task agent from the broker agent increases the scalability of the overall system.* In conventional server-side search agent systems, the only way to handle more in-coming requests is to either upgrade the machine to a faster one or have a faster network connection so access to remote databases could be completed more quickly. The client-oriented distribution architecture, on the other hand, allows easier and cost-effective upgrade path. Just adding more task agents to already existing machines is all that is necessary.

- *The response time from the search agent is shortened.* In the ideal case, the broker agent would assign each connecting client to a task agent that is most closely located to the client. On the Web, client-oriented assignment of task agents reduces the transfer time of the search results and, as we will see in the next section, any images that are present in the query forms.

- *Client-oriented assignment of task agents avoids skewing the logs on the remote databases.* This is because a given task agent acts as a proxy for the subnet it is responsible of. Also, in the next section, we show how our running implementation of the client-oriented search agent passes the client's address to the remote databases in order to avoid skewing their logs.

- *The reliability and availability of the overall system increases.* Having multiple copies of the task agents greatly increases the overall reliability and availability of the system since any one unavailable task agent would not bring down the entire system. There will be other task agents ready to handle the client's query. When one of the machines running the task agent needs a maintenance job or is unavailable for some other reason, the agent system as a whole do not need to be

brought down as well. The task agent that is unavailable just needs to be temporarily marked as unavailable.

- *Agents do not have to be transported.* Compared with mobile agents, such as those using General Magic's Telescript Agents [7], the actual code of the agent does not have to be transported to a machine located close to the client every time a client wants to access its service. All that has to be done by the broker agent is to pass the *address*, not the code, of the task agent.

3.3 The Implementation

We have developed a client-oriented Web search agent, called *Ms. DaChanni*, that queries and gathers information from multiple search engines distributed over the Web. It makes use of the client-oriented distribution architecture to alleviate the deficiencies associated with conventional centralized server-side search agents. The principal design goal of our client-oriented Web search agent is to assist the user in locating information resources available on the Web while minimizing network usage and load on the server running the agent.

In our implementation, the agent running on Ms. DaChanni's home page[3] acts as the broker agent. This is the agent that provides the customized query interfaces to every connecting clients. A task agent that is located closer to the user is given preference over other agents in order to reduce the network traversal of the messages and minimize average response time from the agent. We could predict the network traffic along with server load during a particular time of the day from past access patterns. In order to reduce the number of search engines to contact simultaneously, the search engines are selected based on the keyword and the information domain selected by the user. Log skewing on remote sites are minimized by clustering the clients according to their locations. In addition, we provide a mechanism for passing the client's address to the search engines in order to completely abolish the log skewing problem. Successful query results obtained from the remote databases are cached on the local machine in order to avoid performing the same query to the same search engines when some of the queries could not be completed within the time-frame set by the user. Caching the search results has the added benefit of offering quicker response time since the dominant factor in the speed of Web search agents comes from the communication

---

[3]http://zec.kaist.ac.kr/dachanni/search.cgi

costs associated with accessing the remote search engines. Ms. DaChanni can, if need be, route queries via proxy servers in order to increase the reliability of the system during partial network failure and to also avoid possible network saturation on foreign links during busy hours.

Broker Agent

The broker agent is where all the clients initially connect to in order to begin a search transaction. Its main function is to provide a customized query form to the client with the address of the task agent to contact. Once the client receives the form, any further transactions, including searches, occur between the client and the task agent. Figure 2 shows a sample query form obtained from the broker agent.

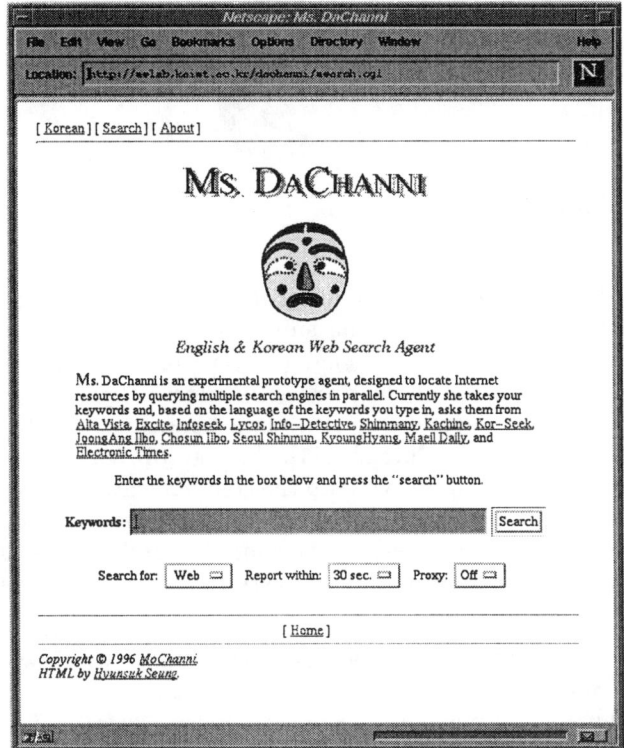

Figure 2: Query interface form of Ms. DaChanni

In our implementation, there are three task agents to select from. Ideally, the broker agent should select a task agent that is beneficial to both the client and the agent system as a whole. We have implemented the broker agent to select the task agent according to the client's location. That is, if there exists a task agent for a given subnet (the subnet the client belongs to), then that task agent is selected for the client. Otherwise, the task agent is selected randomly from a pool

of available task agents. This is a quick and simple way of balancing the overall load of the participating task agents. We are studying ways to ensure load balancing among remotely located task agents and handle situations such as partial network or server failure that may occur at any one of the task agents.

Each distributed task agent has its own unique URL address that identifies itself. It consists of the name "`search.cgi`" and the name of the machine it resides on. A task agent that resides on a host named "`selab.kaist.ac.kr`" under the directory "`dachanni`" is represented by the following URL address:

`http://selab.kaist.ac.kr/dachanni/search.cgi`

Once the task agent is selected, its URL address is passed on the query form, such that when the user submits the query, it would be sent to the task agent instead of the broker agent. A simple method to achieve this action is by using an HTML "form" tag with "action" field set to the address of the task agent. A better approach is to use the "base" tag, however. The base tag allows every anchors or hypertext links referenced within the HTML page to be based from the URL address specified in the base tag. Even images that appear on the query page are retrieved from the task agent. Indeed, this is the approach used by our broker agent to force the clients to download Ms. DaChanni's logo from the assigned task agent.

Task Agent

The task agent is the agent which queries multiple search engines on behalf of the user. Currently, it could handle a total of 14 search engines: 4 for handling English keywords, 4 for Korean keywords, and 6 for Korean newspaper articles. Internally, the task agent consists of three subagents. They are *dispatch agent*, *communication agent*, and *display agent*. Figure 3 shows the internal structure of the task agent.

Dispatch Agent

The dispatch agent is responsible for parsing the user's query and translating it to the appropriate syntax used by each search engine. First, it looks to see if the user is searching for Web documents or newspaper articles. If the user is searching for Web documents, then the dispatch agent inspects the keywords and see if they are written in Korean or English. We detect this by looking at any Base64 [1] encoded Korean characters in the QUERY STRING environment variable submitted by the client's Web browser. If the keyword is written in Korean, then the dispatch agent selects

303

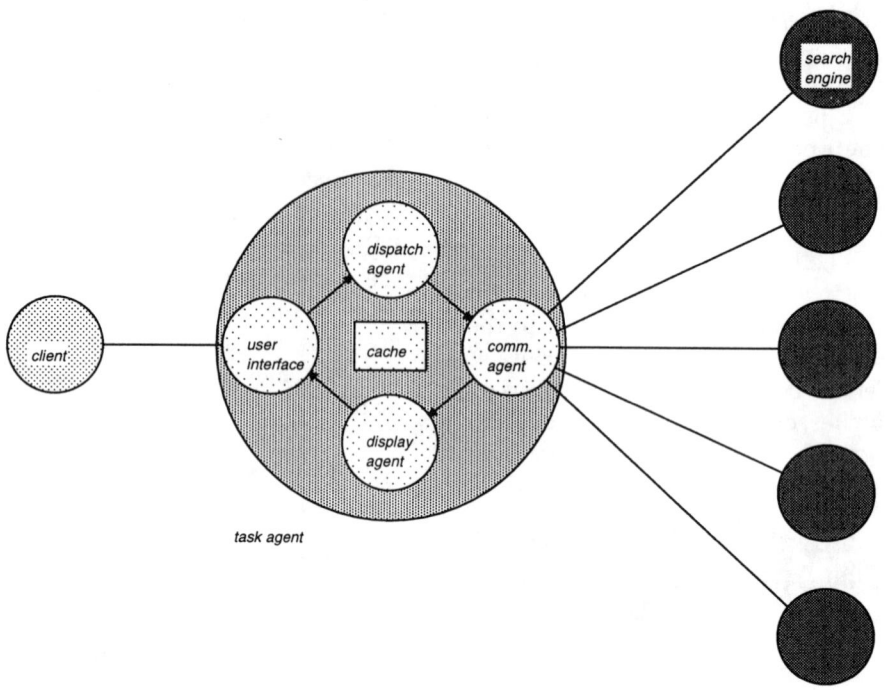

Figure 3: Internal structure of the task agent.

search engines that support Korean searches. If not, then the dispatch agent selects the search engines that support English searches. If, on the other hand, the user is interested in newspaper articles, then the dispatch agent selects the six newspaper sites. This way, the number of search engines to contact simultaneously could be reduced.

Once the search engines are selected, the dispatch agent checks to see if any of the search results for the particular keyword and search engine are available from the local cache. If a particular result from a certain search engine is available, then that search engine is removed from the list of search engines to contact.

Finally, the dispatch agent translates the original query submitted by the user into a form suitable for each of the remaining search engines on the list. When the queries are generated, the list is passed on to the communication agent.

Communication Agent

The job of the communication agent is to submit the queries generated by the dispatch agent to the selected search engines and retrieve their results. Accessing remote search engines is where most of the time is spent on a typical search agent system. To speed up the process, our communication agent makes heavy use of multithreading in order to exploit the potential parallelism available in accessing multiple distributed databases. That is, the submissions of the queries to the selected search engines occur in a concurrent manner, as well as their respective retrieval processes. The communication agent is implemented using IEEE POSIX 1003.1c Threads [11]. Also, to prevent a slow search engine from tying up the communication agent, there is a user-configurable expiration time (from 30 seconds to 3 minutes). This comes in handy for cases where three of the four queries arrive within 30 seconds but one does not.

To avoid skewing the logs on the remote search engines, the communication agent passes the address of the client's machine during the query process. The information is passed in the HTTP's "Client" field. Figure 4 shows a sample HTTP GET session of our communication agent. Using this method, the server could extract the address of the client who is responsible for the query from the environmental variable, HTTP CLIENT.

Another feature our communication agent has is the ability to query via a proxy server. The network connection within Korea is less than perfect. There are times when it is much faster to access the search engines through a proxy server that is not located in the default routing path. Also, we have experienced partial network failures where the queries could still

304

```
GET /cgi-bin/search?query=multimedia HTTP/1.0
User-agent: DaChanni/2.15
From: hseung@selab.kaist.ac.kr
Client: cheese.waist.ac.kr

...
```

Figure 4: Sample HTTP GET session with the "Client:" field set to the client's address.

be made through a different proxy server.

Display Agent

The display agent processes the query results retrieved by the communication agent in order to convert them into a uniform format. Many search engines display their results in their own unique way. Fortunately, however, there exists some regularities among the output generated by different search engines [3]. Our display agent takes this advantage of the regularities present among various result formats produced by the search engines to transform them into a single coherent structure. Currently, it extracts the title and the URL and presents them to the user. This way, the users would not have to deal with different layouts and formats used by each search engine. Also, successfully retrieved results are stored in the local cache so resubmission of queries can be avoided and various sorting options can be conducted locally without making additional network retrievals.

To increase the general reliability of the system and to guide the users in obtaining a complete report, the display agent reports informative error messages and suggests possible actions to take, such as increasing the search time (i. e. timeout value) or forcing the communication agent to query via a proxy server. Future versions would suggest possible actions to take by redisplaying the query interface that reflects the search parameters that would help complete the search process successfully.

4 Evaluation

In this section, we present various statistics and measurements obtained from Ms. DaChanni. It is our goal to show that client-oriented distribution architecture as applied to the Web search agent indeed help to solve the problems that are present among conventional search agents.

4.1 Running Environment

The three task agents participating in our system and their respective running environment are shown in Table 1.

4.2 Client-Oriented Distribution

In this section, we first show the actual distribution of the queries in our system utilizing the client-oriented distribution architecture for a period of four weeks. Following that, the percentage of the clients that benefited from the client-oriented distribution during those period is presented. We also demonstrate the advantage of client-oriented distribution by showing how fast the query forms could be retrieved by clients who have a dedicated task agent for their subnet.

We have logged the queries submitted by the users on the Internet. Figure 5 shows the number of queries processed per week by the three participating task agents. It covers the period of four weeks, starting from April 26 and ending in May 23, 1996. The line at the top represents the total number of queries processed by our system for that week. The three lines at the bottom represent the number of queries processed by each task agent.

As you can see, the queries are evenly distributed among the three agents. The slight drop in Week 4 for $Agent_3$ occurred because the system running that task agent had to be brought down for maintenance purpose for two days. However, $Agent_1$, which is only available when there is no available task agent in the subnet of the connecting client, should have a lower number than the other two. The problem is attributed to two reasons: first, some of the popular meta-search sites within Korea have started including Ms. DaChanni's query form in their All-In-One search page. Users who query through these forms submit their query directly to the broker agent and hence cannot take advantage of the distribution mechanism offered by Ms. DaChanni. The other reason is that once such user query through these search pages, they are bound to $Agent_1$ for the duration of their search transactions. We could have removed such skewing by not bounding the client to the same task agent after each query, but that seemed like an overkill, for the majority of the users who access our system query via authentic forms obtained from the broker agent. Not only that, assigning each client to a single task agent has the added benefit of increasing the cache hits, as we will see at the end of this section.

Currently, clients who are located within the same network as $Agent_2$ and $Agent_3$ benefit from our client-oriented distribution architecture. The percentages of the queries that were submitted by those clients are 11.6% and 18.5% for $Agent_2$ and $Agent_3$ respectively. The reason the percentage for $Agent_3$ is higher than that of $Agent_2$ is attributed to the fact that $Agent_3$

Agent I. D.	Type	Subnet Responsible	O. S.	Hardware
Agent$_1$[4]	Broker / Task	None	Linux 2.0	Pentium 120 MHz, 32 MB RAM
Agent$_2$[5]	Task	.kaist.ac.kr	Solaris 2.4	SuperSPARC 60MHz, 64 MB RAM
Agent$_3$[6]	Task	.dacom.co.kr	Solaris 2.5	UltraSPARC 167 MHz, 260 MB RAM

Table 1: The list of task agents and their running environment.

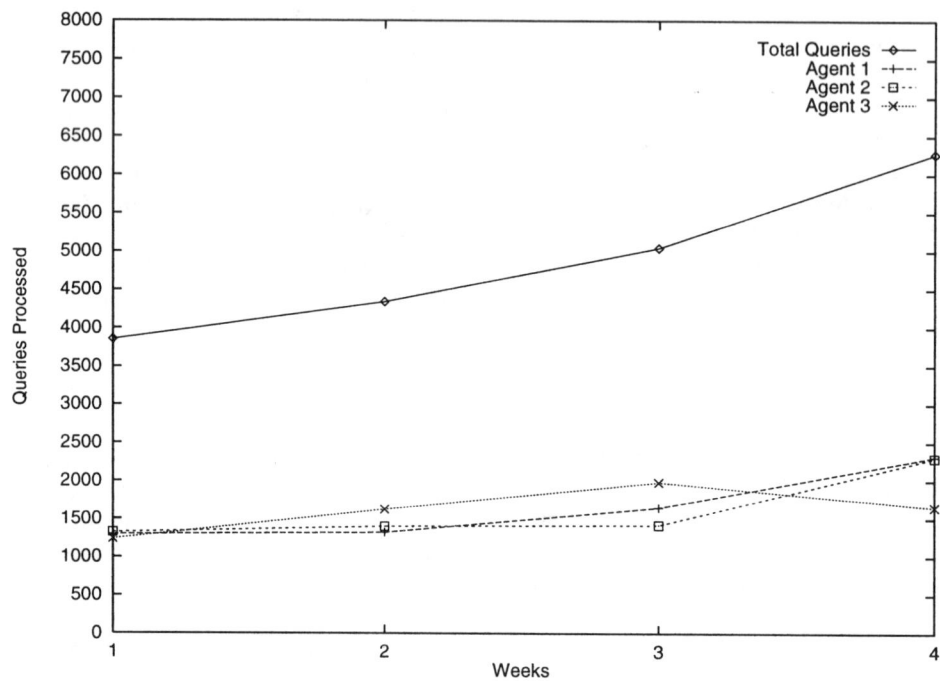

Figure 5: Number of queries processed per week (April 26 to May 23, 1996)

is operating on a major Internet service provider in Korea with many users accessing the Web, as compared to Agent$_2$ which is located on an academic network. Furthermore, to achieve higher number of client-oriented assignment of task agents, more task agents must be deployed and distributed on the Web.

To test how fast query forms can be transferred using the client-oriented distribution approach, we used the tcpdump packet tracer for measuring the retrieval times of the query form (see Figure 2) using the conventional approach and the client-oriented distribution approach. For the measurement, we configured Agent$_3$ to act as the broker agent. Agent$_3$ is reachable through a T1 connection from the client machine, which is located in the .kaist.ac.kr subnet. Hence, the task agent that was assigned to the client was Agent$_2$ located within the .kaist.ac.kr subnet. The query form

consists of a 2608 bytes text file, and two images files that are 2458 bytes and 2638 bytes respectively, bringing the total size of the query form to 7704 bytes. The client program that was used to initiate the request was a Netscape Navigator Version 3.0b4 running on a Pentium 120 MHz Linux system. The Netscape client was configured to the default values of 4 simultaneous network connections and 64 KBytes of network buffer. Disk caching was disabled, and memory cache was erased after each request. Table 2 shows the result of the measurement for ten trials.

In our environment, the speedup gained from utilizing the client-oriented distribution architecture was about 2.19 times the conventional approach. Note that actual speedup may vary depending on the network condition between the client and the broker agent, and that between the client and the task agent

306

Trial	Transfer Time (sec.)	
	Conventional	Client-Oriented
1	3.22	1.41
2	3.67	1.41
3	2.92	1.39
4	2.90	1.53
5	3.28	1.32
6	3.13	1.48
7	3.03	1.32
8	2.83	1.33
9	3.05	1.39
10	3.03	1.63
Average	3.11	1.42

$$\text{Speedup} = \frac{3.11}{1.42} \approx 2.19$$

Table 2: Query form transfer times.

that is assigned to it.

4.3 Communication Benchmarks

We present performance benchmark of the communication agent along with query response times from the search engines Ms. DaChanni accesses.

The execution speed of Ms. DaChanni fluctuates greatly according to the amount of time that is spent by the communication agent in accessing the remote search engines. On the other hand, the time spent by the dispatch and display agents are relatively small and constant compared with the time spent by the communication agent. The average processing time consumed by the dispatch and display agents totals 0.51 seconds[7] for four search engines, as in the case of English and Korean Web searches, and 0.73 seconds for six search engines, as in the case of newspaper searches. The average query response times of the six search engines and the communication agent is shown in Figure 6 according to the time of the day.

As can be seen from the graph, the query response times fluctuate greatly throughout the day. The horizontal line present on the graph is the default timeout value of 30 seconds. It will give you the idea on the number of partial retrievals made by the communication agent. A number of factors affects the query response times of various search engines. They could be a combination of the network load between the task agent and search engines, or the server load on the remote search engine or task agent.

4.4 The Need for Cache

Ordinarily, queries submitted to search engines have very low probability of being overlapped. Selberg and

[7]Measured on $Agent_1$.

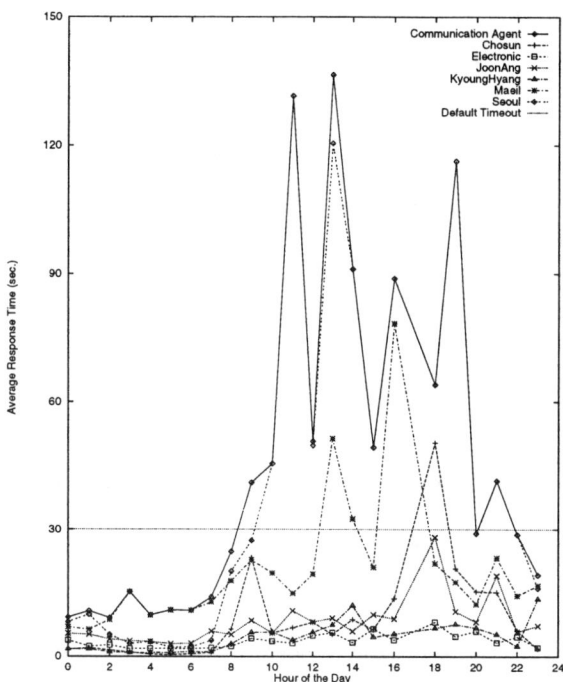

Figure 6: Average query response times of the six search engines supporting Korean news wires and the communication agent throughout the day.

Etzioni [13] report that the combined top ten queries amount to only 3.37% of the total queries and that 46.67% of the queries were unique. Query statistics published by numerous search engines confirm their finding. However, we have found that in our system, a high number of queries were overlapped. Figure 7 show the average number of queries processed per hour by $Agent_3$.

The dotted line represents the number of queries that were redundant *within that particular hour*. This is attributed to the fact that, because the network condition within Korea is not as fast as those in other well-connected countries, such as the United States, our communication agent failed to receive all the query responses within the time-frame set by the user (the default on Ms. DaChanni is 30 seconds). As shown in Figure 6, the default timeout value for our agent is insufficient for retrieving all the responses when a lot of people are using the network. Consequently, unsuccessful queries have forced the users to resubmit the same query either with a longer timeout value or through a different proxy server. Caching successful search results will avoid performing costly network retrievals during retries.

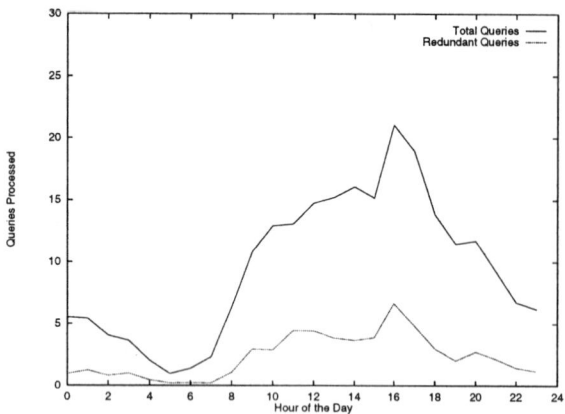

Figure 7: Average number of queries processed per hour by Agent$_3$ throughout the day.

5 Conclusions

We have presented the client-oriented distribution architecture for Web search agents. The architecture provides a platform for building scalable agent systems that reduce network consumption and server load on the server running the agent through the introduction of broker and task agents. Having a broker agent allows intelligent distribution of task agents to each connecting client.

Our Web search agent, Ms. DaChanni, reduces the transfer times of query forms and search results by assigning either a closely located task agent to the connecting client or from a pool of available task agents. To remove the log skewing problem, Ms. DaChanni passes the client's address to the remote search engines. The reliability of the system during partial network failures is enhanced by the ability to route queries via different proxy servers. Ms. DaChanni also reduces the network resources by selective querying. Selective querying is achieved by generating search plans based on the language used in the keywords, information domain selected by the user, and querying only those services that are not in the cache.

References

[1] N. Borenstein and N. Freed. Mime (multipurpose internet mail extensions) part one: Mechanisms for specifying and describing the format of internet message bodies. *RFC 1521*, September 1993.

[2] P. M. E. De Bra and R. D. J. Post. Information Retrieval in the World-Wide Web: Making Client-based Searching Feasible. In *Proceedings of the Second International World-Wide Web Conference: Mosaic and the Web*, pages 137–146, October 25-27 1994.

[3] Robert B. Doorenbos, Oren Etzioni, and Daniel S. Weld. A Scalable Comparison-Shopping Agent for the World Wide Web. Technical report, Department of Computer Science and Engineering, University of Washington, 1996.

[4] Daniel Dreilinger and Adele E. Howe. An Information Gathering Agent for Querying Web Search Engines. Technical report, Colorado State University, Computer Science Department, 1996.

[5] David Eichmann. Ethical Web Agents. In *Proceedings of the Second International World-Wide Web Conference: Mosaic and the Web*, pages 3–13, October 18-20 1994.

[6] Susan Gauch, Guijun Wang, and Mario Gomez. ProFusion*: Intelligent Fusion from Multiple, Distributed Search Engines. *Journal of Universal Computer Science*, 2(9), September 1996.

[7] General Magic. Telescript.
http://www.genmagic.com/Telescript/.

[8] Martijn Koster. Why Simultaneous Search Engines are Not So Great.
http://pubweb.nexor.co.uk/public/cusi/doc/simultaneous.html.

[9] Martijn Koster. Robots in the Web: Thread or Treat? *ConneXions*, 9, April 1995.

[10] Martijn Koster. Guidelines for Robot Writers. *Nexor Corporation*, 1995.
http://info.webcrawler.com/mak/projects/robots/guidelines.html.

[11] Christopher Angelo Provenzano. Pthreads: General Information. 1996.
http://www.mit.edu:8001/people/proven/pthreads.html.

[12] Quarterdeck Corporation. WebCompass.
http://arachnid.qdeck.com/qdeck/products/webcompass/.

[13] Erik Selberg and Oren Etzioni. Multi-Service Search and Comparison Using the MetaCrawler. In *Proceedings of the 1995 World Wide Web Conference*, 1995.

[14] Scott Spetka. The TkWWW Robot: Beyond Browsing. In *Proceedings of the Second International World-Wide Web Conference*, October 1994.
http://fang.cs.sunyit.edu/Robots/tkwwwpaper.html.

Session 4C

Applications:
Train Control and Management Systems

Chair

Maarten Boasson

Distributed Management for Software Maintenance
in a Wide-Area Railway System

Fumio Kitahara*, Humihiro Katano*, Tsutomu Ono*
Yoshiki Kakumoto**, Kuniyuki Kikuchi***, Manabu Shinomoto***

*East Japan Railway Company
1-6-5 Marunouchi, Chiyada, Tokyo 100, Japan, Tel: +81-3-3212-0079, Fax: +81-3-3212-6358
** Systems Development Laboratory, Hitachi, Ltd.
1099 Ohzenji, Asao, Kawasaki, Kanagawa 215, Japan, Tel: +81-44-966-9111, Fax:+81-44-966-6823
*** Omika Works, Hitachi, Ltd.
5-2-1 Omika, Hitachi, Ibaraki 319-12, Japan, Tel: +81-294-53-1111, Fax:+81-294-53-8404

Abstract

In a large-scale real-time control system, functions of subsystems differ slightly although they seem to be identical. Functions change when hardware is replaced or when the function itself is upgraded. If the standard system is generated by merging the differences and is installed then to each subsystem, subsystems become different with maintenance being carried out in each subsystem to fit its operational conditions. Furthermore, subsystems are distributed in a wide area. Software maintenance therefore becomes very difficult. This paper describes a distributed management for software maintenance in the Tokyo metropolitan-area railway system. This management has two phases: generating programs for making maintenance easy and executing programs for increasing the reliability of the system. There it contributes to the effectiveness of maintenance and increases the reliability of the total system.

keywords: The Tokyo metropolitan-area railway system, distributed management, software maintenance

1 Introduction

In a large-scale distributed system, distributed software development in which several groups develop software separately is inevitable. Distributed software management carried out per subsystem is very important because the subsystems differ and each subsystem also has been changing with maintenance being continuously done. Furthermore, the subsystems are distributed in a wide area. All of these things make software maintenance very difficult.

An example of a large-scale distributed system is the Tokyo metropolitan-area railway system, which covers 17 train lines and 250 stations on these lines. Its scale is far beyond the conventional train-traffic control system. This large-scale real-time control system was not fully constructed at once but is being developed step-by-step over 10 years.[1] In this system, the operational conditions of subsystems like railway structures are different, so their functions are a little different. Each function is also continuously changing because hardware including computers and equipment is being replaced or because the function itself is being gradually upgraded. In software development, since we can not generate common

programs for all parts of every subsystem, we generate standard programs and install them in each subsystem. Maintenance must be carried out in each subsystem to fit its operational conditions, so the software of each subsystem becomes different. Software management per subsystem, or distributed management, is necessary for maintenance.

Subsystems execute train-traffic control by communicating with each other; their operation is affected by the data they exchange and by the rate of transmission. The function of the subsystem can only be truly checked during on-line operation.

Distributed development is common in large-scale system development. Several developers have joined in the development of the Tokyo metropolitan area railway system. Some distributed development tools have been already developed.[2][3] With respect to checking functions in on-line operation, test tools[6][7] based on the autonomous decentralized concept[4][5] have been developed and applied to railway traffic-control systems[1], steel production systems[8], and factory automation systems[9]. However, the distributed management for maintenance has not been discussed yet.

This paper describes a distributed management for software maintenance in the Tokyo metropolitan-area railway system. The distributed management has two goals: the management in generating programs for making maintenance easy and the management in executing programs for increasing the reliability of the system. This technique therefore consists of two phases--generation process and execution process. In the generation process, the management technique for programs and subsystems are described. In the execution process, the technique to identify the software modules in the execution environment and the recovery technique to bring back the previous subsystem version are described. The proposed management contributes to the effectiveness of maintenance and increases the reliability of the total system

2 Background

In train-traffic control systems, although functions of subsystems seem to be identical, they are slightly different according to their operational conditions like railroad structures or hardware performance. As construction

311

progresses--connecting station-subsystems, constructing the train-line system, connecting the train-line systems--the total system becomes composed of different subsystems with maintenance being continuously carried out. This section describes the reason why functions of subsystems are different and the maintenance process in this development.

2.1 Features of Stations

Functions of subsystems differ and each function also gradually changes. The reason is explained here by taking a train-traffic control function as an example.

(a) Basic Function

Figure 1 shows the basic function of train-traffic control. When the train-number detector identifies the train departing the previous station, the train-number is passed on to the route control computer. The electrical-route-switching devices collect data on train position, signals and train-route-interlocking devices per 600 msec and transfers them as tracking-data to the route control computer .

When the route control computer receives the train number, it starts its execution by reading the tracking-data. When the train arrives at the control-position indicated in the tracking data, the route control computer sends the control signal for routing to the electrical-route-switching devices. The electrical-route-switching devices receive it and executes fail-safe logic checking and changes the state of signals and the train-route-interlocking devices.

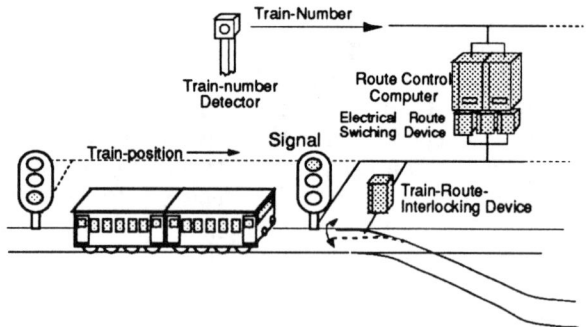

Figure 1: Basic Function of Train-Traffic Control

(b) Individual Function

Figure 2 shows railroad structures in Hachioji and Shinjuku Stations. In each station, two train-lines meet and their railroads intersect complicatedly in up and down lines. The route control computers has to route trains on these complicated railroads. The minimum rush-hour interval between trains is only two minutes. Therefore, besides the tracking and the routing functions, it is necessary for the train-traffic control to define the routing order of trains. The railroad structures of two stations being plainly different as shown in the Figure, their functions of defining the routing order of trains--for example, the routing order of superior and inferior trains, the routing order of some trains coming together from different train-lines, and the starting order of the first trains--become different .

The train-traffic control function is also changed by the gradient of railroads or the position and the performance of the train-route-interlocking device. Namely, even if two stations have the same railroad structures, the train-traffic control functions may be different.

Thus, the train-traffic control functions come to differ slightly although their basic functions are identical.

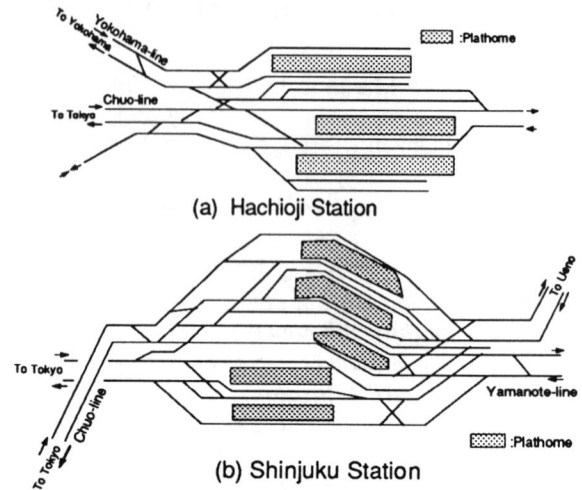

(a) Hachioji Station

(b) Shinjuku Station

Figure 2: Railroad Structures

2.2 Software Maintenance

This section describes step-by-step construction process, maintenance process, and examples of program modification.

(a) Construction Process

As shown in Figure 3, construction process is not to construct all stations at once but to repeat constructing a new station and connecting it to others. Namely, the first step of construction is constructing a station-subsystem and connecting it to others (Figure 3 (a)). The second is constructing the train-line system by repeating the first step (Figure 3 (b)). The third step is connecting train-lines (Figure 3 (c)) because there are some specially-arranged routes passing through parts of several different train-lines, creating a relationship between train-lines.

As construction process is progressing, a station-subsystem upgrades its execution-mode: The first is a single operational mode where a station-subsystem executes the train-traffic control by itself. The second is an adjacent operational mode where a station-subsystem, connected to adjacent subsystems, executes the train-traffic control by communicating with each other. The third is a train-line operational mode where a station-subsystem, connected to management systems in the train-line center, executes the train-traffic control according to the schedule or commands from the management subsystems.

The completed system will cover 17 train lines and 250 stations on these lines with a total line length of approximately 1,200 km. It will include 250 real-time control computers, 250 real-time device controllers, and 3800 workstations and personal computers.[1] Programs of the route control computer amount to 600 Mbytes. The fully automated train-traffic computer-control system is planned with a development period of over 10 years.

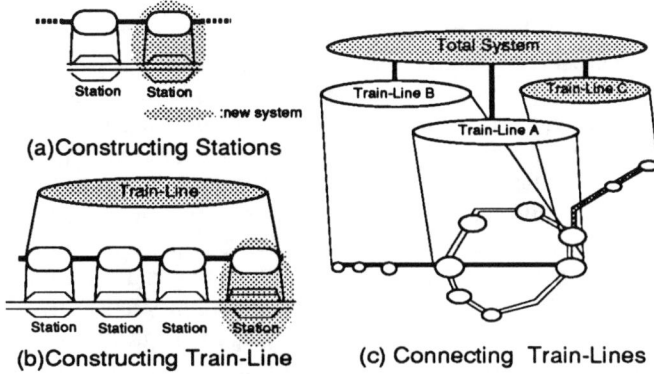

(a)Constructing Stations

(b)Constructing Train-Line   (c) Connecting Train-Lines

Figure 3: Construction Process

(b) Maintenance Process

As mentioned in Section 2.1, although the basic function of train-traffic control is identical, the functions of station-subsystems vary slightly. These differences should be incorporated into one system, and common programs (if these programs are only installed, every subsystem can start its execution) should be generated for software maintenance. However, there are some problems:

(1) Checking the characteristics of over two hundred stations and generating the common programs between them is time-consuming work.

(2) The function of each station itself is so gradually changing that generating the common programs will become endless.

(3) In a real-time control system, functions can not be fixed until the real operation is executed. Hence, the function of a station-subsystem can not be completely understood in advance.

Therefore, the guideline for software development was defined as follows:

(1) Basic function such as the tracking function or the routing function is generated as standard programs which cover the standard types of the railroad structures

(2) Information about the equipment such as the railroad structure or the position and the performance of the devices are generated as individual data in each station.

(3) Individual function such as the function of defining the routing order is generated as individual programs in each station.

(4) Common function such as schedule management or communication, which does not depend on the railroad structure, is generated as common programs

(5)The software structure and program names are unique in all station-subsystems for making maintenance easy.

These programs are installed and maintenance are carried out in each subsystem according to the following two procedures (Figure 4). One is installing the programs to subsystems sequentially. The other is updating the installed programs in each subsystem to fit its own operational conditions. In software maintenance, since revising programs in the actual operation-computer are not permitted owing to

the restriction of time for maintenance, revising and testing programs are done in a simulation system in advance and then the undated programs are installed to the target computer. Testing in the real operation is executed by on-line test technique.[1]

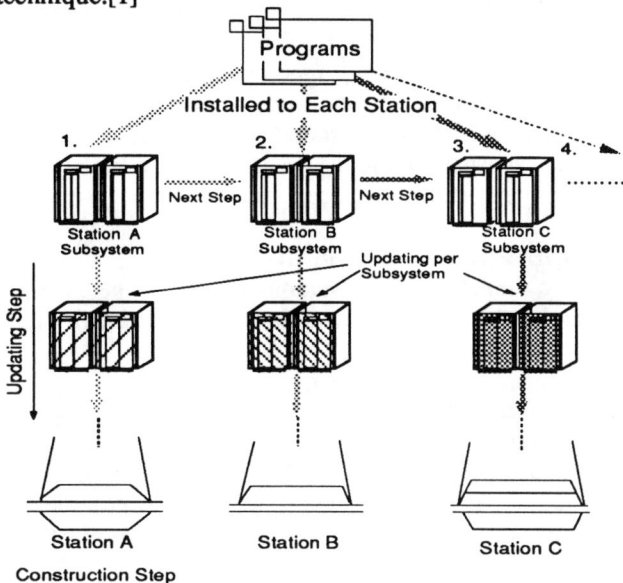

Figure 4: Maintenance Process

Examples of program modification in the maintenance process are described as follows.

(1) Regulation

The functions of defining the routing order of trains differ among stations, so these functions are generated as individual programs in each station. Standard programs have to be checked as if they match the individual programs in each station and may have to be modified if necessary. Of course, a peculiar function for its own station may be added to the programs.

(2) Test in the Real Operation

In the real operation, a station-subsystem executes the train-traffic control by receiving tracking data from the equipment. So the interface programs between them may be adjusted. Also,when the schedule is disordered in the real operation, unexpected operational condition or unexpected changes of schedule may occur. In this case, the additional logic for defining the routing order may have to be necessary or the routing function may have to be modified.

(3)Construction Step

The train-traffic control function has been changing as the construction has been progressing, such as constructing a station-subsystem and connecting station-subsystems. For example, when a station-subsystem is connected to others, a sophisticated, high-speed processor of communication traffic or adjustment of the execution priority of programs is necessary. The train-traffic control function has been also changing as hardware, including computers and equipment, is gradually replaced.

313

(4) Change of Schedule

When the data base for train-operation is replaced, programs utilizing them are sometimes changed. Not only schedules printed in timetable but also schedules for railway-maintenance trains and special arranged trains are written in a train diagram. The basic train diagram is revised per half a year. Train-service data and functions of subsystems may be replaced or added, and once in a while a new station is opened. The function of defining the routing order may also be changed when a new pattern of routing occurs. The train-traffic control function has been changing periodically.

3 Requirements for Maintenance

Requirements for the station-subsystem, the train-line, and the total system in software maintenance are discussed here.

3.1 Station-Subsystem
(a)Generating Programs

As described in section 2.2, the standard programs are installed in each subsystem and the installed programs are updated in each subsystem. There are three kinds of program for updating itself in one subsystem.
(i) Common between all subsystems, for example, common programs for communication with other subsystems, or for schedule-management
(ii) Special for its own station-subsystem, for example, individual programs for defining the routing order
(iii) Not clear whether related to other subsystems or not, for example, standard programs for tracking or routing(Figure.5)

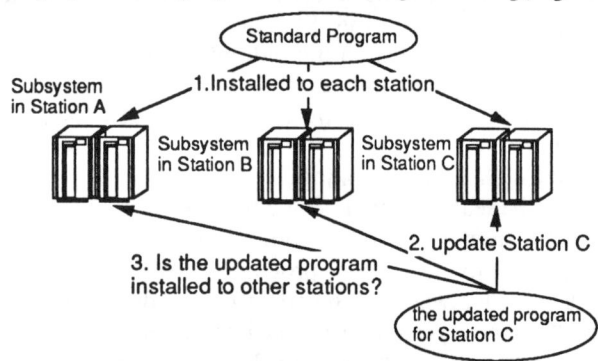

Figure 5: Updating Program

In case (i), it must be installed to other subsystems. This program, an upper version of the previous one, remains the common program. In case (ii), it is managed as the individual program. In case (iii), it is necessary to check whether it can be installed to other subsystems or not. If the revision of the program is due to logical missing--no relation with the operational condition--it must be installed to other subsystems. But if the revision is related to the operational conditions, namely if the revision is done to fit the individual programs, installation in other subsystems may cause them to go out of order. If this program is freely installed to others, irregular operation may occur one after another.

In order to maintain such a program as standard, it is necessary to examine the same programs in other operated subsystems and to keep this program available for others in revising, then to test it in other subsystems. It is time-consuming work and this work may have to be done whenever a new station is constructed. Furthermore, it will have not been certain that whether this program remain standard or not until the total construction will be completed. Therefore, this program tentatively have to be managed as individual and must be kept separated from standard programs to avoid installation in other subsystems. Trying to merge the tentative individual programs in the standard program is done at a construction period such as the completion of the train-line.

As a result, the standard programs, the individual programs, and the common programs coexist in subsystems and their configurations are different among subsystems. As software maintenance is carried out, the configuration itself also changes. In this maintenance environment, in order to do software maintenance effectively and correctly, management for programs(common, standard individual) and subsystems is necessary.
(b) Executing Programs

Requirements in executing programs are as follows.:
(1) Some kinds of programs including subroutines are statically registered in memory in a real-time control system. The subsystem can not be executed as a "system" until these kinds of programs are fully registered in memory, or the execution environment. In software maintenance, these programs have to be checked per registration unit (such as program, subroutine) to see if they are correctly registered in the execution environment.
(2) Even if the subsystem should be put out of order by updating it, recovery to the previous subsystem version must be able to be done. To do this, checking programs registered in the execution environment is necessary.
(3) The operation of the subsystem may be altered by the data exchanged between subsystems or by the rate of transmission. Of course, functions are mostly tested in the simulation system. But since an unexpected operation may happen by exchanging data, checking the function in on-line operation is necessary.

3.2 Train-line and Total System

Subsystems have to work by themselves and cooperate with each other. The requirements for construction / maintenance of the train-line and the total system are briefly mentioned.[1]
(1) The total system or some parts of it must not be stopped while a subsystem is being updated or constructed.
(2) Subsystems do not have to be simultaneously updated but must be able to be updated independently.
(3) Even if a subsystem should be out of order, others must not be affected by the disabled subsystem.
(4) Because subsystems are distributed in a wide area, software maintenance must be able to be done by remote operation.

4. System Architecture

Autonomous Decentralized System (ADS) architecture[4][5] is utilized in order to satisfy some requirements described in the section 3. This section describes the ADS system architecture.

ADS architecture based on the ADS concept in which the subsystems are linked only to the Data Field (DF), where subsystems communicate with each other (Figure.6). Each subsystem has its own management software module called ACP (Autonomous Control Processor) to manage itself and cooperate with others. A subsystem including its application software modules and its ACP is an autonomous unit called "Atom."

System structure may vary depending on construction in the system. To protect the operation of the subsystems from variation in the total system, messages broadcast by the subsystem into the DF has a content code instead of the receiver's address. The content code specifies the data depending on the applications. The ACP decides whether or not to accept a message on the basis of the content code. The ACP executes an application module when all of the necessary data for it are received.

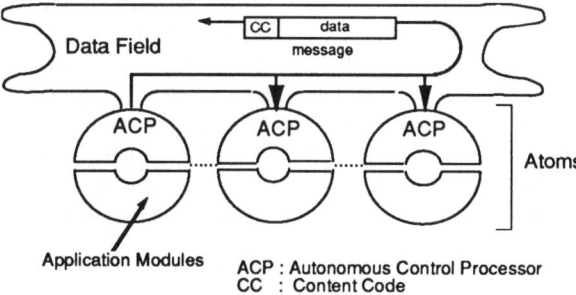

Figure 6: System Architecture

Owing to the content code communication in which each subsystem does not have to understand its relationship to others and executes its applications under its own judgment, subsystems have autonomy in sending and receiving data.

Thus the following is realized.[1][6][7]

(1) Because subsystems are "loosely connected," subsystems can be independently updated or constructed without stopping others.

(2) By checking messages broadcast by subsystems, the function of the subsystem can be checked. This technique can check the updated programs in on-line operation.

(3) Even if application modules generate unusual data, the ACP, checking the content code and its data, does not send or receive these data. Hence, it can prevent malfunction.

5. Distributed Management

ADS architecture enables subsystems to be independently updated or constructed. This section discusses the maintenance technique for the subsystem itself. The first part of technique is distributed management in the software generation process and the second is distributed management in the software execution process.

5.1 Generation Process

This section describes the distributed management technique for programs and subsystems.

(a) Program

As mentioned in section 3.1, when the standard programs are updated in one subsystem, some of them can not be transferred to other subsystems. These programs have to be managed as "individual versions" for this subsystem. Therefore, two kinds of versions--standard and individual--come to coexist as shown in Figure 7. The management must distinguish between these two versions.

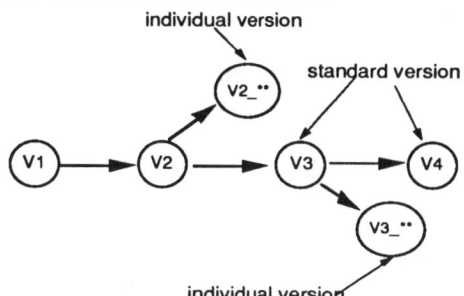

Figure 7: Change of Program Version

Table 1 shows which version is assigned to the standard or the individual. The standard version is only version/revision number. The individual version consists of version/revision number and the subsystem code number. The individual version and its history of changes are managed independently of the standard version. Even if some subsystems have identical versions, each one is named differently in each subsystem because these subsystems will not necessarily continue to have identical version.

Table 1: Program version

program	version(example)
standard	V1.5
individual	V1.5_XXX XXX: subsystem No.

(b) Subsystem

When the individual version is generated, the software configurations of subsystems become different. Maintenance being carried out, the software configuration of the subsystem also changes.

Figure 8 shows changes in the software configurations in two subsystems. In this figure, each Subsystem I, II is composed of Program A, B, C. In first step (95.10.1), standard programs were installed in each subsystem. In next step (96.2.10), Program B was revised. This program was updated as a standard version (V1.2) and installed in two subsystems again. In the third step (96.4.15), Program C was updated as a standard version (V1.2) in two subsystems. On the other hand, when Program A was updated, this program was divided into two versions. In Subsystem I, the individual version (V1.2_I) was generated and installed. In subsystem II, the standard version (V1.2) was installed. Their software configurations were different then; thus the management in each subsystem becomes necessary.

Other requirements for constructing subsystems are described as follows.

(i) Programs are not necessarily independent of each other, for example, the relation between a main program and subroutines. The newly updated programs depending on each other must be installed at once. If not, the consistency of the subsystem can not be maintained.[10] In this case, interface between programs being illegal, the subsystem can not execute its operation regularly even if the programs are correctly executed in the simulation system.

(ii)The subsystem could not work as a system until all necessary programs are complete. Before standard or individual versions are installed, the management must check whether the subsystem will have all necessary programs.

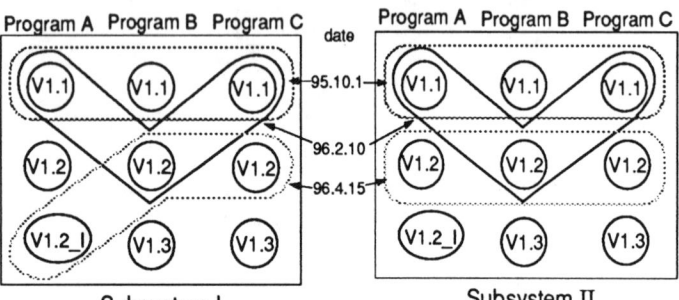

Figure 8: Change of Software Configuration

The conventional tools[11] which install independent and complete program-products like Operating System(OS) can not update these subsystems because they do not construct a "system" but only install common programs uniquely.

In order to satisfy the above requirements, the software configuration is previously determined in each subsystem and programs are absolutely installed according to this definition. The system definition file describing the software configuration of the subsystem is shown in Figure 9.

As shown in Figure 9(a), the subsystem version is set under subsystem No. and the system definition file is made per subsystem version to check changes between the old and the new easily. As shown in Figure 9(b), program group name, program name, version number, and kind of program are written per program in each definition file. Programs are installed according to this definition file.

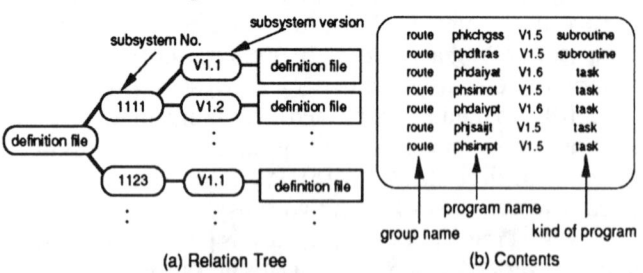

Figure 9: System Definition File

Thanks to this management technique which determines the complete subsystem in advance, the software configuration

can be managed in each subsystem. The consistency of the subsystem can be also guaranteed and the lack of programs can be also prevented if the registration of programs does not fail.

5.2 Execution Process

The defined programs in the definition file are not necessarily registered in the execution environment of the subsystem because registration may fail. So the registered programs must be checked for increasing the reliability of the system. In this section, the loading procedure is described, and the technique to check programs registered in the execution environment and the recovery to the previous subsystem are discussed.

5.2.1 Loading

Program resources are saved in the program management computer. Programs are installed to each subsystem from this computer (Figure 10).

1. Downloading

According to the system definition file, programs are downloaded from the program management computer to the subsystem using a FD (Floppy Disk) or through a network, and are temporarily saved in the disk area. During downloading using a FD, new version programs in the definition file are selected by the operator. During network downloading, as the program management computer compares the contents of the selected definition file with the registration data of the present subsystem version recorded in the target subsystem, only new version programs are downloaded.

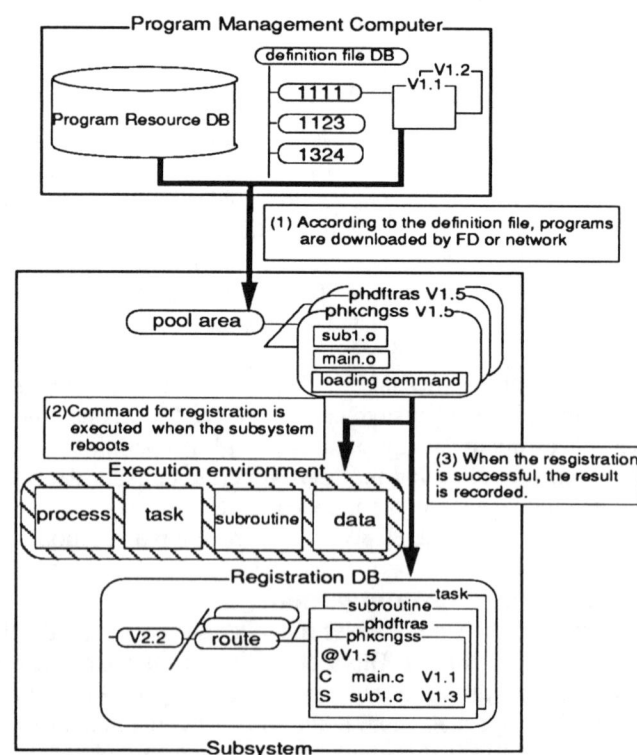

Figure 10: Loading Process

2.Registration

Downloaded programs are registered when the subsystem reboots. The registration can thus be completed before applications starts their executions.[10] The special command for registration, defined for each program, registers its program. When the subsystem reboots, the commands are subsequently executed. The command registers its program to the execution environment and records the registration result to the registration DB if successful.

5.2.2 Checking

Programs registered in the execution environment are recorded in Registration DB by the commands. The execution environment can be checked by referring to this DB. Recovery to the previous subsystem version can be done by referring to this DB.

(a) Registration DB

The command for registration does not only register programs to the execution environment but also records the registration result to the registration DB. It does not record the registration result to the data base until it recognizes whether the program registration is successful. Hence, since it is guaranteed that the execution environment and the registration DB are identical, the execution environment can be checked by referring to the registration DB.

Figure 11 shows the structure of the registration DB in a subsystem. The registration results of the present subsystem version are registered for each program group such as rout control, tracking, and schedule management. Each group has four files according to the kinds of programs (execution program, subroutine, file, data). Program name, program version, etc. are written for each program in each file.

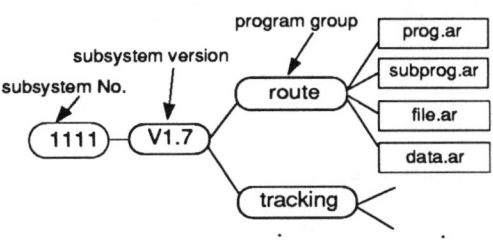

Figure 11: Registration Data Base

(b) Recovery

If the subsystem is put out of order after updating to the present subsystem version, it is necessary to restore the last version to the subsystem. Recovery of the last version can be done by comparing the present subsystem version with the last version in the registration DB, and installing the differences in the last version to the subsystem. In this way, the consistency of the subsystem can be maintained.

6 Execution

This section introduces the execution of software maintenance based on the proposed management technique. Figure 12 shows the simplified overall system of the Tokyo metropolitan area railway system. This system comprises station-computer subsystems in each train-line, the subsystems monitoring the train-line, and management subsystems governing the overall system.

As mentioned before, this system includes real-time control computers, real-time device controllers and WS/PC. ADS architecture is realized in this open computer environment. The network connecting subsystems is dual. Information data like schedule and control data like tracking data are separately transferred in each network.

In the station-computer subsystems, the route control computer and the passenger information control computer, etc. are connected by the Ethernet. The Ethernet is physically divided into two network-lines for the information data and the control data. The route-control computer is replicated.

In the monitoring subsystems, the train-traffic rescheduling management computer and the train-traffic management computer, etc. are connected by the Ethernet. The monitoring subsystems distribute train-traffic schedules to station-computer subsystems, monitor the positions of trains, communicate with crew, etc. The train-traffic rescheduling management computer is replicated. The system for one train line is composed of the station-computer subsystems and the train-line monitoring subsystems, connected by a train-service-line network (Hitachi, TN-100:100Mbps). This network is logically divided into two kinds of networks. There will be 17 train-service-line networks.

The management subsystems consists of the general management computer, the track maintenance computer, the program management computer, and so on. The management subsystems and the monitoring subsystems, which cooperate

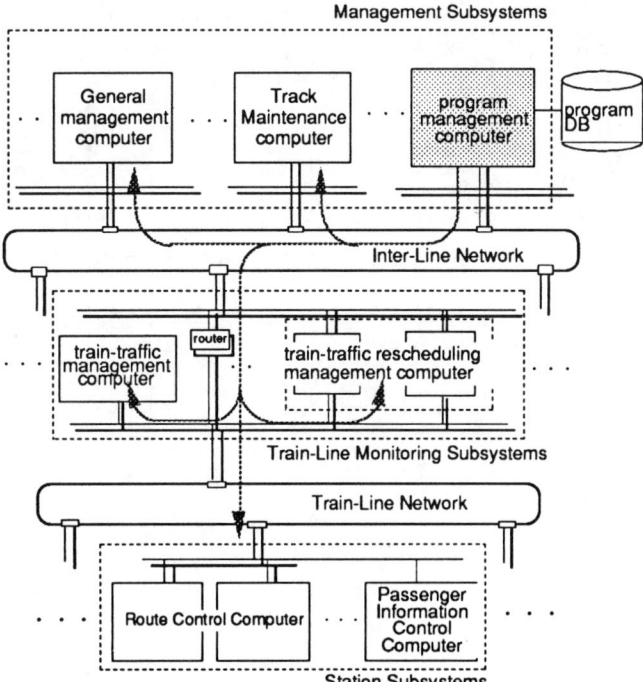

Figure 12: The Simplified Total System

317

to produce train-traffic schedules and monitor the total system, are connected by an inter-line network (TN-100).

The program management computer in the management subsystems downloads programs to the target computers according to their definition file. When the management subsystems have not been constructed yet, programs are installed using a FD. Software maintenance can be independently carried out in each subsystem, thanks to ADS architecture.

Figure 13 shows the program management computer (Hitachi, V90 series). Figure 14 shows its operational display. In the Figure 1, network loading and subsystem version(51.19) are selected (the target computer has already been selected). Comparing the definition file of V51.19 with the registration DB in the target computer, only the differences are to be downloaded. The group name, the program name, the present version (written in the registration DB), and the new version (written in the definition file) of the differences are indicated in this picture. When the operator inputs "OK", these new version programs are downloaded to the target computer.

Figure 13: Program Management Computer

```
線区名:中央本線/駅名:東京/駅バージョン番号:U51.19/装置番号:03010101
をローディングしますか? [y/n]: y

ローディング方法は?
(リモートロード:remote  FD経由ロード: fd): remote
実機 (装置バージョン:U51.20) に登録されている
タスク/IRSUB/プロセス/ライブラリと比較しますか? [y/n]: y
比較中...
..........................................................

相違バージョンリスト
サブシステム名称        名称        変更バージョン   実機バージョン
ダイヤ管理             kihon          U1.70          U1.71        *
ダイヤ管理             libcom.a       U1.70          U1.71        *
ダイヤ管理             pd6b400t       U1.71          U1.73        *
ダイヤ管理             pdmakedt       U1.70          U1.71        *
ダイヤ管理             pds540ht       U1.70          U1.71        *
進路制御              ptrtekot       U1.5           U1.74        *
(EOF):
```

Figure 14: Operational Display

7. Conclusion

A distributed management technique for software maintenance in a large-scaled distributed system has been proposed. The proposed technique consists of two phases-- generating programs to make maintenance easy and executing programs to increase the reliability of the system. In the generation process, the individual programs and the standard ones are managed separately, and the subsystem is constructed according to the system definition file. In the execution process, software modules in the execution environment are recorded in the registration DB and the previous subsystem version is recovered by referring to this DB. The proposed management technique has been applied to the Tokyo metropolitan area railway system and it contributes to the effectiveness of maintenance and increases the reliability of the total system.

The construction of the Central-line (the first train-line) was finished. The Yamanote-line (the second train-line) is under construction and several stations in it have been operated. The second, the third, and next train-line being constructed, requirements for connecting train-lines, in addition, connecting other transportation systems will occur. We will make an effort to settle these challengeable requirements.

References

[1] F.Kitahara, et al. "Widely-Distributed Train-Traffic Computer Control System and Its Step-by-Step Construction", Proc. of ISADS95, Phoenix, U.S.A., 1995, pp.93-102

[2] J.Kramer, et al. "The System Architect's Assistant for Design and Construction of Distributed Systems", Proc. of the 4th. Workshop on Future Trends of Distributed Computing Systems, Lisbon, Portugal, 1993, pp.284-290

[3] M.Sloman, et al., "An Architecture for Managing Distributed Systems", Proc. of the 4th. Workshop on Future Trends of Distributed Computing Systems, Lisbon, Portugal, 1993, pp.40-46

[4] K.Mori, et al. "Autonomous Decentralized Software Structure and its Application", Proc. of FJCC '86, Dallas, U.S.A., 1986, pp.1056-1063

[5] K.Mori, "Autonomous Decentralized Systems: Concept, Data Field Architecture and Future Trends", Proc. of ISADS93, Kawasaki, Japan, 1993, pp.28-34

[6] K.Kawano, et al. "Autonomous Decentralized System Test Technique", Proc. of COMPSAC89, Florida, U.S.A., 1889, pp.52-57

[7] H.Yamamoto, et al. "On-line Software Test Techniques Based on Autonomous Decentralized System Concept", Proc. of the 4th. Workshop on Future Trends of Distributed Computing Systems, Lisbon, Portugal, 1993, pp.291-296

[8] T.Kondo, et al., "Application of Autonomous Decentralized System to the Steel Production Computer Control", Proc. of 3rd. Workshop on Future Trends of Distributed Computing Systems", Taipei, Taiwan, 1992, pp.419-423

[9] M.Omura, et al. "Hi-Cell System Architecture for Manufacturing Systems", Proc. of ISADS95, Phoenix, U.S.A., 1995, pp.154-161

[10] H.Kobayashi, et al. "On-line Software Expansion in Large-Scale Widely Distributed Systems", Proc. of ISADS93, Kawasaki, Japan, 1993 pp.300-304

[11] OSF: "The Whole of Distributed Computing Environment DCE", Computer Today, July, 1991, pp.17-55

New trends of train control and management systems with real-time and non real-time properties

Shigeo Shoji*, Akio Igarashi**

* East Japan Railway Company,
1-6-5 Marunouti, Chiyoda-ku, Tokyo 100, Japan   Tel: +81-3-3212-2108, Fax: +81-3-3212-2109

** Japan Telecom Co.,
4-7-1 Hatchobori, Chuo-ku, Tokyo 104, Japan,   Tel: +81-3-5540-8050, Fax: +81-3-5543-1951

Abstract

The conventional train traffic control system has been focused on safety and punctuality in normal operations. Recently, however, more flexible and adaptive management has been required not only for train regulation, but also for maintenance, and in train-delay situations. An autonomous decentralized train control and management system is proposed to attain both the real-time property for train control such as train traffic and non real-time property for train management such as scheduling. This subsystem has been developed to make these two heterogeneous subsystems to coexist by isolating or coordinating mutually. This system has been applied to the train traffic control and management system for Japanese bullet train line (COSMOS : Computerized Safety Maintenance and Operation system of Shinkansen). The proposed system has been in operation since Nov. 1995, and the effectiveness has been evaluated.

1. Introduction

Train services on the Touhoku and Joetsu Shinkansen (Japanese bullet train) were started in June, 1982 by Japanese National Railways, between Ohmiya and Morioka, Niigata. Since the operations of these services were transferred to East Japan Railway Company (JR East) in 1987, the line has undergone steady growth as a high-speed transportation system offering safety as well as comfort. This bullet train system has an operating distance of 840 km (including the distance between Tokyo and Morioka as well as that between Ohmiya and Niigata), 27 stations, 6 rolling stock bases, and approximately 900 cars. Each day 230 trains made up of 5 to 16 cars carry a total of 220,000 passengers, at a maximum speed of 275 km/h, to their destinations with almost no delays.

JR East's revenue from these services is 4220 billion

Yen, which amounts to almost one fourth of its total earnings from the transportation business. The transportation volume in terms of millions of passenger-kilometers per year has grown remarkably: from 8.254 in 1983 to 16,300 in 1995. During the same period, the number of train runs per day has more than doubled from 102 to 230. Since the inauguration of Shinkansen services, The total number of passengers transported has reached approximately 800 million without a single accident involving loss of life.

Behind this achievement is the COMTRAC/SMIS system (Computer-aided Traffic Control system /Shinkansen Management Information System), which is a fully computerized train traffic control system having a centralized structure to attain safety and punctuality. Recently, the demand for improved management has led to the replacement of the conventional centralized computer system with a distributed system to attain more economical and a efficient passenger service. The new Shinkansen (Japanese bullet train) system, COSMOS (Computerized Safety Maintenance and Operation system of Shinkansen) was constructed on this background ([1],[3]).

2. The requirements and the objectives for train-control and management systems

2.1. Requirements

The situations surrounding the Shinkansen are gradually changing. A number of problems need to be solved so that the conventional traffic control system can be modified to satisfy the following requirements ([1]):

(1) More effective train control and management.

(a) Improved maintenance of trains and facilities.
Conventional facilities maintenance management

authorizes maintenance work plans for Shinkansen ground equipment and controls train routes for maintenance wagons from station offices. The new Shinkansen train traffic control and management system should integrate facility maintenance and train traffic control to reduce the maintenance operations from station offices.

(b) Expansion of the train-line network and computer system.

The situations surrounding the Shinkansen are gradually changing. An official decision was recently made to connect a sector of the Shinkansen to the conventional train-line network in order to offer direct access to local train-lines. The conventional line from Morioka to Akita will be linked with the Touhoku Shinkansen for through service in the spring of 1997, and the Hokuriku Shinkansen will begin service between Takasaki and Nagano in the autumn of 1997.

(2) Passenger service

(a) Flexible and adaptive train traffic control

To increase the number of passengers, Shinkansen trains are connected to conventional train-line networks in order to offer direct access to local train lines. The continual expansion of the Shinkansen network, however , has created problems with certain aspects of its operations. The expansion requirements of the number of trains in the metropolitan area is expected to increase substantially.

Train traffic control should be flexible to support complicated traffic patterns.

(b) Passenger information service

Coping with disturbances requires more than just dynamic scheduling to recover normal operations, it also requires real-time information service to meet the demands of passengers even in the event of sudden disturbances.

2.2. The objectives for the computer system

The system requirements are summarized here for each of the following four objectives. Each of the objectives is to be reached by implementing the proposed techniques shown in **Fig.1**.

(1) Integration of the real-time and non real-time subsystem

Train traffic control system consists of real-time and non real-time data. The system required to integrate of these two different property. The real-time data, such as control signal, require fault-tolerance and real-time

response for the safe and timely train traffic control. This data should be isolated and protected by the unstable disturbances transaction. While the non real-time data such as the information service and maintenance data has the unstable property, which disturb the traffic load of the CPU and network. These data are treated by the human operation. And it's required easy operation and safety maintenance by monitoring the real-time condition of the train traffic control.

(2) Maintainability

Maintainability can also be attained by using the ADS system architecture and construction techniques based on it. For more effective maintainability, remote maintenance was supported in the new Shinkansen system. The system is widely distributed, and the cost of field maintenance is high in time and money. To reduce these costs in times of emergency, a remote maintenance system that includes system diagnosis, software distribution, and configuration management is very effective.

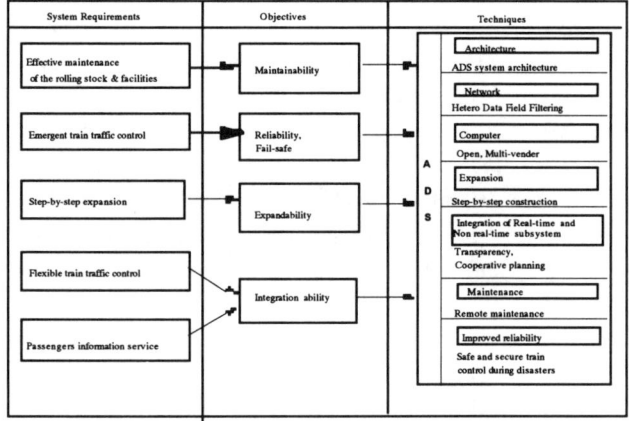

Fig.1.The proposed techniques for the new Shinkansen system

(3) Reliability and Fail-safe

The new system provides safer and more secure train control which integrates weather information with the train traffic control function. The train control function is linked with information from the disaster detectors. This linked system provides automatic train control (speed reduction control) and decision-making support.

These system features and the proposed technique will be explained in detail in the following sections.

(4) Expandability

The step-by-step expansion of the new train network and the system function can be attained by the step-by-step construction techniques based on the Autonomous Decentralized System (ADS) architecture. This architecture provides for on-line expansion without stopping and altering the already operating subsystems. Moreover, the construction techniques make possible on-line testing without disturbing the on-line operating subsystems.([6])

3. The proposed techniques

3.1. Autonomous decentralized system architecture ([4])

(1) Data field

A system with the basic features of non hierarchical structure and local autonomous control can be created by using a new system architecture, in which there is no central operating or coordinating system.

Each subsystem has its own management system, the Memory-base Data Field (MDF), to manage itself and to coordinate its operations with these of other subsystems. A subsystem, including its application software modules and its MDF, is an autonomous unit called an "Atom".

The self-contained subsystems (including their respective MDFs) are integrated into a super system. In the MDF, the subsystems are linked only through the Data Field (DF)—in practice, network, a common memory, etc. The data move around the application programs in each Atom, and the part of the DF in that Atom is called the Atom Data Field (ADF). In the DF, each datum is identified by a "content code" which is uniquely defined with respect to the datum contents.

System structure may vary, depending on expansion or reduction of the partial faults in the system. Therefore, to protect the operation of the subsystems from variations in the super system, messages transmitted by each subsystem contain a content code instead of a receiver's address. A subsystem decides whether or not to accept a message on the basis of that code. Thus, receivers select the data to receive and senders never specify receivers. The content codes to be received by each Atom are determined according to the applications within it.

This content-code communication enables every subsystem to send and receive autonomously. That is, each subsystem does not need to know the relationships between sources and destinations.

This feature of such communication ensures that each subsystem needs only "local" information .

(2) Data-driven application modules

Each Atom selects the message it needs to receive for its application-software modules on the basis of their content codes. After all the necessary data are received, the application module in the Atom starts to execute. This feature shows the autonomous "data-driven mechanism" of the application software modules—no module ever drives another or directs others to receive or process data. This mechanism makes the modules "loosely coupled". No module (or Atom) controls the others—rather, each independently judges and controls itself; that is, all Atoms are equal.

The content codes needed by each application program are listed in the Atom's MDF, which dynamically assigns the content codes its lists, according to changes in the application software modules. Even if the content codes assigned in the MDF are changed, it is unnecessary for the Atom to inform other Atoms or MDFs.

3.2. The step-by-step construction techniques

The system has expanded in two parallel phases: subsystem construction, and the addition/revision of application programs in station-computer subsystems.

The first step in the subsystem construction phase is constructing a station-computer subsystem. In the next step, another station-computer subsystem is added to already operating station-computer subsystems without stopping the previous subsystems' operation. In this step, the newly added subsystem is tested using on-line data. After the operation of the added subsystem is verified to be normal, the subsystem starts actual operation. On-line testing is essential in mission-critical real-time computer-control systems. There are already application programs installed in station computer subsystems. When additional application programs are added to them, on-line tests are inevitable ([4], [6]) .

3.3. The cooperation of the heterogeneous data field

The integration of real-time and non real-time data should be attained by the data-filtering techniques of MDF explained in **Fig.2**. MDF attaches the flag which indicates the real-time or not. When MDF sends and receives the data, it selects the data field that is divided into the real-time and non real-time.

The real-time data, such as control signal, require

fault-tolerance and real-time response for the safe and timely train traffic control. This data should be isolated and protected by the unstable disturbances transaction of the non real-time data. MDF exhibits to send non real-time data to the real-time data field.

While the non real-time data such as the information service and maintenance data has the unstable property, which disturb the traffic load of the CPU and network. These data are treated by the human operation. And it's required easy operation and safety maintenance by monitoring the real-time condition of the train traffic control. Therefore, MDF sends the real-time data to the non real-time datafield, and MDF pass it to the non real-time module such as the information module.([4], [5], [6], [7])

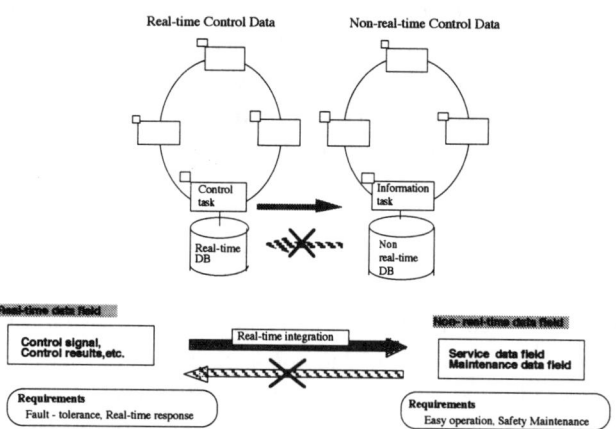

Fig.2.Hetero Data-Field Filtering

The non-real-time data field includes both the information service data and the real-time data. The information subsystem (Transportation planning, Maintenance control and scheduling) enables the real-time control data to be use without disturbing the real-time response or the reliability of the real-time control data field.

3 . 4 . Features of the new system
The outline of the new Shinkansen system is shown in **Fig.3**. The following sections describe the role of the new Shinkansen with respect to the four major subsystem.

(1) Transportation planning subsystems
Function: This is a system to support train operation scheduling for the Shinkansen. Train operation diagrams are made, car operations scheduled, and train

drivers and conductors are assigned by means of this system. It consists of a main system and local systems that are installed at each station and at each crew depot and that are linked to the former via a high-speed general-purpose network (packet- switching network).

The main system is also capable of searching for and retrieving data on train operations. It will transmit this data to each of the departments concerned, including stations, crew depots, and rolling-stock bases.

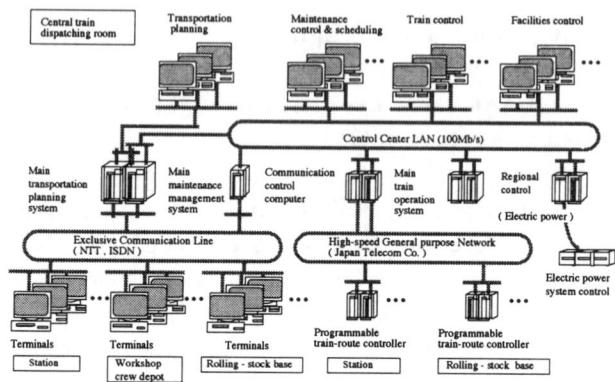

Fig.3.Outline of the new Shinkansen System.

Effectiveness: Supporting human effort in complex transportation planning.This is achieved by automating routing tasks and incorporating the following features:
(a) Automatic checking for consistency among train diagrams, car operation schedules, and crew assignments.
(b) Eliminating data selection and checking tasks now
   being conducted individually at each field site by the
      staff concerned, and transmitting data to each department
   as required and in the format desired for output.

(2) Train-operation-control subsystems
Function: This subsystem is designed to support the operation control of daily train services. It consists of a main system and a number of PRCs (Programmable Route Controllers) that are installed at each station and rolling-stock base, linked with an exclusive network. The main system supports diagram control and schedule adjustment for daily train services. The PRC systems installed at each station route, which can be controlled by the maintenance wagons through a wireless portable terminal. Service information can be displayed or announced to passengers, based on the real-time data and

the rescheduling plan.

Effectiveness: By creating a route-control system with a number of subsystems distributed among the various stations, the risk of a total breakdown can be reduced.

(3) Facilities-and maintenance control subsystems

Function: The role of the facilities-control subsystems are to handle data concerning the inspection and maintenance of trains and facilities. It also monitors the conditions of facilities and takes necessary measures in case of an emergency. The system is of the control-through-regional-center type, comprising a main control center and five regional control centers.

The maintenance control and scheduling subsystem is designed to support planning and management of the maintenance work to be carried out at night. It is composed of a main system and terminals installed at each maintenance depot and linked through a high-speed general-purpose network (packet-exchange network). The work plan is forwarded from a terminal at the maintenance depot to the main system, which in turn controls the commencement and completion of the daily work and checks the entrance into the new Shinkansen facilities according to this work plan. Execution of the work will be checked and authorization given by the system through a wireless portable telephone which maintenance personnel will carry.

Electric Power Control System DECS
(Shinkansen Denryoku Keito Control System)

The power substation control system for the Shinkansen is commonly known as DECS. Basically made up of the same components as those of the present system, it consists of a main system and five regional systems linked to the former via an exclusive communication line. The substations are controlled by the regional DECS installed at each regional control center. The function of the main DECS is simply to approve requests forwarded from regional centers and to monitor the subsequent handling.

Centralized Monitoring System (CMS)

The CMS is a system that monitors information from disaster detectors installed along the railroad lines and that also controls the wireless train telephone system. All the information is transmitted from terminals at each station to a main system via an exclusive line. With the

new system, however, all the information will be fed to the main system.

Effectiveness: Improving the efficiency in total facilities control and maintenance operations.
1) It has achieved automatic checking of the commencement and completion of maintenance work of Shinkansen facilities. With the new system, the commencement and completion of work are reported to the main control system for registration and confirmation. This subsystem is able to support the planing and to readjust the plans at maintenance depots. It also provides centralized monitoring and remote control of the condition of the equipment and the machinery.
2) Improved efficiency through the construction of a comprehensive database in which the equipment data, inspection data, and work records are combined to eliminate manual tabulation and information processing.
3) With the new system, however, the main DECS will be capable of directly controlling substations located in certain regions. This capability will eventually be extended to cover all regions.

3.5.The integration of the different subsystems in new Shinkansen system

The Shinkansen system integrates the different subsystems: transportation planning, train control, maintenance control and scheduling, shunting planning, rolling stock maintenance control, facilities control, electric power system control, etc.. The conventional system was very effective in supporting was safe and punctual traffic control, but each subsystem was not able to use information from the other subsystems for its own needs. That has been the major problem leading to ineffectiveness. The new Shinkansen system, in contrast, provides effective and transparent information sharing among the different task groups. Three examples of this are the schedule information sharing (integrated scheduling) in the transportation planning subsystem, the real-time information sharing on the facility maintenance, and the train traffic control under increment weather.

(1)*The integrated scheduling in the transportation planning subsystem*

The transportation planning subsystem calculates the basic plan of all the other subsystems, which is the first step in the integration of the scheduling. The calculation of the integrated schedule is based on the transportation plan. The routing tasks involved in transportation planning are automated and are performed in the following sequence.

(a) The basic transportation plan and the alternative plan are merged in each of the four seasons. The transportation planning subsystem makes a fifth plan based on these merged plans, which contain train diagrams, train operation schedules, crew assignments, etc.

(b) These schedules are sent to the maintenance and scheduling subsystems, train control subsystem, stations, workshop crew depots and rolling-stock maintenance depot.

Forecast Rescheduling-Checking Method

Basically, instead of carrying out the work on paper, the dispatcher now does the job on the screen of a work station computer (WS). The schedule is represented in the form of a line type diagram on which the completed and predicted portions are displayed in real-time.

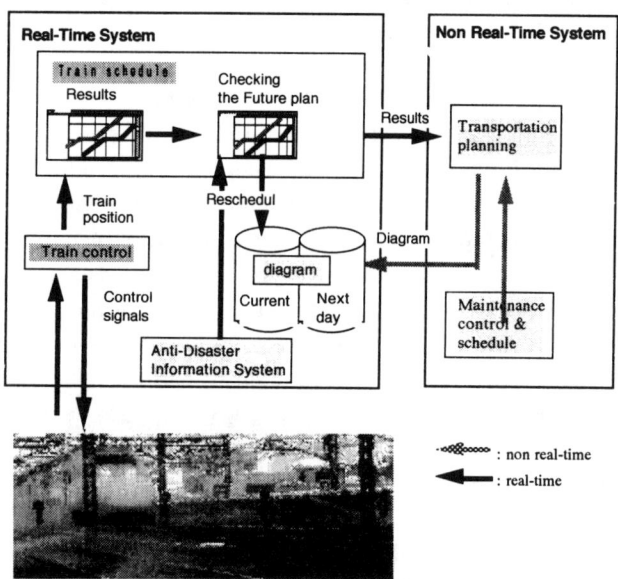

Fig.4 *Forecast Rescheduling-Checking*

Inconsistencies within the line diagram are shown on the lines along with their causes to assist the dispatcher in the regulating operations. While it is possible for the computer to automatically solve these inconsistencies to some extent, because there are slight variations in the methods of each dispatcher, changes made by the computer could be the cause of confusion. For this reason, the system has been designed to provide powerful support to the dispatchers who still make the final decisions.

The control of which line is used for arrivals and departures at each station is controlled to suit the line diagram displayed on the dispatcher's WS by sending the forecast results from the WS to the Programmed Route Control (PRC) at each station. As well as giving the dispatcher a sense of assurance, this also simplifies the control reasoning involved in automatically deciding the points locally.

Revisions made by the dispatcher have to be transmitted rapidly to trains, stations, other dispatchers, and all related field offices. Thus real-time information sharing is achieved.

(2) *Real-time information sharing in the facility control subsystem*

The role of the facilities control subsystem is to handle data concerning the inspection and maintenance of the train and facilities and to elaborate effective maintenance plans. It also issues pertinent instructions and controls daily work. Outline of the Facility and Rolling Stock Maintenance indicated in **Fig 5**.

(a) It stores inspection and maintenance data to be used by all concerned (via an exclusive high-speed network).

(b) It enables each maintenance depot to retrieve and process inspection data collected by a "Doctor Yellow" (Electricity and Track General Test Train) .

(c) To make the system user friendly and keep data from becoming obsolete, it simplifies data input at each maintenance depot terminal, make instructions for maintenance work available as output from terminals , and makes it possible to control overhead wires by using images on terminal screens.

(d) It provides the headquarters and branch offices, as well as each of the maintenance depots, and all the departments concerned, with the result of data processing.

Outlines of the systems to be introduced for the purposes as described above are as follows:

Rolling Stock Maintenance Control System

This is a system designed to handle records of rolling stock, e.g., inspection data and damage or failure information. It consists of a main system and regional systems that are installed at each rolling stock maintenance depot and linked to the former via a high-speed general purpose network (ISDN).

Record control will be conducted for every train using a database stored in the main system. The train maintenance will be carried out with another database accumulated in each regional system so that any train can undergo maintenance work at any rolling stock maintenance depot. Data can be fed by staff members directly into any of the terminals interconnected via a LAN within a rolling stock maintenance depot. This data can be processed or used there, and stored in the main system.

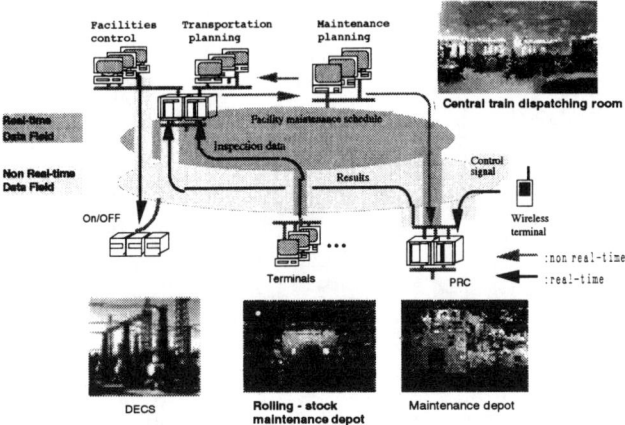

Fig. 5. Outline of the Facility and Rolling Stock Maintenance

3.6. Maintenance

Remote software maintenance is essential in this widely distributed system. Faults and failures have widespread effects on society, and rapid fault diagnosis and recovery are provided by the following remote maintenance functions.

(1) Remote diagnosis

The major purpose of this function is speedy system diagnosis in case of fault or failure. The error message and the logging data are collected and sent to the maintenance control & scheduling subsystems along with the alarm messages. Analysis of these messages and this data is performed in order to determine the cause of the fault or failure as soon as possible.

(2) Software distribution

The remote distribution of software reduces the cost (both money and time) of software maintenance. It also provides the secondary advantage of enabling the remote collection of the loading result, the download site, the loading software (name, version, attribute,.....), the

loading time, etc. This remote confirmation of the loading result prevents discrepancies between the field results and the recorded results.

The new Shinkansen system provides for realtime monitoring of the information about the software installed at each site. It displays two software versions: the existing software version (former version) and the revised software version. The operator can easily change these two software versions simply by selecting one of the other version windows. This operation makes it easy to revise the software and confirm that the correct software versions are installed at each of the sites.

(3) Software management

(a) Software configuration management

The software configuration management function provides the version management for all the configuration files for the computer system and the network. The configuration management is especially important for the safety operation of the widely-distributed system.

The management of each subsystem, computer, and network is carried out by the individual maintenance managers. The installation and repair of the software are also managed in a distributed fashion. To ensure the safety of the system operation, these distributed configuration managements should be integrated. The applied techniques provide a transparent monitoring function for all of this configuration information.

(b) Site installation management

The site installation management function supports the checking of the uninstalled software version and the revision on a field computer using, network elements. This function confirms that the correct software is installed at each site.

All these functions can also be processed during the on-line operation as part of the on-line maintenance of this autonomous decentralized system.

4. Evaluation of the new system

This new Shinkansen system integrates the entire system consisting of the different task groups, as well as transportation planning, train operation control, facilities control, rolling-stock inspection, and maintenance control and scheduling. The supporting task groups are larger and more distributed than the conventional Shinkansen system. The autonomous decentralized system structure enables information to be shared effectively and transparently among the different task groups, thus improving the workability of the whole system. The proposed system has been in operation since Nov. 1995, and the effectiveness of this system has been evaluated as described below.

1) The time needed to revise diagrams and schedules is now only two thirds what it was with the conventional system. All schedules can easily be changed in response to changes in traffic conditions, and this improves the services provided to passengers.

2) The highly computerized station-system reduces the amount of time required to perform routing maintenance operations. For example, during the middle of the night when there are no trains in operation, maintenance is carried out without assistance from the station office.

3) Efficient management was achieved by the information sharing among train diagrams, crew assignment plans, maintenance work plans, etc.

5. Conclusions

The introduction of the autonomous decentralized system architecture has facilitated the expansion of the Shinkansen network. As a result, the conventional line from Morioka to Akita will be linked with the Touhoku Shinkansen for through services in the spring of 1997, and the Hokuriku Shinkansen will begin its services in the autumn of 1997. The introduction of the new traffic control system has already transformed the handling of the Shinkansen into a series of highly efficient controlling tasks centered around decision-making by human intelligence. This will enable us to cope with a high-speed, high-capacity transportation system offering a large variety of services. That is, it will make possible the future Shinkansen.

References

[1] Akio Igarashi, "The new Shinkansen system", Rail International, Feb.1885, p.18~p.34.

[2] Wataya,K. : "The cooperating Autonomous Decentralized System Architecture", Proc. of ISADS 95, Arizona, USA, 1995, pp.40-47

[3] Kitahara, F. and Kera, K.: "Widely Distributed Train Traffic Control System and Its Step-by-Step construction ", Proc. of ISADS 95, Arizona, USA, 1995, pp.93-102

[4] Mori, K. : "Autonomous Decentralized Systems: Concept, Data Field Architecture and Future Trends", Proc. of ISADS 93, Kawasaki, Japan, 1993, pp.28-34

[5] "Special Issue on Distributed Computing Systems", COMPUTER, Vol.24, No.8, 1991

[6] Eiji Nishijima: "On-line Testing for Application Software of Widely Distributed System", Proc. of 15th Symposium on Reliable Distributed Systems, 1996. p.54-63

[7] Yau, S.S., et al.: "An Object-Oriented Approach to Software Development for Autonomous Decentralized Systems", Proc. of ISADS 93, Kawasaki, Japan, 1993, pp.37-43

Future Framework for Maglev Train Traffic Control System Utilizing Autonomous Decentralized Architecture

Takashi Kawakami

Construction Dept., Central Japan Railway Company, 1-6-6 Yaesu, Chuou-ku, Tokyo 103, Japan

Abstract

A future framework for the traffic control system of an ultra high speed and huge transportation capacity Maglev (magnetically levitated) linear motor line is presented. This Maglev system must have the ability to recover from disrupted traffic scheduling situations and to provide electrical energy savings. The proposed system consists of the CSS (Central Commander Support System), RPCCs (Regional Power Control Centers), TPCSs (Traffic and Power Control Substations) on the wayside, TSS (Train Supervisor Support System) on board, and IFN (Information Field Network) joining all the facilities.

The delay recovery control function has three layers. The top is a strategic rescheduling layer. The CSS helps to plan recovering strategy. The middle layer is a headway minimizing layer including the TPCSs which direct the Maglev train toward a 3-D target so as to minimize the train headway. The bottom layer is a temporal speed control layer.

The energy saving function has a two-layered structure. The top is a total load control layer coordinating the powering/braking schedule of each train and the bottom is the same as in the traffic control function.

The proposed architecture is expected to present quite controllable circumstances to the central commander, and also to realize the desirable overall power dissipation characteristic.

Keywords

Autonomous Decentralized System, Control System, Linear Motor Car, Magnetic Levitation, Maglev, Superconducting, Automatic Train Operation, Train Traffic Control

1. Introduction

Experts are well able to foresee the consequences of an emergency or abnormal situation and to make a good judgement of action to take through a decision process which may be conscious or not. A large scaled realtime system such as ultra high speed intercity transit has to provide the same expert ability[1] free from functional omissions.

Today, in Japan, a superconducting maglev train system is being developed which is projected to appear in a 5 million man•kilometer/hour capacity transportation system with a 500km/h speed to run between Tokyo and Osaka.

This system will have to face the following problems.

•Provide sufficient variety in the train time table patterns
Short transit time commuters and long journey travelers use the trains together. Therefore, various time table patterns must be possible.

•Have a powerful ability to reschedule disrupted traffic[2]
The maglev trains must be operated at the limit of the headways to maintain the transportation capacity and to serve a variety of time tables. So, rescheduling becomes far more severe than encountered with today's trains.

•Have energy saving characteristics
Maximum power for each train will be about 5 times that of an ordinary Shinkansen train. In view of the speed ratio, this means the maglev system will have good efficiency. But, the absolute value will not be small, so informational control is expected to become the next power saving technique.

The distributed architecture of today's railways is the result of the efforts of many researchers and many developments. Although the central traffic control, stations, trains, route controller, interlocking system, wayside and

0-8186-7783-X/97 $10.00 © 1997 IEEE

cab signals, power substations, etc. and employees working for each department are combined through only minimum width information channels, the resultant arrival/start time of trains are quite exact in short headway and high speed operation.

But more time is being required to handle disrupted traffic situations. This results from the current role apportionment of train operation[3]. It is impossible for a central train dispatcher to command the movement of all trains at every moment exactly.

Therefore, the future maglev system must have a new architecture for train operation.

This paper proposes a future framework for the maglev train traffic control system utilizing an autonomous decentralized architecture[4][7]. Section 2 overviews the hierarchical architecture and cooperative relationship between items. Section 3 introduces the methods for delay recovery control, consisting of three layers. Section 4 describes the energy saving function which has a two-layered structure.

2. System Architecture

2.1 Overview

Fig.1 shows an imaginary structure for the future Maglev transportation line and traffic control items discussed in this paper. It is assumed that the number of stations (STn) including two terminals ST1 and ST8, on a 500 km length line, is eight, and the average distance between Traffic and Power Control Substations (TPCSs) is about 25 km. The Commander Support System (CSS) assists the commander to observe, judge and dispatch the train traffic at the central traffic control center.

In an ordinary railway traffic control center, the person who reschedules the time table is called a "dispatcher". The train dispatcher adjusts the time table, including the start time, home track assignment and starting order.

Although the dispatcher recommends the speed deceleration to a train driver via a wireless channel, it is difficult for the dispatcher to control exactly the arrival time, even though it is the most important parameter in the dispatching job.

On the other hand, using an issued system, movement of each train is freely controllable since central personnel are helped by the user-friendly characteristics of the system. Therefore, "dispatcher" is renamed "commander".

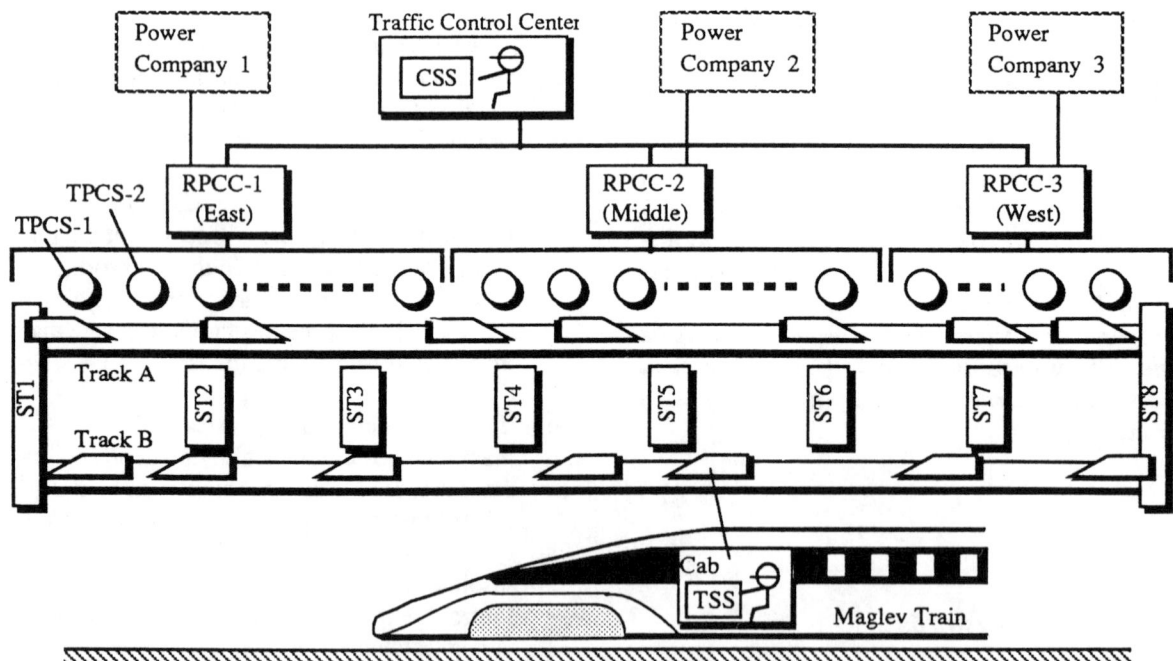

Fig.1 Items of Future Maglev Traffic Control System

There are three Regional Power Control Centers (RPCCs) corresponding to three electric power companies. Each RPCC treats the TPCSs in its administering zone. Each TPCS controls trains running in both directions in its control district.

During a normal situation, the train supervisor's role will be different from that of today's driver, because there is no need for a supervisor to operate the train. Train Supervisor Support System (TSS) displays the state of the train and traffic, and it helps the supervisor in the cab to drive the Maglev train manually in an emergency.

All the items on ground and on board are connected through a high speed transmission system which has a sufficient communication capacity to compose the information field network (IFN). Then from the central traffic control center, daily and temporary data of train schedules and regional power demands are broadcast to the IFN. Each RPCC plans its regional tactics to meet the power demand and to watch the energy consumption trends. If the RPCC detects the possibility of an over-demand, it recommends power accommodation control between TPCSs. The power commander (someone at RPCC) negotiates with the power company if it becomes necessary to feed more energy, or to exchange the source power feed line as protection from a partial failure in the TPCS group.

The TPCS manages the train schedule and propulsion power directly. The train schedule includes departure times, arrival times, passing times, stopping times, route control plans and speed profiles.

Propulsion power control means speed control and electric power accommodation between trains in the local area and neighboring TPCSs.

2.2 Hierarchy of the system

Hardware and functional hierarchy of the future Maglev traffic control system are shown in Fig.2. At the top of the hierarchy is the management system. This layer decides train time table according to judgements from convenience for passengers, energy demand level, profitability, maintenance capability , etc. The roles of the management system are almost the same as those of ordinary railway system.

Functionally, the Maglev traffic control system is separated into two parts. One is for the delay recovery control facilities, and the other is for energy saving. The former has three layers, and the latter, two. The delay recovery control function is effective in both normal and disrupted situations.

If a small delay is detected, the temporary speed control layer suppresses the expansion of delay under the level that the departure time is on schedule. The function of this layer is performed by cooperative action of the TPCS group. When the starting or passing time schedule is delayed too much, it is necessary to recover the lost time as soon as possible to keep the planned schedule and to avoid unexpected deceleration or stopping motion of following trains before reaching a station.

The headway minimizing layer generates a suitable three-dimensional (3-D) target defined by (x,v,t) for each train aiming to recover its schedule. Here, x, v, and t are position, velocity, and time. If a train succeeds in passing

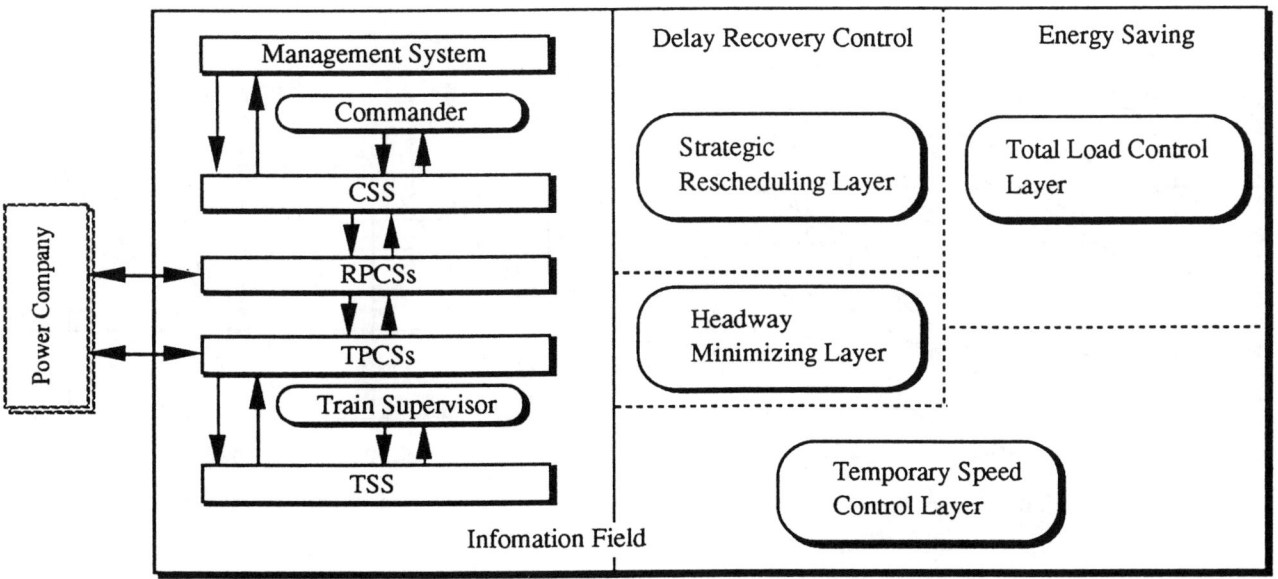

Fig.2 Hierarchy of Future Maglev Traffic Control System

the target, unnecessary braking is clipped out and at the same time, then, headway becomes the shortest. This layer is also processed by the TPCS group.

If traffic is severely disrupted, in the strategic rescheduling layer, the commander makes a dispatching plan comprised of changing the order of departure, shunting, and shopping out from the stock yard. In almost all cases, trains are controlled exactly by IPCs. So, it is easy for the commander to get accurate perspectives of traffic all over the 500km Maglev transportation line.

There are two layers in the energy saving function. These are most active in the normal state. The future Maglev system will have a high precision train traffic simulator[6] which will calculate the exact power consumption. So all train time tables planned in the management layer will have good characteristics for energy saving, given from the power simulation.

But, in practice, various factors change a train speed pattern. Basically, the temporary speed control layer adjusts the velocity of each train, according to the time keeping rule, and energy minimizing theory[5]. The total load control layer is composed of RPCCs. It balances the electric energy between trains and TPCSs so as to minimize the regional total amount of power consumption.

3. Delay Recovery Control

3.1 Strategic rescheduling layer

Fig.3 is a conventional train schedule diagram. Four express trains and four way trains start from and arrive at each terminal (ST1 and ST8) every one hour. The express train services the two terminals only. At stations ST2 and ST6, the express train gets ahead of a way train. Train time table is designed with a planned stopping time, running time from station to station, and operative headway which is the sum of static headways.

Usually, these factors include some time margin. This is a well known method and, possibly the future Maglev railway will also use the same way to make train schedules. But, the future CSS will display traffic information in different style from today's dispatching console. Time schedule, and locus of trains (result and scope) are drawn with straight lines on an ordinary central train dispatcher's screen (Fig.3) while the future CSS will represent a perspective of loci of trains, using many types of safety zones(Fig.4). Vertical width of the safety zone means an assured braking length, which is necessary for a following train to decelerate from speed at each time to zero.

Therefore, the real safety zone varies according to the speed of the following train and other parameters, such as gradient of track, and deceleration rate of speed.

Fig.4 shows two example patterns of a safety zone. In case-1, both the preceding way train and following express train stop and start at station STn. The departure of the express train occurs at t2, so that, just at that time, the safety length for the preceding train is zero. According to

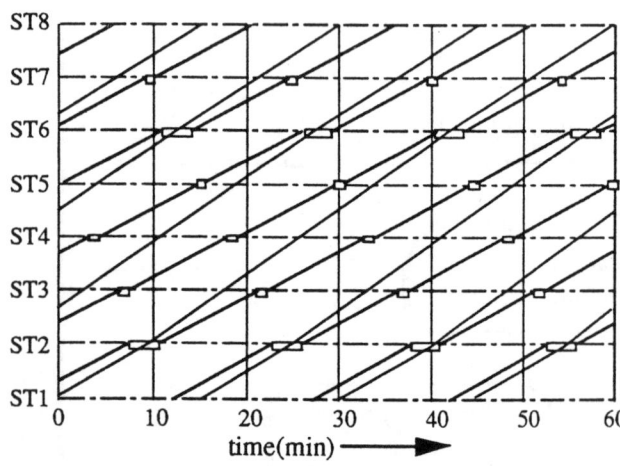

Fig.3 Conventional Train Diagram

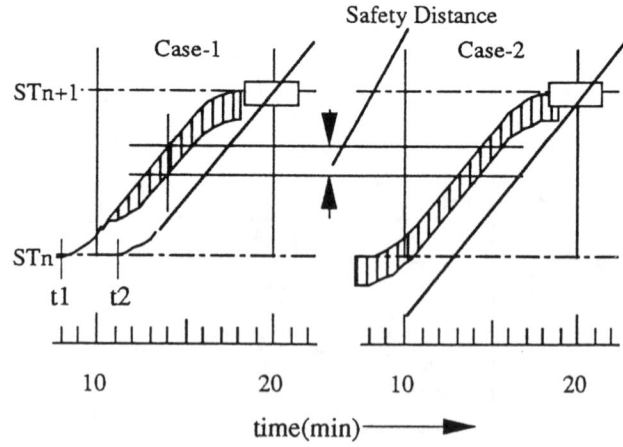

Fig.4 New Items for Train Diagram

330

acceleration of the express train, the zone becomes wider.

After the arrival of the way train at STn+1, the traverser (equal to the switch of an ordinary railway) moves and provides a route for the passing express, then, the safety zone disappears. Case-2 indicates the situation in which the following train passes STn and STn+1 at full speed. So, the safety zone width is constant. There are many other possible situations between trains, or train and track facilities. Various types of safety zones will be drawn, and their actual shapes will change dynamically, depending on the velocity of an approaching train.

Safety zone representation, which CSS will provide is done conveniently for the central commander. By observing the displayed traffic state on the CSS screen, the commander easily finds a "channel" or "space" on the distance-time plane which is important for strategic dispatching. So as to provide accurate perspectives, each TPCS calculates the scope of the movement of a train , then checks its rationality among the TPCS group. TPCSs cooperate with each other, so that, even if one TPCS is lost due to some trouble, neighboring IPCs, on both sides, separate the remaining substations about half and half, and continue to control them.

From the commander's standpoint, the issued traffic control system is tough, controllable and friendly.

3.2 Headway minimizing layer

Fig.5 shows a case in which the traverser at station STn+1 experiences some trouble moving. The following express train, scheduled to go ahead, like the dotted line in Fig.5, can not escape from the unexpected stopping just in front of the entrance side traverser. From the standpoint of the central commander, two dispatching methods are possible. One is, not to exchange the scheduled departure order (express starts first). The other is the opposite, i.e. that way train 1 will be passed by the express at the next, or subsequent station.

In Fig.5, the commander selects the former. Coming after way train 1, way train 2 approaches ST n+1, and its head position draws a curve close to the safety zone extended from the tail of the preceding way train 1. At a 500km/h speed, the safety distance will be 12km or over.

Fig.6 is a more detailed diagram of the safety zone between way trains 1 and 2. Generally, it is known that to shorten headway, one-stage braking (assured braking) is effective, when used together with moving block type protection system. Supposing that the minimum unit of a propulsion facility is set along a Maglev guideway, it will function the same as a blocking unit. Then, the ability to shorten the headway is quite similar to that of an ideal moving block system. Safety zones in Fig.5 and 6 are shaped, based on this assumption.

Dotted lines in Fig.6 are safety boundaries in the one-stage braking pattern. Headway in this case can be calculated from static factors, i.e. deceleration and acceleration rates of trains, approaching speed of a scheduled train, gradient of track, and margin of safety distance. These give "Headway Level 1". A train, approaching through a suitable 3-D point.

P (x,v,t) gains the minimum headway. Two braking occurrences and one accelerating occurrence are done before the train is stopped. If the following train attends the optimum "Level 2" point P (x,v,t), the resultant headway (Headway Level 2) is shorter than Headway Level 1.

TPCS estimates the time when the preceding train starts, and calculates the points, Level 1 and Level 2, then broadcasts the data of those points onto the IFN. Other TPCSs, which are directing Maglev trains, receive the headway minimizing point data, sent from the forward direction.

Once the train schedule is heavily disrupted, the central commander begins to restore the traffic. During dispatching work, the commander often wantes to move a train too close to another one. Then, there is a problem that, from the conclusions on the console display, it is hard for the dispatcher to judge if the recommendation is possible or not.

Even now, drawing safety zones as described here, on the display equipment in a traffic control center, is not so difficult, if each train's speed and position at every

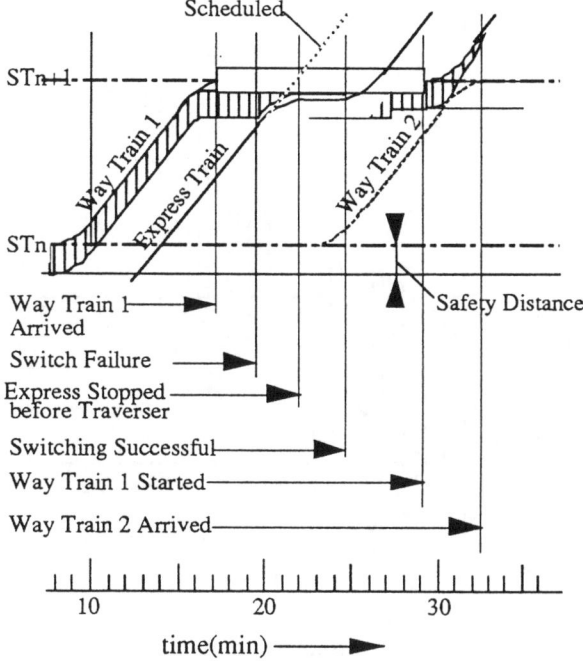

Fig.5 A Case Preceding Train Delayed

331

moment are known. But, such a display is meaningless, because the traffic situation in the near future is uncertain. In the future Maglev system, the TPCS automatically selects what level of headway is fit for the situation at that time, and predicts the shape of safety zones which are influenced by the commander's request. Therefore, the shape of the safety zone on the CSS screen varies continuously, and the commander is freed from worrying whether the perspective of traffic, and the expected result of dispatching, will be accurate or not.

4. Energy Saving

4.1 Background

As mentioned in sections 1 and 2.2, the Maglev system has good energy efficiency, comparing its transportation capacity with the amount of electric power consumption, but, because of the largeness of propulsion power per train, it is necessary to have some energy control methods from the view points of operating reliability, passengers' convenience and profitability. In present railway systems, information channels between transportation departments, such as the traffic control center, driver or automatic train

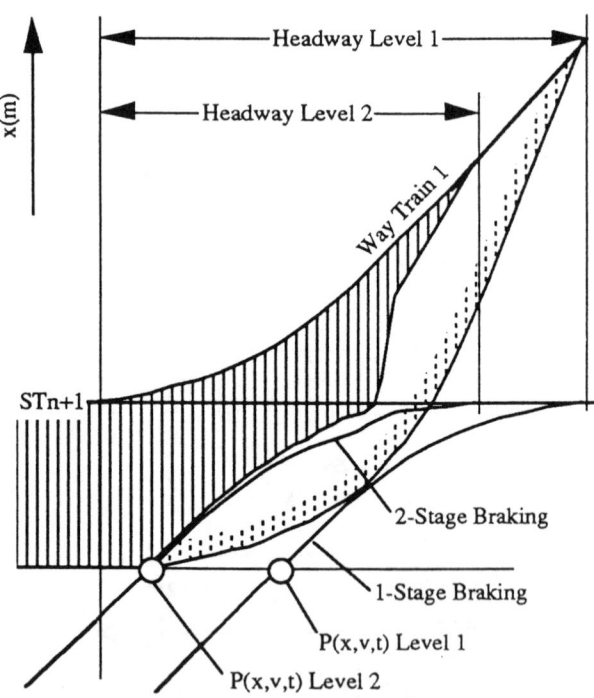

Fig.6 Effect of Headway Minimizing Control

operation equipment on board, and power substations, are very narrow. And, there are not enough number of substations which make up the continuous voltage/ regenerative power control facilities. So it is not easy for today's electric railway to distribute the transportation energy ideally.

But, quite accurate power simulation technology for railway systems has already been developed[6]. The future Maglev system is expected to have a wide and high speed IFN, while the TPCS has powering/regenerating functions to drive the Maglev train. Therefore, enough facilities to realize the power accommodation control must be provided.

In the management layer, an energy saving train time schedule will be planned, using the high precision train traffic simulator which calculates exact predictions of power consumption, even though there are many parameters which forcibly shift the train velocity from the planned pattern.

4.2 Temporary speed control layer

Considering the train speed pattern from station to station as a kind of waveform, in velocity mode, each train has phasic relations between other trains. It is thought that, even if substations on the wayside have power feedback equipment, the sum of power dissipation is sensitive to the phase. This phenomenon is caused by ohmic loss in the power line. It is useless to distribute regenerative energy beyond too long a distance. The keypoints of temporary control of train speed to decrease the sum of the power, are as follows.

1) To maintain an energy saving pattern
   Every train velocity pattern between two service stations has at least one period including one acceleration and one deceleration. The most economical method is to make the speedup as rapid as possible within each period.

2) To provide regenerated energy to the best item at anytime
   So as to efficiently distribute kinetic energy from a decelerating train, each TPCS broadcasts the predicted powering and regenerating energy pattern of trains under its control, and decides the tactics to minimize the dissipation energy.
   The best way to give regenerative energy is sometimes by the TPCS itself, and other times it is given to a train running nearby.

3) To suppress the speed of a train having much time to spare
   Suppressing top speed is quite useful for decreasing power consumption.

The TPCSs also combine the above methods, according to self judgement and/or to recommendations from the RPCCs

4.3 Total load control layer

The future Maglev railway will run within the boundaries of three electric power companies. To satisfy demands with each power company, the respective RPCC watches the trend of the sum of the power. Each TPCS will be fed electric power directly from the power company, and contracted values of demands differ. The RPCC assesses power demand of each TPCS. When train traffic is disrupted, the central commander plans a recovering strategy, then broadcasts it onto the IFN. Generally, more energy is necessary in one zone than others to restore the train schedule. If necessary, in such a case, the RPCC allow some TPCS to use more energy over the demand contracts, or the RPCC generates an alarm to personnel to negotiate about the demand with the power company.

5. Conclusions

The basic architecture for a future Maglev train traffic control system was described. Distributed and cooperative Traffic and Power Control Substations (TPCSs) broadcast accurate predictions of traffic, including dynamically changing safety zones onto an Information Field Network.

The Commander Support System (CSS) displays the information to help make the dispatching strategy. TPCSs realize the commander's recommendations cooperatively.

At the same time, Regional Power Control Centers (RPCCs) issue power demands, and the TPCSs accommodate electric energy so as to minimize the sum of power. Thus energy saving and an autonomous decentralized architecture will be established for the ultra high speed, high density Maglev system.

Acknowledgements

The author would like to thank Mr. Masanori Ozeki, President of Railway Technical Research Institute, for his continuous support and many helpful suggestions. The author is also grateful to Mr. Sakio Masunaga, General Manager, Construction Dept., Central Japan Railway Company, for giving him the opportunity to work on the Maglev development.

References

[1] T. Kawakami:"Concept of Friendly Autonomous Decentralized System for Next Generation Train Traffic Control", ISADS93, pp.316-322, 1993

[2] T. Kawakami, M. Ohtani:"Realization to 'Maximum Number of 1500 Trains'", JR-Tokai Confidence and Review, Vol.2, No.9, pp.7-9, 1992 (Japanese)

[3] S. Sone:"Intellectualization of the Traffic System", J.IEE Japan, Vol.112, No.11, pp.837- 840, 1992 (Japanese)

[4] K. Mori:"Autonomous Decentralized Loop Network and Its Application to Train Traffic Control System", IECON'84, Tokyo, pp.663-666, Oct. 1984

[5] S. Murata, et al.:"Train Control Method for High Speed and High Density Railway Systems", COMPRAIL94,435, 2, (1994)

[6] K. Tashiro:"Train Operation & Traffic Simulation", IEEJ National Conf. on Industrial Application, S2-5,1995 (Japanese)

[7] T. Kawakami:"The New System Architecture for Next-Generation Shinkansen Based on the Friendly Autonomous Decentralized System Concept", ISADS95, pp.86-92, 1995

Panel Session P2

Standards for ADS

Panel Chair

Kane Kim

Panelists

Donald S. Pieronek

Seiichi Shin

Masaharu Oku

Volker Tschammer

Bhavani Thuraisingham

Standards for ADS

Kane Kim
Panel Chair
University of California, Irvine, USA

Background

As the autonomous decentralized system (ADS) philosophy is meeting increasing acceptance by the system engineering community, the type and number of highly autonomous subsystems will grow rapidly. Standardization of the methods and protocols by which autonomous subsystems interconnect or interact among themselves is an important issue with respect to further enhancing the economic advantages of the ADS and enabling broader acceptance of the ADS approach.

Such standardization is a political challenge as well as a technological challenge. Not only a sound and concrete form of the ADS philosophy but also various proven and promising ADS architectures must be reflected in the standardization process. The trends in the requirements imposed by the system customers must also be reflected. Already a number of major standardization efforts dealing with various building-blocks of ADS's are under way. The two ADS application communities which have been the most active in such standardization efforts are the telecommunication application community and the computer-based control application community.

A team of distinguished panelists will review those major standardization efforts under way and discuss some major issues not yet resolved as well as near-term prospects.

Discussion Topics:

1. Emerging ADS architectures
2. Needs and targets for standardization:
   Needs created by new user requirements
   Interaction protocols at various levels: WAN, LAN, device networks, blackplane
3. Emerging standards:
   WAN for information services
   LAN for control applications
   Device networks for control applications, etc.

Chair:
Kane Kim
University of California, Irvine, USA

Panelists:
1. *Donald S. Pieronek*
Principal Technologist, Corporate Research and Development of Eaton Corporation, Milwaukee, WI, USA
Representative of ODVA (Open DeviceNet Vendor Association)

2. *Seiichi Shin*
Associate Professor, Course of Mathematical Engineering and Information Physics, Graduate School of Engineering, University of Tokyo, Japan
Representative of IROFA (International Robotics and Factory Automation Center), Japan

3. *Masaharu Oku*
Department Manager, Automation Controls and System Development Department, Bridgestone Corporation, Japan

4. *Volker Tschammer*
Assistant Research Department Manager, GMD FOKUS, Berlin, Germany

5. *Bhavani Thuraisingham*
Senior Principal Engineer, The MITRE Corporation, Bedford, MA, USA

Organizers:
Katsumi (Kevin) Kawano,
Systems Development Laboratory
Hitachi, Ltd., Japan
Tel: +81-44-966-9111, ext 3111
Fax: +81-44-966-6823
E-mail: kawano@sdl.hitachi.co.jp

Kane Kim
University of California, Irvine, USA

ADS Technology in the
Telecommunications and IT Environment
(Position Paper)

Volker Tschammer
GMD FOKUS, Berlin, Germany

Future telecommunications and IT services are to be realised by the combination and co-operation of distributed components, such as agents, controllers, managers, and communication units. These components will be distributed across multiple technical and organisational domains and will have to contribute to different services in sequence or even in parallel. The number and characteristics of components will change dynamically as new components will be introduced, existing components will fail or be withdrawn for maintenance, and peak loads will impose stringent requirements on the availability and performance of components.

The ADS technology has been designed for applications in large distributed systems which include many autonomous entities that have to co-operate for a common task. An ADS system is integrated from a set of components which can undergo constant changes due to failures. modifications, and addition or removal of components. These characteristics are the reason why there is an increasing interest to apply ADS technology to the telecommunications and IT environment in order to solve problems of scalability, online-expansion, online-maintenance, and fault tolerance, as well as to integrate the ADS standards with those of IT and telecommunications platforms, services, and applications.

Issues on Real-Time Object Request Brokers (Position Paper)

Bhavani Thuraisingham
The MITRE Corporation, Bedford, MA, USA

Introduction

For many applications such as command and control, tele-communications, and process control, it is critical that they meet timing constraints and ensure predictable computation. Furthermore, some of these applications are also distributed in nature and utilize distributed object management technology such as object request brokers (ORB) for interoperability. Therefore, it is necessary to integrate real-time systems technology with distributed object management technology for many distributed real-time applications. This position paper describes some of the issues that need to be investigated in order to develop real-time data processing extensions to the common object request broker architecture (CORBA). In particular, issues on extensions to ORBs are given. A discussion of some real-time services and applications hosted on real-time ORBs will also be provided. Finally, an overview of the work of the Object Management Group's Real-time Special Interest Group will be provided.

Real-Time Extensions to ORBs

The major issues on developing a real-time ORB and object adapter involve the time constraints and criticalities attached to requests. That is, when clients request services from object implementations, time constraints are attached to the requests, such as start times, start events, deadlines, and periods, to be imposed on the execution of the operations. These ideas have been proposed in the RTSORAC model of [PRIC94]. Techniques to handle these timing constraints have to be developed. This is a heavily-analyzed problem in real-time scheduling which needs to be evaluated for applicability to the ORB environment. For instance, if an object server is to have some responsibility in controlling the order of execution of its requests, the entire request queue needs to be available for reading and writing, not just the top of the queue. Also, real-time communication is also a major issue that needs to be addressed for a real-time ORB.

An example of how CORBA can be used to advantage in a real-time problem: an operation may have multiple implementations which are differentiated by the accuracy of their computed result and their time to execution (worst-case, av-erage, etc). When one implementation is too time-consuming to meet the required deadline, the best implementation which is feasible can be selected by the ORB/object adapter.

Real-Time Services and Applications

As stated in [THUR96], while a real-time ORB is intended to provide the underlying mechanisms for time constrained communication between different objects, services such as real-time scheduling for period, aperiodic, and sporadic tasks have to be built on top of real-time ORBs. That is, it is desirable top keep the ORB as simple as possible and to hosted the various services. One approach here is to develop a real-time infrastructure library to encapsulate and extend the use of the operating system and ORB real-time features. For example class libraries can be developed to extend the functionality of threads by allowing them to have their priorities set by a server that calculates the priority based on the period of a task, have timing constraints based on CPU usage or other timing criteria enforced, and to be controlled by events generated by other processes or threads. The library may support features that are important that are directly supported by operating systems or ORBs. Examples may include semaphores, mutexes and condition variables which support priority inheritance. This enables for greater portability of application code that uses the features provided by the library.

Real-time data manager is a example of an application hosted on an ORB. A real-time data manager manages persistent and temporal data, ensures that the queries and transactions meet the timing constraints, enforces temporally correct serializable schedules, and combines techniques from data management and real-time scheduling. Some implementations of real-time data managers relevant to the discussion here is given in [WOLF94] and [BENS95]. In [WOLF94], a discussion of real-time extensions to Texas Instruments' Open Object-Oriented Database System (Open OODB), which is currently stated to be consistent with CORBA specifications, is given. These extensions follow the RTSORAC model and thus would be consistent with our proposed changes to the Object Model. In [BENS95], the design and implementation of an infrastructure and a data manager, which is based on a real-time object model, for next generation command and control systems is described. Future plans with this latter effort is to use a real-time ORB to integrate the infrastructure services, the data manager, and the application.

OMG's RTSIG

To address the needs of many real-time distributed applications utilizing object technology, the Real-time Special Interest Group (RTSIG) has been formed as part of the Object Management Group (OMG). This group was formed in February of 1996 and has been conducting meetings since then. The group has so far focused on two aspects. One is the Request For Information (RFI) that was posted sometime in late 1996. The other is the white paper that the group has been working on. The white paper addresses many issues for real-time ORBs. In particular, it defines what is meant by real-time, has explored issues on real-time scheduling, and also addresses other topics such as integration with real-time data managers, fault tolerance, and security. The white paper will serve as a guidance document for the vendors interested in developing real-time ORB products. A copy of this white paper which is in draft form can be obtained from the OMG.

Directions

While the current challenge is to develop real-time ORB products that can be used for various real-time applications such as those in command and control, telecommunications, and process control, we believe that with the significant progress made by the RTSIG and the interest with commercial vendors, this will be a reality within the next year or so. Following this, the next big challenge is to integrate this technology with some of the emerging technologies such as Intelligent Agents. Agent technology is now being used for various applications including command and control and telecommunications. Furthermore, agent technology is a key component to Internet technology. Therefore, work should be directed toward integrating distributed object management, agents, and real-time technologies. That is, we need agent-based real-time ORBs for future distributed real-time applications.

References

[BENS95] Bensley, E., et al., Evolvable Systems Initiative for Real-time C3, Volume II: Design and Implementation of the Infrastructure, Data Management, and Application, MTR 95B 0000116, September 1995.

[PRIC94] Prichard, J., et al., RTSORAC: A Real-time Object-Oriented Database Model, Proceedings of the International Conference on Database and Expert System Applications, 1994.

[THUR96] Thuraisingham, B., et al., On Real-Time Extensions to the Common Object Request Broker Architecture, Proceedings of the WORDS 1996 panel, Laguna Beach, CA, February 1996.

[WOLF94] Wolfe, V., et al., The Design of Real-Time Extensions to the Open Object-Oriented Database System, Proceedings of the IEEE Workshop on Object-Oriented Real-time Dependable Systems, October 1994.

Session 5A

Multiagent Systems III

Chair

Feng-Jian Wang

Adaptive Environment Observation for Distributed Agents Moving in a Lattice World

Takashi Watanabe   Masuhiro Mizuno   Tadanori Mizuno
Department of Computer Science
Shizuoka University
Hamamatsu, JAPAN 432

Abstract

In a distributed cooperative system, each agent has to observe its environment in order to create an adequate subgoal. This paper discusses tradeoff between cost of observation and achievement of the shared final goal of agents moving in a lattice world. The model is referred to as the restricted Tower of Babel. A cooperation mechanism called LM-DMax is evaluated through simulation against observation interval, the number of agents and moving area size. We find that the model has an optimal observation interval, an optimal area size to minimize the total cost. Then three adaptive policies which modifies the observation interval of each agent are proposed. They are geometrical policy, arithmetical policy and slow-start policy. Slow-start policy shows the best result among them. Though the discussion is restricted to the specific model, it is directly applicable to a vehicle scheduling problem and a robot planning problem.

1 Introduction

An autonomous agent is an independent intelligent activity, which acts on its own decision making policy. A multi-agent system contains quite a few loosely coupled autonomous agents achieving a common goal. Among problems of the multi-agent systems [6] are inference on interaction between agents [5], cooperation protocol KQML [3], organization [10]. In a multi-agent system, agents cooperate with each other, through searching [1] or plan recognition [2], and cooperative agent can achieve a goal efficiently. Thus , agents should exchange its beliefs, knowledge, intention to keep up with the world, which gives agents a significant amount of time to create subgoal. Most of research on multi-agent systems assume agents can observe the environment anytime, even time for observation and deliberation is mostly assumed to be ignored[6].

If we solve an actual problems with multi-agent paradigm, time for observation of environment, communication with other agent, and deliberation for an agent to create its subgoal can be significant for its performance.

Specifically there are two issues with respect to consumed time. Firstly, if an agent try to create an optimal plan, it has to observe the environment precisely. It means every change in the environment should be informed to the agent. Secondly, even if an agent could obtain the exact environment, creation of its subgoal consumes time, which makes the information out-dated. So, for a practical use of multi-agent system, an agent have to act on inaccurate environmental information.

For agents to get the environment information, we have two policies [2]:

1. observe the environment via communication (explicit policy)[3], and

2. estimate the environment using information obtained so far (implicit policy)[5] [4].

In this paper, we investigate criteria for agents' environment observation and achieving its subgoal, where agents interact explicitly.

To the similar problem Huber and Durfee [2] tackles from a view point of early versus late commitment. But, the fundamental discussion on the environment observation has not been done, especially for practical use.

We use here an extended Tower of Babel which is one of typical cooperation models. Though it is quite simple, the model contains crucial aspects for agents as follows:

- As agent action is very simple, we can focus on tradeoff between observation and achievement of agents.

- The model can be widely applied for real systems such as AGV (auto guided vehicles).

We have made fundamental discussion on subgoal creation of autonomous agents [12], where some cooperation mechanism to achieve the final goal of the Tower of Babel, and we have studied creation and autonomous modification of agent's subgoal.

Several experiments give us criteria for effective movement of agents. We show optimal observation interval does not depend on number of agents, but on space size and cost of agent observation.

And then we propose adaptive observation mechanism which enables agents to establish their observation frequency autonomously.

343

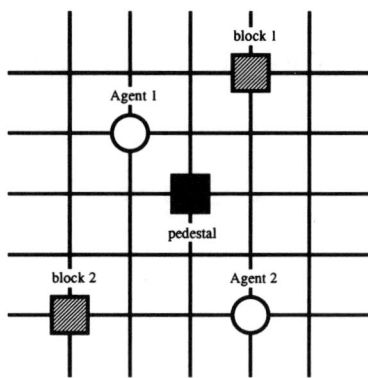

Figure 1: The Tower of Babel

2 Evaluation Model

2.1 The Tower of Babel

Typical model for studying distributed cooperative problem solving includes Tileworld[7], Pursuit Problem[8], MICE[9], the Tower of Babel[10], and so on. The Tower of Babel is proposed by Ishida towards self-organization mechanism. The model of the Tower of Babel is defined as follows (Fig.1): There are some blocks numbered serially and scattered in a two-dimensional lattice world. The final of the model is to carry all blocks to a central place, called pedestal, in an order of the block number. An agent can carry a block on it at a time. An agent can move to an adjacent intersection in a unit step. An agent cannot move an obstacle that might be a block or another agent.

In the model, there are two kinds of interrelated subgoals; block subgoal and intersection subgoal. Block subgoal means decision making of the next block to carry to the pedestal. And intersection subgoal means decision making of the next intersection to move to in order to avoid conflict or to achieve its own block subgoal efficiently.

2.2 Restrictions and Assumptions

In the model, agents can move and observe all over the lattice world. As a bus location system or a physical distribution system, it is done that agents are assigned their territory. Then, we assume that agents stop during their observation and periodically.

To recognize another agent's subgoal, agents communicate each other. In this paper, we assume that agents can communicate all over the lattice world, and we don't take communication cost into account.

There are 3 types of cost to be considered. (1) Moving cost (2) Observation cost (3) Communication cost.

2.3 Observation Cost and Moving Cost

In a multi-agent environment, the environment varies dynamically since multiple agents act autonomously and parallelly. In this environment, they must observe their territory to recognize blocks in the world. Then, agents can choose one block as their

subgoal. After that, they move to a temporary destination according to the subgoal. Thus, to achieve the final goal the total cost includes observation cost and moving cost.

Specially, observation cost depends on the following factor:

1. Territory size
   Larger territory of an agent requires more cost, but enables agents to create an efficient subgoal, to reduce moving cost.

2. Observation frequency
   Frequent observation requires much cost, but enables agents to create an efficient subgoal.

3. Observation cost per intersection
   We define that Observation cost per intersection is A if an agent can observe a intersection as much as move A step.

3 Cooperation Protocol

The cooperation protocol consists of cooperation mechanism and communication protocol.

3.1 Cooperation Mechanism

We have developed some algorithms to solve the model. One of them, DMax (q) algorithm (Delay Maximing Algorithm), is the most efficient algorithm in some algorithms that agents can create a subgoal when they arrive at the pedestal[12]. LM–DMAX (q) algorithm (Limited Movement–Delay Maximizing Algorithm) [11] is a revised version of DMax (q) when agents have their territory. Appendix A gives a brief explanation of LM–DMax(q).

3.2 Communication Protocol

Agents must know another agents' subgoals. We designed a simple communication protocol which assumes a point-to-point communication between agents, and enables agents to get minimum information. We have designed a simple communication protocol containing two messages, *ask_subgoal (source_agent, destination_agent)* and *reply_subgoal (source_agent, destination_agent)*.

An agent can ask subgoals of other agents only when it observes its territory.

4 Evaluation for the Observation

We evaluate observation interval, territory size, the number of agents against the total cost. At first, we make the following definitions.

m:	the total number of agents
d_j:	round trip distance between block b_j and pedestal
s_k:	steps at sample number k
l:	the total number of samples

a rate of overlapping M (%): This is a overlapping ratio of area which multiple agents occupy together to all over the lattice world. If the territory size becomes larger, a rate of overlapping becomes also larger.

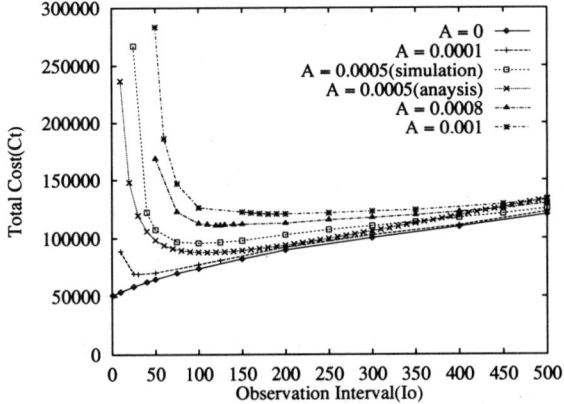

Figure 2: Evaluation of observation interval

LB: LB equals to steps when all agents act in a completely parallel fashion and they carry all blocks without latency.

$$LB_k = \frac{\sum_{j=1}^{l} d_j}{m} \quad (1)$$

denotes LB at sample No. k.

Inefficiency (I): I is greater than 1 and that I approaches m with large m because of the degradation of inherent parallelism of the model.

$$I = \frac{\sum_{k=1}^{l} s_k}{\sum_{k=1}^{l} LB_k} \quad (2)$$

4.1 Simulation Assumption

Table 1: Simulation Assumption

Blocks (n)	1000
Pedestal	center of world
Agent initial position	pedestal
Block initial position	random
Trials (l)	50
q	10
c	10

We assumed that no conflict of intersection exists because we study the influence of observation interval and territory size in detail. If an agent observes its environment, it stays there for a time in proportion to the product of territory size S and observation cost per intersection A.

4.2 Evaluation for Observation Interval

Fig. 2 shows the simulation result for observation interval. Here the territory is assumed to be all over the lattice world. And the total cost when $A = 0$ is the lower bound because of no observation cost. The figure also shows an analytic result obtained in Appendix B.

In case that $A \neq 0$, we found the optimal observation interval that minimizes the total cost for each observation cost per intersection. Total cost increases suddenly if observation interval becomes shorter. This causes agents to require much more time for observation, and to prevent from achieving their subgoal. Conversely the larger observation interval is, the larger the total cost is. As observation interval becomes larger, environmental information with agents are different from actual environment. Because agents create their subgoals based on the inaccurate information, they cannot create appropriate subgoals, as a result the total cost increases.

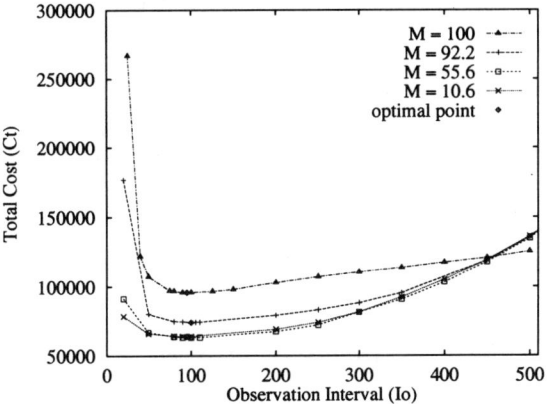

Figure 3: Evaluation of agent's area size (Total Cost, $A = 0.0005$)

4.3 Evaluation of the Number of Agents

In order to evaluate observation interval (I_o) and the total number of agents (m), we assume that space is 200×200, observation cost per intersection (A) equals to 0.0005 besides Table 1.

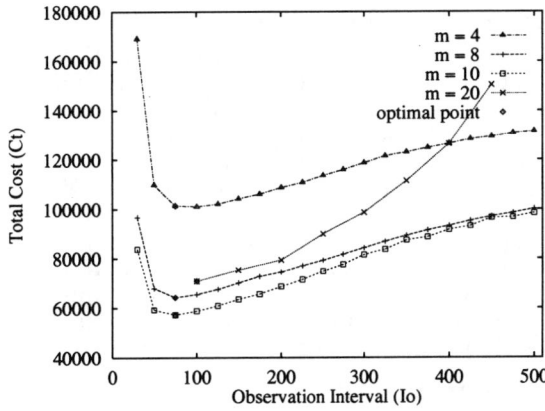

Figure 4: Evaluation of the number of agents (Total Cost, $A = 0.0005$)

From Fig. 4, the optimal observation interval (I_o) is around 75 in spite of the number of agents. In other words, the optimal observation interval is independent parameter of the number of agents. In order to minimize the total cost, a policy can be considered that while fixing observation interval $I_o = 75$, the number of agents varies according to the environmental transition.

4.4 Evaluation of Space Size

To evaluate observation interval (I_o) and the total number of agents (m), we assume that the number of agents (m) equals to 4, observation cost per intersection (A) equals to 0.0005 besides Table 1.

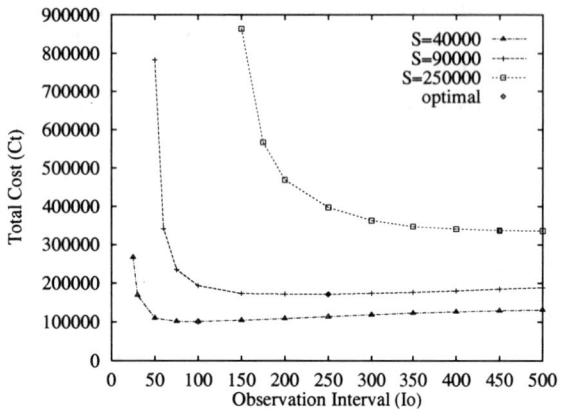

Figure 5: Evaluation of Space Size

As shown by Fig. 5, we found the optimal observation interval for each space size. This optimal observation interval is monotonically increasing against space size. If space size is bigger, observation cost is increasing because agents have larger territory. Furthermore, if observation interval becomes smaller, observation cost increases much more because agents have their territory to observe more frequently. In this case, since agents take less time for achieving their subgoal, the total cost increases.

5 Adaptive Observation Mechanism

As shown the previous section, on condition that observation cost cannot be ignored in multi-agents environment, agents must decide their observation interval taking observation cost per intersection, territory size, and space size into account in response to environmental transition. However, if observation interval is given to an agent initially, the agent cannot create an appropriate subgoal in case of sharp fluctuation in its environment or observation interval being too large. So, agents require an adaptive observation mechanism in which agents establish observation interval autonomously according to environmental change.

5.1 Design Policy

For agents to observe their environment adaptively, agents can examine the following parameters of observation:

Observation Interval

If an agent can modify its observation interval, the agent can establish the balance of moving cost and observation cost well.

Number of Agents

As shown in section 4.3, the optimal observation interval is almost fixed value in spite of the number of agents. It is expected that agents can achieve the final goal if agents estimate required the number of agents, and then they spawn or die by themselves.

Viewing Area of agent

Environmental transition which occurs far from an agent may not affect on making its decision. If it can detect an event only in a circle with a radius around an agent as the viewing area of the agent, the agent can ignore far transition. Or a case may exist which physical condition restricts viewing area. By varying the radius dynamically, agents can save time for observation and concentrate on creating an efficient subgoal.

In this paper, we described observation interval periodically in section 4.2. In order to discuss advantages and disadvantages of periodical observation and adaptive observation, we consider the policy that agents adapt environment by varying observation interval autonomously.

5.2 Observation Interval Decision Algorithm

An agent a_i have observed its territory at time t_n, t_{n+1} (obviously, $t_n < t_{n+1}$), when it observe its territory at time t_{n+3}, a_i decides observation interval I_o as follows:

$$dif_1 := dif(E(t_{n+1}, a_i), E(t_{n+2}, a_i));$$
$$dif_2 := dif(E(t_n, a_i), E(t_{n+1}, a_i));$$
$$if\ (|\frac{dif_2}{t_{n+2}-t_{n+1}} - \frac{dif_1}{t_{n+1}-t_n}| > I_t)\ then$$
$$\quad if\ (\frac{dif_2}{t_{n+2}-t_{n+1}} - \frac{dif_1}{t_{n+1}-t_n} > I_t)\ then$$
$$\quad\quad I_o(a_i) := max(decrease(I_o(a_i), v_d), L)$$
$$\quad else$$
$$\quad\quad I_o(a_i) := min(increase(I_o(a_i), v_u), U);$$

where,

$I_o(a_j)$:	observation interval of agent a_j
$E(t, a_j)$:	environmental information that a_j obtains of its area at time t
$dif(E_1, E_2)$:	difference between environmental information E_1 and E_2
U:	the upper bound of the interval
L:	the lower bound of the interval
$decrease(I_o(a_i), v_d)$:	a function that decreases a_i's interval $I_o(a_i)$ by v_d
$increase(I_o(a_i), v_u)$:	a function that increases a_i's interval $I_o(a_i)$ by v_u
I_t:	threshold to change interval
v_d:	decremental factor of *decrease*

v_u: incremental factor of *increase*

In the above algorithm, when an agent observes its environment, to decide observation interval, an agent uses the recent two environmental information, and finds a degree of environmental transition by function dif, which converts environmental transition into a scalar value. If the term which denotes acceleration of environment, $(|\frac{dif_2}{t_{n+2}-t_{n+1}} - \frac{dif_1}{t_{n+1}-t_n}|)$, is bigger than I_t, the agent decreases its observation interval because environmental changes rapidly. Whereas, if the acceleration is smaller than I_t, the agent increases its observation interval. In this way, agents establish their observation frequency autonomously in response to environmental transition, and make efforts to reduce their observation cost.

The function, *decrease* and *increase*, include an operation which decreases or increases observation interval, respectively. Note that simple function is preferable to ignore time for varying observation interval. In the following section, we examine the geometrical, arithmetical, and slow-start observation interval modification policies.

Observation interval may be much smaller or bigger. If observation interval is much smaller, agents tend to take much time for observation and cannot act for achieving their subgoals. If observation interval is much bigger, it takes much time to achieve their goal because agents don't observe their environment. To avoid these situations, we use an upper limit (U) and an lower limit (L) of observation interval.

5.3 Modification Functions

We consider the following modification functions:

Geometrical policy

     An agent a_i modifies its observation interval using the following function, *decrease* and *increase*, i.e., geometrically:

$$decrease(I_o(a_i), v_d)\{\ I_o(a_i) := I_o(a_i)/v_d\ \}$$
$$increase(I_o(a_i), v_u)\{\ I_o(a_i) := I_o(a_i) \times v_u\ \}$$

Arithmetical policy

     An agent a_i modifies its observation interval using the following function, *decrease* and *increase*, i.e., arithmetically:

$$decrease(I_o(a_i), v_d)\{\ I_o(a_i) := I_o(a_i) - v_d\ \}$$
$$increase(I_o(a_i), v_u)\{\ I_o(a_i) := I_o(a_i) + v_u\ \}$$

Slow-start policy

     This policy uses geometrical decrease function and arithmetical increase function, as TCP dynamic sliding window mechanism[13].

5.4 Evaluation for Modification Functions

     We evaluate the previous modification functions by simulation.

5.4.1 Simulation Assumption

Here U and L are assumed to be 400 and 20, respectively.

5.4.2 Geometrical Policy

Fig. 6 shows an evaluation for geometrical modification under the condition that v, $v_d = v_u = v$, $I_o\_init = 100$. Fig. 7 shows a locus of an agent's observation interval, $v_d = v_u = 1.5$, $It = 3.0$, $I_o\_init = 100$.

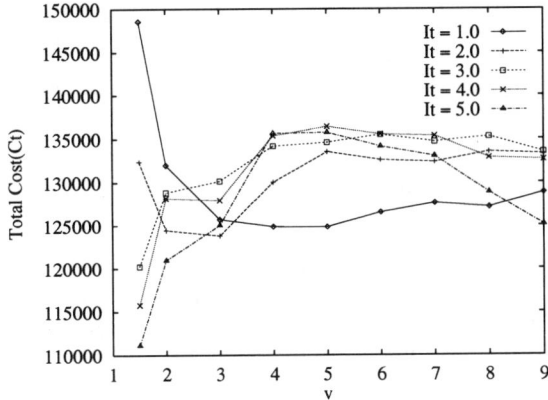

Figure 6: Evaluation for geometrical modification ($I_o\_init = 100$)

As shown in Fig. 6, if I_t is smaller ($I_t \leq 2$), the total cost is concave for v, and if I_t is bigger ($I_t > 2$), the total cost is convex for v.

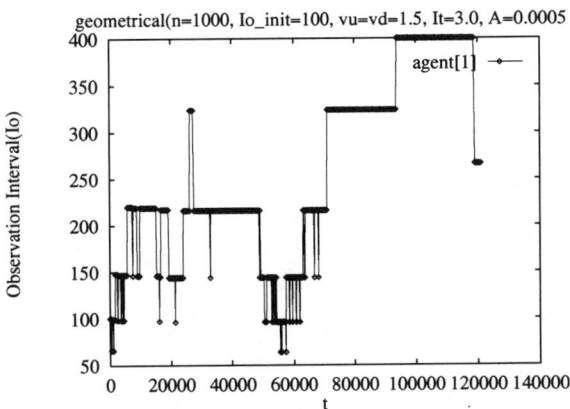

Figure 7: Locus of observation interval (geometrical modification)

We found the total cost converges on a value as I_t is bigger. The value is the total cost (101292.76) in periodical observation, $I_o = 100$. If I_t is bigger, agents cannot modify their observation interval unless the environment is drastically changed (Fig. 7). So, with large I_t, (and more if v is large) observation interval of each agent tends to oscillate.

5.4.3 Arithmetical Policy

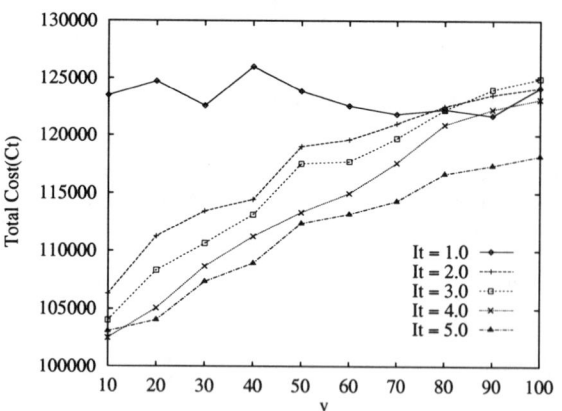

Figure 8: Evaluation for arithmetical modification ($I_o\_init = 100$)

Fig. 8 shows the evaluation for v in arithmetical modification, $v_d = v_u = v$, $I_o\_init = 100$. From the figure, as v increases and/or I_t decreases, the total cost increases almost monotonically.

5.4.4 Evaluation for Slow-start Policy

Fig. 9 shows an evaluation for slow-start modification, $I_t = 5.0$, $I_o\_init = 100$. Fig. 10 shows a locus of an agent's observation interval, when $It = 3.0$, $v_u = 50.0$, $v_u = 3.0$, $I_o\_init = 100$.

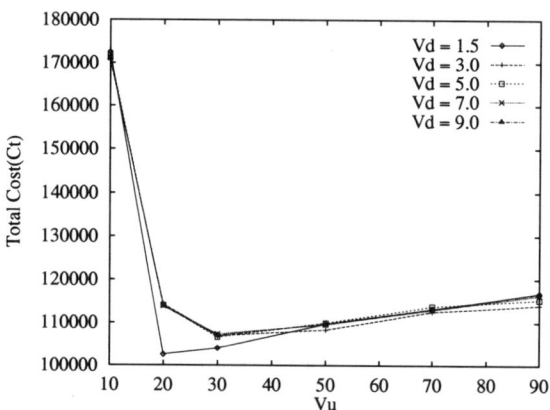

Figure 9: Evaluation for slow-start modification

As shown by Fig. 9, there is the optimal v_u that minimizes total cost. And this value hardly depends on v_d. In addition, the total cost does not depend on v_d, either. And the cost achieved when $v_d = 0.5$ and $v_u = 20$ is the lowest cost among the experiments.

When $v_u = 10.0$, the policy suffer large cost. The reason is that the modification policy tends to denote similar characteristic to geometrical and arithmetical modification. When $v_u > 50.0$, the total cost is gradually increasing. This is because agents set observation

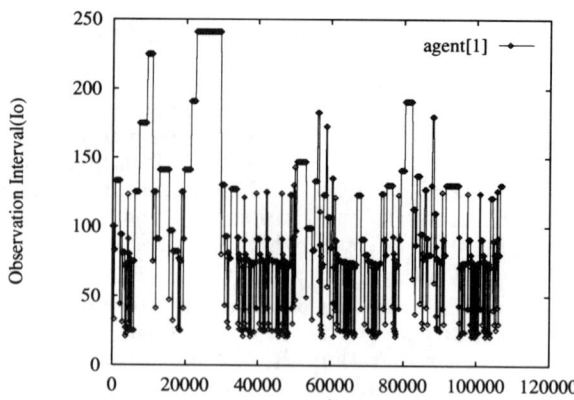

Figure 10: Locus of observation interval (slow-start modification)

interval to bigger value than the optimal observation interval of periodical observation.

From Fig. 10, observation interval is scattered around 75, i.e., the optimal interval from a log term viewpoint.

5.4.5 Comparison of Adaptive Observation Policy with Periodical Observation

We make a comparative evaluation of adaptive observation mechanism and periodical observation because we examine whether adaptive observation mechanism is available.

We assume that observation cost per intersection (A) equals to 0.0005.

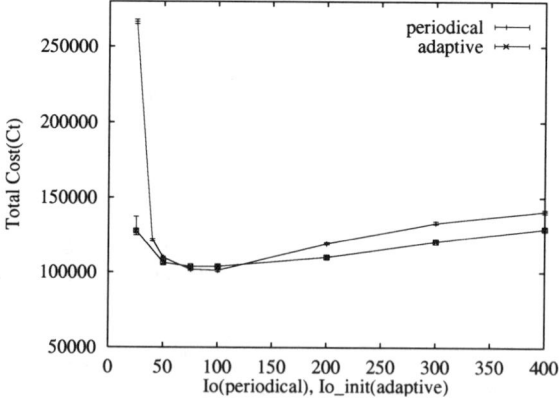

Figure 11: Comparison adaptive observation mechanism with periodical observation

Fig. 11 shows the comparative result of periodical observation and adaptive observation mechanism. As adaptive observation mechanism, we use the result on condition that agents use arithmetical modification in $v_d = v_u = 10.0$, $I_t = 1.0$ as a sample case. In Fig. 11, the x axis is observation interval for periodical observation, and initial observation interval for adaptive observation mechanism, respectively.

From shown by Fig. 11,

- when initial observation interval is smaller ($I_o\_init \leq 50$),

  adaptive observation mechanism is better. In periodical observation, all agents take much time to observe their territory but they cannot modify their observation interval, so the total cost is increasing. In adaptive observation mechanism, agents recognize that the environment hardly changes, so they modify bigger observation interval. In this way, they can dissolve the circumstance, so that the total cost is decreasing.

- when initial observation interval is middle ($75 \leq I_o \leq 200$, $75 \leq I_o\_init \leq 200$),

  periodical observation is better. Using adaptive observation mechanism, the total cost is bigger than periodical observation because agents tend to establish smaller observation interval. However, these differences are almost 2000 step, and we think small difference in comparison with $I_o \leq 50$ ($I_o\_init \leq 50$) or $I_o < 200$ ($I_o\_init < 200$).

- when initial observation interval is bigger ($I_o < 200$, $I_o\_init < 200$),

  adaptive policy is better.

6 Conclusion

In this paper, we discussed the tradeoff between time for environment observation and achievement of subgoal.

We considered territory size, observation cost, time for achieving a final goal on the Tower of Babel. We applied LM–DMax (q) algorithm as subgoal creation algorithm. In case that observation cost be under consideration and agents observe their territory periodically, we show that the optimal observation interval exists.

And we showed that the optimal agent's territory size existed, which enabled agents to achieve the final goal efficiently. Then this paper proposes three adaptive environment observation policies. Through simulation we found the slow-start policy for modification of observation interval is relatively preferable.

As the model is quite simple, the results obtained here are directly applicable to a vehicle scheduling problem and extensible to a real world problem such as automobile guiding vehicles, automatic warehouse, factory automation, bus location systems.

References

[1] Toru Ishida: "Two is not Always Better than One: Experiences in Real-Time Bidirectional Search," *International Conference on Multiagent Systems*, June(1995).

[2] Marcus J. Huber and Edmund H. Durfee: "Deciding When to Commit To Action During Observation-based Coordination," *International Conference on Multiagent Systems*, June(1995).

[3] Smith, R. G.: "The Contract Net Protocol: High-Level Communication and Control in a Distributed Problem Solver", *IEEE Trans. Comput.*, Vol. 29, No. 12, pp. 1104-1113(1980).

[4] Genesereth, M.; Ginsberg, M.; and Rosenschein, J.: "Cooperation without communications," Technical Report 84-36, Stanford Heuristic Programming Project, Computer Science Department, Stanford University, Stanford, California 94305(1984).

[5] Gmytrasiewicz, P. J.; Durfee, E. H.; and Wehe, D. K.: "A decision-theoretic approach to coordinating multiagent interactions," *Twelfth International Joint Conference on Artificial Intelligence*(1991).

[6] Victor Lesser (General chair): *ICMAS-95: First International Conference on Multi-Agent Systems*, The AAAI Press, June(1995).

[7] M. Pollack and M. Ringuette: "Introducing the Tileworld: Experimentally Evaluating Agent Architectures," *The Eighth National Conference on Artificial Intelligence*, pp.183-189(1990).

[8] M. Benda, V. Jagannathan, and R. Dodhiawalla: "On optimal cooperation of knowledge sources," *The 1988 Workshop on Distributed Artificial Intelligence*, May(1988).

[9] E. H. Durfee and T. Montgomery: "MICE: A flexible testbed for intelligent coordination experiments," *The 9th AAAI Distributed Artificial Intelligence Workshop*, pp. 25-40(1989).

[10] Toru Ishida: "Towards Organizational Problem Solving," *IEEE International Conference on Robotics and Automation*, pp. 839-845(1993).

[11] M. Mizuno, T. Watanabe, and T. Mizuno: "Environment Observation Cost of Distributed Autonomous Agents", In Proceedings in *Multimedia Communication and Distributed Processing System Workshop*, Information Processing Society of Japan, pp.39-46, October(1994) (in Japanese).

[12] T. Watanabe, T. Yamazaki, and M. Mizuno: "Dynamic Subgoal Generation of Autonomous Agents Moving in a Lattice World," *The Second International Symposium on Autonomous Decentralized Systems*, April(1995).

[13] Douglas Comer: "Internetworking With TCP/IP Vol I: Principles, Protocols, and Architecture," 3rd ed., Prentice-Hall, (1995).

A Brief Explanation of LM–DMax(q)

In the following, an agent has just put a block numbered k (b_k) in the environment that m agents and n blocks exist.

LM–DMax (q) algorithm enables agents to create subgoals which has the largest interval after other agent puts a block at the pedestal. If an agent only can move the blocks, the agent moves the smallest numbered block immediately. That is, agent x creates a subgoal b_j as follows:

1. If no other agents are moving b_{k+1} and the agent, x can move b_{k+1}, agent x takes b_{k+1}.

2. If the agent x is only one which can move a block among the remaining q blocks.

3. Among the remaining q blocks which aren't carried by another agents, the agent x selects a block which has the most largest interval between the time which agent x can put the block and the time another agent have put b_{k+1} at the pedestal.

If the pedestal number is k and an agent, which carries a block numbered more than $k+2$, approaches the pedestal, the agent has to wait around the pedestal because it cannot put the block. For this problem, if an agent have considered that it cannot achieve the current subgoal immediately, it modifies its subgoal autonomously and puts the block there, and creates the next subgoal. We call this scheme self-regeneration scheme.

Furthermore, in case of intersection conflict, another agents and blocks are obstacles. If some blocks surround the pedestal, the final goal cannot be achieved. At that time LM–DMAX makes a room for other agents to evade placed blocks around the pedestal. The following terms are defined:

number of the top block of the tower:	a
agent subgoal number:	b
number of vacant intersection:	c

An agent put a block at a intersection at a distance r from the pedestal;

$$r = \lfloor \frac{-1 + \sqrt{1 + 2 \times (b - a - 1) \times c}}{2} \rfloor \quad (3)$$

B Evaluation for Observation Interval

In this paper, the total cost is approximated by the sum of moving cost C_m and observation cost C_o. Therefore,

$$C_t = C_m + C_o. \quad (4)$$

In $A = 0$, C_o equals to 0. So observation interval influences the total cost only. From Fig. 2, we can approximate that the specific characteristic in $A = 0$ is a linear function of I_o. Thus,

$$\begin{aligned} C_t|_{A=0} \\ = C_m \\ = \alpha \times I_o + \beta. \end{aligned} \quad (5)$$

Equation (5) shows while agents observe their environment taking some time, early observed information gets out-dated, and agents may create inappropriate

subgoals. So we can consider that a linear relationship is established between the total cost and increased cost which inappropriate subgoals may cause.

And observation cost is inversely proportional to observation interval, and linearly increases to observation cost per intersection. Further similarly as equation (5), it is considered that observation cost increases in proportion to observation cost per intersection. From Fig. 2, given constants, γ, δ, the equation for the observation cost is

$$C_o = \frac{(\gamma A + \delta) A}{I_o}. \quad (6)$$

Hence the total cost, C_t, is

$$C_t = C_1 I_o + C_2 + \frac{(C_3 A + C_4) A}{I_o}. \quad (7)$$

From Fig. 2, $\alpha, \beta, \gamma, \delta$ are approximated as follows, respectively.

$$\begin{aligned} \alpha &= 147 \\ \beta &= 56086 \\ \gamma &= 2.8 \times 10^{12} \\ \delta &= 2.2 \times 10^9 \end{aligned}$$

By partial differentiation with respect to I_o, an optimal observation interval I_{opt} is derived.

$$I_{opt} = \sqrt{\frac{(\gamma A + \delta) A}{\alpha}} \quad (8)$$

Fig. 2 shows the equation (7) in $A = 0.0005$.

A Negotiating Agents Model for the Provision of Flexible Telephony Services

M. Rizzo
Dept. of Computer Science & A.I.
University of Malta
Msida MSD 06
Malta

I. A. Utting
Computing Laboratory
University of Kent
Canterbury, Kent CT2 7NF
England

Abstract

Current telephone systems suffer from a service interface bottleneck problem, wherein network resources are under-utilised and customer requirements are often not met, in spite of these resources' ability to satisfy such requirements. This bottleneck is primarily due to a coarse-grain service interface, coupled with the inability to support arbitrary terminal types, and the inability to inter-operate with other systems.

This paper outlines a new model for telephony services, based on the concept of negotiating agents. In this model, functionality is not made available to users in the form of services, as has been the case traditionally. Instead, users specify policies that describe how they wish their calls to be handled. These policies are used to guide agents appointed to act on behalf of users. The paper also describes a prototype that was built to demonstrate the capabilities of the model.

1 Introduction

Developments in telecommunications technology have, in recent years, enabled network operators to introduce a range of supplementary services. These first appeared as extensions to the traditional Public Switched Telephone Network (PSTN), then in the Integrated Services Digital Network (ISDN), and more recently in the Intelligent Network (IN).

Supplementary services modify the normal call setup process to provide some useful function. Popular examples are *Freephone*, *Call Forwarding on Busy*, and *Call Waiting*. In Call Forwarding on Busy, for instance, call setup is modified so that if the destination number is busy, the call is routed to another number.

Many supplementary services allow some degree of customisation, thereby enabling subscribers to tailor services to their specific needs, e.g. Call Forwarding on Busy allows subscribers to specify a forwarding number. However, the extent to which such services may

be customised is generally rather limited, offering little expressive power to the subscriber, who is only allowed to specify values to service parameters and turn services on and off. Consequently, these services cannot always meet customer demands, even if network resources could, in principle, be used to provide the required functionality. This problem is termed the *service interface bottleneck*.

This paper describes NAT (Negotiating Agents for Telephony), an alternative model for the provision of telephony services which aims to deliver increased functionality to end users. Rather than package functionality into a set of rigid, inflexible services, NAT allows users to express policies which describe how they would like their calls to be managed. These policies are used to guide users' agents, which negotiate and take action on behalf of their users. The aim of negotiation is to find a suitable course of action that is acceptable to all users concerned.

The paper is structured as follows: section 2 describes the service interface bottleneck, showing how it is related to the well-known problem of *feature interaction* [1]; section 3 reviews relevant research activities; section 4 introduces the NAT model; section 5 describes a prototype system that was built to demonstrate the capabilities of the model; section 6 concludes with an evaluation of the NAT approach.

2 The service interface bottleneck

Consider a telephone user X who wishes to have all his/her calls forwarded to a user Y if Y is not busy at the time of the call, or to another user Z otherwise. Clearly, a stored program control exchange is capable of doing this, and any necessary control messages could easily be carried over a modern digital transmission signalling network. Yet, such a service is not defined for PSTN and ISDN supplementary services, nor is it listed in existing IN capability sets.

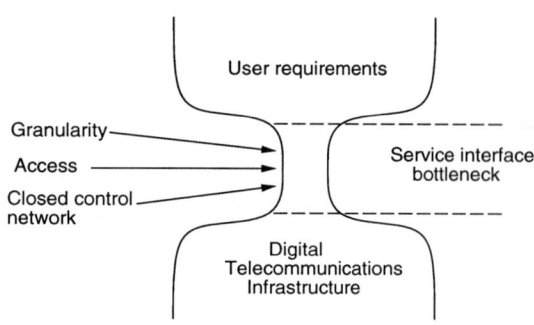

Figure 1: The service interface bottleneck

Unfortunately, the problem highlighted by this simple example is characteristic of today's telephone systems: the service software that sits between the transport network and the user represents a *service interface bottleneck* (figure 1) that often results in the under-utilisation of network resources and the inability to meet subscribers' needs, in spite of the fact that the resources *per se* are able to meet these needs. Three principal causes of this bottleneck are:

- *service interface granularity*: the service interface offered to users is too coarse-grained, offering a limited number of highly specialised, inflexible services which suffer from feature interaction;

- *service interface access*: user terminal equipment is rather restrictive, and support for introduction of new terminal types is cumbersome;

- *closed control network*: service logic must execute on network operator equipment and cannot utilise sub-systems that are available on customer premises equipment.

NAT was designed with all the above causes in mind. However, this paper will focus primarily on the interface granularity aspect, as this turned out to be more interesting than the others. For further discussion of the service interface bottleneck problem see [2]. Essentially, the interface granularity problem prevents users from indicating how they would like their telephone calls to be managed. The interface made available to them is not flexible enough to allow them to specify their requirements. Furthermore, evolution of this interface is controlled by network operators and is made difficult because of feature interaction.

The remainder of this section illustrates the interface granularity problem (2.1), and discusses how it affects service deployment and scalability (2.2).

2.1 Illustration

The interface granularity problem will be illustrated using the Call Forwarding on Busy/No Answer (CFC) service feature as described for Capability Set 1 (CS-1) in clause B.2.8 of ITU-T Q.1211:

> *Call forwarding on busy/no answer* (CFC): This service feature allows the called user to forward particular calls if the called user is busy or does not answer within a specified number of rings.

Throughout this illustration, *scenarios* will be used to illustrate problems and concepts. A scenario is described in terms of a list of participating subscribers, a description of initial conditions, and a sequence of events relating to a particular call. Our first scenario illustrates the basic operation of CFC:

Scenario 1: Basic operation of CFC
Parties: A, B, C.
Setting: B has enabled CFC to C.
Sequence: A calls B; B is busy.

The outcome of this scenario is fully determined by the status of C. In the event that C is busy then A hears busy tone. Otherwise A hears ring tone until either C picks up the phone, in which case A is put through to C, or A gives up and replaces the handset.

Now consider the following scenario:

Scenario 2: Feature interaction in CFC
Parties: A, B, C, D.
Setting: B has enabled CFC to C. C has enabled CFC to D.
Sequence: A calls B; B is busy.

In this scenario the outcome is not necessarily determined by the status of C. If C is not busy, then A is put through to C, but if C is busy then there are two possible outcomes: either A hears busy tone (as in scenario 1), or C forwards the call to D. As things stand, this choice is non-deterministic because the description of CFC given earlier is incomplete: it makes no explicit statements about what should happen if the user to whom the call is forwarded is also busy, or does not answer within a specified number of rings.

Thus, two possible interpretations of the description of CFC in the context of this scenario are: (i) that the attempt should fail and the calling user should hear busy tone, and (ii) that the user to whom the call was forwarded should be viewed as the new called user, so that CFC may be invoked again if required.

In fact in CS-1, CFC is actually a feature of an IN type A service (Q.1211, clause 5.1), which effectively dictates that the correct interpretation of the description of CFC is (ii). However, this does not necessarily correspond to what is required: some subscribers may require that their incoming calls are not forwarded to any numbers other than those which are explicitly named, in which case the behaviour described in interpretation (i) is desirable. This exemplifies how the use of *static resolution rules* to combat feature interaction contributes to the service interface bottleneck.

Next, consider a subscriber requiring the ability to specify two forwarding numbers for CFC. This might be provided in the form of a 'slightly' generalised form of CFC, say *Double Call Forwarding on Busy/No Answer* (DCFC). The following scenario shows how DCFC might be used:

Scenario 3: Double call forwarding with DCFC
Parties: A, B, C, D.
Setting: B has enabled DCFC to C and D in that order.
Sequence: A calls B; B is busy; C is busy.

The outcome of this particular scenario is unambiguous: B attempts to forward the call to C but fails and so re-attempts to forward the call to D, which accepts the call. However, along with the introduction of this new feature comes at least one new feature interaction type, which is exemplified by the next scenario:

Scenario 4: DCFC interacting with CFC
Parties: A, B, C, D, E.
Setting: B has enabled DCFC to C and D in that order. C has enabled CFC to E.
Sequence: A calls B; B is busy; C is busy.

In this example there are two possible courses of action once the call has reached C. The call might be forwarded to either D or E, depending on whether B's or C's feature is invoked! Either behaviour might be acceptable in the right circumstances; it would be wrong for the definition of DCFC to stipulate that paths offered by the first forwarding number should or should not be tried before trying the second DCFC forwarding number. Ideally, the decision should take into account the preferences of all the parties involved, with those of the caller A being given priority.

On to the next example! Along comes another subscriber asking for CFC with three forwarding numbers, and yet another asking for five. The CFC service provider decides to extend the coarse-grain service interface by providing *Multiple Call Forwarding on Busy/No Answer* (MCFC) with n numbers where n is a number deemed sufficiently large to satisfy all subscribers' demands.

Scenario 5: Multiple call forwarding with MCFC
Parties: $A, B, C_1, C_2, \ldots, C_n$.
Setting: B has enabled MCFC to $C_1, C_2, \ldots, C_n$ in that order.
Sequence: A calls B; B is busy ; $C_1, C_2, \ldots, C_j$ $(1 \le j \le n)$ are busy.

If $j < n$, then the outcome can be described as follows: B cannot accept the call and forwards the call to C_1; C_1 cannot accept the call so B forwards the call to C_2; $\ldots$; C_j cannot accept the call so B forwards the call to C_{j+1}; C_{j+1} accepts the call. The remaining case i.e. $j = n$, results in A hearing busy tone.

MCFC represents a generalisation of CFC that is somewhat better able to meet customer requirements in the sense that the range of possibilities it offers is slightly larger. But it must be stressed that it is only *slightly* larger, because the improvement is not that significant in the context of the range of call management requirements that subscribers might have! Consider a subscriber requiring a combination of MCFC and *Call Waiting* (CW) which behaves such that if all MCFC forwarding numbers are busy, CW is invoked. The service provider can provide yet another new service but clearly there are limits to this approach, because of the infinite number of ways in which service features can be combined[1].

2.2 Service deployment and scalability

The illustration above highlights the principal problem with current approaches to service provision, namely that the granularity of the service interface provided to subscribers is too coarse. Functionality is packaged into simple features designed around what the service provider believes that subscribers need, offering very limited expressive power to the subscriber.

The service interface granularity problem cannot simply be solved by adding more services, for at least two reasons. Firstly, it does not make sense to deploy network-wide services which, because they are so specialised, will only be used by a small number of people. This problem can be avoided by the IN architecture to some extent, but such services still have to be subjected to regulation by the network provider. Even if this heavyweight service acceptance procedure was removed to allow unconstrained addition of new IN services by third parties, this would easily overwhelm the network's resources: it would be better to utilise third-party equipment to handle such specialist services. A second reason is that with each service that is added,

---

[1]The notion of creating new services by combining others does exist in CS-1 e.g. *Selective Call Forwarding on Busy/No Answer* combines CFC with *Terminating Call Screening*.

the feature interaction problem is compounded. The more services that are added, the greater the potential for interaction and subsequent undesirable behaviour.

The service interface granularity problem can only be overcome by providing a lower-level service interface to subscribers, giving them more expressive power that allows them to specify call management requirements in terms of finer-grain primitives than service features. But doing this requires careful thought and consideration. Giving subscribers more flexibility could lead to abuse or incorrect use of this flexibility. Moreover, a significant increase in expressive power could very well overwhelm subscribers with simpler requirements and limited know-how.

3 Related work

This section briefly reviews other work which is relevant to the issues highlighted in the previous section.

3.1 The IN architecture

The IN architecture offers more service control flexibility than ISDN and its predecessors by providing a wider range of supplementary services. However, as we noted earlier, a finite number of pre-defined services can never be enough to satisfy all users' requirements, regardless of the number of services available.

The IN architecture's main improvement is that it makes service creation by third-party service providers possible. Unfortunately the approach is cumbersome and requires updating of components that fall under the auspices of the network operator. This does not permit lightweight deployment of behaviour tailored to one user's specific needs.

Another problem with IN is that the all-important Basic Call Processing (BCP) module still represents traditional circuit-related control. IN service invocation can only take place after the BCP has started, with the result that the IN services appear as an extension of the traditional telephone service instead of services in their own right.

Feature interaction continues to be the dominating problem within the IN community. Several solutions have been proposed, but none seems to be general enough to capture the problem in its entirety.

3.2 Velthuijsen & Griffeth model

Velthuijsen and Griffeth [3] propose a model which aims to avoid the need to study the behaviour of features taken in combination each time a new service is introduced. Users, terminals, and the bearer network are all represented by agents which negotiate with each other in order to establish a course of action which will

achieve some specified goal, the assumption being that there may be several ways to satisfy a goal, not all of which may be acceptable to the parties concerned.

This approach is useful in resolving feature interactions where a pre-defined set of service features is deployed across an entire network. However, because agents must share a common goal hierarchy, which essentially determines the service features available, it is not possible to support lightweight introduction of new services, meaning that the model still suffers from the interface granularity problem.

Nevertheless, this negotiation model certainly represents a step in the right direction with respect to the feature interaction problem, because it avoids the need to check all possible combinations of features. The NAT model we will be describing attempts to go a step further by avoiding the need to have standardised, network-wide features.

3.3 ODP and TINA

There has recently been much activity in the area of *Open Distributed Processing* (ODP), including standardisation of a reference model (RM-ODP) [4], which addresses issues relating to the development of heterogeneous, large-scale distributed systems. RM-ODP centres around an object-based paradigm and makes extensive use of the principles of encapsulation and abstraction. Two fundamental concepts in RM-ODP are: *viewpoints* for separation of concerns, and stream bindings to support communication of continuous media. Both are used extensively in the NAT model.

One of the first applications to which ODP technology is being put is that of telecommunication services over broadband networks, the main initiative here being the *Telecommunications Information Networking Architecture* (TINA) [5]. TINA has no notion of a basic call process as in IN and its predecessors. Telephony is simply one kind of application, for which several access interfaces may be provided. Clearly there is a lot of potential for reducing the granularity aspect of the service interface bottleneck [6]. However, because TINA is a general architecture, it does not prescribe too much detail relating to the structure of services, and consequently it does not provide any insight as to how the granularity problem should be tackled.

4 The NAT model

NAT is a framework for the provision of flexible telephone services. It is described in terms of concepts and notations of RM-ODP.

NAT provides modelling concepts for the design of telephone systems, and is particularly geared towards

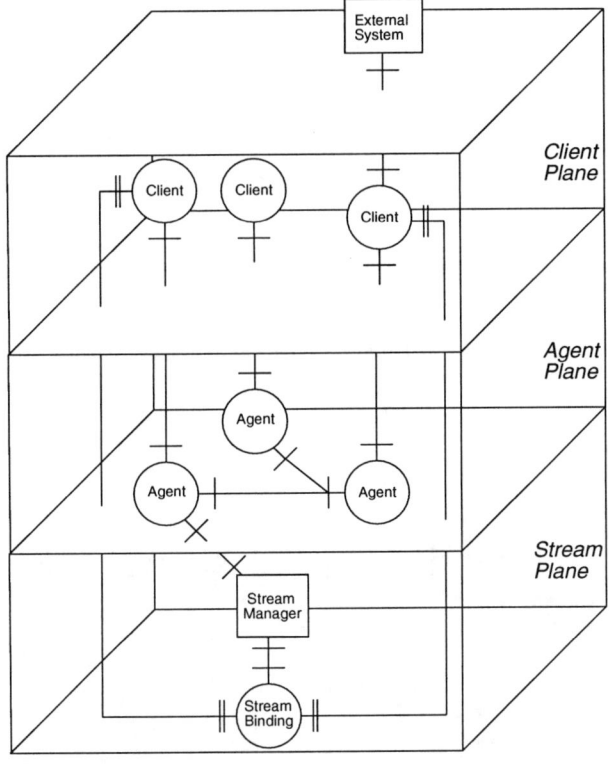

Figure 2: The NAT computational model

When a client wishes to pursue some task that necessitates participation of other clients, its agent negotiates with these clients' agents in order to find a course of action that is acceptable to all involved. In the course of the negotiation process, agents may exchange questions, answers, proposals and counterproposals until some consensus is reached, or until it is established that no consensus can be reached. New agents may be brought into, and others may be withdrawn from, the process at any time before the negotiation is completed. Once a consensus is reached, transmission paths may be set up over an underlying transport network. The transport network may actually be capable of carrying several types of multimedia traffic, and is not necessarily restricted to voice.

The computational viewpoint for NAT-based systems may be specified in terms of three planes, namely the stream plane, the agent plane, and the client plane. Each of these has a static aspect and a dynamic aspect. The static aspect defines relationships between object and interface types, whereas the dynamic aspect describes relationships between instances of such types. Figure 2 depicts a particular configuration of the dynamic aspects of all three planes of a NAT system at some point in time.

The stream plane defines a multimedia transport service in terms of a set I of stream interface types, and a set B of stream binding types. I determines the kinds of subscriber terminals and teleservices supported by the network with respect to the information flows that may be carried over the lifetime of a call. It does not, however, have any say as to how such terminals or teleservices should be controlled. B determines the kinds of information communication that can take place over the course of a call. However, creation of and subsequent control over bindings is not the responsibility of the stream plane.

The agent plane contains agents which act on behalf of clients in the client plane. Each agent communicates with its client via a *client interface*. The static aspect of the agent plane defines a set of agent types and a set of client interface types. In the dynamic aspect of the agent plane, each client interface instance provided by a client is paired with exactly one agent instance and vice-versa. Agents also host a negotiation interface via which they negotiate amongst each other according to some *negotiation protocol*.

The client plane defines a number of client types, each of which supports exactly one client interface type. The same client interface type may be supported by more than one client type, e.g. for a given client interface, one client may provide an implementation

eliminating, or at least widening, the service-interface bottleneck by (i) making a finer-grain interface accessible to end-users, (ii) facilitating the connection and management of arbitrary third-party terminals, and (iii) supporting integration with other systems. Thus well-designed telephone systems based on NAT are better able to meet customer requirements, and can make use of network resources more efficiently.

NAT is not a design for any one particular telephone system, but is a framework that provides concepts which may be used to construct several such designs. In a particular NAT-based system, the degree to which customer requirements are met and network resources are utilised ultimately rests on the design of the system. Thus simply basing a design on NAT does not necessarily constitute a guarantee for customer satisfaction and efficient resource utilisation.

NAT centres around the concept of a *negotiating agent*. A negotiating agent is an object that engages in and conducts negotiations with other such agents on behalf of some *client* for which it has been appointed. Several kinds of client may be represented by a negotiating agent, including users, terminals, teleservices, administrators, user groups, and organisations.

355

strmgr	Stream manager: manages bi-directional bindings between audio stream interfaces.
usterm	User terminal: provides an audio stream interface, a user interface, and a client interface for interacting with a user agent.
usagent	User agent: provides a client interface for interaction with a user terminal, a negotiation interface for negotiating with other agents, and a policy specification language for expressing user requirements.
stardev	Bridges the workstation network with the PABX network to enable exchange of voice across the two (based on the StarT32 [7]).
starman	Manages a pool of stardev components.
stagent	Gateway agent: translates between negotiation messages in the workstation domain and signalling messages in the PABX domain.

Table 1: NEMO-2 components

based around the standard telephone, whilst another might provide an implementation based on a multimedia workstation.

Further details on the NAT model can be found in [2]. The next section gives a clearer indication of NAT's capabilities by describing a more concrete demonstration system.

5   NAT demonstration system

NEMO-2 (NAT dEMO 2) is a NAT-based system that makes use of various resources available at the University of Kent at Canterbury (UKC) including the Computer Laboratory's computer network, an active badge network [8], and the university's internal telephone network. The system was implemented using ANSAware [9] as an engineering platform.

In NEMO-2, users communicate via workstation-based terminals attached to the computer network, and via telephones connected to UKC's internal telephone network. Each user is represented by an agent, and a policy specification language is defined to allow expression of call management policies.

Table 1 lists the components that make up NEMO-2. Figure 3 illustrates how these components would interact in a call between a workstation terminal and a telephone.

5.1   Negotiation protocol

The negotiation protocol used in NEMO-2 centres around the notion of a *negotiation session* between an initiator agent and a responder agent. Within a session, agents may ask each other questions about their intentions. At some point, the initiator may is-

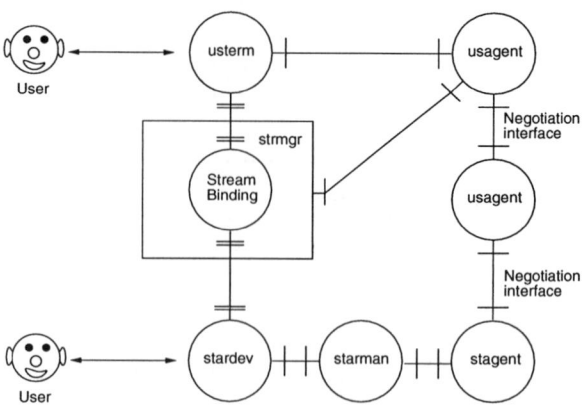

Figure 3: Call between workstation and telephone

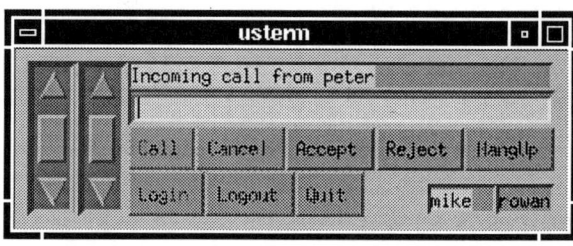

Figure 4: Workstation-based terminal for user agents

sue a request for connection. The responder can react by meeting the request or by offering an alternative course of action, referred to as a *fallback*. In the latter case, the initiator is allowed to start another exchange of questions and answers in order to discover information about the proposed fallback. The initiator might then choose to pursue the fallback or check whether the responder has other fallbacks to offer.

5.2   User agent terminal

The user agent terminal makes use of workstation audio hardware and a graphical user interface (GUI). The audio hardware is used to provide a bi-directional audio stream interface, and the GUI enables users to communicate with their agents.

Figure 4 shows the GUI presented by the user agent terminal, comprising a message display, an input window, sliders for controlling input and output volume, and a number of control buttons. In order to use the terminal, a user must first log in. Subsequently the user may use the terminal to place outbound calls and to receive incoming calls. The figure depicts an incoming call from a user *peter* for the logged user *mike*. At

```
String coffeeroom := "SW101"
String lab := "SW103"
StringSet donotdisturb := [coffeeroom,lab,"S120"]
StringSet pricallers := ["ian","peter","john"]
BhvExpr dcf := forward("peter") {
    nolocate,noanswer,busy,reject,
    fallback,other -> forward("ian") {
        nolocate,noanswer,busy,reject,
        fallback,other -> giveup
    }
}
BhvExpr con := connect() {
    nolocate,noanswer,busy,reject,other -> dcf
}
BhvExpr incoming := callerid(pricallers) {
    true -> con
    false,dontknow -> timein(09:00,17:00) {
        true -> location(donotdisturb) {
            true -> dcf
            false,dontknow -> con
        }
        false -> dcf
    }
}
InitiatorForwApproval(on)
ResponderPolicy(incoming)
```

Figure 5: Policy specification for user Mike

this point, user *mike* can press either of the **Accept** or **Reject** buttons to accept or reject the call respectively.

5.3 Policy specification language

The policy specification language is particularly geared towards supporting flexible call forwarding policies, which may make use of users' identities, the time of day, and users' locations (obtained via the UKC active badge system).

A policy consists of a list of *constant definitions* and *policy statements*. Constant types available include: strings, string sets, booleans, and *behaviour expressions*. The latter can be used to descibe a *behaviour tree* which may then be used to guide an agent.

Figure 5 shows the policy, containing two policy statements, for a user Mike. The first statement enables a feature of the NEMO-2 agent called **InitiatorForwApproval**. This causes Mike's agent to ask for Mike's approval before any of his calls are forwarded. The second describes how Mike's incoming calls are to be handled in terms of the behaviour expression **incoming**. This specifies that if the caller is in the list of priority callers, then a connection attempt

```
BhvExpr incoming := timein(9:00,17:00) {
    true -> connect() {
        nolocate,noanswer,busy,
        reject,other -> giveup
    }
    false -> giveup
}
InitiatorForwApproval(on)
InitiatorForwScreen(["peter"])
ResponderPolicy(incoming)
```

Figure 6: Policy specification for user Dave

should be made, otherwise the action to be taken depends on the time of day.

5.4 Example scenario

Consider a user Dave, whose policy is shown in figure 6, calling Mike at 10:00 whilst Mike is in the coffee room. Dave's policy makes use of another agent feature, namely **InitiatorForwScreen**, which indicates that Dave's calls should *never* be forwarded to Peter. Dave also has **InitiatorForwApproval** enabled, so that forwarding attempts other than those to Peter may be approved interactively. The ensuing sequence of events is described in figure 7 (D denotes Dave's agent, M denotes Mike's agent).

6 Discussion

The contribution of this paper is twofold: (i) it views current approaches to telephone networks from a new perspective, exposing problems relating to satisfaction of customer requirements and efficient use of network resources, as well as throwing some light on the problem of feature interaction, and (ii) it suggests an alternative approach to telephone networks, emphasizing the need for de-centralised control and increased communication amongst the individual components that make up a system.

It is clear that the current philosophy of making functionality available as a tightly-controlled set of rigid, inflexible service units is not a good long-term paradigm for the telephony services of the future. This raises serious questions about whether it is worth spending so much time and effort investigating the feature interaction problem, when this problem arises only as a consequence of a particular choice of service provision paradigm. Indeed the feature interaction problem may be unsolvable, and the only way around it might be to make it disappear by adopting an altogether different approach to service provision.

1. D requests a stream interface reference from M.
2. M asks for the identity of the caller.
3. D obliges.
4. M determines that the caller is not in the list of priority callers.
5. M determines that the time of day is between 09:00 and 17:00.
6. M determines that Mike is in the coffee room.
7. M rejects D's request.
8. D asks M whether it has any fallbacks to offer.
9. M indicates that it is willing to forward the call to another user.
10. In view of Dave's policy statements on call forwarding, D asks M for the identity of this user.
11. M obliges, supplying the identity *peter*.
12. D immediately informs M that it is not interested in this course of action, and asks whether another fallback is available.
13. Again M indicates that it is willing to forward the call to yet another user.
14. Again D asks M for the identity of this user.
15. M obliges, supplying the identity *ian*.
16. D displays a message on Dave's terminal, asking Dave to approve the forwarding attempt.
17. Dave presses the `Accept` button to approve the forwarding attempt.
18. D informs M of Dave's decision, and M then tries to set up a call with Ian.

Figure 7: Sequence of events in example scenario

The NAT model represents one such alternative approach. It allows users to express their requirements in terms of policies rather than service features, making it possible for users to specify a much wider range of requirements. Moreover, the negotiation mechanism enables *all* participants in a call to have a say in the way the call is processed, and conflicts are resolved dynamically by taking all policies into consideration, rather than statically by a central service provider.

Of course the approach suggested by NAT also introduces a number of new problems, many of which require further research before the approach can be adopted in large-scale systems. For instance, new approaches to charging and marketing are required, because functionality is no longer packaged into convenient tarriffable, easily-understandable units. The choice of negotiation mechanism is also extremely important, and should be designed to allow evolution and interworking with legacy systems or systems utilising different negotiation mechanisms. Care is needed to ensure that the increased flexibility offered by the approach does not compromise safety and reliability.

Last but not least, dimensioning to meet performance guarantees is made harder by the proposed approach, as it is not possible to predict utilisation of the network in advance of negotiation.

The importance of each of these problems may vary, depending on the applications to which the model is put: whilst charging may be an important concern in a public network, it is not likely to be an issue in a localised office communications system.

The NAT model is seen to be very much in the spirit of TINA. Work is currently underway to integrate NAT concepts into the TINA framework.

Acknowledgment

The authors are grateful to Olivetti Research Limited (UK) for donating an Active Badge system.

References

[1] E. J. Cameron *et al.*, "A feature-interaction benchmark for IN and beyond," *IEEE Communications Magazine*, vol. 11, pp. 64–69, Mar. 1993.

[2] M. Rizzo, *A Model for Flexible Telephony Services based on Negotiating Agents*. PhD thesis, Computing Laboratory, University of Kent, UK, 1996.

[3] N. D. Griffeth and H. Velthuijsen, "The negotiating agents approach to runtime feature interaction resolution," in *Proc. of 2nd Int. Workshop on Feature Interactions in Telecommunications Systems*, pp. 217–235, IOS Press, 1994.

[4] P. F. Linington, "Introduction to the basic reference model of open distributed processing," *IFIP Transactions C, Special Issue on Open Distributed Processing*, vol. C-1, pp. 3–13, 1992.

[5] M. Chapman and S. Montesi, *Overall Concepts and Principles of TINA (TB_MDC.018_1.0_94)*. TINA Consortium, 331 Newman Springs Road, Redbank, NJ 07701, USA, Feb. 1995.

[6] R. Minerva, "TINA service architecture: some issues in service control," in *Proc. of TINA '95, Melbourne, Australia*, pp. 97–112, Feb. 1995.

[7] Staria Ltd, 13 Market Place, Ross-on-Wye, Herefordshire, UK, *StarT32 Programmer's Manual*.

[8] A. Harter and A. Hopper, "A distributed location system for the active office," *IEEE Network*, vol. 8, pp. 62–70, Jan. 1994.

[9] Architecture Projects Management Ltd, Poseidon House, Castle Park, Cambridge, UK, *RM.099.02: An Overview of ANSAware 4.1*, May 1992.

Architectural Considerations about
Open Distributed Agent Support Platforms

M. Mendes[1/2], O. Falsarella[1], I. Fontes[3], S. Krause[5], W. Loyolla[1/2],
C. Mendez[4], P.S. Silva[3], C. Tobar[2]

[1] *PUCCAMP -Campinas Brazil,* [2] *UNICAMP-Campinas, Brazil,* [3] *UNESP-Bauru, Brazil,*
[4] *UMAG- Punta Arenas, Chile,* [5] *TU-Berlin, Germany*

Abstract

Agents are becoming one of the most important topics in massively distributed and autonomous decentralized systems. Here, agents are considered as software computational entities, created and supported through Agent Platforms, and acting in an Agent Environment.

Based on some recognized agent skills and roles, this paper presents a discussion about the capabilities required in order that a standardized platform supports the action of agents in distributed heterogeneous systems.

After posing some architectural goals and principles, a proposal is presented for an Agent Platform Architecture that offers a set of configurable functionalities aiming to support generic agent necessities within their execution environment. The proposal follows the concepts of agent facilities described for Corba environments.

1. Introduction[1]

The imperative necessity for improving work efficiency, due to current world competitiveness, is leading enterprises to review their working styles and to create work integration and work cooperation with many partners. Organizational improvement has been driven by evolutionary results in several areas, such as new management techniques, new operational methods and new computational technologies. However, in the field of Information Technology, a diversity of computing platforms, information structure standards, software development techniques and local and wide area networks causes discrepant tendencies to the way enterprises integrate and cooperate.

Nevertheless, efforts to develop and standardize distributed information processing, storing, and exchanging have been major elements to allow enterprises to develop work integration and cooperation {iso94a} {iso94b}

{hars96}. Advances provided by new computer and telecommunication technologies are promising "information at any time, at any place, in any form" through the development of new technologies as ubiquitous and mobile computing and of new abstract concepts such as the Virtual Environment (VE) {niii96}. VE is an open "electronic" place where new flexible and powerful forms for enterprise integration and cooperation could take place, with or without direct interference and control by human beings.

Agent technology will play a major role for the realization of VE applications, as well as for a wide range of other problem domains, since the trends for modern enterprises point to largely autonomous, decentralized and geographically dispersed computing systems. Generally, an agent can be any entity (physical or logical) that is responsible for the execution of a set of tasks, delegated to it by an user human-being or another entity. *Software agents* differ from other software systems because of their broad skills, particularly the abilities of autonomy and itinerant execution combined with a certain degree of intelligent and cooperative behavior {page96}. Issues related to multi-agent systems {dalg95}, intelligent agents {cacm94}, information agents {nwana96}, interface-agents {maes94}, and mobile agents {chess95} are not treated here as separate aspects, but as complementary and combined aspects, that result in very flexible software systems.

A computational infrastructure is necessary to support the activities of agents. This infrastructure varies according to the characteristics and design of the agents that need to be supported. An *Agent Platform* (or *Agency*) is related to the infrastructure necessary in each computer node of the distributed system, in order to support all the necessities of an agent during its existence. Because there are so many types of agents, there may exist many types of agencies {page96}. An agent should normally require to be executed in functionally compatible platforms in differ-

---

[1]The work presented in this paper received several research grants: CNPQ and CAPES/PIDIC (fellowships), CNPQ/GMD nr. 91/96., FAPESP P. T. nr. 92/357-0

ent nodes, belonging to different domains. At the present time, several projects {mole95}, {ara95}, {agl96} treat this subject in different incompatible ways as, for example, the support to agent mobility. Recently, some proposals have been made {omg96b} in order to define standard Object Oriented API's for mobile and secure agents, while KIF, KQML and Ontolingua Tools and standards {kse94} allow the sharing of knowledge, communication and cooperation between intelligent agents.

This paper describes some initial steps in the scope of the PAGE (Prototyping an Agent Facility Environment), and MAGNA {kraus96} Projects, where prototypes of agent platforms are being implemented. The main goal here is to propose an architecture to a flexible platform that offers support to create agents, to let them migrate, to execute, and to be managed in distributed, heterogeneous and autonomous decentralized systems. Section 2 explores the extent of agent potential use through a summarized presentation of some areas of interest related to VE. Section 3 discusses what computational capabilities are necessary in order to adequately support agents performing their roles. Section 4 proposes a functional architecture for an agent platform, as an Agent Facility as proposed by OMG for a CORBA environment {omg95b}. Finally, in Section 5, some conclusions are presented.

2. Agent Applications in Virtual Environments

The application domain of our current interest, related to the use of agents, comprehend the high level application area known as *Virtual Computing Environment*. Two special Virtual Environments, the Electronic Commerce and Virtual Enterprises, were analyzed, as well as three supporting areas, Workflow Systems, Flexible Information

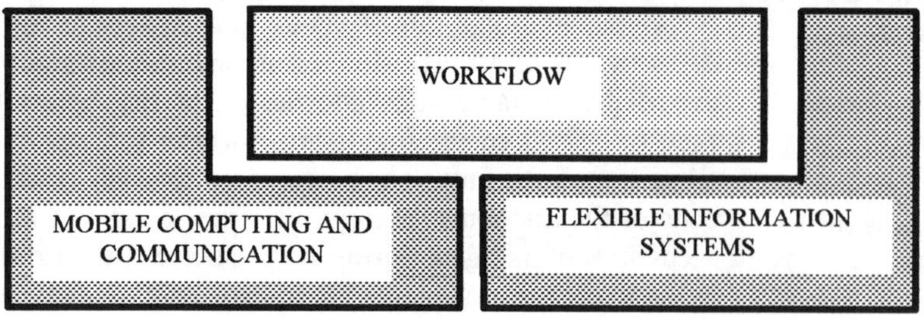

Figure 1. Agent Application Environment

Systems, and Mobile Computing. Here it is important to emphasize that together they constitute *an open cooperative system* and that many other application areas may exist in the same context.

Electronic Commerce Applications may be considered as hybrid systems, where people and computers are working together in order to reach certain commercial results through a generic commercial transaction that is "a finite sum of interaction processes between members of different roles" {klel94}. The aim of a commercial transaction is a trading agreement between the actors (supplier and customer, or intermediates, like dealers and brokers), and the following settlement for the delivery of goods and/or services.

Agents may be useful in the electronic commerce realm by searching potential suppliers to buyers or potential buyers to suppliers, asking about or offering marketing conditions, receiving and analyzing invoices, performing whole or part of a negotiation, providing electronic payment, and accompanying processes as packaging, storage, shipping, insurance, and customs clearance.

A **Virtual Enterprise** corresponds to a temporary alliance of companies formed to share costs and skills, regardless of the diversity of their computing systems and applications {niii96}.

Agents may contribute by searching for potential partners according to some conditions, negotiating partners admittance and workflow templates, searching, installing and configuring software, monitoring status of multiple and related remote resources and services, searching, filtering and translating data, monitoring particular responsibilities, controlling the use of resources, offering support to presenting personal positions and summarizing voting, searching for unfinished tasks or negotiating responsibilities when dissolving the VE.

Workflow Systems are related to the automated execution, coordination, and management of any type of work, that include almost all activities of any enterprise. By the very nature of Virtual Enterprises and Electronic Commerce, they are closely related to the concepts evolved from Workflow {joos95}, {schj94}. Agents may be useful in the workflow realm with several functions: searching for different entities (workflow templates, interfaces, applications, resources, different types of information and workers), configuring working applications, monitoring tasks execution status, and balancing and relocating tasks.

Flexible Information Systems constitute the basis for any information manipulation requirement, considering open cooperative sys-

360

tems for all application domains. Agents may be useful in browsing and exploring for information according to particular requirements, adapting information to the user characteristics, or suggesting, trading and delivering specific information to the user.

Mobile Computing establishes a set of basic support facilities for the mobility of people from place to place, accessing a wide range of networking services through wired or wireless access technology, in addition to the implicit support for the access of heterogeneous computer platforms. These aspects constitute basic requirements to the analyzed high level application areas and to almost any collaborative environment in the near future. Agents may be used in: device configuration (specially lightweight ones); communication reduction; communication optimization; interaction adaptation; and user service customization.

3. Agent Platform Capabilities

An Agent Platform is responsible for the infrastructure that supports all the life-cycle necessities of an agent. A standardized agent platform has to present managerial, processing, storage, communication, and security capabilities to support basic agents skills and roles.

3.1 Agent Skills and Roles

An agent presents skills in order to be able to perform one or many of the tasks in VE environments. An agent skill is a capacity that an agent presents, in some graduate level, that enables the agent to accomplish its goals. Several major skills are mandatory{mend96}:

- *intelligence* or the degree of reasoning and learning ability of the agent;
- *mobility* or the degree of migration power of the agent;
- *autonomy* or the degree of auto-control ability of the agent;
- *communicability* or the degree of communication ability of an agent with other agents or its environment;
- *cooperation* or the degree of an agent ability to collaborate and to do some joint work with other agents or entities;
- *trustability,* or the degree of an agent that makes it secure and believable in the environment where it exists.

Otherwise, depending on the applications where they live, agents specialize in certain types of tasks and may be classified, in respect to the *roles* they play (e.g. information agents, user agents, resources agents, and service agents).

The capabilities of an agent platform are related to the services and tools that support the agent skills and roles. An agent may not necessarily require all the possible capabilities of an agent platform and it is possible to deploy

agent platforms that support only one or some types of agents. It is thus important to implement the agent platform in a modular way, that will allow different levels of configurable functionality.

Due to the distributed nature of agents, it is necessary to use a reference model that allows the understanding of the computational requirements imposed by the different agent skills, in an orderly fashion, as well as the definition of common functions necessary to compose an agent platform. The different computational aspects can be divided into several categories as explained in Fig 2.

Based on these categories, several functions were identified. A complete description of these functions is presented elsewhere {page96}. Here only some aspects are discussed, representing a typical sequence of required functionalities, that begin when an agent is created at a certain node.

3.2 Managerial Aspects

Functionalities for fault, configuration, accounting, performance and security (CAPS) management have to be considered. Here, as an example, only some aspects of configuration management are discussed.

First of all, it is important that the agent platform supports the dynamic construction and creation of agents. This may happen through the use of an ***agent creation service*** driven by an ***agent authoring tool*** together with the ***agent life cycle service***.

The construction of agents should be conducted to allow the intensive reuse of patterns and agent frameworks or class-libraries in object oriented systems.

Static Agents may be programmed with classical languages such as C++ and SmallTalk. In this way, during their execution, they will be processes for the Operating System {ara95}. In this context, intelligent agents are very representative, and other languages (e.g. KIF) may be used for knowledge representation and communication, providing the means to express beliefs, goals and tasks in appropriate vocabularies for various domains. Mobile agents present more problems regarding the portability and security. This is one of the reasons to the initial success of script languages (Telescript, Tcl, etc.) that are either directly interpreted or compiled to a portable intermediate interpreter-based language, not depending on the platform, since the script execution engine has been ported to all necessary platforms. In {mole95} some considerations are made to the systematic use of an intermediate interpretable code (e.g. JAVA ByteCode).

Moreover, it is essential that agent developers safely ***test*** and ***debug*** their products before deploying them. This may pose new problems because simulation techniques may be necessary in order to emulate the interaction envi-

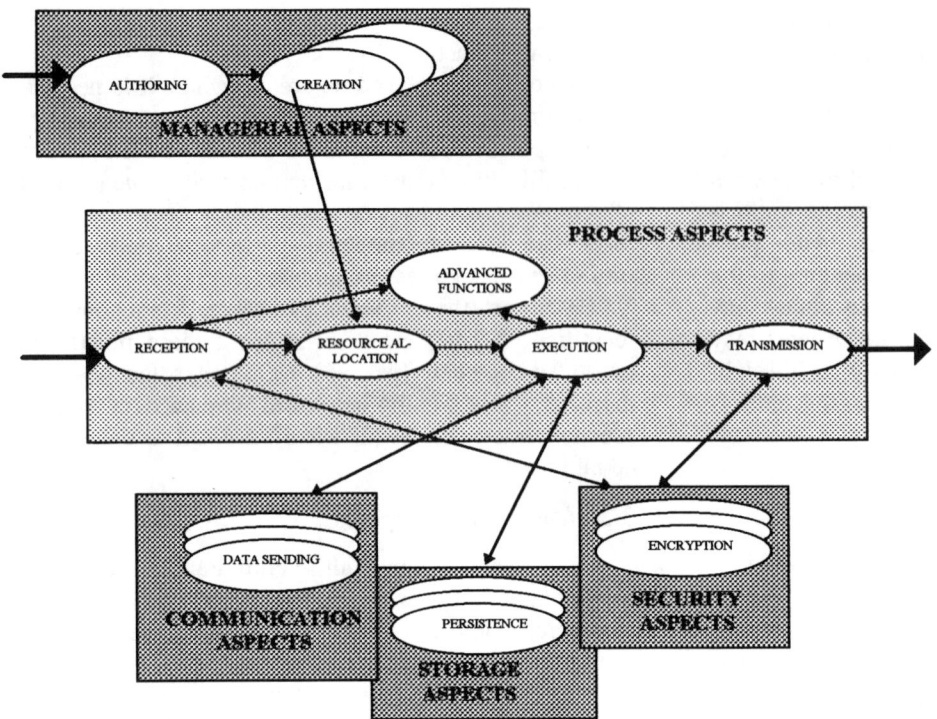

Figure 2 - Functional classes of Agent Platform computational aspects

ronments of agents as well as the dynamic behavior in distributed systems.

The *agent life cycle service* has to support a mechanism of agents "cold-boot", e.g. an *agent installation service*, responsible for the activation of agents on a specific machine, or even for their destruction (*agent removing service*). Also, an *agent naming service* must be available in order to assign a name to new agents. Since agents can move between different name domains, naming them in a location dependent way is counter productive and facilities to assign names that are globally unique may be necessary. After its creation and naming, the agent is delivered to the execution phase.

3.3 Process Aspects

In order to describe the different issues raised by process aspects, we consider a sequence that begins when an agent moves to another platform. This sequence encompass five phases: Reception, Resource Allocation, Execution, Transmission and Advanced Functions (figure 2).

3.3.1 Reception Phase

The agent credentials, issued by the agent owner (see the transmission phase later), are first analyzed by the *security service*. During the process of authentication, if an error occurs, a message refusing the agent reception is sent to the sending node.

When a platform receives an incoming agent as an encoded packet a *decoding and unmarshalling service* must be considered. The platform receives the agent into a temporary location, where the agent implementation must be stored. The existence of an agent implementation with the same name in the platform must be verified and, if necessary, a new name allocated. This functionality must be provided by an *agent naming service*. After that the agent is moved to a definitive location.

3.3.2 Resource Allocation Phase

Before running an agent, some actions should be performed to verify if the platform has the required resources. This is carried out by a *resource allocation service*. At this point several aspects should be analyzed: does the platform has the required resources and are they available?, does concurrence apply to those resources?, are there partial resources available and are they allowed to be used by the agent?, if not, can the agent wait for them?. After this preliminary verification the available resources are allocated or scheduled to the agent and concurrence control is used for the execution.

3.3.3 Execution Phase

This phase begins when the agent is sent to the appropriated execution environment using an *execution control service*. If the agent is in source code, it should be compiled and, in the case of compilation errors, the agent should be deleted (*life cycle service*) and an error message sent to the agent owner.

The first agent action after its *starting* (activation) is to initialize its state through an *agent state restore service*. When the agent is executed a message will be sent to the owner to inform about the agent situation. The agent execution can be *suspended* and *resumed* by the platform for some reason (e.g., if the platform needs to do maintenance and changes the agent work place). The agent execution will also be *stopped* when the agent finishes its work or when the platform decides to stop it for some reason (e.g., the agent wants to destroy its work place). If an error occurs during the execution phase, the agent and its work place will be deleted (*life cycle service*) and an error message will be sent to the agent originator.

If the agent wants to migrate, its state must be stored using an *agent state store service* and the migration process starts again.

The concurrent execution of several agents may be an important issue. Support for *time establishment and synchronization* between agents as well as support for *establishing and recognizing priorities* may be needed.

3.3.4 Transmission Phase

Before sending the migrating agent, the origin platform should ask the destination platform for permission, by sending the agent credentials, generated using an *agent credentials generation service*. The information that composes the agent credentials (global agent name, sender, receiver, resources needed, coding type, agent programming language, etc.) is based on pre-defined formats. Another alternative is to send the agent with its agent credentials without previous announcement.

In order to transport agents, two other services must be considered: *encoding* and *marshaling services*. The agent is transmitted to the remote platform by a *transport service* that uses, for example, the established data transmission protocols of the native platform (HTTP, TCP/IP, OSI, etc). When the agent is sent, it can travel with or without state. In the former case, the transmission of state must be considered.

Others services as *negotiation of QoS transmission, transmission recovery*, etc., could be considered depending on the particular platform complexity.

3.3.5 Advanced Functions Phase

When the agent task is very specific and the information that will be manipulated is previously known, the user may insert instructions to deal with them directly into the agent code. In this case the agent will be very limited. A more flexible agent could present some level of intelligence skill. In this case, the agent should possess a knowledge base about the task domain (e.g. electronic commerce) and a description of the task to be performed (its goal). During its execution, the agent will use this knowledge to analyze and decide what should be its next action in order to accomplish its task. To process this knowledge the agent may use some reasoning mechanism offered by the agent platform through some *intelligence services* (reasoning capabilities, as rule processing, problem solving and planning, etc.).

An even more sophisticated agent may be able to learn and adapt itself to its dynamic environment, both in terms of the user's objectives, and of the resources available to the agent. Like an assistant, this kind of agent might discover new relationships, connections, or concepts, independently from the human user, and exploit these decisions making in anticipating and satisfying the user needs

{maes94}. To perceive its environment, the agent should register its desire about the user or other agent actions and be notified. In order to support this level of intelligence, the platforms should offer to the agent *notification services* and semantic *communication services*.

3.4 Storage aspects

Persistence may be mandatory and implemented, for example, by special repositories, in order to allow agent restart and recovery in the case of faults and crashes

Agents have the need to query and discover services offered by agent environments. *Query services* arise because agents may wish to query services offered by its platform at run time, or it may query (and compare) services offered by other platforms before moving to the appropriated one. *Discovery services* are related with an agent directory service that supports discovery at run time. Some kind of directory services may be addressed (e.g. a *trading service*), with the support of sophisticated agent facility discovery.

3.5 Communication aspects

During the execution phase the agent might require to communicate with other agent(s). There are two possibilities of communication among agents: *agent-to-agent communication* or *agent-to-group communication*.

In agent to agent communication two types are again possible: *synchronous* and *asynchronous*. In *agent-to-group communication* scheme there are essentially two possibilities: *broadcasting* and *multicasting*.

3.6 Security aspects

This issue considers two subjects: the *platform security*, that refers to the ability of the agency to prevent any action of the agent that could damage the runtime environment, and the *agent security*, where *agents* must be protected against any kind of attack during its travels or execution.

The Platform must provide an inherently secure environment for analyzing the incoming agent through a control structure by determining the agent owner and its original location. *encryption* and *decryption* functionalities should be provided as part of the management of the local infrastructure. The most significant security building block is related to digital signatures generation and authentication, task typically associated with a public key cryptography.

Although even a difficult subject, some aspects of security may be implemented {ches95}, like, origin authentication, agent admission and host security, data integrity, agent privacy, and the integrity of gathered information.

The security services should provide functionality to allow, at least, two kind of interactions: *application-agent interaction, and platform-agent interaction*. The first one is needed when an user wishes to startup an application that is going to make use of agents at its platform. From the point of view of security, this application must have authorization granted by an *agent authentication service*. This procedure will prevent any intruder to gain access to the platform resources.

Platform-Agent Interaction is a more complex case, that involves several actions related to security for mobile agents. This may be subdivided into two parts, one for actions taken at the agent departure point and the other for actions taken at the arriving station. At the departure point, when an agent requests for travel, the platform, based on this request, provides the creation of agent credentials through an *agent credential generation service* that must be sent to the remote agency. Before this step, if requested, the private information carried by the agent is encrypted through an *agent encryption service*.

At the arriving station, the reception of the agent encompasses two steps: first the agent credentials must be validated by an *agent authentication service* and after reception, decrypted by an *agent decryption service*.

Other aspects are related with agent to agent and agent to operating system interactions. In the first case aspects related to malicious actions that agents may perform over one another are taken in account, and this functionality should be integrated during the development of the agent itself. For the agent to operating system interactions, like system calls, certain functions calls should be granted to the agent depending on its functionalities.

4. Proposals for an Agent Platform architecture

The next step is the definition of an Agent Platform Reference Architecture, the Software Architecture that will enable applications to use Agents efficiently.

4.1 Architectural Guidelines

The Platform Architecture should be distributed and open, exploiting object-oriented technologies. It is not necessary to repeat here the convenience of Object Orientation: Agents , in our work, are objects or object collections (like clusters).

The Architecture Goals include reuse, interoperability, extensibility, easy introduction of new services, and portability.

The Architectural Principles include the adoption of the main OMG recommendations, such as the use of ORB implementations as a basis for operation, and the support of the main adopted CORBA Services. We propose to de-

fine the Agent Platform as an Agent Facility in the OMA-context, taking into account the RFP published by OMG {omg95a}, dedicated to mobile agents.

Several other societies and companies, like TINA-C and NIIP, are defining solutions in those areas, so that our proposal also aims to adopt these technologies when they become available.

4.2 Preliminary Reference Architecture

As a result of the analysis made in this document, a preliminary architecture is proposed, which is to be refined in the next phases of the PAGE project.

Figure 3 represents the main building blocks of the proposed Agent Platform Architecture. We are building a collection of services that will enable us to develop and test hypotheses about agent mobility, security, management and intelligence aspects in an OMA environment. A brief description of these modules follows below.

Concierge Module

Reception services: Unmarshalling, Admission control/registration, Store and Forwarding, Decoding, Security (Origin authentication, Agent admission, Decryption, Virus detection) Services;

Transmission services: Storing the agent (persistency), Generation of agent credentials, Marshalling, Encoding, Encryption, Negotiation of QoS transmission, Transmission recovery, Support for multiple transport means (FTP, E-MAIL, HTTP, etc.);

Execution module

Checkpointing: State construction, State restoration, State storage;

Execution control: Execution management (Instantiation, Starting, Suspending, Resuming and Stopping the execution), Multi-language, Time synchronization, Priority handling;

Repositories: Persistency support (non-transparent by the user, transparent by the agency), Support to agent repositories;

Resource protection during execution of an agent: Mediators, Script execution supervising.

Communication module

Basic communication: Asynchronous peer-to-peer, Data Sending, Notification, Broadcast communication, etc.;

Advanced communication features: Semantic communication (KQML, etc.), Multicast communication;

Management Module

Fault management: Tracking, Execution failures, Transport failures, Logs , etc;

Configuration management: Life cycle services (Creation, Installation, Deletion, Copying, Equivalence, etc.), Auto maintenance (Modification, Append, Deletion

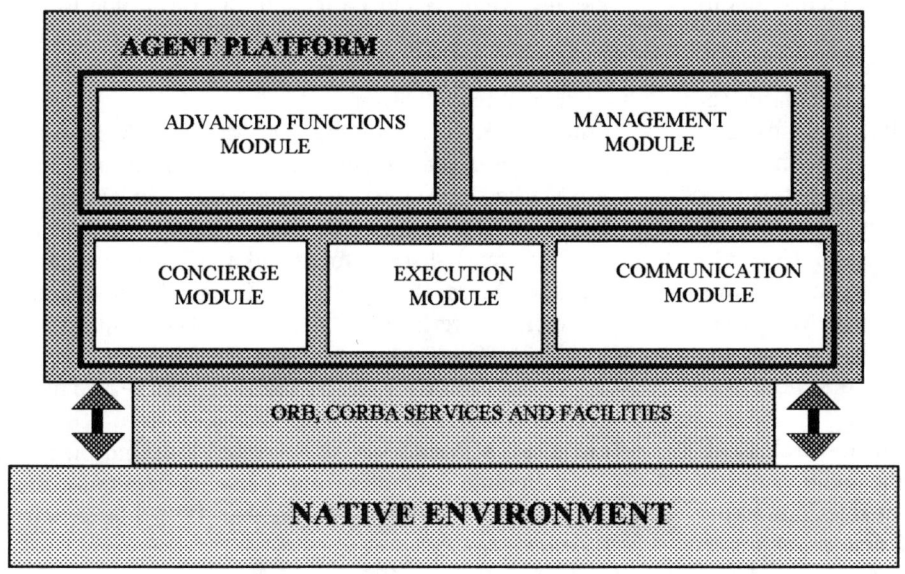

Figure 3. Agent Platform Architecture

of data), Naming and identification (Naming, Query, Discovery), etc;

Accounting management: Accounting, Banking, etc;

Performance management: Load balancing, Quality of service, etc;

Security management: Security monitoring/auditing, non-repudiation, etc.

Advanced Functions Module

Special Migration: Forced migration, Event-driven migration, Migration blocking, Migration scheduling, Invitation, Co-migration, etc;

Intelligence support: Preferences, Planning, Reasoning, Learning, etc;

Agent cooperation: Negotiation (of cooperation and conditions, of dictionaries, etc.), Meeting points, etc;

Coordination: Support to concurrence control, Support to recovery, Support to the coordination of multiple agents via transactions, Specification and control of complex activity control flows, etc;

Storage: Query, discovery, trading,etc.

Authoring Module

Agent programming: Support to different programming languages, Editing, Visual programming (support to class construction, object relationships, temporal diagrams), Reuse (frameworks, class repositories, browsers, etc.);

Debugging and testing;

Creation of executable agents: Generation of executable intermediate script code, Late binding and polymorphism, Testing of executable agents, installation, removing.

4.3 OMG Infrastructure

The interfaces already standardized by OMG (e.g. Corba Services) will be used (and perhaps extended) even if their implementations are not yet commercially available.

The *notification service* will be based on the *CORBA event service.* The *agent state storage/restoration services* will be based on *CORBA externalization/internalization services*, respectively. The *directory service* that supports the *query and discovery services* will make use of the name structure provided by the *CORBA naming service.*

The *agent naming service* may use the *CORBA naming service* but, to ensure a globally unique name, each agency must do the naming for each one of its agents. *Corba security services*, like authentication, authorization, credential generation and registering are the essentials that must be extended to attain the secure environment posed by mobile agents.

5. Conclusions

The applicability of mobile agents working as specialized elements in important business areas has been introduced. Mobile agents that can present some level of intelligence when performing their tasks certainly will influence the way people works in the highly automated future Virtual Environments. Particularly, the fields of Electronic Commerce and Virtual Enterprises offer a multitude of opportunities to the use of agents. This use may save human time when exploring, trading, adapting, and delivering the desired information. Agents may also perform efficiently profitable automated transactions (e.g. during negotiation of goods and services) in an electronic market. In a virtual enterprise agents are efficient in either controlling the use of shared resources or optimizing the flow of automated work.

However, agents are quite different from usual software in many aspects, as the autonomy and mobility skills. This poses some new challenges. Agents are itinerant and need new programming tools: languages like JAVA and its virtual machine are an immediate response to this need. The movement of pieces of code during execution, in a distributed environment poses severe security restrictions: however very promising results in cryptography have been reached, the authentication of network entities may be standardized, and new secure execution environments are under development. The mandatory autonomy of agents demands a standardized platform functionality in the form of transport and execution API' s. And above all, agents are a promise to the implementation of non trivial intelligent functionalities, like learning and reasoning, and cooperating in agent societies, without becoming heavy cumbersome processes. In the paper, such a platform is proposed, based upon the standardization efforts from OMG, ODP and TINA-C. Basic CORBA-Services may be reused and the future evolution of Corba-Facilities is promising. At present time, other Midleware Architectures like DSOM/OLE-ActiveX don't deliver the needed functionality and general support to several features of object orientation.

Agent technology is a very interesting research field. As it always occurs with technologies, the solution of basic aspects, as some concepts about agent roles and skills and platforms functionalities, produced new problems to be solved and new important issues to be standardized.

References

{ara95}, ARA Project (Agents for Remote Actions), University of Kaiserslauten, Germany, 1995.

{agl96} IBM Aglets Workbench Home Page, http://www.trl.ibm.co.jp/aglets/

{bagc95} Bagrodia, R., Chu, W.W., Kleinrock, L. and Popek, G., "Vision, Issues, and Architecture for Nomadic Computing", IEEE Personal Communication, dec/95.

{blad92} Blattner and Dannenberg, Multimedia Interface Design ACM Press, 1992.

{cacm94} Communications of the ACM Journal, Intelligent Agents, 37 (7), jul/94.

{cheg95} Chess, D., Grosof, B., Harrison, C, Levine, D., Parris, C. and Tsudik G., Itinerant Agents for Mobile Computing, IEEE Personal Communication Magazine, 2(5), oct/95.

{dalg95} Dalmonte, A. and Gaspari, M., Modeling Interaction in Agent Systems, http://www.cs.umbc.edu/kqml/papers/gaspari-ijcai95.ps.

{hars96} Hardwick, M., Spooner, D.L., Rando, T. and Morris, K.C., Sharing Manufacturing Information in Virtual Enterprises, CACM, 39(2), feb/96, pp. 46-54.

{iso94a} ISO/IEC Reference Model of Open Distributed Processing - Part 1., 1994.

{iso94b} SO/IEC Reference Model of Open Distributed Processing - Part 3, 1994.

{joos95} Joosten, S., Conceptual Theory for Workflow Management Support Systems Technical Paper, Center for Telematics and Information Technology, University of Twente, 1995.

{klel94} Klein, S.and Langenohl, T, Electronic Markets: An Introduction, In Proc. of Information and Communication Technologies in Tourism, Springer Verlag, Wien, 1994.

{kraus97} Krause, S. and M. Mendes, A DPE-based Platform for Mobile Agents in Electronic Service Markets", see the proceedings of ISADS97.

{kse94} Knowledge Sharing Effort Home Page, http://www-ksl.stanford.edu/knowledge-sharing/

{maes94} Maes, P., Agents that Reduce Work and Information Overload , CACM, 37(7):31-40, jul/94.

{mazk95} Mazer, M. et al, Issues in Mobile Computing, Guest Editor's Note in IEEE Personal Communication, dec. 1995.

{mend96}, M.Mendes et.all , Agents skills and their roles in mobile computing and personal communications, IFIP 96, Camberra, September 1996 (accepted invited paper).

{mole95}F. Hohl, MOLE Project, Konzeption eines einfachen Agentensystems und Implementation eines Prototyps, Diplomarbeit, Stuttgart University, October 1995.

{nwana96} Nwana, H.S, Software Agents: An Overview, The Knowledge Engineering Review Vol.11(3), 1996.

{niii96} National Industrial Information Infrastructure Protocols Consortium, Reference Architecture: Concepts and Guidelines, Stanford, CT, USA, 1996.

{page96}PAGE Project - Prototyping an Agent Platform Environment, Informatics Institute - Puccamp, Brasil, 1996, available at www.puccamp.br/II/PAGE/Page.html.

{omg95a} Object Management Group, CORBA services: Common Object Services, USA, 1995, available at www.omg.org.

{omg95b} Object Management Group, Common Facilities Request for Proposals 3, available at www.omg.org.

{omg96} Object Management Group, CORBA 2.0 Specification, 1996, available at www.omg.org.

{schj94} Schuster, H., Jablonski, S., Kirsche, T. and Bussler, C., A Client/Server Architecture for Distributed Workflow Management Systems, www.informatik.uni-erlangen.de:1200/publ/sjkb 94. ps.Z.

Session 5B

Management and Coordination

Chair

Charles Jung

An Autonomous Agent-Based Infrastructure for Inter-LAN Systems Management

Shyh-horng Jou and Shang-Juh Kao

Department of Applied Mathematics
National Chung Hsing University
Taichung, Taiwan, 40227 R.O.C
E-mail: {shjou,sjkao}@flower.amath.nchu.edu.tw

Abstract

Along with the growth of distributed services and applications in a networked system, to monitor, interpret, and control the behavior of the distributed resources become relatively complicated. A centralized management could be efficient in a privilege-owned LAN system. However, without a privileged owner, to manage multiple LANs requires self-determined management within a LAN, and cooperating operations between LANs. This paper presents an agent-based architecture distributed framework which has the advantages of centralized control. This approach relies upon several well established technologies including objected resources, intelligent agents, and cooperated communication. The framework addresses the hierarchical domain management and functionalities of layered modules within a domain agency. A simplified application of printer management is also presented in the paper.

1 Introduction

Management of a networked system deals with supervising and controlling the activities of the system to fulfill the requirements of efficiency and productivity [1]. Along with the relentless growth in complexity, diversity and heterogeneity of distributed resources and applications, the behaviors of a system can no longer be managed by a labor-intensive and platform-centered paradigm. Consequently, proactive and decentralized control mechanisms must be utilized to enforce management more comprehensively.

The problems of extending a centralized model involve frequent negotiations between the manager and the management agents in a distributed environment. Some degree of centralized administrative control and local autonomy, i.e, partial decentralization[2], is necessary for multi-vendor networks. In order to efficiently manage an inter-LAN system, task-oriented.

domain hierarchy is necessary. Additionally, management policies from above should be carried out with self-contained intelligence. In this approach, integration of both the management domains and policies within the intelligence agent-based model creates a more flexible and decentralized management paradigm.

1.1 The Aspects of Management Domains

When distributed applications and services need to be managed efficiently, it is impossible to manage these resources individually in a large scale management system. Therefore, introducing a mechanism for grouping resources, the concept of the management domain, is required.

A management domain[3] is a collection of managed objects grouped together explicitly to meet the requirements of an organization or administration. Because a domain may also be a member of another domain by representing itself as a managed object, and because managing large numbers of objects in a purely linear way is impossible, a hierarchical skeleton of domains is logically constructed to serve the management structure. The complexity of management is reduced by refining management tasks through the different granulations of the domain's hierarchy. Lower layer domains provide their services, which could be fulfilled by subordinate layers, to those of the upper layers and cooperate with coordinate peer-domains to achieve the global objectives.

A domain is not only a concept to describe separate administrative-controlled environments. It is a building block for grouping resources and specifying boundaries of management responsibility and authority. It also provides an integrated and uniform view of the distributed and heterogeneous managed environment.

1.2 The Motivation for Management Policies

Monitoring for mal-functional activities in a system and making management decisions for modifying the behavior of the system is the responsibility of a manager in any scale networked platform. Monitoring and control can be accomplished by dictating management operations to and accepting notifications from managed objects. Because making decisions for control actions from a set of management goals can be a very complicated task, the concept of management policies is introduced as an intermediate step [4]. As mentioned in [5], policies are derived from the goals of management to define the desired behavior of distributed heterogeneous systems and recognized as a concept to support this complex management task by specifying conditions that enable a manager to enforce this behavior. From the view point of organizational structure, a high-level policy may be used as the basis for initiating control actions. By refining and partitioning the goals, delegating responsibility to other managers, multiple lower-level policies can be derived. We can define these recursively constructed policies as policy hierarchy[6].

By adopting delegation mechanisms and hierarchial relationships, the input of Nth layer domain is a set of goals specifying the objectives of management from the upper domain(i.e, (N-1)th layer), these goals turn out to be policies which define what task of management must be accomplished, from these policies plans are derived which state the objectives must be achieved. If derived plans cannot be executed in the local domain, these plans will be passed to subordinate domains in the lower layer to be refined again, or control will be delegated[7] to peer-domains to achieve the goal. The automation for management is possible through a step by step refinement within each autonomous domain and the domain hierarchy of neighboring layers.

1.3 Intelligent Agent-Based Framework

Due to the increasing computation power of management agents, the role of the servant will be replaced with a proactive entity that functions as both agent and manager. For the purpose of reducing the load on the system manager and utilizing network bandwidth efficiently, problem solving and reasoning will require decentralized control and intelligent agents. Hence, the potential impact of intelligent agents will have a great influence on network management as observed in [8].

In order to coordinate the activities for solving a task among peer agents, situation assessment and inter-agent transaction parameters must be determined for creating a multi-agent system environments as presented in [9] and [10].

We propose a framework based on intelligent agents for autonomous system management to resolve management requests. Initially, the translation of a high-level abstract management goal into a number of policies is done either by a policy translator (automatically) or by manager interference (manually). These resultant policies will be formulated into plans which may be executed by the domain itself or passed on to subordinate domains for further refinement as a new subgoal. That is, a policy could be divided into subpolicies to be carried out recursively by coordinate domains (or, subdomains). The situation assessment will play an important role in decision making and reasoning applications.

2 Autonomous Decentralized Agent-Based Architecture

A group of tasks and services is known collectively as an agency[11]. Drawing upon the principle, pragmatic sense in hierarchical domain management, we will treat an agency as a dominant management-platform which controls its subordinate domains and provides access to the world outside its supervisory boundary. Two types of communications can be derived from this organization: loosely coupled negotiation or tightly coupled control. In the former case, a domain agency can communicate with peer domain agencies and negotiate to determine what services they can provide in order to cooperatively solve a management task. The other type of communication is concerned with managing the services that the servant agent is executing within its domain. Note that this domain agency model can be organized into flat, hierarchical, or hybrid architectures to meet different functional requirements. Furthermore, a virtual domain can be formed by selecting a set of services which are provided by different servant agents. We can also designate a virtual domain agency to manage all the activities occurring within a virtual domain. Hierarchical relationships between domains and agencies are depicted in Figure 1.

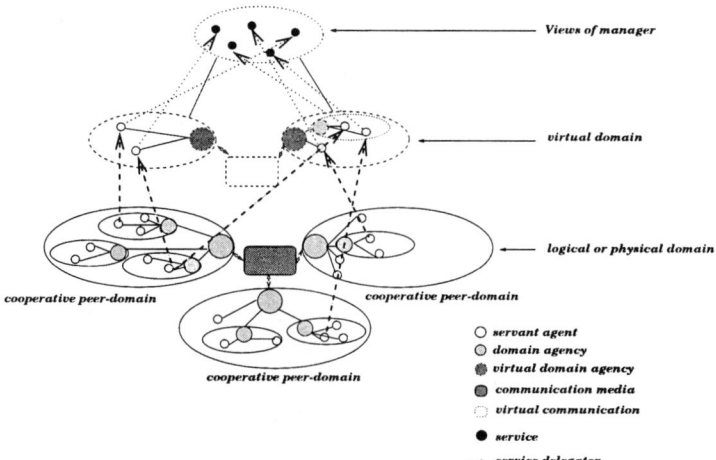

Figure 1: Hierarchical relationships between domains and agencies

Having identified the basic concept and function of agency, we will now discuss the infrastructure within a domain agency. A domain agency comprises multiple layers: control, inference, knowledge, and communication. Figure 2 depicts the modules within a domain agency and shows the relationships between them.

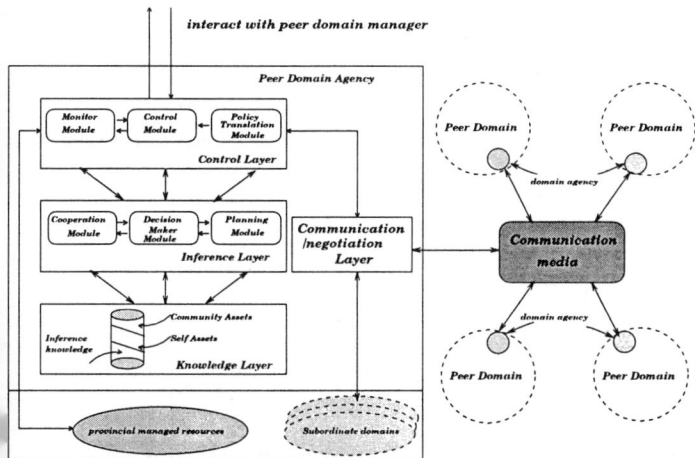

Figure 2: Autonomous Decentralized Agent-Based Infrastructure

2.1 Modules of Knowledge Layer

Two important factors which should be accounted for in designing a cooperation framework are directing local problem solving and coordinating activities with others within the community[9]. For either situation, adequate information about the requests from upper layers provides the bases for autonomy. Therefore, the basic knowledge about how, where, and who should solve a problem must be identified. The fundamental data-repository in this layer includes inference knowledge, self assets, and community assets.

- **Inference Knowledge:** It supports the inference layer a knowledge base of past experiences for decision making under uncertainty, for realizing plan decompositions, and for cooperating problem solving. These past decision experiences provide useful information when an agency is confronted with novel but similar situations. Based upon the situation, problems can be solved by deductively retrieving previous problem operations and creating appropriate operations to solve the new problem.

- **Self Assets:** Contain the knowledge about what kind of activities can be performed locally, information of the current state, and an abstract description of services it can provide. Through the hierarchical structure of a domain agency and its subdomains, a mechanism for encapsulation and abstraction of services can be achieved. To facilitate cooperation, each agent must share information regarding available services during negotiations with its peers.

- **Community Assets:** Information about all available shared services which can be supplied by other agencies in the community. Community assets can be derived from parts of domain agencies within the community, and collected through importing assets that have been exported by coordinate agencies.

2.2 Modules of Inference Layer

It is impossible for a system to predict what situations it will face in an uncertain environment and to estimate the amount of information that will be required to deal with each event, consequently the system must be able to infer. Processing of inferences are categorized into submodules concerning how to plan, make a decision and cooperate with others.

- **Planning Module:** A local scheduler inputs the self assets, local commitments, and cooperative commitments to produce several possible schedules for the decision maker module to evaluate. A scheduled plan derived from a policy may contain a set of actions which can be executed immediately with managed resources. For a complicated plan, decomposing into subplans may be necessary.

- **Cooperation Module:** When a plan(or subplan) must be performed through cooperative means, the cooperation module will ascertain where the

371

commitment can be best satisfied, how to establish and adapt cooperative activities, and determine if cooperation has been successful. In addition to facilitating cooperation, the module must also account for the negotiation process that achieves cooperation.

- Decision Maker Module: The purpose of the decision maker module is

  1. Choosing an appropriate schedule based on suggestions from the planning module to accomplish subplans/actions through locality or community.

  2. Monitoring commitments under changing circumstances and solving conflicts during negotiations between agencies.

  3. Choosing additional system adaptations or error-reporting whenever an expected objective cannot be achieved.

  4. Deciding what actions should be taken when accepting notifications are from underlined infrastructures.

Considering the factors of inadequate information, dynamic, and uncertain environment, we seek a method to allow a management system to utilize previous inference experiences to formulate solutions to new problems. Among the available reasoning mechanisms, case-based reasoning[12] [13] could be the best candidate for developing an inference based management system.

2.3 Modules of Control Layer

After having introduced the essential functions of two the baseline layers, we come to the top layer, the control layer. The control layer has a direct interaction with the domain manager. It is an interface between the manager and the managed resources. There are three modules inside the control layer. They all perform together to accomplish the necessary interactions.

- Monitor Module: The various of scenarios with which the monitor module must deal are summarized as follows.

  1. When accepting a notification emitted from underlined managed resources, the monitor module uses the decision maker module to filter, analyze and process it. It then issues an alarm to drive the control module to take corresponding action.

  2. Monitor managed resources according to filter criteria predefined by the control module, and report result to the control module.

  3. Monitor the status of operations in both local and cooperative problem solving, and inform the control module what is happening.

- Control Module: The activities of the control module can be classified into three divisions:

  1. Supervision of managed resources - initiates data monitoring by providing filtering parameters to the monitor module, and takes action on managed resources if necessary.

  2. Decision making for plan apportioning - makes use of services which are provided by the inference layer to decide how and where to initiate these plans.

  3. Activity execution control - directs the monitor module to manage the schedule execution for local and collective activities, and takes appropriate action on incoming notifications.

- Policy Translation Module: This module translates high-level abstract goals into concrete plans.

2.4 Modules of Communication/Negotiation Layer

A management task can be satisfied through one of the following operations:

1. Combining the information of self assets and decentralizing problem solving to its local community.

2. Collecting the information of community assets and delegating cooperative problems to its coordinate community.

In either case, the communication/negotiation layer is assuming control over operation requirements. In fact, we may encapsulate various communication protocols and hide the exact format for data transfer within this layer.

The obligation of this layer involves packing the content of dialogue messages into an appropriate format which is meaningful to its recipients, and conveying the resultant messages to the recipients. A common agent communication language, such as KQML[14], can be used to format dialogue messages, but conveying messages to the recipients requires a specific mechanism to provide transparent message routing.

3 Application to Quality of Service for Printer Management

In order to depict the application of the proposed framework for inter-LAN management, we present the automation of Quality of Service(QoS) printer management problem in a 2-layered Ethernet architecture in the following subsections.

3.1 Experimental Environment of two-layered Ethernet Architecture

The experimental environment comprises two subnets. Each subset has a proprietary printer server and consists of two individual management domains. The dual-domain for the 2-layered Ethernet environment is represented in Figure 3.

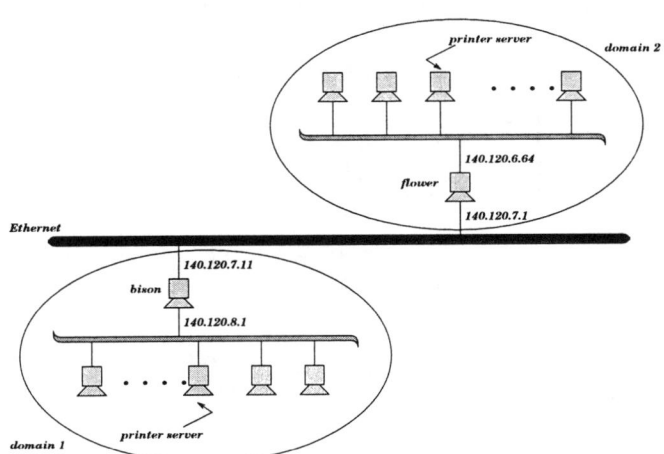

Figure 3: The dual-domain of 2-layered Ethernet environment

Conceptually, we created a virtual domain which contained two printer servers, and designated a virtual domain agency to manage all printing activity. Through the aid of an autonomous agent-based infrastructure, the virtual domain agency not only provided the transparent service to users, but also supported high quality service by balancing the printing load between printer servers. Figure 4 shows the structure of the virtual domain from the perspective of service management.

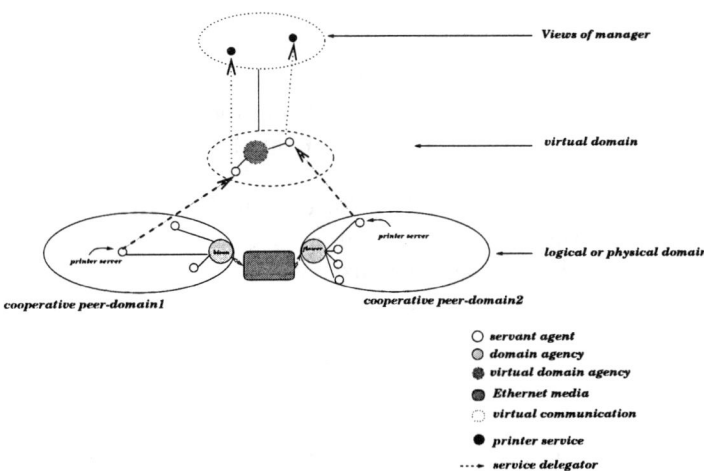

Figure 4: Structure of the virtual domain

3.2 Application Scenarios

Having introduced the experimental environment, let's demonstrate a specific example to exhibit how cooperative agents achieve the goal of load balancing. The premise of our hypothesis is that the agency records necessary proprietary information about the printers, for instances, type, speed, quality, postscript cartridge, etc. The procedures followed within the virtual domain agency, from accepting a goal to delivering the derived subplans, are described as follows.

1. Upon accepting the goal, the policy translation module within the control layer will translate the abstract description of the goal into plans using intermediate policy templates. These derived plans will be fed into the control module for further processing.

2. The control module makes use of services which are provided by the inference layer to decide where, who, and how to carry out these plans. For example, the question about the destination of printing servers can be solved easily using policy templates, self assets and community assets within the knowledge layer. The difficult work is to determine how to distribute the overall workload to all available printers, in other words, to determine the job loads each printer will service. The capability of the adaptive service agent(printing server) is a limiting factor. We can simplify and formulate the evaluation of the workload[1] distribution by factoring of speed, and

---

[1] $wload_i\% =$

$$\frac{\sum_{printer \in subdomain_i} Speed(printer)}{\sum_{printer \in subdomain_i} Speed(printer) + \sum_{printer \notin subdomain_i} Speed(printer)}$$

loading percentage in each subdomain.

Therefore, as specified in our example, the expected workload in $subdomain_1$[2] and $subdomain_2$[3] can be estimated respectively.

3. After the subplans have been derived from the original goal, they can be submitted to the proper subordinate domains through the communication/negotiation layer.

During the progressing of a specific goal, the monitor module may be required to supervise execution and the control module may initiate appropriate actions(e.g. reestimate the distributed workload or issue a alarm to notify the manager) whenever the goal can't be met.

The control flow derivation from an expected goal to subplans can be shown in Figure 5.

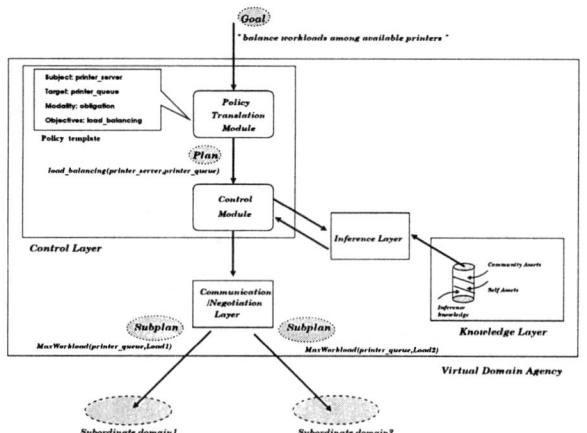

Figure 5: The progress of derivation from a goal to subplans

Now, the problem is how an autonomous servant agent adopts its internal assets for inference and adaptation in order to accomplish the expected goal. This comes to the processing of the adaptation and inference as described in the following.

1. When a printing request arrives, a servant agent evaluates what percentage of its current loading can be formulated. The formula of calculating the percentage is as following.

$$Workload\_of\_agent_i\% =$$
$$\frac{Size(current\_request\_job) + \sum_{job \in queue\_of\_agent_i} Size(job)}{Size(current\_request\_job) + \sum_{job \in queue\_of\_agent_i} Size(job) + \sum_{agent \neq agent_i} Workload(agent)}$$

From the community workload, acquired by the communication/negotiation layer, a servant agent can calculate its percentage of workload easily.

2. In order to alleviate the burden of printer overloading, the next step a servant agent has to deal with is determining if this current request should be carried out locally or migrated to a appropriate servant agent within its cooperative community.

Accordingly, the planning module within the inference layer retrieves a similar case(past experience) stored in inference knowledge bank to supply a suggestion for the decision maker module to determine how a servant agent can satisfy this current request.

The contents of the retrieved case may include:

> subject: CPU
> trouble: Workload of Adaptation
> resolution:
> Heavy = $\mathcal{F}$(current_load, max_load, overload_tolerance)
> Migration = $\mathcal{G}$ (Heavy)
> status: OK

The fuzzy function of $\mathcal{F}$ [15] within the resolution is given in the following:

$$Heavy = \mathcal{F}(current\_load, max\_load, overload\_tolerance)$$
$$\equiv \begin{cases} 0.0 \\ \quad current\_load \leq max\_load \\ (1 + (\frac{current\_load - max\_load}{max\_load * overload\_tolerance})^{-2})^{-1} \\ \quad otherwise \end{cases}$$

The boolean function of $\mathcal{G}$ within the resolution can be determined as :

$$Migration = \mathcal{G}(Heavy) \equiv \begin{cases} False \\ \quad Heavy \leq 0.5 \\ True \\ \quad otherwise \end{cases}$$

---

[2] $wload_1\% = \dfrac{\sum_{printer \in subdomain_1} Speed(printer)}{\sum_{printer \in subdomain_1} Speed(printer) + \sum_{printer \in subdomain_2} Speed(printer)}$

[3] $wload_2\% = 1 - wload_1\%$

According to the similar characteristic of loading adaptation, the planning module can create a new case for the printer:

subject: Printer
trouble: Workload of Adaptation
resolution:
Heavy = $\mathcal{F}$(current_load, max_load, overload_tolerance)
Migration = $\mathcal{G}$ (Heavy)
status: OK

In this case the planning module can reason whether the current request should be migrated or not, and then sends the result to the decision maker module for further resolution.

The decision maker module may have various preferences as in the following:

- Accept the proposal which the planning module has suggested.

- Migrate a job to be executed cooperatively to facilitate efficiency and on time delivery.

- Notify the planning module to reevaluate the alternative by changing the parameter *overload_tolerance* in fuzzy function $\mathcal{F}$ in order to scale the degree of tolerance for load balancing efficiently.

The conception of inference described previously is depicted in Figure 6.

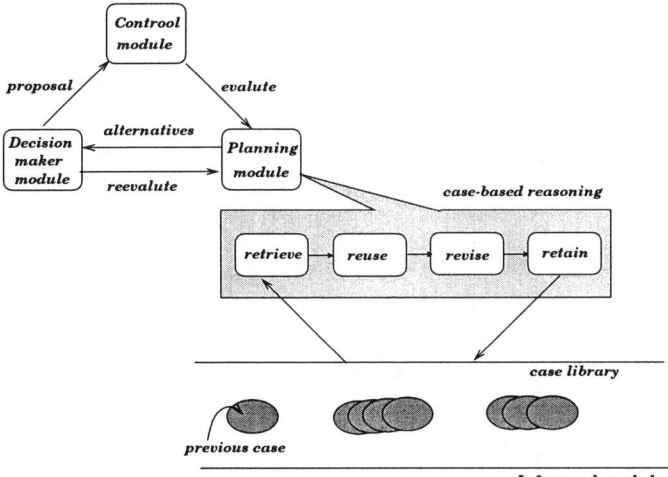

Figure 6: The conception of inference

3. The Control module sends the request to local queue or migrates the request to the remote queue

of cooperative agents through the communication/negotiation layer, and then triggers the monitor module to supervise the activities of execution.

4 Summary and Future Works

Despite the fact that most current management systems fall in the paradigm of centralized control and labor-intensive, we propose a decentralized approach to inter-LAN systems management. In this paper, a partial decentralization-like infrastructure is presented which takes the advantages of efficient management from centralization even further through a locally autonomous agent structure. Meanwhile, the approach of hierarchical management domains enables solving management faults through cooperation and negotiation among domains. Consequently, the inter-LAN management system is more flexible and easier to scale.

The infrastructure consists of cooperative peer-domains which are interconnected through communications links. A virtual domain can be formed by logically incorporating several physical domains. Within a domain, there are four layers: knowledge, inference, control, and communication/negotiation layer. Upon receiving requests from the domain manager, the policy translation module is in charge of converting requests into plans. As soon as this process is completed, the control module feeds the planning module these plans to distribute further. The decision maker module takes the results of the distribution as input to take on appropriate action, which can be tackled either locally or cooperatively. In the former situation, the requests will be accomplished internally or in concert with its subordinate domains. The other case, cooperative problem-solving is ongoing through out the communication/negotiation layer. Triggering the monitor module to observe immediately after a decision is determined will occur if necessary. All underlined information and inference during processing is provided by the knowledge layer. No matter where the requests are executed, the process of using historical data can be applied recursively until the original goal is satisfied.

Through out the presentation, We have focused on the management system structure. Nevertheless, there are still several technical issues left for further development. Among these issues, the following are important:

- Communication protocol. The two most important communications are between managers and agents, and between agents themselves. Here, we

375

temporarily hide all the communication issues in the communication module. In the future, integration of SNMP(or CMIP) with multimedia data transfer(for instance, HTTP) is anticipated.

- Managed Resources. Object definitions which are beyond the standardized managed objects are necessary for inter-LAN spectrum. Both data storage and database interface require explicit specifications.

- Inference rules. Computer intelligence needs human-like preprocessing. In addition to knowledge that has to be collected in advance, inference logic requires carefully design in order to present the intelligence. Furthermore, self assets(managed resources) and community assets play the role of managed objects which can be integrated into the management information base(MIB). In order to use the MIB for more sophisticated management, the guidelines for the definition of managed objects(GDMO) must be enhanced.

- Management overhead. Efficient management usually pays the costs of system overhead. The study of polling frequency, for instance, deserves further exploration.

References

[1] Kornel Terplan, "Communication Networks Management", Prentice-Hall, Inc. 2nd, 1992.

[2] Germán Goldszmidt, Yechiam Yemini, "Decentralizing Control and Intelligence in Network Management", in Proceedings of the 4th International Symposium of Integrated Network Management, Santa Barbara, CA, May 1995.

[3] "Information Technology - Open Systems Interconnection - Systems Management Overview", ISO/IEC 10040, 1992.

[4] R.Boutaba, S.Znaty, "An Architecture Approach for Integrated Network and Systems Management, ACM Computer Communication Review, October 1995.

[5] René Wies, "Policies in Network and Systems Management - Formal Definition and Architecture -", In Journal of Network and Systems Management, Plenum Publishing Corp., Vol 2, pp. 63-83, March 1994.

[6] Jonathan D. Moffett "Policy Hierarchies for Distributed Systems Management", In IEEE JSAC Special Issue on Network Management, Vol 11, No. 9, Dec. 1993.

[7] Germán Goldszmidt, Yechiam Yemini, "Distributed Management by Delegation", In Proceedings of the 15th International Conference on Distributed Computing Systems, June 1995.

[8] T. Magedanz, K.Rothermel and S. Krause, "Intelligent Agents: An Emerging Technology for Next Generation Telecommunication ?", INFOCOM 96, March 24-28, San Francisco, CA, USA, 1996,

[9] N.R.Jennings, E.H.Mamdani, I.Laresgoiti, J.Perez and J.Corera, "GRATE: A General Framework for Cooperative Problem Solving", IEEE BCS Journal of Intelligent Systems Engineering 1(2), 1992.

[10] T.Witting, N.R. Jennings and E.H. Mamdani, "ARCHON - A Framework for Intelligent Cooperation", IEE-BCS Journal of Intelligent Systems Engineering - Special Issue on Real-time Intelligent Systems in ESPRIT, 3(3): 168-179, 1994.

[11] J.L.Alty,D.Griffiths,N.R.Jennings,E.H.Mamdani, "ADEPT - Advanced Decision Environment for Process Tasks: Overview and Architecture", In Proceeding BCS Expert System Conference(Applications Track, ISIP Theme), Cambridge, UK, 1994.

[12] Agnar Aamodt, "Case-Based Reasoning: Foundational Issues, Methodological Variations, and System Approaches", Artificial Intelligence Communications, 7(1):39-59, 1994.

[13] Lundy Lewis, "A case-based reasoning approach to the resolution of faults in communications networks", Integrated Network Management,III(editors: H.-G. Hegering and Y. Yemini). Elseiver Science Publisher(North Holland). 1993.

[14] Tim Finin and Rich Fritzson, "KQML - A Language and Protocol for Knowledge and Information Exchange", Knowledge Building in Knowledge Sharing, Ohmsha and IOS Press, 1994.

[15] L.A.Zadeh, "Fuzzy Sets", Information and Control, vol 8, pp. 338-353, 1965.

Redesigning the Web: From Passive Pages to Coordinated Agents in PageSpaces

Paolo Ciancarini*   Robert Tolksdorf[†]   Fabio Vitali*   Davide Rossi*   Andreas Knoche[†]

Abstract

Currently, Web does not support distributed applications well. Existing approaches are oriented towards centralized applications at servers, or local programs within clients. To overcome this deficit, the PageSpace platform was designed for distributed, coordinated agents in the Web.

We take a specific approach to coordinate agents in PageSpace applications, namely variants of coordination language Linda that support rules and services to guide their cooperation. This technology is integrated with the standard Web technology and the language Java.

Several kinds of agents life in the PageSpace: User interface agents, personal homeagents, the agents that implement applications, and the kernel agents of the platform. Within the architecture it is possible to support fault-tolerance and mobile agents as well.

Keywords: Java, Linda, Coordination, Internet, Web Applications, Open Distributed Systems

1 Introduction

The Web has evolved into the dominating platform for information systems on the Internet. There is increasing demand to use it as a platform for distributed applications in which processing of information occurs. For example, the application domains groupware and workflow management require distributed access and processing due to the distributed nature of the work these applications support. Still there is no widely accepted platform for implementing distributed applications on the Web.

PageSpace is a platform that has the potential to provide sufficient functionalities to do so. It is based on the core Web technology for access and presentation, on Java as the execution mechanism, and on coordination technology to manage the interaction of agents in a distributed application. This paper describes the rationale for our platform, its design, and the implementation strategy currently applied.

This paper is organized as follows. In the next section, we review approaches to implement applications that require active processing on the Web. We then describe the technology, on which our specific approach to coordination in distributed applications is based. The next section describes the PageSpace platform and its agents. Then, the current approach taken in engineering and implementing PageSpace is outlined.

2 Existing Approaches for Web Applications

At its core, the Web is a static hypertext graph in which multimedia pages of information marked up in HTML are offered by servers, retrieved by clients with HTTP and displayed in a graphical interface that is very easy to use.

Because of its high availability, it becomes more and more desirable to use the Web as a platform for dynamic, distributed applications. The support of the core Web platform for applications is rudimentary – only the CGI mechanism allows for processing of information that is entered by the user in forms, or retrieved from auxiliary systems.

A number of mechanisms has been proposed and implemented to make the Web a platform for distributed applications. The following classification is structured according to the loci of activity possible with such mechanism.

- **Activity located at Web servers.** The CGI mechanism can be used to access other application servers from the Web. An example is database access, where a query is formulated in a form at the browser and a CGI script at the server passes that query – probably in a translated form – to some database server. The results of the query then are sent back to the users browser as HTML. Figure 1 shows that structure.

  However, with respect to the aspect of distribution of an application, this approach turns out to have nothing in common with distributed paradigms like client-server interaction, or others. In fact, interfacing an application via CGI to the Web does not mean to offer a distributed application. There is no processing at the

---

*Dept. of Computer Science, Univ. of Bologna, Pza. di Porta S. Donato, 5, I-40127 Bologna, Italy. *mailto:{cianca|vitali|rossi}@cs.unibo.it http://www.cs.unibo.it/{~cianca|~rossi}*

[†]Technische Universität Berlin, Fachbereich 13, Informatik, FLP/KIT, FR 6–10, Franklinstr. 28/29, D-10587 Berlin, Germany. *mailto:{tolk|knoche}@cs.tu-berlin.de http://www.cs.tu-berlin.de/~tolk/*

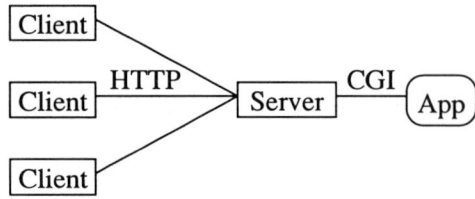

Figure 1. Application within one Web server

client besides displaying results. Moreover, there is only one central location of activity. Thus, such an application is basically a mainframe/terminal system at a large spatial scale. The Web server is like the mainframe – the only location of processing. The Web browsers are nothing but easy to use and graphical terminals, that use HTML as the display language.

- **Activity located at Web clients.** With execution mechanisms within Browsers, like the Java virtual machine, JavaScript, or plug-ins, activity can be performed at clients. If a Java applet is executed within the browser, it usually forms no distributed application. The applet is like a program that is run locally on the users machine. There is no generally accepted way to connect applets; the remote method invocation mechanisms lead to security problems that are not solved yet. If two or more applets communicate with some mechanism, they introduce an application specific protocol. Some applets and plug-ins connect to other proprietary servers and again leave the core Web. Figure 2 shows this structure of activity focused on clients.

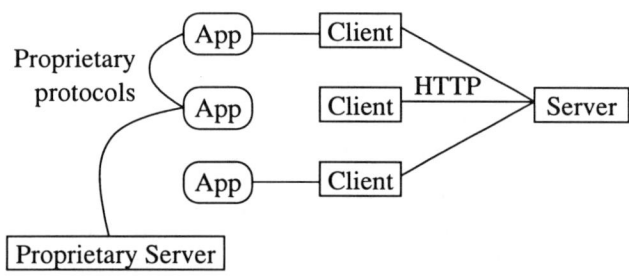

Figure 2. Activity at Web clients

- **Activity located in midware.** The previous two approaches are based on core Web technology, probably extended with proprietary communication schemes. A third approach to distributed applications on the Web is to use middleware to connect the active parts in an application, which can be located in clients and/or servers. Here, the Web technology takes the rôle of

providing a uniform access and presentation mechanisms. Figure 3 depicts that structure.

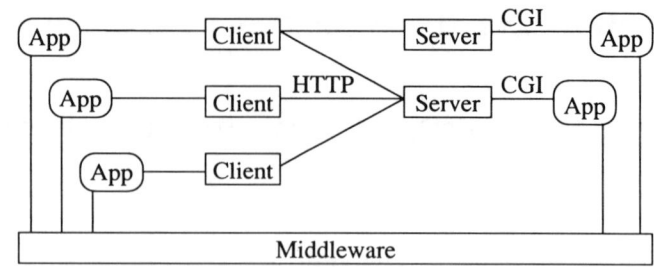

Figure 3. Activity within middleware

Examples for such approaches are [BIV$^+$95], [HK94], or [AJ95], where the Web is an access mechanism to CORBA- or DCE-based applications, or integrated with middleware to profit from its services, such as secure communication. Client side middleware access is enabled by Java-CORBA embeddings like Sun's JOE.

The PageSpace platform falls into the last category, as only here activity is really distributed. We provide a middleware platform that is smoothly integrated with the dominating core Web technologies and also addresses issues of integrating the user interfaces of applications with the Web. The key conception is the use of coordination technology to manage the interaction amongst PageSpace agents.

3   Coordination Technology for the Web

The PageSpace platform ([CKTV96]) is based on the notion of agents that use coordination technology for their interactions. We use the term agent reflecting that processing is performed in such an entity. Applications are composed by a set of distributed agents. Each user has a *home-agent* that provides the interface to the PageSpace and its agents. We rely on Java as the main implementation language for our agents. In the main focus of the PageSpace is the issue of coordination amongst these distributed, concurrent agents, and we explored the use of Linda-like coordination technology to solve that coordination problem.

3.1   Basic Coordination Technology

Three issues are important in a distributed application: How do agents synchronize their work, how do they communicate, how is activity started? Amongst the various approaches to solve this coordination problem, is one line of research called *coordination technology* that is based on the concepts introduced by the language Linda ([CG89b]).

Linda introduces an abstraction for programming concurrent agents and defines a very small set of coordination operations. In a program based on Linda, a set of agents work on a task within a shared environment, called the *tuplespace*. It is a collection of tuples that contain information relevant for the application. Variants like distributed, or hierarchically structured ones, have been studied.

Linda's primitives provide means to manipulate that shared tuplespace, thereby introducing coordination operations. A tuple can be emitted to the tuplespace by an agent performing the out-primitive. As an example, out(<" amount",10,a>) emits a tuple with three fields, that contain a string, an integer, and the contents of the program variable a. This operation is non-blocking.

Two blocking primitives retrieve data from the tuple-space: in and rd. Both take a *template* as argument – for example in(<"amount",?int,?b>). A *matching rule* governs the selection of a tuple from the tuplespace: The template and the tuple must be of same length, the types of the fields must be the same, and – for a constant field (an *actual*) – values of fields have to be identical.

The example pattern retrieves a tuple that contains the string amount as the first field, followed by an integer, followed by a value of the same type as the program variable b. The notion ?b means that the retrieved value is to be bound to the variable b after retrieval. The difference between in and rd is that the former removes the matching tuple, while rd leaves it untouched in the tuplespace. Both operations are blocking as long there is no matching tuple found in the tuplespace. Linda makes no further guarantees on the selection of matching tuples and waiting operations.

It has been demonstrated ([CG89a]) that Linda is capable to express all major styles of coordination in parallel programs. in is a very powerful operation – it combines synchronization (the operation blocks until a matching tuple is found) with communication (the binding of values to program variables). Linda's operations together form a so-called *coordination language* ([GC92]). Combined with a sequential programming language, a new language for concurrent systems is generated. This combination is called *embedding* and can be implemented by changes to the programming language, by preprocessing source code, by libraries, or can be provided an extended operating system.

The following characteristics make Linda-like coordination attractive for distributed applications on the Web:

- **Uncoupling of agents.** The tuplespace as the coordination media uncouples agents in space and time. An agent can perform an out even when the retrieving agent does not yet exist, and can terminate before the out-ed tuple is retrieved. The tuplespace is a very high-level abstraction from locality issues.

- **Associative addressing** by using a template to retrieve a tuple means to state what kind of tuple is sought, not what tuple. This addressing is more abstract than retrieving a specific message.

- **Asynchrony and concurrency** are implicit notions of the tuplespace abstraction.

- **Separation of concerns.** The conception of coordination languages focuses on coordination only. It is not influenced by characteristics of the used programming language and leads to a clearer coordination model.

3.2 Coordination Technology in PageSpace

Coordination technology based on Linda uses repository of shared elements and operations for the addition and withdrawal as its core. To use this basic coordination mechanism for the Web, a Linda embedding into Java was defined for PageSpace and implemented. This system – derived from *Jada* ([Ros]) – forms our coordination kernel.

However, the pure data oriented style of coordination as in the original Linda-conception is not suited to support open distributed applications. It can well be used to keep state within one application, but it becomes difficult to support multiple applications that share one tuplespace.

With operations from two other coordination languages, we introduce two additional flavors of coordination styles:

- **Rule oriented coordination**

  The plain coordination mechanisms in Linda can be raised to a higher level by allowing declarative rules on coordination. ShaDe ([CCR96]) is an object-based coordination language. It offers a basic abstraction called the Object Space, that is similar to a tuple space with the difference that it contains objects. In fact, the Object Space is a distributed collection of objects and messages. Each object encapsulates a state in form of multiset of tuples and methods as rewriting rules.

  ShaDe objects are *active* units of computation with the ability to react to messages sent by other objects with an internal activity defined by methods. The state of an object is a multiset of tuples, so that the object itself can be considered as a tuple space, whereas the object space is a meta tuple space supporting inter object associative communication. Objects can use unicast, multicast, and broadcast communicate.

  For PageSpace, the most interesting feature of ShaDe is that coordination is expressed by rules. We intend to exploit such a feature to build "coordination" services enacting declarative cooperation laws. In the first prototype ShaDe was matched with Prolog, to obtain a distributed logic programming language. For PageSpace ShaDe is implemented on top of Jada.

- **Service oriented coordination**

  The notion of services is well adapted in open distributed systems. If can also form the basis for service oriented coordination languages that support the basic interactions in service-usage and provision.

  Laura ([Tol96]) is a coordination language designed for open distributed systems. Here, the tuplespace containing data is replaced with a *service space* containing forms describing service offers, requests, and results. The respective coordination primitives are serve, service, and result. Matching is performed on the service interface that is included in each of these kinds of forms. A subtype relation amongst these interfaces guides the matching routine in the selection of offers matching a service request.

  With a Laura reimplemented in Java, PageSpace agents are able to use and offer services at interfaces with Laura's coordination operations.

4 Kinds of Agents in PageSpace

In the PageSpace architecture, we distinguish several kinds of agents, denoted by Greek letters:

- **Alpha** agents are the interfaces of applications. They are manifested as a display in the users browser and can be "written" in HTML, Java, JavaScript etc. According to the capabilities of the users browser, there may be different instantiations of Alphas.

- **Beta** agents (also called *homeagents*) are the persistent representations of users in the PageSpace. They are responsible for the generation of Alphas upon the users request, for providing the user access to the agents and applications within the PageSpace, and for collecting incoming messages from other agents for the user.

- **Delta** agents offer and use services, interact by shared data, and implement applications within PageSpace. We distinguish application Deltas that also provide a user interface for interaction.

- **Zeta** agents are gateway agents to other coordination environments, or wrappers for legacy applications.

- **Epsilon** agents act as the kernel on each machine connected by the PageSpace platform. Epsilons manage the access to the PageSpace with an HTTP server, control Delta agents and offer coordination for agents.

Figure 4 shows an application in the PageSpace. There is an alpha in the users browser which is generated by a Beta. A set of Deltas implement the functionality of applications, and a Zeta provides access to a CORBA based coordination

environment. The PageSpace environment Gamma space is established by a set of Epsilon agents on different nodes. In

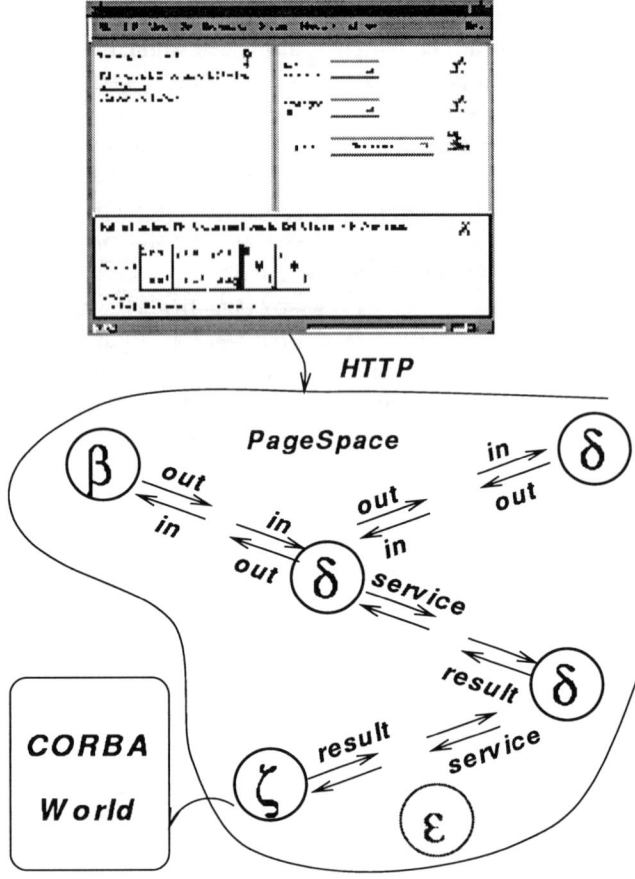

Figure 4. An application in PageSpace

the following, we describe the kinds of agents in detail.

4.1 Applications, GUIs, and Homeagents

PageSpace and its applications are accessible from any Web browser. This browser can be located at a different machine than the actual agent that performs an application. Also, the user can move during that interaction from one browser and machine to others. Thus, it is necessary, to deal with the user interface of an application separately.

As the interface has to be displayable by a Web browser, it is written in HTML. Due to the different characteristics of browsers, it can come in different formats. A text-based browser requires an interface without graphical components. Thus, we conceptually foresee that an application provides multiple representations of the interface.

The interface is moved from the application to the user, where the browser displays it and offers interactions. The processing of these interactions can take place at different

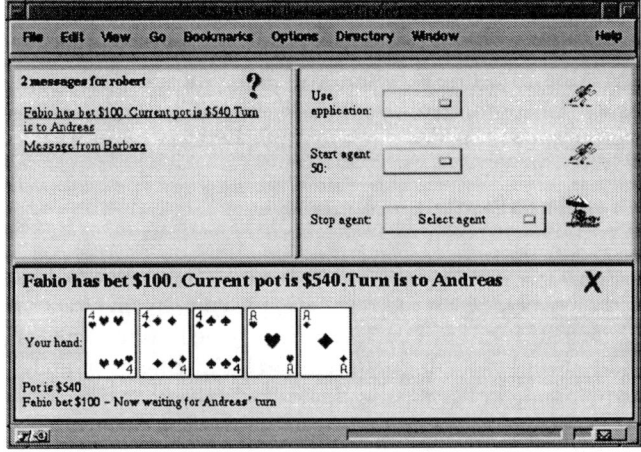

Figure 5. The user interface to PageSpace

locations – within the browser, if it is enabled by some mechanism like Java, or at the server, if it is form-oriented.

PageSpace has the potential to support any of the structures for applications on the Web as outlined in section 2:

- **Server located activity** can be supported by moving a static interface to the browser, which interacts with the application via a Beta agent located at a Web server.

- **Client located activity** by including a link to an applet within the HTML interface which does processing of inputs also. In PageSpace terminology, this means to transfer Delta agents to the browser.

- **Midware mediated activity** by transferring an applet with the interface that makes use of the coordination operations provided by PageSpace to interact with Delta agents distributed somewhere in PageSpace.

Each user of PageSpace has a persistent representation in the net, called the *homeagent* or "Beta". It has two faces – one to the user, one to the PageSpace. For the user, it provides the interface to PageSpace and the applications and agents therein. Figure 5 shows that interface for an example Poker application, which we call "Alpha", when it is manifested in the browser. This display is shown, when the user contacts his or her homeagent by retrieving a specific URL.

Alpha consists of multiple sections. One part of the user interface does provide operations of Beta – to use applications, and to start and stop own agents in the PageSpace. A list of messages shows the results of interactions with applications in the PageSpace. If the browser supports this feature, the queue is updated by a client-pull mechanism. When a message is selected, the user interface of that application is displayed in the third section of Alphas interface.

The other "face" of Beta is that it is a persistent representation of the user in the net. From Alpha, a user can use applications and start agents. However, he or she does not have to be online, while the application is running. Consider as an example a groupware application in which users all around the world participate in some work.

It is unacceptable to force users to be logged into the PageSpace all the time, as most distributed work is asynchronous in nature. Thus, while there is no connection to the user, Beta can still receive messages from a joined application. Beta looks like a complete agent to the PageSpace. However, it only stores incoming messages in a persistent store until the user retrieves them and reacts to them. In a future PageSpace, we will explore mechanisms to instruct Beta to automatically react to incoming messages.

4.2 Applications and Agents

Applications in the PageSpace are composed of agents, called "Deltas". There are three sorts of Deltas:

- An application Delta, which provides the specific user-interface of the application. As described in section 4.1, we separate the user interface and the application in order to access to it from Web browser. Still, some agent has to be contacted by Beta to get that interface. Any Delta returning such a GUI as the result of a method invocation is an application Delta.

- Deltas that implement the actual application specific functionality. A subset of a distributed application consists of agents that share an application specific context, which is of no use for other agents.

- Deltas used by an application, but which offer services to any application. These services are generic in that they can be used in the context of any application.

The agents that offer generic services can be started by a user within the PageSpace. They remain therein and answer to service requests by other agents until they are withdrawn.

The integration of legacy applications and gateways to other coordination environments can be achieved by wrapping and gateway agents that are called "Zeta". Like Deltas, they offer services to the PageSpace, but implement them by interacting with a closed application or via some middleware protocol to other middleware specific object.

5 Implementing the PageSpace Platform

The PageSpace is currently implemented as a prototype used for demonstration purposes and for experiments. This prototype follows the implementation strategy outlined in the following. Work remains to be done on the engineering of the platform, however, we believe that the main principles of our architecture can remain unchanged.

On each machine participating the PageSpace, one kernel Epsilon agent is running. Each Epsilon runs on a Java virtual machine and manages multiple threads. Figure 6 shows the logical outline of the Epsilon kernels. The several objects that run in threads are connected by streams for purposes of communication and management.

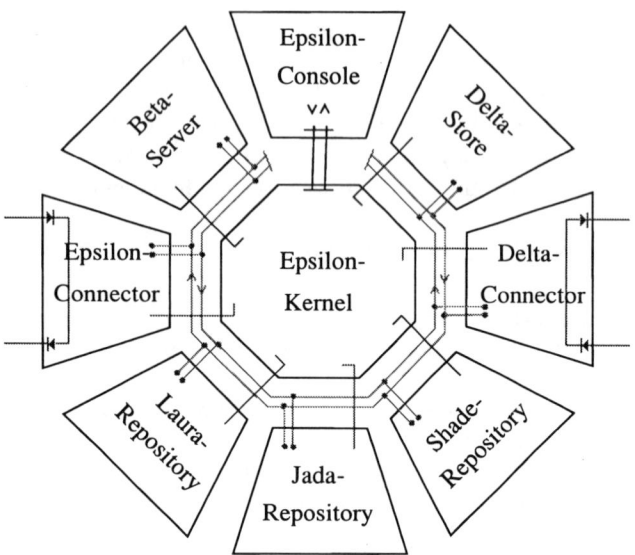

Figure 6. The logical structure of Epsilon

5.1 Provision of access to the PageSpace

Users access PageSpace via their homeagents that are contacted from a browser. Thus, a Web server has to be colocated with an Epsilon. As there are implementations of Web servers in Java – like Jigsaw from the W3C –, we integrate one of them as a thread in Epsilon. Thereby, interfacing of Beta to HTTP becomes much easier – instead of the CGI mechanism which only passes the CGI environment to a Java process, a call to a Beta object suffices.

5.2 Management of Beta agents

Betas are implemented by a single object within Epsilon. They are parameterized with the identification of a PageSpace user. After passing a login form in which a user name and password is entered, each user receives Alphas with the same components, but based on a different message queue.

The message queue in Beta is stored persistently in a database. Currently, we set up an mSQL ([Hug]) server for this purpose. Future databases written in Java, or interfaced with JDBC fit more smoothly in the implementation.

Besides the interaction with messages, the user can use applications, and start agents from Beta. Both result in the execution of a thread within the Beta object. That thread issues the appropriate coordination operation, waits for the results, stores it in the database and terminates.

5.3 Management of Delta agents

Each Delta agent is executed as a thread in Epsilon. This is reasonably, as we can make use of the native interaction mechanisms within one Java virtual machine for threads, and to avoid executing multiple virtual machines on nodes participating in PageSpace.

All Delta agents have the same kernel structure as depicted in figure 7. The main purpose of it is to pass invocations of the coordination primitives on to the Epsilon kernel. The specific coordination style can be supported by prefabricated handlers, for example for the dispatching of methods when the service-based Laura operations are used.

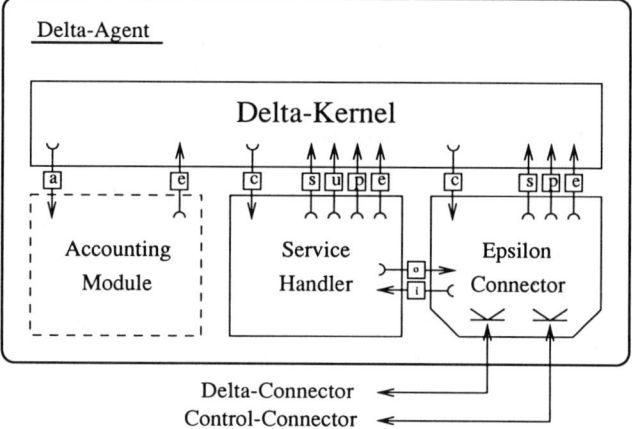

Figure 7. The logical structure of Deltas

Epsilon can easily manage its exception and monitor the operation of Delta threads. In a sense, Epsilon establishes the kernel of an operating system for Delta agents within a PageSpace coordinated Java virtual machine. We inherit thread management from Java, and add "process" interaction mechanisms by coordination technology.

5.4 Implementing the coordination operations

As outlined in section 3.2, we include several flavors of coordination technology in PageSpace. All of these have in common that they are centered around the use of a shared space of element of some kind, and that a matching rule guides the coordination primitives.

Thus, Epsilon contains instances of a generic basic component, the *repository*. These are collections of elements of some type. Each such repository implements the specific operations of a coordination language with a specific matching routine, thus is may be optimized, but still is based on the management of a pool of elements of some type. Within the Epsilon architecture, multiple repositories can be integrated to the internal control and data streams.

5.5 Identifying agents

Accessing services and naming are central issues in middleware. In PageSpace, we do not enforce a registry of agents, but take a different approach similar to how pages are accessed in the Web. As Epsilon knows about all agents it is managing, it is able to provide lists of these and their interfaces. The natural way to access such a list is by the built-in HTTP server. Thus, the "name" of an agent for the outside world is a simple URL.

Users that want to use agents, keep their personal list of known agents – just as one does for known Web pages with a bookmark-list. This list is used by Beta to offer the use of agents and applications to the user. It can be extended by the user, and be the subject to public catalogs of agents – resembling search engines and index services on the Web.

For a service-based use of PageSpace, the interface of a requested agent has to be passed to the coordination operations. This interface is accessible from the Epsilon that manages an respective agent. Beta retrieves it via HTTP, and constructs the appropriate coordination operation.

5.6 Distributed PageSpaces

The Epsilon kernels manage and coordinate agents on one machine. For distributed applications, these kernels have to have a distribution architecture and according protocols. A special concern with such a protocol is scalability – the ability to provide efficient coordination for a platform involving a large number of machines.

Establishing a shared repository of information can lead to scalability problems due to the amount of overhead for replication. We can take a flexible approach to structuring the system to overcome these problems. We follow the approach of the Internet to scalability: The set of machines that participate in the PageSpace is organized in a loose hierarchical fashion: Locally connected machines follow a replication schema in a logical sub-PageSpace and one machine is defined as the gateway to other sub-PageSpaces. Thereby, we imitate interconnected LAN the Internet.

The specific organization of Epsilons within one sub-PageSpace is a local decision. Known architectures for distributed implementation of Linda-like systems include full replication of a repository to all nodes, no replication with a centralized repository, or a partial replication as in [CG86]. As long as there is one defined node that runs a gateway protocol to other sub-PageSpaces, our architecture supports all of them. In fact, the current Jada implementation uses a centralized or fully replicated repository, whereas Laura implements a partial replication scheme.

For a gateway, a "routing-table" exists that instructs the gateway to which other sub-PageSpaces requests for matching elements shall be forwarded. Thus, the distribution structure can be statically or dynamically configured. This configuration will be based on the structure and behavior of the agents within a sub-PageSpace, and supports them in their coordination requirements. The several flavors of coordination employed in PageSpace give way to several intelligent optimizations, that are to be evaluated.

6 Features of the PageSpace Platform

The PageSpace architecture has several features that are yet to be explored. In this section, we point to two of them, namely fault tolerance and mobility. We show how these features can be introduced to the platform, and how they are enable by the design of the platform.

6.1 Fault Tolerance

The architecture of PageSpace opens perspectives to satisfy the needs for fault tolerance. Failures of the Alpha agents – because of a crashing browser, or a fault in the users machine – do not affect the PageSpace at all. The failure of a Beta agent does not introduce problems, as the queue of messages for a user is kept persistent.

The Beta, Delta and Zeta agents are managed by Epsilon. Thus Epsilon can keep a log of their external interactions and to request state information from that is stored persistently. Epsilon thus can monitor the managed agents, and restart them in case they crashed with a given state. We foresee that any of the managed agents can provide a method that transfers state information to Epsilon. The log of the external interaction can be used to keep the repositories within Epsilon fault-tolerant. The state information can be used to keep the managed agents tolerant to failures.

In the case of an Epsilon failure, the kernel and all managed objects are lost. The log of external interactions can be used to reestablish the repositories after restart; the managed agents can be restarted accordingly. An alternative would be to make the repositories themselves persistent, and the coordination operations transactional – however, the overhead involved has to be evaluated.

6.2 Mobile PageSpace Agents

As stated above, Delta agents interact location transparent. This fact, and the technical characteristics of the Java

383

platform makes them candidates to establish a notion of mobility of agents within PageSpace.

To do so, agents have to pass their internal state to an Epsilon, and Epsilon has to be able to start an agent with a specific state. The use of agents compiled into code for the Java virtual machine together with its run-time-linking capabilities makes the code of the agents portable within the PageSpace environment.

Deltas may want to be moved because they detect that they interact with each other and try to make the coordination more efficient by "meeting" at a specific location. They can be asked to move by an authoritative Epsilon, because a specific policy applies to their current location.

In any case, the notion of location has to be introduced. We foresee, that a Delta can ask its Epsilon about the current location and that it is able to communicate it to another Delta. We do not foresee any operations on location representations available to Deltas. Epsilons can stop agents, transfer them to another Epsilon, which in turn restarts them. The state of Delta has to be passed along with the byte code of Delta. The access to that state is provided by Deltas, as foreseen in the fault tolerance mechanisms.

It has to be evaluated, what protocols are most efficient to perform such operations, and what strategies for mobility should be followed by Deltas and Epsilons.

7 Conclusion

The PageSpace is a platform to support distributed applications on top of the Web. We provide a framework that is based on the core Web technologies and Java, and add a specific approach to coordinating distributed agents in applications, namely Linda-like coordination technology. Our approach is generic towards the usage of several variants of coordination technology, as demonstrated with data-oriented, service-based, and rule-driven coordination styles. The design of the platform is enabling for a straightforward implementation of several desirable features, such as fault tolerance and mobility of agents.

The first phase of project PageSpace was concerned with the development of our approach, a prototypical implementation, and a demonstration of its potential. Now the focus is on engineering the platform, and on validating our conception with applications in the field of electronic commerce. Information on PageSpace can be found on the Web at http://www.cs.tu-berlin.de/~pagespc.

Acknowledgments. PageSpace has been supported by the EU as ESPRIT Open LTR project #20179.

References

[AJ95] G. Almási and V. Jagannathan. Integrating the WWW and CORBA-based Environments. In *Proceedings of the Fourth World Wide Web Conference*, 1995. (Web* Home Page: *http://webstar.cerc.wvu.edu/lpi/*).

[BIV+95] Ashley Beitz, Renato Iannella, Andreas Vogel, Zhonghua Yang, and Tak Woo. Integrating WWW and Middleware. In RS Debreceny and AE Ellis, editors, *Innovation and Diversity - The World Wide Web in Australia. AusWeb95 - Proceedings of the First Australian World Wide Web Conference*, 1995.

[CCR96] S. Castellani, P. Ciancarini, and D. Rossi. The ShaPE of ShaDe: a coordination system. Technical Report UBLCS-96-5, Department of Computer Science, University of Bologna, 1996.

[CG86] Nicholas Carriero and David Gelernter. The S/Net's Linda Kernel. *ACM Transactions on Computer Systems*, 4(2):110–129, 1986.

[CG89a] Nicholas Carriero and David Gelernter. How to Write Parallel Programs: A Guide to the Perplexed. *ACM Computing Surveys*, 21(3):323–357, 1989.

[CG89b] Nicholas Carriero and David Gelernter. Linda in Context. *Communications of the ACM*, 32(4):444–458, 1989.

[CKTV96] Paolo Ciancarini, Andreas Knoche, Robert Tolksdorf, and Fabio Vitali. PageSpace: An Architecture to Coordinate Distributed Applications on the Web. *Computer Networks and ISDN Systems*, 28(7–11):941–952, 1996. Proceedings of the Fifth International World Wide Web Conference.

[GC92] David Gelernter and Nicholas Carriero. Coordination Languages and their Significance. *Communications of the ACM*, 35(2):97–107, 1992.

[HK94] Edwin E. Hastings and Dilip H. Kumar. Providing Customers Information Using the WEB and CORBA. In *Proceedings of the Second World Wide Web Conference '94: Mosaic and the Web*, 1994.

[Hug] David J. Hughes. Mini SQL – A Lightweight Database Engine.

[Ros] Davide Rossi. Jada: multiple tuple spaces for Java á la Linda. *http://www.cs.unibo.it/~rossi/jada/*.

[Tol96] Robert Tolksdorf. Coordinating Services in Open Distributed Systems with Laura. In Paolo Ciancarini and Chris Hankin, editors, *Coordination Languages and Models, Proceedings of Coordination '96*, LNCS 1061, pages 386–402. Springer, 1996.

Incorporating Business Process Management into Network and Systems Management

Lundy Lewis and Jim Frey

Cabletron Systems
486 Amherst Street
Nashua, New Hampshire 03060 USA

Abstract

Integrated network and systems management for communications networks is relatively mature. In addition, techniques for off-line simulation and analysis of business processes is relatively mature. Since many of the components of a business process now include network communications, devices, and applications, the time is ripe for integrating business process (BP) modeling methods and network management into real-time BP monitoring and control. In this paper we describe our direction and results towards this end, where we employ the Spectrum network management platform, the AppControl systems management agent, and the NerveCenter process modeling tool in an integrated, decentralized architecture.

Keywords: Integrated Network and Systems Management, Business Process Management

1: Introduction

A communications network is an element in a larger picture which includes all aspects of an enterprise's business processes (BPs). These BPs depend on the network and its various nodes in order to successfully complete their stated goals. For example, when a standard process such as "payroll" is described, we discover that required components may include client workstations, computer servers, file servers, database applications, peripherals, and the network connecting them together.

Network Management Platforms (NMPs) for monitoring and controlling computer networks have become relatively mature. Examples of NMPs are Cabletron Spectrum, HP OpenView, and IBM NetView. Spectrum is considered to be the most advanced of these NMPs due to its distributiveness and event correlation capabilities.

In addition, Systems Management Platforms (SMPs) have become relatively mature. Examples are BMC Patrol, Tivoli Inc. Tivoli Management Enterprise, Computer Associates UniCenter, and Seagate AppControl. SMPs provide micro-management of systems, including workstations and servers, with capabilities to monitor and control system operating parameters, user security, and applications and processes running on the systems. SMPs typically consist of agents that reside on the systems of interest, and client consoles are used to configure the agents and to display important events that occur on the system.

The integration of NMPs and SMPs has matured as well -- e.g. Spectrum has been integrated with each of the SMPs listed above. NMPs are very good at overall network management but are not designed to micro-manage systems, while SMPs are good at micro-management of systems but are not designed for network management. Thus, NMP/SMP integration is complementary, contributing a large part to total management of enterprise networks. See Reference [1] for a thorough discussion of integrated network and systems management.

Finally, off-line BP simulation products have become increasingly prominent, e.g Proforma Corporation ProVision Workbench, Meta Software Corporation WorkFlow Analyzer, LogicWorks BPwin, CSA Silverrun, and many others. These products are used typically for re-engineering, streamlining, and improving existing BPs, or designing new BPs.

With the advent of integrated Network and Systems Management, the time is ripe for integrating BP modeling methods and network management into real-time BP monitoring and control [2]. For example, when a network element is not functioning properly, network fault management methods help us isolate and correct the problem, but does not help us in determining how a business process is affected. A router failure may cause a marketing forecast report to fail, or a file server crash might interrupt a nightly software distribution. With proper integration with BP representations, we may also effect on-line BP monitoring and control.

We find that Spectrum's modeling techniques, distributed representations, multi-level reasoning, and alarm roll-up can be applied to real-time BP management. First, we model the systems, servers, applications, peripherals, network devices, operators, and domains in Spectrum in the usual way. This gives us a complete picture of the network. Next, we select and combine a subset of these elements in order to form a representation of a BP.

The objective is to monitor the health of the elements in the BP and to roll the health metrics up to the top level BP representation, whereby we can monitor the health of the BP. Since the elements in the BP are likely to be distributed across multiple domains in a large enterprise, we must consider that the information will be coming from several elements across domains. Importantly, we must consider that it is the task of the BP representation to correlate this information and issue alarms when the BP fails, faces eminent failure, or becomes generally degraded.

The approach we take here is based upon the concept of distributed, multi-level reasoning (DMR) [3]. Section 2 discusses the motivation for distributed reasoning in network management. Section 3 describes distributed reasoning in the Spectrum client/server architecture. Sections 4 and 5 illustrate the architecture in an internal testbed and a customer installation. Section 6 describes the concept of multi-level reasoning with respect to the general problem of inter-domain alarm correlation in network fault management. In Section 7 we argue that real-time BP management is a special case of the former problem, and show how to deploy BP management in similar fashion. Section 8 provides an example. Finally, in Section 9 we discuss future work and outstanding issues regarding the approach.

Note that while the techniques of BP modeling and integrated network/systems management exist today, the method of combining them into a distributed, decentralized model for the purpose of BP management is original with this work.

2: On distributed, multi-level reasoning

Three important truths regarding network management motivate our distributed, multi-level architecture for the enterprise network (EN) management platform:

1. The EN is inherently a distributed, multi-domain enterprise. ENs typically are partitioned in ways that help administrators understand and manage the EN, e.g. with respect to geographical domains, functional domains, or managerial domains [4]. For an ordering process, the functional domains are usually divided into customers, sales, credit/invoicing, production, and assembly/shipping.

2. The data types for analyzing and reasoning about EN behavior come in various forms, e.g. traffic data (e.g. packets and cells, load, collisions), topological models of the network, events, alarms, subjective assessment by humans, et al.

3. The models in an EN likewise come in various forms, e.g. models of systems, workstations, peripherals, communication links, operators, domains, and applications.

Therefore, the tasks involved in managing the activities spread over a distributed EN are too complex for a single controlling agent, and thus the tasks themselves must be partitioned into distributed, cooperative agents making up a decentralized, autonomous (or semi-autonomous) monitoring and control system.

3: Spectrum's distributed client/server architecture

For the reasons above, the Spectrum NM platform was designed as a distributed client/server architecture. See Figure 1. The Spectrum servers, called SpectroSERVERs (SSs)

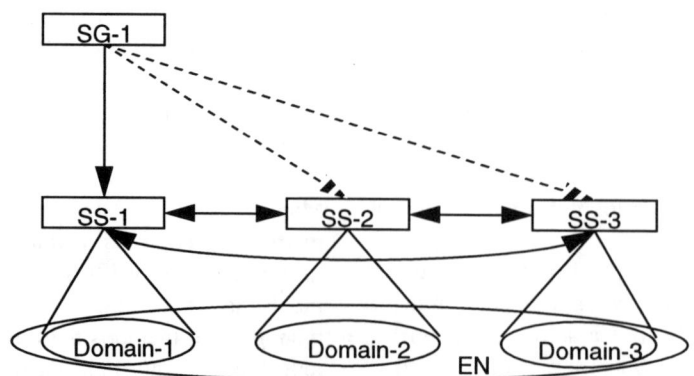

Figure 1. The Spectrum Distributed Client/Server Architecture

386

monitor and control individual EN domains. The Spectrum clients, called SpectroGRAPHs (SGs) may attach to any one SS in order to graphically present the state of the domain, including topological information, event and alarm information, configuration information, and control options.

Importantly, all domains can be managed from a single SG. If SG-1 is attached to SS-1, but the user wishes to manage the domain controlled by SS-2, the user can click on an icon in SG-1 that represents SS-2. Figure 1 shows the primary client/server attachment between SG-1 and SS-1, where virtual attachments between SG-1 and other SSs are indicated by dotted lines.

4: Experiences with the architecture in an internal testbed

We have tested the architecture in an internal testbed to establish operational feasibility. A virtual enterprise network was configured using 250 SSs situated in our U.S. and U.K. facilities. A three-layered hierarchical topology was used, with one master SS connecting to 14 SSs, each of which in turn connecting to 15-20 more SSs. Each end-node SS was given a database with several hundred manageable devices.

We recommend a 1:7 ratio among SSs that are configured hierarchically. This ratio is derived from workstation operating system characteristics rather than communications traffic load among SGs and SSs. Note that load would be a strong, adverse factor if we forced a one-to-one correspondence between SGs and SSs, or if we allowed heavy communication among SSs.

In our internal testbed, we purposely exceeded the 1:7 ratio in order to stress the architecture and shake out the bugs. We monitored performance, accuracy, and reliability. The test was run over a period of several days, during which time a variety of simulated failures were introduced and the resultant behavior analyzed. A few communication flaws at the physical layer were identified and corrected, but as a whole the test was successful and the test bed operated without incident. Subsequent passes through this test plan are being used to further exercise distributed functions and to test fault tolerance.

These initial tests provided a good, empirical argument for the scalability of the distributed, client/server architecture. We consider each SS as an intelligent agent, capable of presenting management data on demand to any client SG. This keeps inter-SS communications to a minimum. Each SS "knows about" its peer SSs, but is prohibited from extensive communication with them. In Section 2.2, we show how SSs may communicate by intermediary agents who reside at higher levels of NM abstraction.

5: Customer installations

The distributed version of Spectrum has been installed at many customer sites, with setups ranging from a few (2 or 3) SSs to several hundred. Active sites are domestic U.S. and International, including Bermuda, Taiwan, Australia, and Germany. For the most part, customer ENs are divided into geographical domains, using an SS at each facility or campus, with a central master SS at a headquarters location.

A particularly challenging current project is the management of a telecommunications network in Eastern Germany, deployed by Deutsche Telekom [5]. This project poses unusual requirements because it is purely non-SNMP, using only a proprietary management protocol. Spectrum was chosen as the management platform because (1) it has a distributed, client/server design, (2) it has APIs for developing non-SNMP management applications, (3) it has APIs for configuring intelligent SS agents, and (4) it enables representation of both devices and services involved in EN management.

Consider (3). In multi-domain ENs with corresponding SS agents, polling-based management can be costly in terms of bandwidth load. By restricting SS polling (i.e. only using it for testing basic element presence/status) and instead having managed elements forward management data to the SSs (e.g. via traps or intelligent agents), in-band management traffic is reduced considerably. This was a requirement for the application. Note that a transition from polling to trap-based management and intelligent agents is considered by some to be the future of EN management.

Consider (4). Data collected via the management system is being utilized in two ways. First, network devices are represented in the usual way in order to monitor and control the operations of the telecommunications network as a whole. Second, a service-based representation is used to monitor/manage usage and repairs so that, for example, large business customers may be given relatively higher priority for repairs than residential customers. This secondary representation is accomplished by giving relative weights to each managed element's alarms.

As of November 1996, the number of SSs deployed for this application was more than 1500. The NM configuration is similar to our internal testbed save that (i) each end-node SS manages devices and services, (ii) there are two tiers (i.e. there is not yet a top-level master SS), and (iii) there are more end-node SSs per second tier SS (up to 40). Performance and capability results thus far have been excellent. The project is on track towards a planned deployment of 4000 SSs (first tier plus second tier). The success of this project has resulted in similar projects begun by teleco providers in several other countries.

6: Multi-level reasoning

Let us stop a moment to think about the management tasks that occur within network domains and management tasks that occur across domains. We will use fault management as our first example.

Fault management consists of event monitoring and filtering, event correlation, escalation of events to alarms, alarm correlation, and diagnosis/repair. The SSs that monitor individual network domains perform these tasks with Spectrum's inductive modeling technology (IMT) (see Reference [6]). We may refer to this as *intra*-domain event/alarm correlation.

With multi-domain ENs, the requirement now is to perform the same function across domains. For example, a failed router in Domain-1 may affect the applications running in Domain-2. Conversely, the cause of an application failure in Domain-2 may be identified as the failed router in Domain-1. We refer to this as *inter*-domain event/alarm correlation. Since we have limited inter-communication among SSs, we need to find some other way to do inter-domain alarm correlation.

We can conceptualize the inter-domain event/alarm correlation task as shown in Figure 2. The figure, however, is somewhat misleading because it is in two dimensions. The bottom-most two levels can be performed by SSs that monitor and control individual domains in the EN. The number of SSs is the (implicit) third dimension. The agent that resides on the top level collects alarms from multiple SSs and carries out inter-domain alarm-correlation, communicating with other SS agents as appropriate. Note that the SS agents communicate indirectly via the intermediary coordination agent on the top level.

What reasoning paradigm is appropriate for the coordination agent at the top-most level? Several reasoning paradigms are at our disposal, including simple look-up tables, expert systems, case-based reasoning systems, state-transition graphs, et al. Several commercial products that incorporate some one or other of these paradigms are available. We have integrated NerveCenter from the SeaGate Corporation (which uses the state-transition graph paradigm) with Spectrum, where NerveCenter is the top-most agent [7].

Figure 3 shows a clearer picture of the integration architecture. (We have left out the SG clients). The Spectrum Alarm Notifier (AN) is a client daemon that collects alarms from all domains in the EN. The AN can be configured to allow only select alarms to be passed to NerveCenter. NerveCenter performs high-level reasoning over multi-domain alarms and communicates with other SS agents via the Spectrum Command Line Interface (CLI). Communications may include requests for further bits of information, or notification of inter-domain alarms.

An inter-domain problem scenario is implemented in NerveCenter in a temporal state transition graph (STG) representation. NerveCenter comes off the shelf with some half dozen standard STGs, but users may model problem scenarios in order to customize the tool for particular ENs.

The classic example of such a problem scenario is when two routers in two domains become overstressed at approximately the same time, a consequence of which is that overall network performance is likely to become degraded. The SS agents in each domain may issue independent alarms when this happens, but neither agent can predict the overall effect of the alarms. However, since each alarm is rolled up to NerveCenter, NerveCenter may reason about the overall effect and issue inter-domain alarms

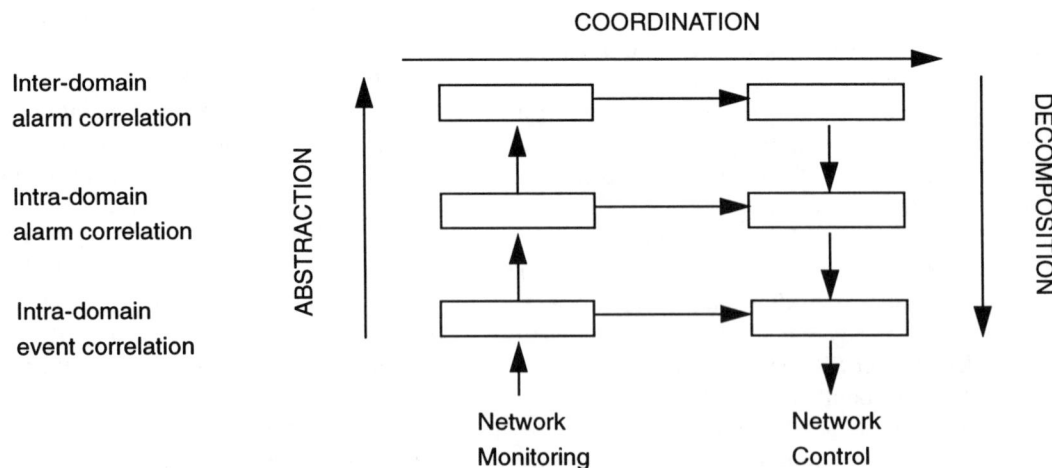

Figure 2. A Multiple-Level Architecture for Inter-Domain Alarm Correlation

388

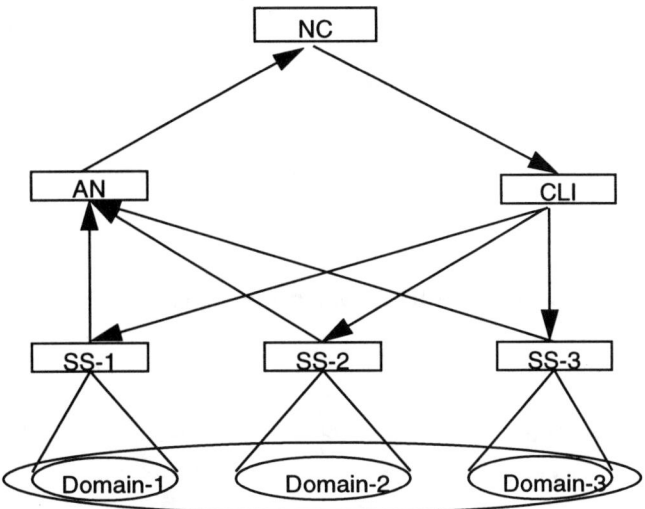

Figure 3. An Integrated Architecture with Spectrum and NerveCenter

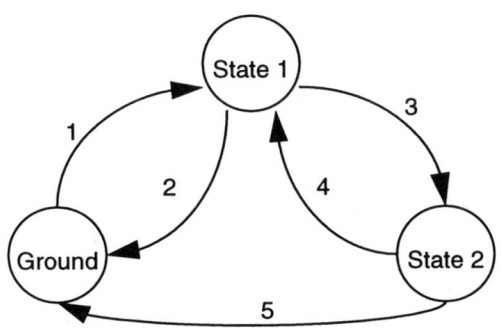

Ground:No activity

State 1: Mild inter-domain alarm issued (yellow)
State 2: Urgent Inter-domain Alarm Issued (red)

Arc 1:   Intra-domain alarm on X at time T1
Arc 2:   Intra-domain alarm on X cleared
Arc 3:   Intra-domain alarm on Y at time T2, X /= Y, T1 = T2
Arc 4:   Intra-domain alarm on Y cleared
Arc 5:   Inter-domain alarm cleared by user

Figure 4. A Sample State Transition Graph

accordingly. An alarm is indicated as the coloring of a state node in the STG, or alternatively in a SpectroGRAPH Alarm View. Figure 4 shows the STG that can accomplish this.

At this juncture we wish to shift views and examine business process management. An important insight in our work is that the problem of BP management is a special case of the problem of inter-domain alarm management.

7: On-line business process management

Thus far we have identified two kinds of representations that may be useful in BP management: eminent process failure models and on-line process models.

Eminent process failure models are quite simple, but useful nonetheless. The idea is that we design an STG that shows the health of each communications device that enters into a BP [8]. Figure 5 shows a very simple STG, where each state corresponds to a specific device in bad health. Note that such an STG does not model the structure of the BP, but rather models the elements that make up the structure. For example, one may look at the representation to insure that all systems are up, before initiating the process. ICS GmbH has worked through this idea with their Continuity product for Spectrum [8].

On-line process models are more-or-less true representations of the process. The simplest kind of BP model is a linear process in which each application in a BP has to finish and report success before the next application is initiated. SMP agents such as AppControl may monitor the applications and issue events upon initiation and completion. These events may be rolled up to NerveCenter for the purpose of monitoring and control. Figure 6 shows such a linear process. The State1x nodes indicate that an application is in progress, while the State2x nodes indicate that an application failed to reach completion.

Importantly, we may combine STGs in NerveCenter so that results of one STG provides input to a second STG. For example, the STGs in Figures 5 and 6 may be combined so that the failure of a system on which an application is running enters into the picture of BP management.

389

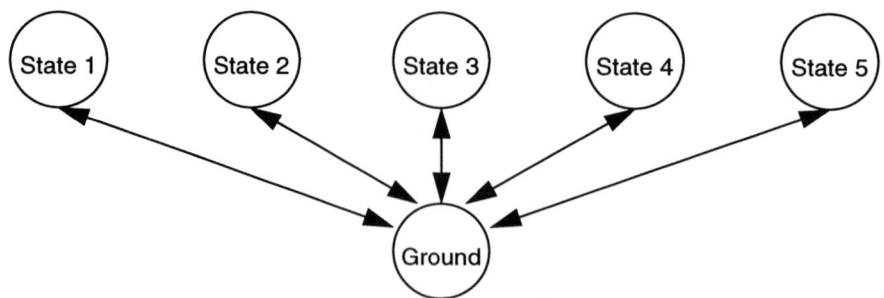

Figure 5. A Sample STG for Eminent Process Failure

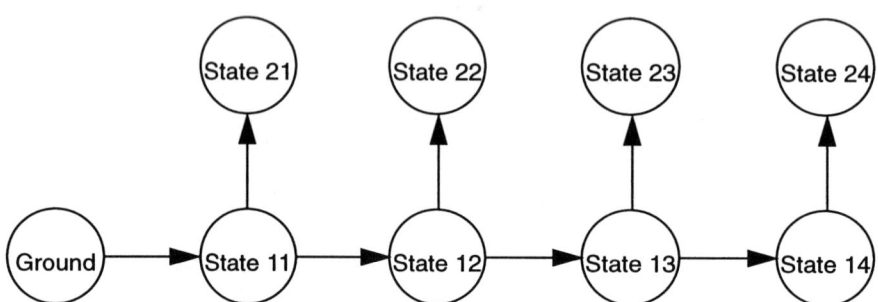

Figure 6. A Sample STG for On-Line, Linear BP

As an aside, let us observe that the integration of BP management into network and systems management will benefit from several useful services already enjoyed by network and systems management. For example, Spectrum and other NMPs have been integrated with several Trouble Ticket Systems such as Peregrine's ServiceCenter, paging and email systems such as SpectroPHONE, and Knowledge-Based Systems such as SpectroRx [9], where alarm information is passed automatically to these peer applications for further processing. From Spectrum's point of view, "an alarm is an alarm is an alarm" whether it be a device alarm, communications alarm, or BP alarm. Thus, a BP alarm and surrounding information may be transformed into a trouble ticket, or may be passed as a phone call to administrators upon occurrence.

8: A hypothetical example

In this Section we provide a hypothetical example of the approach. Figure 7 shows a hand-drawn topology of a rather complex enterprise network for Company X, with various kinds of subnet, satellite communications, file servers, and workstations.

Suppose X has a policy to update File Server FS1 every Wednesday with the prior week's accounts received, which has been logged in File Server FS2 and a home-grown database on Workstation W1. This is an example of a BP.

An eminent process failure model would group the devices upon which the BP depends in a single, logical view. For example, in Spectrum there would be icons that represent these devices. An icon would turn red if an alarm occurred on the device.

An on-line process model would model the applications upon which the BP depends, showing the temporal relations of parts of the process. For example, suppose the retrieval of data D1 from W1 initiates the process, and D1 is sent to Application A1 on FS2. A1's responsibility is to combine D1 with Data D2 retrieved from FS2 (say D3 = D1 + D2), format it, and ship D3 to FS1 for permanent storage. Application A2 on FS2 performs a security check on D3 to make sure it hasn't been corrupted or tampered with in route, and then deposits it in the database on FS1.

Figure 8 shows an on-line process model for this BP.

9: Future work

We have demonstrated the feasibility of the infrastructure required for the management of distributed BPs using the Spectrum NMP, the AppControl SMP, and the NerveCenter process modeling tool. Our tests involved the monitoring and control of simple processes spread across

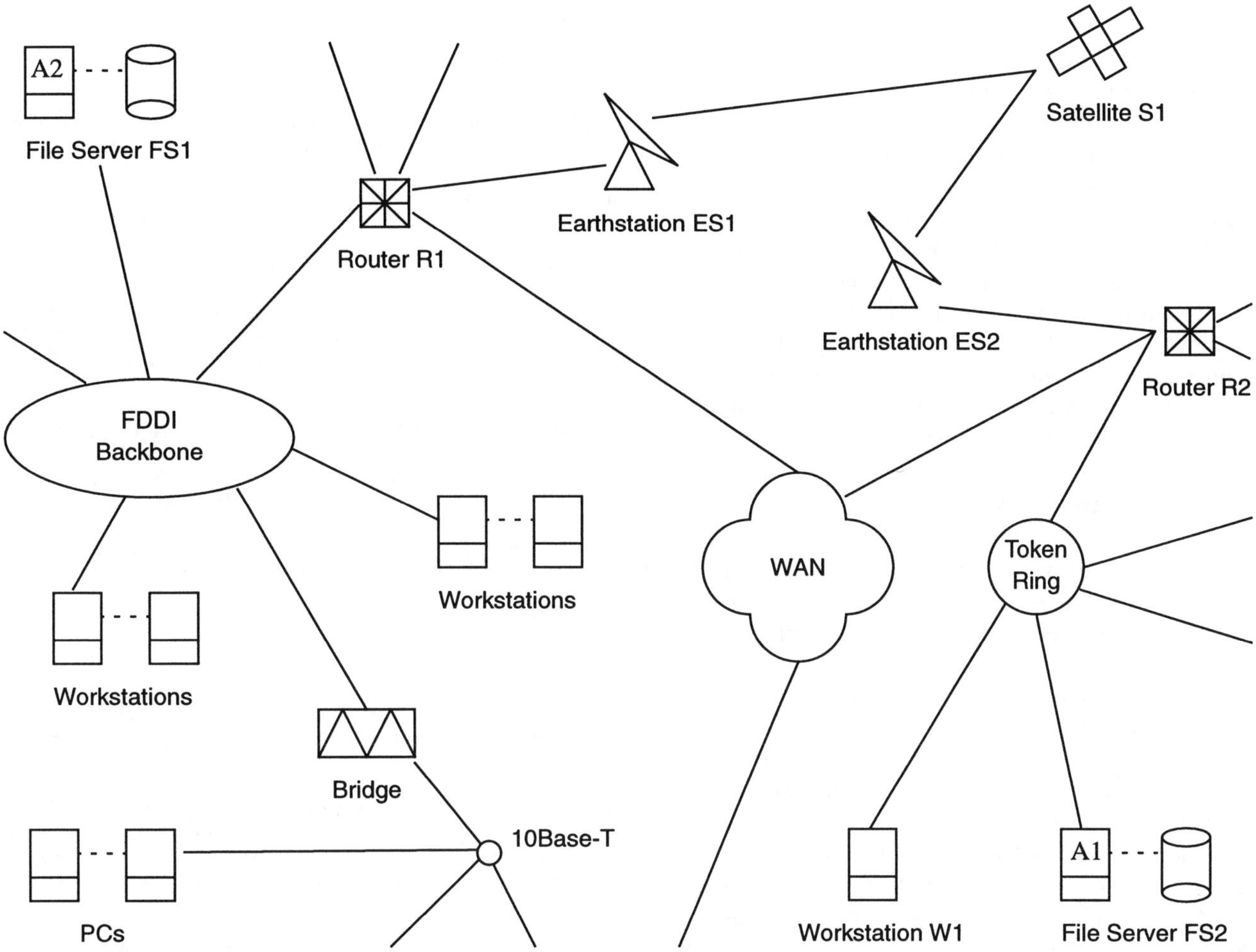

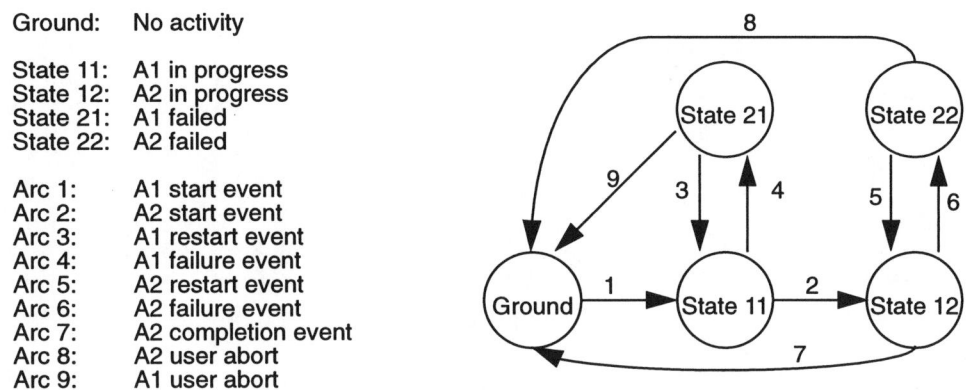

Figure 7. A BP Embedded in an Enterprise Network

Ground: No activity

State 11:   A1 in progress
State 12:   A2 in progress
State 21:   A1 failed
State 22:   A2 failed

Arc 1:   A1 start event
Arc 2:   A2 start event
Arc 3:   A1 restart event
Arc 4:   A1 failure event
Arc 5:   A2 restart event
Arc 6:   A2 failure event
Arc 7:   A2 completion event
Arc 8:   A2 user abort
Arc 9:   A1 user abort

Figure 8. An STG for the BP in Figure 7

geographical domains, designed to show that the mechanics of alarm passing in fact worked in the field.

Our next experiment is to model a "real" BP at a customer site. The classic example of a real BP is a distributed payroll process. Since the infrastructure of decentralized control of a BP is in place, our primary challenge will be to understand the semantics of the BP and to form a representation of the BP in the process modelling tool. Importantly, we require a formal method by which to map a BP over to a STG. At present, this task is more of an art than a science, and requires further work in its own right. An explanation of the situation and results of our future experiments are forthcoming.

10: Summary and conclusion

In this paper we have described an approach towards autonomous, decentralized monitoring and control of BP. As vehicles to realize BP management, we employ existing techniques in integrated network and systems management. The agents that make up the system involve SS agents whose responsibility is to monitor individual network domains, SMP agents whose responsibility is to monitor systems in those domains, an alarm monitoring agent whose responsibility is to collect and forward domain alarms to a BP model, a BP reasoning agent whose responsibility is to correlate alarms, and finally an alarm notification agent who displays BP alarms to users, or takes corrective actions automatically. We described the approach in some detail for the general problem of inter-domain alarm correlation, and argued that the BP modeling problem is a special case of the former problem.

References

[1] H.-G. Hegering and S. Abeck. *Integrated Network and System Management*. Addison-Wesley. 1994.

[2] R. Oliveira, D. Sidou, J. Labetoulle. *Customizing network management based on application requirements*. Proceedings, First IEEE International Workshop on Enterprise Networking (ENW-96). Dallas, June 1996.

[3] L. Lewis and J. Frey. *Distributed, Multi-level Reasoning for Enterprise Network Management*. Proceedings, First IEEE International Workshop on Enterprise Networking (ENW-96). Dallas, June 1996.

[4] M. Sloman and K. Twidle. "Domains: A Framework for Structuring Management Policy" in *Network and Distributed Systems Management*. Edited by M. Sloman. Addison-Wesley Publishing Company. Wokingham, England. 1994.

[5] W. Weipert. The Evolution of the Access Network in Germany. *IEEE Communications Magazine*. February 1994.

[6] W. Hamscher, L. Console, and J. de Kleer (editors). *Readings in Model-Based Diagnosis*. Morgan Kaufmann, San Mateo, 1992.

[7] L. Lewis and A. Noushin. Outline of the Spectrum/Nerve-Center Integration. Technical Note lml-aj-95-10. Cabletron Systems. 1995.

[8] T. Klober. Business Continuity Management with SPECTRUM. Technical Note 1.0/28.3.1996. Intelligent Communications Software International GmbH. Kistlerhofstrabe 111, 81379 Munchen, Germany. (URL http:/www.ics.de) 1996.

[9] L. Lewis. *Managing Computer Networks: A Case-Based Reasoning Approach*. Artech House, Boston. 1995.

Session 5C

Applications: Building Control

Chair

Edgar Nett

Decentralized Autonomous Object-Oriented EMS/SCADA System

Tomomichi. Seki[+],   Hideaki. Sato[*],   Toshibumi. Seki[*],
Tatsuji. Tanaka[#],   Hadime. Watanabe[**]

[+]Engineering R&D Division, Tokyo Electric Power Company, Yokohama, Japan
[*]Systems & Software Engineering Lab., R&D Center, TOSHIBA Corporation, Kawasaki, Japan
[#]Heavy Apparatus Engineering Lab. TOSHIBA Corporation, Fuchu, Japan
[**]Utility Power Systems Engineering Department, TOSHIBA Corporation, Tokyo, Japan

Abstract

Usually Energy Management Systems / Supervisory Control and Data Acquisition (EMS/SCADA) systems are geographically distributed and have hierarchical operational organizations. They are continuously changing in accordance with the various and varying environments, and they should be flexible enough to adopt to those changes quickly. This paper proposes a new architecture called SCOPE (System Configuration Of PowEr control system) to realize flexible and reliable EMS/SCADA systems.

SCOPE makes application programs independent of operational organization and system configuration of EMS/SCADA system, i.e., application programs are not influenced by changes in them. By these properties, EMS/SCADA systems become flexible and reliable, and also development of EMS/SCADA systems become efficient and economical. Through developing and evaluating a SCOPE prototype system, it has been confirmed that the flexibility and maintainability of EMS/SCADA systems based on SCOPE architecture had been improved.

1. INTRODUCTION

Electric power systems are continuously expanding. Power stations, transformer substations and transmission lines are installed in accordance with the growing demands, and new functions are also introduced to improve power supply capability. The operational organization of electric power systems constitute hierarchy of many control offices, that is, central load dispatching office, regional load dispatching offices and local load dispatching offices. This hierarchy is modified frequently in accordance with the change of management policy.

In the conventional systems, modifications of control office's supervisory areas cause significant changes in the database of supervisory control and data acquisition (SCADA) systems and application programs at each control offices[1,2]. Also replicated data and programs at several offices with different interfaces, make it difficult to directly access databases and programs in different offices.

To make development and maintenance of SCADA systems flexible and reliable, redundant data and software should not be localized at individual offices. They should be accessible from all offices. In order to keep operations continuously even under changing network topology, operational organization, computer system configuration and functional configurations, data and programs distributed over the network must be accessed transparently from different locations. For realizing uniform and dynamic access to data and programs in different computers, software structure should be divided into encapsulated layers with well-defined interfaces.

As a result of analysis for conventional SCADA systems, it is confirmed that the software in electric power system classified into three layers, i.e., "computer system dependent part", "power system dependent part" and "task dependent part". The computer system dependent part encapsulates computer system configuration and resource location. The power system dependent part encapsulates equipment configuration and operational organization of power systems. The task dependent part depends on application specification, such as supervision. An modification in a layer is encapsulated into it, and does not influence other layers. Therefore, any modification of application programs is not necessary even if the configuration of computer systems and power systems are changed.

In this paper, SCOPE (System Configuration Of PowEr control system) has been proposed as a basic framework so that changes of environment do not affect application programs. In section 2, the goal of SCOPE is mentioned. Section 3 shows the design concept of SCOPE including the definition of the three layers. In section 4, the technique for establishing desired flexibility of the power system is introduced. In section 5, a SCOPE prototype system is described, and section 6 shows its evaluation results, followed by the conclusion in section 7.

2. GOAL OF SCOPE

The goal of SCOPE is to realize a highly flexible and maintainable SCADA systems which enable incremental software development with the changes of environment, and offer non-stop operation of application programs during system maintenance where configuration of power systems and computer systems are changed.

In order to continue the correct operation in these environments, SCOPE has the following features:

1. Application programs and data are independent of the system operation and configuration.
2. Application programs and data are independent of modification of other programs and data.

When new offices are built in the conventional system, application programs in regional load dispatching offices have to be halted and modified in order to access to programs and data in the new office consistently. On the other hand, in SCOPE, it is intended that application programs in existing regional load dispatching offices can access to the newly added offices without modifications nor operational breaks.

3. DESIGN CONCEPT OF SCOPE

3.1. Base Structure of Future SCADA System

In order to increase the flexibility of the SCADA systems, it is important to isolate subsystems which are influenced by changes of environments. Therefore, SCOPE classifies programs in electric power systems into three hierarchical layers as shown in Figure 1. Interfaces between different layers encapsulate the inner-structure of each layer; therefore, the modification of programs and data in one layer does not affect other layers.

Task dependent part
Power system dependent part
Computer system dependent part

Fig.1 Three Layers of SCADA System

Task dependent part : Programs and data in this part are independent of the configuration of both power systems and computer systems. Since the programs in this layer become general package software independent of system configuration, they can be used in all offices with minimum modification. Namely, their modifications are needed only when application specification is changed.

Power system dependent part: This part encapsulates the hardware configuration and operational organization of real power system, and provides two abstract data models. One corresponds to hardware of power systems, such as a topology of transmission lines, circuit-breakers and transformers. The other corresponds to operational organization of power system, such as office configuration, control areas and order authority. These models make programs of the task dependent part independent of real power system configuration.

Computer system dependent part: This part hides a computer system configuration. It encapsulates network topology of computer system, computer architecture, location of resources etc., and the programs and data of upper two layers can be developed without awareness of computer environments.

In conventional SCADA systems, programs are not classified in this way, and the affects of a program modification may propagate to other programs in a complicated way . By this three-layer structure, it becomes possible to develop application programs independent of not only other application programs but also the configuration of power systems and computer systems. In addition, since configurations of power systems and computer systems are encapsulated in the corresponding layers, modification of their inner-structure can be done independently of other programs. Programs can access to data and functions in a different layer through well-defined interfaces.

3.2. Service of SCOPE Functions

SCOPE constitutes the lower two layers of SCADA three-layer structure mentioned in the previous subsection, i.e., the power system dependent part and the computer system dependent part. It consists of two subsystems, two models and a utility, as shown in Figure 2.

The power system dependent part includes the network information model and the operational information model. They are abstract models corresponding to the real power system.

396

Power System Supervisory Application Power System Operation Application Generator Control Application	Task Dependent Part
Network Information Model(NIM) Operational Information Model(OIM) Model Modification Utility	Power System Dependent Part
Configuration Management Subsystem Fault Management Subsystem	Computer System Dependent Part
Distributed Object-oriented OS(IDPS) : Object-oriented Database	
UNIX OS	

SCOPE

Fig.2 SCOPE Support Function

Network information model: This is an abstract model of power system equipment, such as transmission lines, transformers, circuit-breakers, etc. and it encapsulates their real configuration. It has the information relating to topology and status of power system equipment. For example, transmission lines have attributes, such as starting points, end points, present voltage and current, etc.. These attributes are updated periodically by obtaining them from the corresponding physical equipment. In addition, status of physical equipment can be changed by setting values of the corresponding attribute of the model.

Operational information model : This is an abstract model which encapsulates operational organization of power systems. It has the information relating to control areas, order authorities, necessary application names, etc.. For example, the central load dispatching office maintains names of regional load dispatching offices, power stations and transmission lines as control areas, and also maintains application names such as the supervision and generator control.

In the followings, "Network information model" and "Operational information model" are abbreviated to "NIM" and "OIM", respectively. The *Model modification utility* supports the on-line maintenance of the NIM and the OIM.

The computer system dependent part consists of the configuration management subsystem, fault management subsystem, distributed object oriented OS and database management system.

Configuration management subsystem : This subsystem monitors and manages hardware, software, and communication network. When it detects a fault, it informs the status to the fault management subsystem. This subsystem is installed in each computer.

Fault management subsystem : This subsystem manages the availability of the system. For example, when it receives a fault detection message from the configuration management subsystem, it tries to run a back-up application program on another computer.

The distributed OS and the database system encapsulate the location and inner-structure of each subsystem.

4. FLEXIBILITY OF SCOPE (Model Architecture)

This section explains two models, i.e. NIM and OIM, of the power system dependent part. In order to improve flexibility of SCADA systems, these models encapsulate changes of operational organization and power system topology.

An EMS/SCADA system is geographically distributed, and has hierarchical structure as shown in Figure 3. The hierarchical structure reflects the operational organization of power system. Office in higher levels handles more abstract and aggregate data. For example, although a substation manages the behavior of individual equipment such as CB (Circuit Breaker), LS (Line Switch), a regional load dispatching office recognizes them as the set of transmission lines and buses instead of individual CB and LS status. A central load dispatching office recognizes them as a node in an electric power network by abstracting equipment. In this way, the degree of abstraction and aggregation differs according to the hierarchical level. This hierarchy may change with the modifications of operational organization of SCADA system.

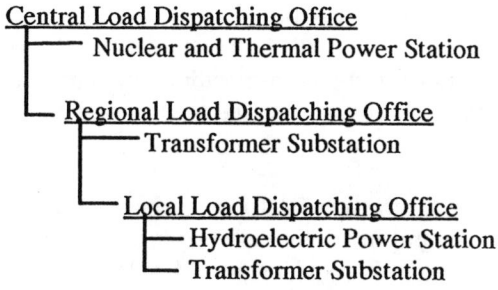

Fig.3 Configuration of Supervisory Control System

NIM makes application programs independent of these modifications of operational organization. That is, NIM in the local load dispatching office automatically reflects these modification to NIM in the higher level offices by abstracting them while maintaining consistency among layers. This removes inconsistency of database

397

among individual layers. In addition, new network information layers can be easily add to without causing modification in other subsystem even when new layer's load dispatching offices are established.

We found that it is enough for NIM to provide the following three hierarchical abstraction levels by evaluating the present EMS/SCADA system.

Level 1 :   A model for equipment which is used for supervising the behavior of the whole area of the power system.

Level 2 :   Models for equipment which are used for supervising the behavior of areas including lines and stations. This layer is used to detect the power failure caused by line accidents.

Level 3 :   Models corresponding to individual equipment. This layer is used for supervising and operating the power station and transformer substation.

NIM is developed based on the object-oriented technology. Figure 4 shows the class structure and the relation among classes of equipment. Each layer has a set of equipment class according to the abstraction level mentioned above. Individual equipment, such as CB and LS, is defined as an object. Figure 5 shows CB class structure which indicates the abstraction level. The top layer has only three abstracted attributes, however the bottom layer has five detailed attributes and it also has four attributes which inherits upper-class. The attribute values are classified into the following three kinds. 1) Fixed individual information such as an equipment name. 2) Status value (SV) having the binary state, e.g. open/close state of CB or LS. 3) Numerical value (NV) such as voltage and power flow.

The equipment in the higher level is the abstraction and aggregation of the corresponding equipment in the lower levels. The abstraction rule can be summarized as follows.

1. Logical Abstraction : Attribute values for the higher level equipment are the result of logical operation of those for lower equipment. SV values are mainly abstracted by this rule.

2. Numerical Abstraction : Attribute values for the higher level equipment are mean, total or representative values of those for lower equipment. NV values are mainly abstracted by this rule.

3. Connection Abstraction : Connections between equipment in the higher level are decided based on the connection states between the corresponding equipment in the lower level.

These abstraction algorithms are implemented as a method in individual objects of each level's NIM. When an optional abstraction network information layer is required, the new NIM is generated by redefining the abstraction rule in accordance with the characteristics of the new layer's attribute values.

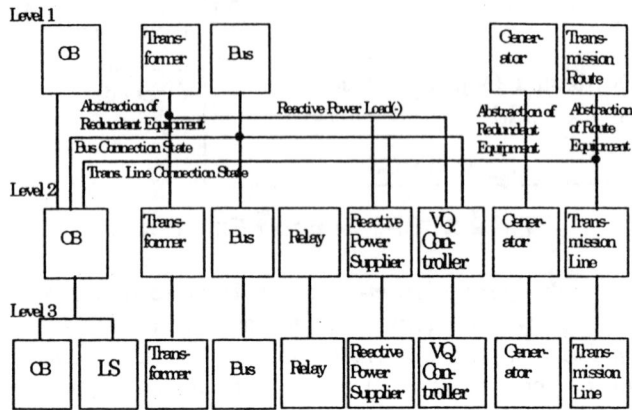

Fig.4 Configuration of Network Information Model

CB1	/* CB class of level 1 */
	name
	kind of change
	on-off status
CB2	/* CB class of level 2 */
	equipment type
CB3	/* CB class of level 3 */
	voltage rank
	rated capacity
	order value
	last break
	re-closing type

Fig.5 Class Structure of Circuit Breaker

OIM is a model to realize independence against the various operational organization. When operational organization is changed, modifications are required only for elements in OIM directly related to the changes. OIM is created corresponding to each office or station. Individual OIMs have the following information.

1. Control Area : Names of power system elements, such as offices and transmission lines, which the corresponding offices have responsibility to control them.

2. Order Authority : Names of functions which the

corresponding office can execute.

3. Application Name : Names of application programs which are necessary for the corresponding office to achieve jobs.
4. Upper Office Name : A name of the office which supervises the corresponding office.

5. SCOPE PROTOTYPE SYSTEM

A SCOPE prototype system is a SCADA system which covers all area of the power system of Tokyo Electric Power Company, and has been developed based on the framework of the design concept mentioned in section 3. The effectiveness of design concepts and functions of SCOPE are evaluated by this prototype system.

This section explains the basic structure of operational organization of the target power system and shows the hardware and software structure of SCOPE prototype system.

5.1. Operational Organization of the Power System

The target power system of SCOPE prototype system consists of five control offices as shown in Figure 6. The relationship among these offices and stations is a hierarchy such as the administration system. The roles of each office are as follows.

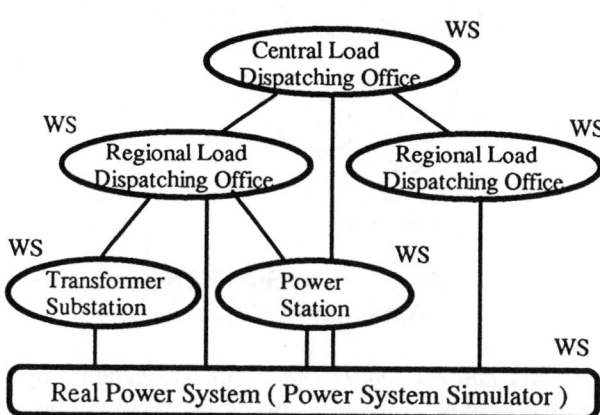

Fig. 6   A Configuration of SCOPE Prototype System

Central Load Dispatching Office : Central load dispatching office supervises the accidents or troubles occurring in bulk power systems, especially power stations and transmission lines. It also decides the output power of each generator in accordance with the power demand, considering the characteristics of each generator.

Regional Load Dispatching Offices : Regional load dispatching offices behave under the control of central load dispatching office. They supervise transformer substations and transmission lines which are located in control areas corresponding to them and also detect accidents happening in those areas. There are three regional load dispatching offices, i.e. West, East, and South regional load dispatching office. Initial configuration of the prototype system consists of the West and East regional load dispatching office which supervise west and east part of the all power system area, respectively. The South regional load dispatching office is an optional office for an evaluation case.

Transformer Substation and Power Station : The transformer substations supervise status (on-off status, voltage, current, etc.) of equipment which are located in them such as CB, LS or tap switch, and control them. The power stations control the output power of generators, and supervise their status.

5.2. Hardware and Software Structure

5.2.1. Hardware Structure

Figure 6 also shows the hardware configuration of the SCOPE prototype system consisting of six workstations connected through the Local Area Network(LAN). Five workstations form the platform where application programs corresponding offices and substations are running. Programs corresponding more than one offices and/or substations can run on one workstation simultaneously. The remaining one workstation is a simulator which simulates the behavior of real power system. Figure 7 shows the photograph of the SCOPE prototype system.

Fig. 7 Photograph of SCOPE Prototype System

5.2.2. Software Structure

Figure 8 shows a software configuration of each office or substation on the SCOPE prototype system. Each

SCOPE function is designed as an object in order to encapsulate its inner structure and improve its autonomy. Because of encapsulation of objects, SCOPE can flexibly adapt to various changes occurring in the system. That is, each object can dynamically bind with other related objects in accordance with environments.

SCOPE uses the IDPS (Intellectual Distributed Processing System) OS [4,5,6] as an object-oriented distributed infrastructure. IDPS-OS has a reliable broadcast communication mechanism to establish the location and replication transparency of each object. An object-oriented database is also used to implement the NIM and the OIM which represent relationships among individual elements in the power system.

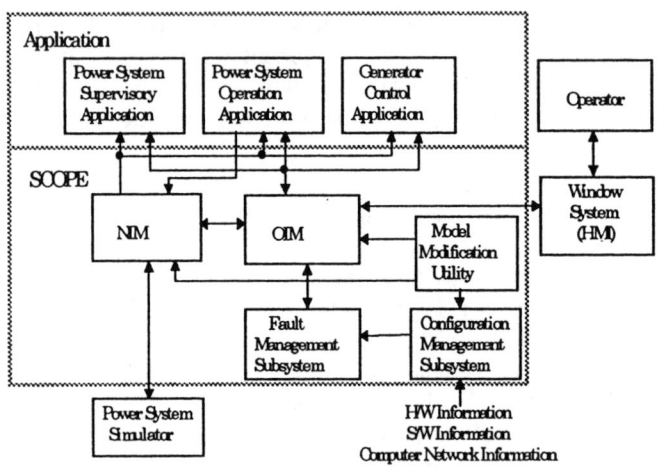

Fig.8 Software Configuration of SCOPE Prototype System

In the remaining of this subsection, three application programs for evaluating SCOPE design concept and a power system simulator are explained. Figure 9 shows the relation among objects in the three applications.

Power system supervisory application : This application program periodically supervises the SV values (CB/LS open-close status) and NV values (voltage, current, power, etc. of equipment) which represent the states of the power system. It also displays the states of the power system and gives warning to operators when the states represent some accident such as excess power, voltage and/or frequency values in the power system.

As shown in Figure 9, the NIM of a substation receives real individual states information of the power system, and abstracts them and then informs the abstracted states to the upper NIM by acquiring its name through the OIM of the same level . Accidents of power

system equipment is detected by NIM in substations. NIM informs the supervisory application of the same level and also informs the NIM in the upper level. When the supervisory application display the power system, it receives their present states from the corresponding NIM. The upper NIM repeats this process.

Power system operation application : This application program controls the CB/LS states (open/close) of substations in accordance with the operator's order. As a function which is not installed in the current SCADA system, the automatic operation which controls substation's CB/LS from the regional load dispatching office is introduced. This function also displays an operation sequence, its executing state and its executed results.

As shown in Figure 9, the automatic operation application confirms the correctness of the order by asking the OIM of the regional load dispatching office before it orders the substation to execute the operation. Only when the order is confirmed, does it control the equipment of the substation through the operation object and the NIM of the substation. The results of the operation are returned to both the operation object and the automatic operation object through NIMs similarly to accident notification.

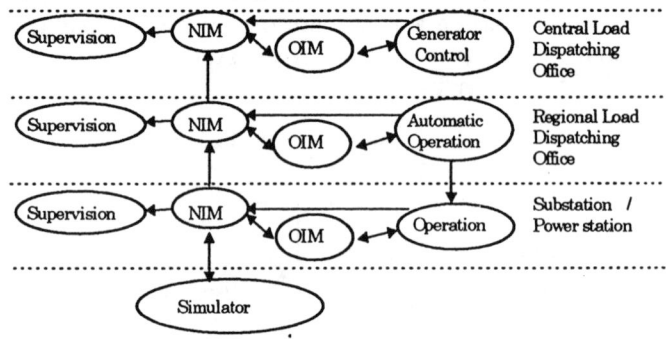

Fig.9 Relation among SCOPE Models and Application Objects

Generator control application : This application program controls the power of generators in the power stations in order to balance the load in the whole system. The execution of control is only permitted by the central load dispatching office, because the power station is under the control of that office.

Power system simulator : The power system simulator has a power system model and generates the states of each of its element. It also simulates the accidents (ex. CB trip) in the power system and the effects caused by them.

400

6. EVALUATION OF SCOPE

The following six evaluation cases have been used to evaluate the flexibility and maintainability of the SCOPE architecture and to confirm the functions of the SCOPE subsystem. These evaluation are done by running the SCOPE prototype system.

Case 1: Construction of the present SCADA system
Case 2: Supervision of remote offices or stations
Case 3: Automatic remote operation of transformer substation
Case 4: Migration of office and/or objects
Case 5: Online equipment installation
Case 6: Introduction of a new office or a station

Evaluation results are as follows.

Case 1: Construction of the present SCADA system : This case evaluates that the present SCADA system can be constructed by using the SCOPE architecture. The evaluation test has shown that the supervision, operation and generator control applications are performed as well as the present system. That is, the supervision of a local office (e.g. display of the network diagram, SV/NV value and warning notification.), operations of the substations and the generator control have been confirmed.

Case 2: Supervision of remote offices or stations : This case evaluates how SCOPE adopts to the changes of operational organization. The function to supervise the equipment of arbitrary offices/stations from remote offices has been tested. By this function, it has been confirmed that the operator at the central load dispatching office had been able to supervise the states of transformer substations which are normally supervised locally.

Case 3: Automatic remote operation of transformer substation : This case evaluates effectiveness of the new operation of SCOPE. An operator of a regional load dispatching office controls and supervises the equipment of arbitrary remote transformer substations. It has been confirmed that orders issued from a regional load dispatching office are automatically transferred to the target office and broken down into a set of orders which are executable at the target transformer substations. The results of operations in the individual equipment have also been sent to the application programs. Orders issued by unauthorized offices have been rejected. Figure 10 shows an operation flow of the automatic remote operation.

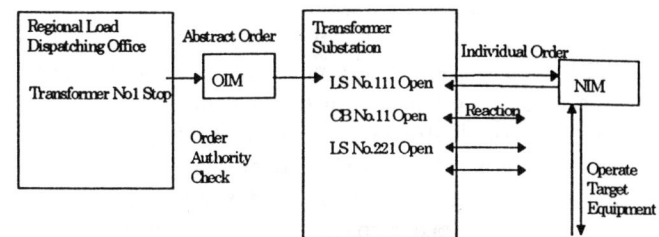

Fig. 10 Operation Flow of Automatic Remote Operation

Case 4: Migration of Office and/or Objects : In this case, application programs corresponding to offices have been migrated from one computer to another computer. It is confirmed that the configuration management subsystem and IDPS-OS successfully migrate objects while continuing normal operations of other objects. In case of computer fault, the configuration management subsystem detected it and the system successfully continued operations through the reloading stand-by objects on other computers by the fault management subsystem. Operation switching to the stand-by objects was very smooth because of the transparency of object's location. More than one office and/or station has been able to run on one workstation.

Case 5: Online equipment installation : In this case, a new equipment has been installed in order to evaluate the online database maintenance ability of SCOPE. Model modification utility has successfully updated the data in NIM corresponding to the equipment, without stopping the system nor modification of application programs. In addition, the modification at a lower level NIM has automatically reflected to upper level NIM. Figure 11 shows an example of an equipment installation to a transformer substation.

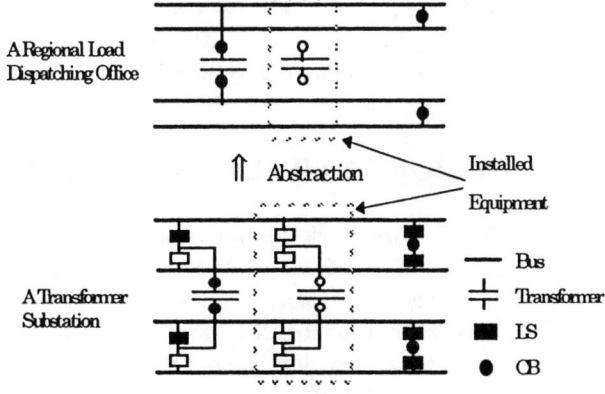

Fig.11 An Example of Online Equipment Installation

Case 6: Introduction of a new office or station : In this case, the ability of SCOPE for dynamic office establishment has been confirmed. That is, NIM and OIM at each layer have been consistently updated to reflect a newly established office/station without stopping normal operations. Figure 12 shows the operation flow of office establishment. The model modification utility establishes the new regional load dispatching office in the specified workstation by using the IDPS-OS function. At the same time, it informs the OIM of each existing offices of the modification of the control area, and also requests the NIM to reset the control area, and finally requests the supervisory applications to redraw the network diagram.

It has been confirmed that computers or programs for the new office/station was dynamically connected to the system and communicates with existing ones without stopping the system nor modification of application programs. The control area has also been dynamically changed while maintaining the consistency among OIMs and NIMs.

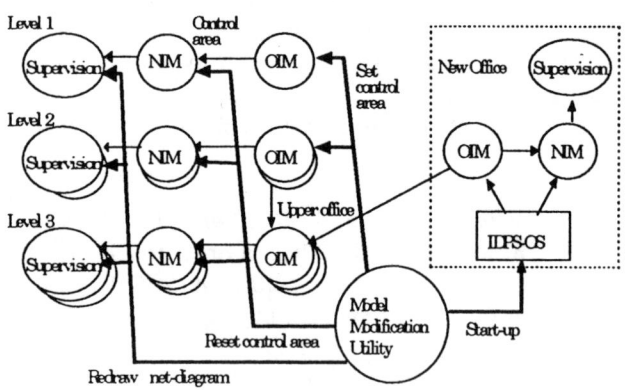

Fig.12 Establish a new office

Through these six evaluation results, the achievement of the goal mentioned in section 2 has been confirmed. Namely, modification of computer system configuration and power system configuration can be accomplished without stopping the system nor modifying other application programs.

7. CONCLUSIONS AND FURTHER WORK

The SCOPE framework for realizing the flexible and maintainable EMS/SCADA systems has been proposed. In order to make application programs independent of various changes, they are classified into three different encapsulated layers. Through the development and the evaluation of the prototype system, it has been confirmed that SCOPE has been able to adapt to changes of computer and power system configuration

without stopping operations nor modification of application programs.

As future work, more precise quantitative comparison of the performance and flexibility between SCOPE and conventional systems is needed. Problems derived from the communication delay in wide area networks also should be analyzed and solved before developing real scale EMS/SCADA systems based on SCOPE concept.

References
[1] Tsuruma, et al. "New SCADA system based on open-distributed architecture," CIGRE Symposium Integrated Control and Communication Systems, 1995

[2] W.Dieterle, el al. "LAN based data communication in modern energy management system (EMS)," IEEE Trans. on Power Systems, Vol.11, No.1, Feb. 1996

[3] Y.Hanawa, et al. "Supervisory control of power systems: Reliability improvement", Proc. of 2nd IFAC Workshop on Safety and Reliability, Nov. 1995

[4] Tamura, et al. "IDPS: Intellectual distributed processing system," Proc. of Pacific Computer Communications Symposium, pp.129-133, 1985

[5] T.Seki, et al. "An Operating system for the intellectual distributed processing system – an object oriented approach based on broadcast communication –," IPSJ JIP 14, pp.405-413. 1991

[6] T.Seki, et al. "A fault-tolerant architecture based on autonomous replicated objects," Proc. of 28th HICSS, pp.506-515, 1995

An Implementation of a Reactive Distributed Air Conflict Resolution System*

Jean-François Bosc
ENAC† Département MI
31055 Toulouse CEDEX 4
FRANCE

Garfield Dean
Eurocontrol, B.P. 15
91222 Bretigny sur Orge CEDEX
FRANCE

Abstract

The fast growth of air traffic that began during the 80's raises some questions about what future Air Traffic Control (ATC) systems will need to be. A major change will probably be required in Europe within a 30 years timeframe. Fortunately, recent developments in air navigation systems provide new opportunities for efficient autonomous navigation. Several studies have been conducted in order to develop automation in ATC, based either on centralized or distributed systems. An important part of the ATC task is to avoid separation losses (which are called "conflicts") between aircraft. This article presents an implementation of a reactive distributed conflict resolution method on a traffic simulator providing a realistic traffic sample. The method is described, and some results regarding its efficiency and its degradation with traffic increase are given. In particular, a critical traffic level appears, above which efficiency drops dramatically. This level depends on the method used; the airspace itself is not saturated.

1 Introduction

Since air travel began to increase rapidly in the late 50's, Air Traffic Control (ATC) has been performed in a similar manner. Aircraft follow airways, which consist in a succession of route segments linking crossing points defined by navigational aids (radio-electrical beacons providing distance or bearing information). Air traffic controllers provide instructions to pilots in order to ensure fluidity and safety of the traffic. In particular, minimum horizontal or vertical separations have to be maintained between aircraft in order to avoid collisions. A potential violation of these separations is called a conflict. One important part of the control task consists in detecting conflict situations and modifying aircraft trajectories to prevent any loss of separation from happening.

Airspace is divided in "control sectors". A sector is a volume of space defined by lateral, upper and lower limits, which comprises a few route segments and crossing points. Generally the size of a sector corresponds to 15 mn flight, or about 100 NM. Each sector is placed under the responsibility of two controllers (one planning controller and one tactical (or radar) controller). The main limitation of the system is the number of aircraft that a controller can handle simultaneously, which is about 15. Therefore there must never be more than 15 aircraft in the same sector at the same time. The division of airspace into sectors is intended to maximize the flow of traffic that can be handled. Reducing the size of sectors beyond a certain limit can't be a solution. A minimal transit time in sectors is required so that the controller can grasp the situation and achieve the necessary maneuvers. Moreover, the transfer of a flight between two sectors generates a specific workload ("coordination" workload).

The whole ATC system can be viewed as a succession of filters with decreasing time horizons, designed to make sure that a a controller will never be overloaded. These filters are :

- Airspace Management (sector and route design)

- Air Traffic Flow Management (ATFM) (attribution of routes and take-off times, ...)

- Medium term (strategic) planning (performed by planning controller)

- Radar control (performed by tactical controller)

- Short-term collision avoidance systems (ground or airborne), which provide warnings and in some cases suggest avoidance maneuvers.

---

*Supported by Eurocontrol Experimental Center

†Ecole Nationale de l'Aviation Civile / National School for Civil Aviation

Some recent technological developments in navigation system, e.g Flight Management Systems (FMS) and Global Positioning Satellite (GPS), offer the possibility of suppressing some restrictions in navigation, in particular the requirement of following pre-defined route paths. Modern FMS can provide precise navigation to any GPS coordinate, without any need for radio beacons. Apart from ATC requirements and restricted areas (military in particular), each equipped aircraft could now fly its own optimal route to its destination airport. These capabilities are not used in the current system. Allowing free navigation to flights (which is sometimes referred to as the "free route" concept) would provide significant benefits in flight time and fuel consumption. A simulation conducted with a 1-day traffic sample including all controlled flights in the French airspace indicated a 6.4% reduction in flight time.

Moreover, the recent increase of air traffic, which is expected to go on during the next 10 or 20 years, has raised some questions. In particular, many people believe that an ATC system based on control sectors where control is performed by human beings will not be able to handle the traffic density expected around 2015-2020 ([2], which considers scenarios ranging from 50% to 500% increase for the 1990-2020 period).

From that perspective, several attempts to develop automation in ATC have been made since the early 80's ([6, 7, 5, 1]). A fully automated ATC system could still be centralized, or totally distributed. In a centralized ground system, aircraft follow instructions, with possibly some freedom for negotiation. In a distributed system each aircraft would have to negotiate its trajectory with surrounding aircraft in order to ensure avoidance. The present article describes an implementation of a distributed conflict resolution system.

2 CATS : Caml All-purpose Traffic Simulator

The LOG[1] Team developed a traffic simulator in order to provide a realistic environment to test various algorithms developped by the team, that deal mainly with ATFM and medium- or short-term conflict resolution. The simulator uses flight plan data from the French ATC system (CAUTRA) archive, and a performance model coming from the ENAC ATC simulator which is used for ATC controllers training. The traffic sample used for simulation includes all controlled flights in the French airspace for a particular day. The

flights can follow their scheduled route (just as they really did) or fly straight to their destination. The simulator provides realistic control-free flight profiles.

It is possible to change several parameters, in particular vertical and horizontal separations. Typical values used here are 6 nautical miles horizontally and 1000 ft vertically. A traffic increase can be simulated by dividing all take-off times by a constant factor. During time periods where traffic is stable, there is a very strict correlation between the number of aircraft in flight and that factor, which therefore represents global traffic density (the density of the original sample being equal to 1).

Results are normally observed in a time window chosen during peak hours, at a time where traffic is reasonably stable.

The simulator has been used for testing resolution algorithms developped by the team. These algorithms are based on genetic global optimization techniques for medium-term control ([4]), and on reactive techniques ([9]) for short-term resolution. In the near future, some tests will be made with ATFM algorithms.

3 Reactive conflict resolution

The first resolution method implemented in the simulator was based on reactive techniques, which consist in altering aircraft trajectories in real-time (without planning) to ensure separation. The main goal was to provide a reference for comparison with genetic techniques, which are expected to be more efficient.

For a particular aircraft, trajectory alterations are derived from the computation of forces induced by surrounding aircraft that may cause a conflict. Karim Zeghal provided a complete theory of interaction between mobiles, for the purpose of avoidance or interception ([9, 8]). In particular, he introduced a tangential force to ensure bypassing of obstacles (either fixed or mobile). The main advantage of this method is simplicity. The minimal set of information concerning an intruder aircraft is distance, altitude, azimuth, relative speed, and the intruder's type of behaviour (avoidance with an identical or different method, interception). A force can then be computed for each intruder aircraft. Forces corresponding to different intruders are summed, and the results define the course change. Some additional information, in particular planned changes in course or vertical speed, can help to distinguish intruders that constitute a real threat.

It is clear that this kind of method can be distributed. Necessary information can easily be exchanged via an interrogation-response process similar to that of secondary surveillance radar (SSR)[2]. Each

---

[1] Laboratoire d'Optimisation Globale / Global Optimization Laboratory

[2] An SSR radar sends an interrogation message that is replied

aircraft can then compute its own force. With a time horizon of 5 minutes, the number of intruder aircraft is less than 10 in almost all cases at current traffic densities. Therefore the computing power required for each aircraft is very limited. On the other hand, computing maneuvers on the ground would require the assignment of a specific portion of airspace to each computing center, and therefore a special treatment near the borders (just like with current control sectors).

4 Implementation

The reactive resolution method used here has been adapted from Karim Zeghal. In a first step some additional limitations have been put on aircraft maneuvers. Normally, forces can alter aircraft motion in three dimensions. However, commercial aircraft performance in cruise flight is optimized for a particular speed and altitude. Flying at a lower altitude is less efficient, and a higher altitude may be unreachable because of engine limitations. Moreover, a climb maneuver near the maximum altitude is costly in time and fuel. Regarding speed, changes are limited to 2 to 3% of normal cruise speed, and must be planned a very long time ahead in order to produce some effect. Karim Zeghal described a way to integrate performance constraints in the computation of forces. However, because of the reasons listed above, altitude changes are not used by ATC services for cruise flights, and speed changes are rarely used. Since our studies mainly deal with en-route traffic, we decided to only allow lateral maneuvers. Even though the main reason was simplicity, this restriction is very close to reality.

However, this raised some difficulties in the case of aircraft flying parallel track, eg for an overtaking. When the two aircraft fly side by side, the forces are longitudinal and tend to increase the speed of the faster one and slow down the slower one. To avoid disruptive reactions, an additional repulsive component has been added to the force vector. The repartition between the two components depends on the conflict geometry. The resulting force for a particular aircraft A and a particular intruder I (see figure 1) is :

$$\vec{F} = F \cdot \{\cos(i).\vec{u_T} + (1 - \cos(i)).\vec{u_R}\}$$

where F is the intensity of the force, $\vec{u_T}$ and $\vec{u_R}$ unary vectors in the tangential and repulsive directions respectively, and i the angle between the intruder's azimuth and the relative speed vector $\vec{V_r}$.

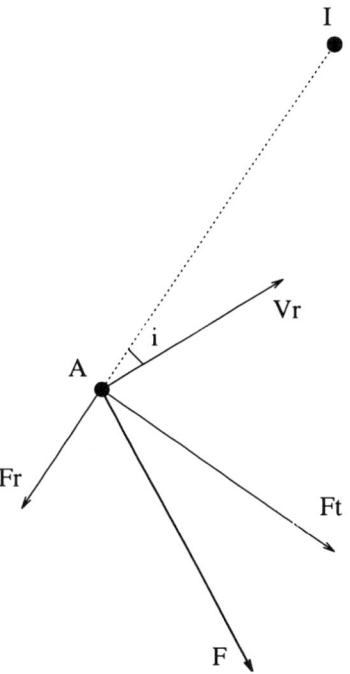

Figure 1: Force vector

Therefore $\vec{F}$ is purely tangential if both aircraft are on a colliding course and purely repulsive if they fly on parallel tracks. The tangential component, which determines the characteristics of the maneuver (which aircraft passes ahead of the other), complies with the initial geometry of the conflict, ie the position of the relative speed compared to the intruder's azimuth.

Moreover, risk criteria have been suppressed. The intensity of the avoidance force is proportional to the inverse of the time remaining until normal separation will be lost, assuming that the closing speed will remain constant.

An attraction towards the aircraft's destination is added to the total force-vector (the current implementation only allows straight route navigation). The aircraft then tries to reach the resulting direction. There's a limit of $3^\circ/s$ on the turning rate. It is moreover assumed that all aircraft in the simulation behave exactly the same way.

In most cases this method ensures avoidance and generates smooth trajectories (which however are not flyable by a human pilot since during maneuvers heading changes slightly at every step of computation). Some problems still occur in cases where the angle between the two trajectories is small and the ratio of the speeds is close to 1. These cases are known to be the most difficult to solve, in the sense that the

405

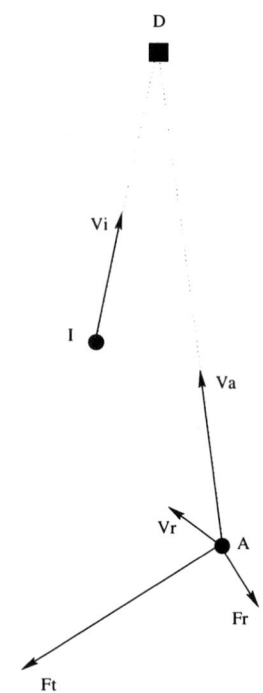

Figure 2: Improper resolution

avoidance maneuver takes more time and generates a greater increase of flight time or distance ([3]).

In figure 2, aircraft A and its intruder I (slower) are heading towards the same destination D. Speeds are Va and Vi, relative speed is Vr. Generated forces are Ft (tangential) and Fr (repulsive). In this case, the crossing geometry given by Vr (A passes behind) is incorrect. After the maneuver, the two aircraft will find themselves in a symetric situation (A on the left, I on the right). Another (similar) maneuver will begin, and so on, until eventually separation is lost.

Two versions of the simulator have been tested. One is truly reactive, all intruder aircraft are taken into account, which includes many aircraft that don't really constitute a threat. The only parameters used to distinguish aircraft that may generate a conflict are distance and closing speed, therefore the selection process can't be very efficient.

In the other version, aircraft trajectories are first simulated (without resolution, ie aircraft fly straight to their destination) to detect future conflicts. Typically the simulation occurs every 1 to 5 minutes, and spans over the next 5 to 20 minutes. An increased horizontal separation (usually twice the normal separation) is used in order to take into account aircraft flying near a conflict, which may interfere with the resolution maneuver. Even so this method dramat-

ically reduces the number of aircraft pairs to consider. Aircraft involved in conflicts are then grouped in "clusters" by transitive closure. The computation of reactive forces is then only applied to pairs of aircraft belonging to the same cluster. Compared to the purely reactive method, the time of computation is approximately divided by 6, despite the time spent in pre-simulation of trajectories.

The information necessary (knowledge of planned future trajectories) is not available in the current system. However, it may be approximated over a small period of time (eg 5 minutes), provided that planned changes in vertical speed or heading are transmitted between aircraft. In the future, aircraft may have a 4-dimension flight plan valid for the next 20 to 30 minutes of flight. It will be desirable to have descriptions of trajectories defined and transmitted between aircraft and/or ground systems. For example, the ARC2000 simulator ([7]) is based on the planning of aircraft trajectories described by 4-dimensional tubes, which are transmitted and negotiated between aircraft and the control system.

Both versions (i.e., with and without conflict pre-detection) give similar results regarding the number of unsolved conflicts. The knowledge of future trajectories reduces the amount of processing (but the volume of data to transmit is increased), and also the number of useless avoidance maneuvers. The average increase in flight time induced by maneuvers is 0.3% (compared to 1.5% with the purely reactive version), which is very low.

5 Results

Only true "en-route" conflicts are taken into account. There's a lower bound on altitude (6000 ft), and the first and last 5 mn of each flight are not considered either. This is to suppress problems due to aircraft entering or leaving the simulated airspace at the same point and time. In the first case separation is lost as soon as the planes are created in the simulator, in the second the proximity of the exit point hinders avoidance maneuvers.

Figure 3 shows the evolution of the number of separation losses versus traffic density, both without (first curve) and with (3 others) conflict resolution. Here resolution is performed with pre-detection of conflicts, with different conditions for each curve : respectively every 5 mn with twice the normal separation, every mn with twice the normal separation, and every 5 mn with one normal separation. A few corresponding values are presented in table 1. Figure 4 shows the percentage of unsolved conflicts relative to the number of separation losses observed without resolution. The

Traffic	std	+50%	+100%	+150%	+300%	+500%
no resol	508	748	962	1266	2071	2953
2 sep, 5 mn	7 (1.4%)	24 (3.2%)	38 (4.0%)	89 (7%)	1487 (72%)	
2 sep, 1 mn	10 (2.0%)		22 (2.3%)		112 (5.4%)	739 (25%)
1 sep, 5 mn	18 (3.5%)		72 (7.5%)		197 (9.5%)	545 (19%)

Table 1: Remaining conflicts.

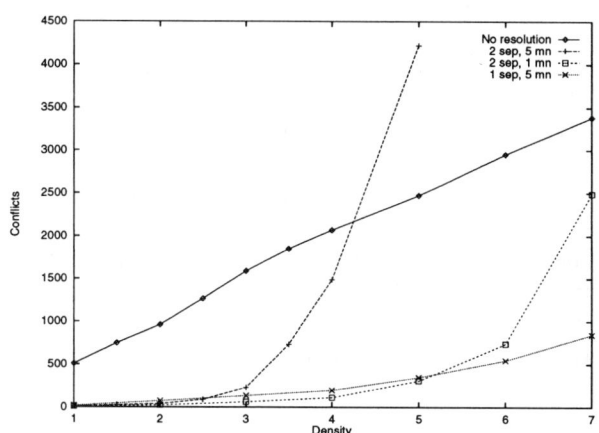

Figure 3: Number of separation losses vs density

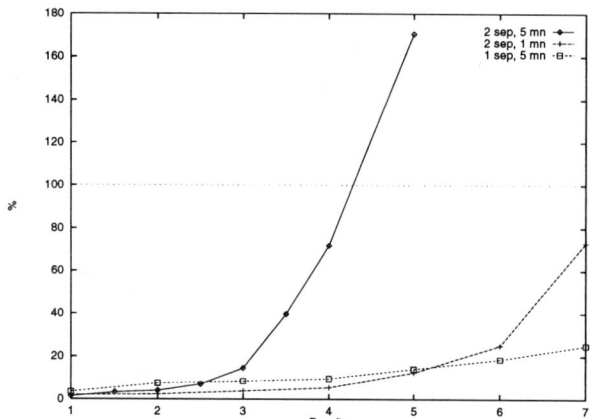

Figure 4: Percentage of unsolved conflicts vs density

results here are observed on the full traffic sample (1 day), which explains the linear shape of the conflict number without resolution. If traffic was stable during the whole simulation, this would be quadratic.

At normal traffic density, all the 7 remaining conflicts either have the same geometry described above (small angle and speed ratio), or occurred near the entry- or exit-point of the flights involved. It appears that density itself is not a problem.

When density is increased, the curve clearly shows the saturation of the resolution method. The percentage of unsolved conflicts is pretty good up to density 2.5, but it increases dramatically when density exceeds 3. At some point the resolution method generates more conflicts than it solves, because of "oscillations" of trajectories generated at high densities. It is interesting to note that with ARC2000 saturation appears at a traffic density that we believe to be similar. However, the ARC2000 traffic sample has been generated manually in order to maximize its complexity, and the resolution process itself is submitted to much stricter operational constraints.

It is possible to delay saturation when density increases. The previous results were obtained with conflict pre-detection performed every 5 mn on a 15 mn period, which is not optimal. A higher frequency

allows more accurate updating of clusters, while a shorter period avoids treatment of conflict that are too far ahead in the future. With a frequency of once every mn and a period of 5 mn, performance remains similar at low density, but saturation appears only at densities higher than 5.

Another possibility is to use a smaller horizontal separation for pre-detection. On both figures, curves "1 sep" are obtained with pre-detection every 5 mn, using the normal horizontal separation. This limits the number of aircraft pairs that are treated simultaneously, which is the cause of saturation. At some point the trajectories of aircraft submitted to multiple forces become so irregular that more conflicts are generated than solved (which doesn't mean that no conflict-free trajectories exist). The curves show that saturation appears much later. However performance is slightly worse at low density (more than twice as many conflicts remain unsolved). This is because with a reduced separation for pre-detection, maneuvering aircraft may create new conflicts with surrounding aircraft not initially involved. Those conflicts won't be treated until the next pre-detection step, which of course is unacceptable in a real system.

6 Conclusion

Some tests should be conducted with different traffic samples. However, the first results obtained are excellent. At low densities, unsolved conflicts are mostly due to conflict geometry (as has been explained above). It is clear that at current traffic density the resolution method is far from being overloaded. Therefore, a special treatment of "difficult" geometries and of flight exits might ensure the resolution of all conflicts in the sample.

With an additional filter to solve possible failures, a simple reactive method like the one used here may even have the capacity to reach the desired level of safety (which is around 2.5×10^{-9} collision per flight hour). For example, the horizontal separation required for such a filter could be only half of the normal separation. This would enable the resolution of remaining conflicts without interfering with those already solved by the previous filter.

Our next step will be to use genetic and reactive techniques in a complementary way. Genetic conflict resolution is performed every 5 mn, and it plans trajectories for the next 20 mn. Only the first 5 mn of each plan are definitive. Since uncertainty on future trajectories is taken into account, the rest of the plan can be computed again at the next step with better results because trajectories will be known more accurately. Tests have been made on particular conflict situations and gave good results with large numbers of aircraft (up to 50).

In case of a failure of the genetic resolution, the reactive method can be used as a secondary filter within a time horizon of 5 mn, and with a smaller horizontal separation. It will then be possible to evaluate the performance of the two methods used separately or complementarily. Considering the performance level of the genetic method, some more significant data should be obtained regarding airspace saturation. Points of interest include the maximum traffic density acceptable in the airspace given some separations and aircraft performance (speed, turning rate), regardless of the resolution method, and whether saturation is created by aircraft density in a portion of airspace or by some particular conflict situations.

References

[1] Jean-Marc Alliot. *Techniques d'optimisation stochastique appliquées aux problèmes du trafic aérien*. Thèse d'habilitation, INPT, Mai 1996.

[2] I. C. Berriman et al. ATLAS Phase II, Volume 2.5 : Feasability of the Operational Requirement and Representative Operational Philosophies. Interim Report 3.0, PA Consulting Group, October 1993.

[3] Nicolas Durand. Modélisation des trajectoires d'évitement pour la résolution de conflits en route. Technical report, CENA, Janvier 1994.

[4] Nicolas Durand. *Résolution optimale de conflits en route*. PhD thesis, INPT, Mai 1996.

[5] Xavier Fron, Bernard Maudry, Jean-Pierre Nicolaon, and Jean-Claude Tumelin. ARC2000 : Automated Radar Control. Technical report, EUROCONTROL.

[6] Lawrence Goldmuntz et al. Automated En-Route Air Traffic Control (AERA) Concept. Technical Report FAA-EM-81-3, FAA, March 1981.

[7] ARC2000 Team. ARC200 Technical Report. EEC No. 274, EUROCONTROL, May 1994.

[8] Karim Zeghal. A Reactive Approach for Distributed Air Traffic Control. In *International Conference on Artificial Intelligence & Expert Systems*, Mai 1993.

[9] Karim Zeghal. *Vers une théorie de la coordination d'actions - Application à la navigation aérienne*. PhD thesis, Université Paris VI, Décembre 1994.

An Autonomous Decentralized System Platform
under Multi-vendor Environments in Building Automation

Akio Orihara
Nikken Sekkei, Ltd.
1-4-27 Koraku, Bunkyo, Tokyo, 112 Japan

Ken Nozaki
Systems Development Laboratory, Hitachi, Ltd.
1099 Ohzenji, Asao, Kawasaki, Kanagawa, 215 Japan

Nobuhisa Kobayashi
Mito Works, Hitachi, Ltd.
1070 Ichige, Hitachinaka, Ibaraki, 312 Japan

Masahiro Oguri
Airconditioning & Refrigeration Systems Division, Hitachi, Ltd.
4-6 Kanda-Surugadai, Chiyoda, Tokyo, 101 Japan

Shinobu Tajima
Inazawa Works, Mitsubishi Electric Corporation
1 Hishi-machi, Inazawa, Aichi, 492 Japan

Masahiro Inoue
Living Environment Systems Laboratory, Mitsubishi Electric Corporation
5-1-1 Ofuna, Kamakura, Kanagawa, 247 Japan

Abstract

This paper proposes an autonomous decentralized building control system to cope with the requirements of building control systems under multi-vendor environments. In the proposed system, building equipment is horizontally distributed on a network and autonomously works by sharing communication data on the network according to objects for building control systems. The proposed system deals with the requirements of building control systems and offers its users, such as building designers, a user-friendly tool for the system construction. In this paper, we call such a communication platform an autonomous decentralized system (ADS) platform. By utilizing the ADS platform, the building control system obtains two ADS properties, i.e., autonomous controllability and autonomous coordinability. Furthermore, the proposed system can realize the on-line expandability, on-line maintainability and fault tolerance of the system by relaying messages between the building system objects on the ADS platform without specifying the sending addresses.

Keywords: *Autonomous decentralized systems, building control systems, object oriented approach, communication platform, on-line expandability, on-line maintainability, fault tolerance*

1. Introduction

Recently, building control systems have been often employed not only to efficiently control building equipment such as air conditioning and lighting in a building but also to make living rooms comfort. Building control systems are classified into two types, *i.e.*, centralized and distributed building control systems. A centralized building control system has a large-scale computer that directly controls equipment in a building. On the other hand, a distributed building control system is composed of subsystems that control building equipment. Each subsystem can independently work and is controlled by a small-scale computer. The installation of distributed building control systems into a building reduces development costs, offers sensitive controls of building equipment and enlarges the number of products from which building owner can choose.

Since equipment in a building control system is supplied by different vendors, a building control system is constructed under multi-vendor environments. In such environments, it is difficult to connect building equipment with each other and to realize cooperative control among subsystems. To cope with this difficulty, there have been several standardization activities of interfaces among subsystems [1][2][3][4]. By standardizing interfaces, it is easy to realize cooperative control among subsystems. We summarize the users' benefits of using the standard interfaces in a building as follows:

vendors:
- market expansion due to improved connectivity

designers:
- low-cost and flexible system construction due to equipment integration
- improved efficiency in building design due to the transparency of interface specifications

owners/residents:
- energy savings
- flexible responses to improve habitability

To construct such a distributed building control system based on the standard interfaces, there are some system requirements which need to be addressed, that is, how to define applications in building control systems as building system objects and the kind of system architectures to be used. To realize the system requirements, a concept of horizontal autonomous decentralized building control systems has been proposed in [1], where each building subsystem is connected on a network and communicates information by using messages. In [1], the introduction of object oriented concepts has been suggested to facilitate the communication of messages via a network in building control systems.

Recently, an autonomous decentralized system (ADS) architecture has been proposed to realize a low-cost and flexible system construction corresponding to user requirements such as the on-line expandability, on-line maintenance and software productivity [5]. The effectiveness of the ADS techniques has been verified in the factory control industry and so on [6][7][8]. In ADSs, communication messages can be sent without specifying sending addresses by assigning content codes to messages. ADSs that can be flexibly constructed with low costs have been paid attention in the building industry. To construct a building control system based on the ADS architecture, we define objects for building control systems and propose an object oriented platform for building control systems. The introduction of object oriented concepts makes the system construction environment of the ADS based building control systems more user-friendly.

2. Requirements of Building Control Systems

In general, a building control system consists of air conditioning, lighting, power distribution, security subsystems and so on. These subsystems supplied by different vendors are integrated on a network to realize the cost reduction of system integration, energy savings and so on. Fig.1 shows building subsystems integrated on a network. By integrating the subsystems on a network, it is possible to realize cooperative control between building subsystems. For example, as shown in Fig.1, when the disaster prevention subsystem detects a fire in a building, the fire alarm is communicated to the air conditioning subsystem through a network and then the air conditioning subsystem stops the operation of air conditioners (see (A) in Fig.1). Furthermore, when the power distribution subsystem broadcasts an electric power output as a demand message through a network, the air conditioning subsystem utilizes the demand message for energy savings of the operation (see (B) in Fig.1).

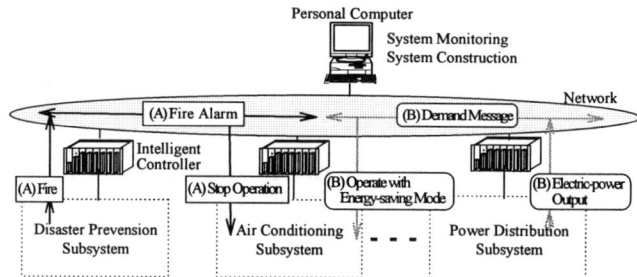

Fig.1　Cooperative control in building control systems

Thus, when constructing a building control system under multi-vendor environments, it is important to have the communication between subsystems. However, because of the difference of vendors supplying subsystems, it is necessary for users to develop communication interfaces case by case. It requires a great deal of effort to develop them. This effort is a difficulty to construct a distributed building control system under multi-vendor environments. Table 1 summarizes system requirements for constructing building control systems under multi-vendor environments.

It is essential for users of building control systems to utilize common communication interfaces between subsystems to improve the efficiency in designing, installing and adjusting a building control system under multi-vendor environments. We summarize some merits of constructing a building control system by using common communication interfaces as follows.

Table 1 Systems requirements of building control systems

	System Requirements
System Architecture	System construction under multi-vendor environments
	Easy connectivity among building subsystems
System Expandability	Easy attachment/detachment of building equipment
	Reduction of initial investigation
Maintainability	Improvement of maintainability by supporting online test capability
Fault Tolerance	Localization of faulty expansion

(1) Reduction of engineering efforts to adjust the specifications among multi-vendors. Users can reduce the consultation time and effort necessary to adjust interfaces and functions among subsystems.

(2) Flexible system construction corresponding to users requirements. Due to the reduction of restrictions for interfaces, it is possible to realize cooperative controls by directly communicating with subsystems.

(3) Elimination of interface imperfections. By using standard interfaces, users can eliminate communication errors and the imperfections of communication interfaces that used to be specified case by case.

(4) System expandability and maintainability. By increasing the amount of building equipment based on the standard, users can easily expand and maintain it because of its interoperability.

Thus, construction of building control systems under multi-vendor environments require the following:

(i) Connectivity should be improved by standardizing communication interfaces among subsystems.

(ii) Each subsystem can autonomously work to prevent the effects of a fault from spreading from the faulty subsystem to the rest of the subsystems.

To realize this, this paper suggests the following:

(i) A standard communication interface that makes communication among subsystems transparent.

(ii) An ADS such that a system consists of autonomous subsystems that require the intelligence to manage themselves without directing to and being directed by the other subsystems and to coordinate with the other subsystems.

In this paper, to cope with the above-mentioned problems of building control systems, we propose an ADS based building control system.

3. Autonomous Decentralized Building Control Systems

3.1. Building Control System Structure

There have been a lot of successful application examples of ADSs in fields such as the steel production, factory control, traffic control and so on [6][7][8]. An ADS consists of subsystems having autonomous controllability and autonomous coordinablity (see, for example, [5]).

These two properties assure the on-line expansion, on-line maintenance and fault tolerance of the system. They suggest that every autonomous subsystem requires intelligence to manage itself and without directing to and being directed by the other subsystems and to coordinate with the other subsystems. An building control system having the above ADS properties is called here an ADS based building control system. Fig.2 shows an ADS based building control system structure. In Fig.2, a building control system consists of intelligent building controllers and a man-machine interface. Intelligent building controllers control and monitor equipment of building subsystems such as air conditioning, lighting, power distribution subsystems and so on. These subsystems are integrated on a network, communicate with each other and then are distributed according to locations or functions. In this building control system, we employ Ethernet as a network and consider a system structure that can be connected to a device level network for the future.

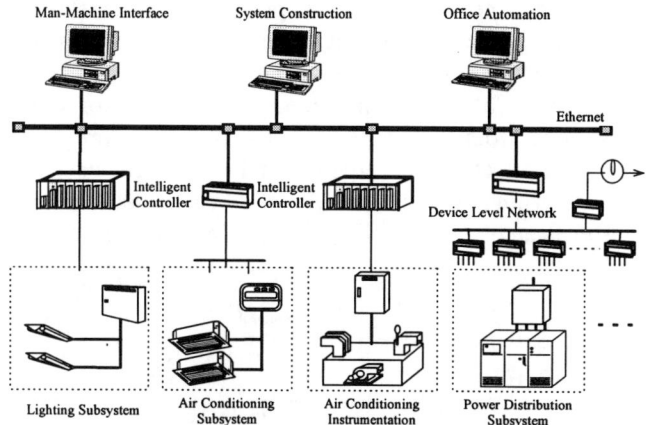

Fig.2 Building control system configuration

Each intelligent building controller has autonomy with the ADS firmware and is equally connected to the network. Then, it shares data on the network by broadcasting messages to the network and autonomously selecting them to receive only the necessary messages from the network. By constructing a building control system based on the ADS architecture, it is easy to realize the on-line

411

maintenance and the on-line expansion of the building control system due to ADS properties such as autonomous controllability and autonomous coordinability.

3.2. Object Oriented Building Control System Model

3.2.1 Object description of building control systems.
The proposed ADS based building control system consists of subsystems corresponding to several building equipment in which there are application programs and objects. These objects can be considered as interfaces for exchanging information between application programs in subsystems. By utilizing objects to represent functions of subsystems and communication services to access the objects, the proposed system can realize the interoperability among subsystems under multi-vendor environments without considering any physical building equipment.

This paper proposes how to construct a building control system under object oriented environments based on objects and communication services specified by "a Data Communication Protocol for Building Automation and Control Networks (BACnet)," which was developed and approved by the American Society of Heating, Refrigerating and Air-conditioning Engineers (ASHRAE) [4].

BACnet models building equipment as a set of objects by using object oriented concepts in order to realize network-transparent data communication with the standardized data representation. In BACnet, objects are defined as the collection of properties by abstractly representing data in internal memories of sensors for analog input, binary output and so on. Application programs in a subsystem can get information of other subsystems from their objects. Thus application programs can control other subsystems and exchange data by reading/writing the properties of their objects. Fig.3 illustrates an object model of a building control system.

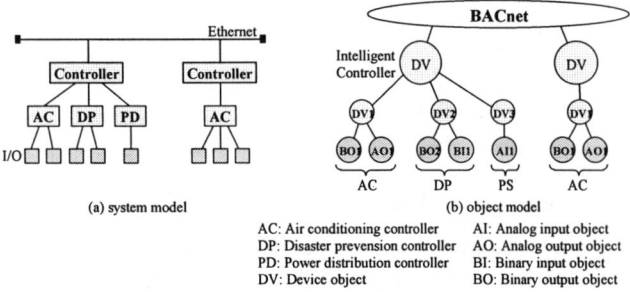

(a) system model   (b) object model

AC: Air conditioning controller   AI: Analog input object
DP: Disaster prevension controller   AO: Analog output object
PD: Power distribution controller   BI: Binary input object
DV: Device object   BO: Binary output object

Fig.3   BACnet based object model

To maintain the interoperability to the BACnet standard, objects in the input/output (I/O) device level defined in BACnet are employed as basic elements.

However, it is more useful to employ interfaces in the level where controllers have functions instead of those in the I/O device level. Thus, this paper proposes a hierarchical representation of objects consisting of four levels, that is, facility, sub-controller, equipment and I/O device levels. Fig.4 shows the hierarchical structure of objects. Objects in the facility, sub-controller and equipment levels are defined by the description forms of non-standardized objects in the BACnet standard and has a property with the list of objects in the lower level.

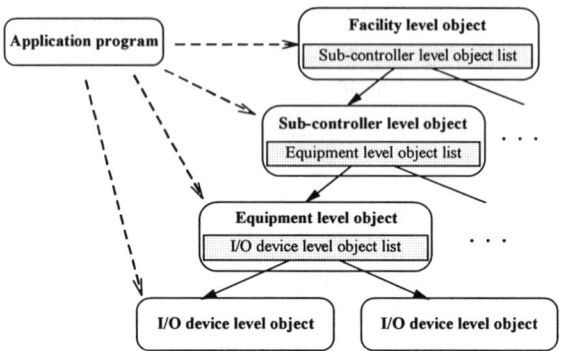

Fig.4   Hierarchical structure of objects

3.2.2. Application services.
The proposed system utilizes BACnet-compatible services that are based on Application Layer Protocol Data Units (APDUs) formats and encoding rules defined in BACnet. In this paper, four types of application services are used, that is, Notifying, Writing, Reading and Operating services. These services are realized by application services defined in BACnet. Notifying services are realized by broadcasting a message and thus several subsystems can simultaneously receive the same message. Writing, Reading and Operating services can also send a message to several objects at the same time. In the proposed system, communication between application programs is specified by an application service, the identifier of a message sending object and the identifier of the device object of the intelligent building controller having the message sending object.

3.3. BACnet based Autonomous Decentralized Building Control Systems

In BACnet based building control systems, functions in each subsystem are described by using the BACnet standard objects and data communication between objects are realized by the BACnet standard services. Thus by using the standard objects and services, each subsystem can be easily connected with the others under multi-vendor environments.

In order to communicate data between objects in BACnet, however, users such as building designers must know the address of every object. This means that when users design applications based on BACnet, they explicitly specify the address of every object. This leads to be difficult to attach new building equipment to a building or detach it from a building, and be complex to modify application programs when addresses of objects are changed. Moreover, since sending and receiving objects are explicitly connected with their addresses, when a fault occurs in a subsystem it easily influents other subsystems.

3.3.1. Autonomous Decentralized System Platform. This paper proposes a platform that users can design data communication between objects without considering addresses of objects by using content codes in the ADS based protocol [9]. Content codes are message tags that represent contents of communication messages. The proposed platform is called an ADS platform. It can be easier to attach/detach building equipment and to realize the localization of faulty influence.

In the conventional ADS based factory automation systems, it is necessary for users to pre-specify some control information of the system such as data field numbers, multicasting group numbers, content codes and so on [6][7][8][9]. However, Users are not familiar with such control information of the system. The proposed ADS platform provides a user-friendly system construction approach without considering control information in the ADS based protocol by automatically translating BACnet into the ADS based protocol.

The ADS platform is illustrated in Fig.5. As shown in Fig.5, each object can communicate with other objects by using object oriented application program interfaces (APIs). By utilizing the ADS platform, each object can be equally treated and one message can be simultaneously transferred to several objects. Users can design building control system by only focusing on message communication between objects without knowing a network. Moreover, users can easily realize the on-line expansion, on-line test and system monitoring in building control systems by communicating messages with objects on the ADS platform.

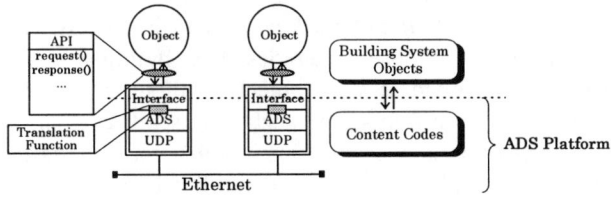

Fig.5 ADS platform

Fig.6 shows a protocol architecture used in the ADS platform. The protocol architecture is based on a seven layer OSI reference model. The application layer is defined in the building control system interface that employs the objects and services described previously. The presentation and session layers are defined in the ADS firmware that realizes the filtering function and attaching/detaching functions of building equipment. Fig.7 shows a message configuration used in the communication of the system. In the message configuration, the APDU is based on the BACnet standard and ADS heading on the ADS based protocol [9].

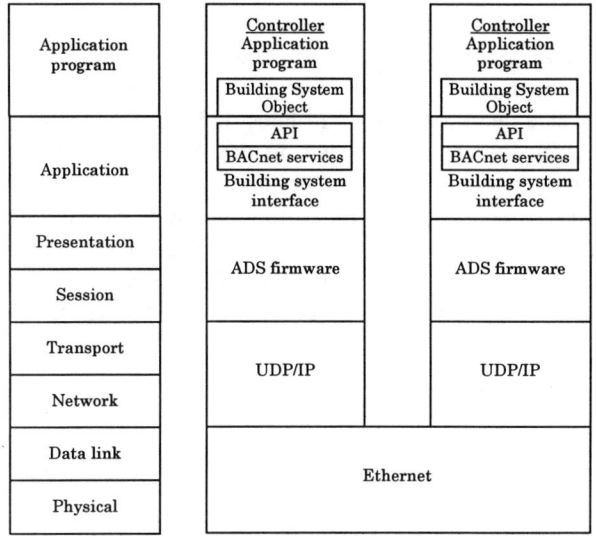

Fig.6 Protocol architecture

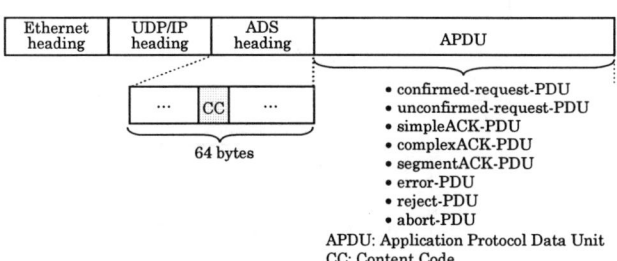

Fig.7 Message configuration

3.3.2. Basic functions of autonomous decentralized building control system.

A) Filtering function. Message communication between application programs in the ADS based building control system is specified by API messages defined in the BACnet standard. The building system interface shown in Fig.6 translates the API messages into the content codes in the ADS based protocol and vice versa. That is, to realize the communication between application programs by using application services, the services and other information in the API messages are automatically translated into the content codes by the system. This is illustrated in Fig.8.

As shown in Fig.8, an application program transfers data with a service and other information in the API message to the building system interface. Next, the

413

building system interface automatically translates the information in the API message into a content code and transfers data with the content code to the ADS firmware. Then, the ADS firmware broadcasts the data with the content code to the network (see Controller #1 in Fig.8). The data with the content code is selectively received by the ADS firmware that requires the data and is transferred to the building system interface on the ADS firmware. Finally, the building system interface interprets the API message in the data and then transfers the data to the application program on the building system interface (see Controller #2 in Fig.8). Thus, since the building system interface automatically translates information in a API message into a content code in the ADS based building control system, users can define the communication of messages among application programs by using the object oriented interface without being conscious of any message transactions of the ADS firmware. This selective message receiving function based on content codes is called here the filtering function.

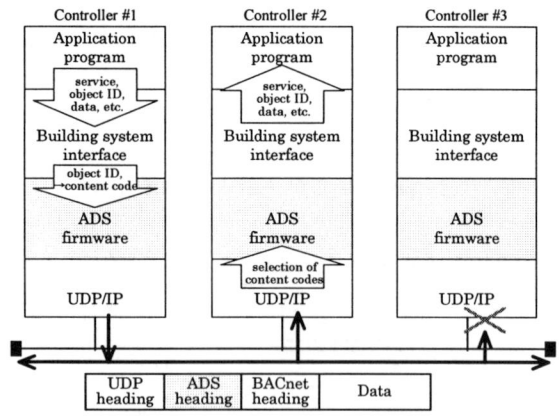

Fig.8 Software configuration

To realize the filtering function, translating API messages in BACnet into content codes in the ADS based protocol should be uniquely defined. This makes possible to specify sending objects with content codes. API messages in BACnet consist of the PDU type, service, device object identifier, object identifier and property number and so on. However, the total size of these parameters in a API message defined in BACnet is more than 16 bits that is the size of a content code in the ADS based protocol. Thus the information in the API message should be compressed to 16 bits.

B) Attaching/detaching functions of building equipment.
The ADS firmware makes easier to attach building equipment in a building and detach it from a building by weakly connecting application programs in intelligent building controllers on the ADS platform. The following procedures are employed in the ADS based building

control system to attach/detach building equipment.
(1) Attachment. Each intelligent building controller opens UDP ports for sending and receiving messages to attach building equipment, when starting up and initializing the ADS firmware. The attachment of building equipment to intelligent building controllers using the socket interface is as follows:
  (a) Open a socket for sending/receiving messages in IP and UDP.
  (b) Bind the socket with the wildcard address (IP address = 0) and a UDP port number.
  (c) Adjust the buff size of the socket for receiving messages according to the communication amount.
(2) Detachment. The detachment of building equipment is done by removing the bind to the above UDP port number. The detachment of building equipment from intelligent building controllers using the socket interface is just to close the socket.

C) System Cooperative Control. By sharing system information with subsystems connected on a network, these building subsystems can be cooperatively controlled, which we call system cooperative control here. This is realized by using system mode object.

To realize the system cooperative control, the status and information of building control systems (system mode), such as earthquake information, fire information, security status information and power demand information and so on, are communicated between subsystems by using application services. This makes it possible that equipment in subsystems can be appropriately controlled by referring the system mode objects. Let us consider that the security subsystem detects a system mode that specifies existence/non-existence status information denoting whether or not someone is in some area. In this case, lighting and air conditioning subsystems can appropriately control building equipment under their controls and realize energy savings by receiving the status information.

3.3.3. Features of autonomous decentralized building control systems.
The features of the ADS based building control system are here summarized.
(1) Efficiency to design a building control system. By using the ADS platform, users can design a building control system and operate it with objects without considering communication information such as sending/receiving addresses. This reduces the time to develop a building, easy to newly add and to modify a building control system.
(2) Toughness of building control systems. By using the ADS based protocol for data communication between application programs in subsystems, several merits are obtained, that is, the localization of faulty influence, the expandability, the maintainability and the responsibility of building control systems. Fig.9 shows the online

expandability of a building control system. Fig.9(a) illustrates the online expansion by newly adding building equipment and Fig.9(b) denotes the online expansion between heterogeneous subsystems, for example, by integrating a control subsystem such as an air conditioning subsystem with an information subsystem such as an office automation subsystem. Moreover, the online maintainability of a building control system is shown in Fig.10 where it is easily to realize to trace alive messages of subsystems by using the alive message reporting function in the ADS based protocol.

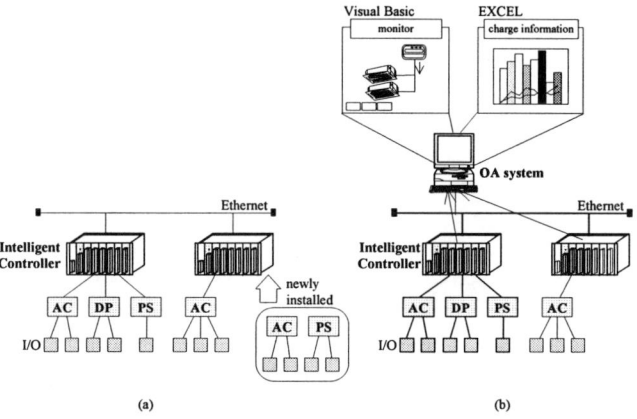

Fig.9　Online expandability in building control systems

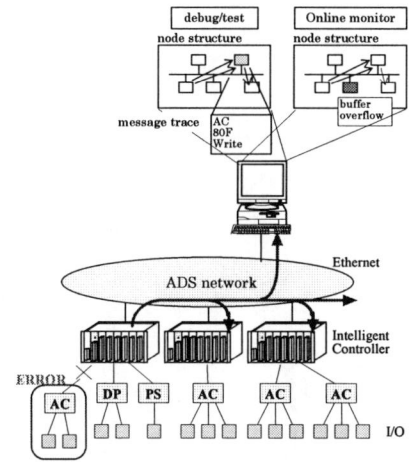

Fig.10　Online maintenance in building control systems

4. Conclusion

In this paper, we discussed the system requirements for building control systems and proposed an ADS based building control system to cope with the requirements. The proposed system employed the ADS architecture and the building system objects and had the features such as the on-line expandability, on-line maintainability and fault

tolerance of the building control systems. Moreover, the proposed system easily realizes the cooperation between building subsystems and improves the efficiency of operation and designing of building control systems. For the future, we will apply the proposed system approach to the cooperation between buildings and the development of urban infrastructure management systems.

References

[1] The Institute of Electrical Installation Engineers of Japan, "The Proposal of Communication Protocol for Autonomous Decentralized Building Systems," (in Japanese) 1995

[2] The Japan Refrigeration and Air Conditioning Industry Association, "Interface for Air-conditioning Systems for Building Use," (in Japanese) 1995

[3] The Japan Machinery Federation, "The Survey for the Standardization of Building Management Systems," (in Japanese) 1994

[4] American Society of Heating, Refrigerating and Air-Conditioning Engineers, Inc., "A Data Communication Protocol for Building and Control Networks," ANSI/ASHRAE Standard 135-1995, 1995

[5] K. Mori, "Autonomous Decentralized Systems: Concept, Data Field Architecture and Future Trends," Proc. of the First International Symposium on Autonomous Decentralized Systems, pp.28-34, April 1993

[6] M. Omura and M. Oku, "Hi-Cell System Architecture for Manufacturing Systems," Proc. of the Second International Symposium on Autonomous Decentralized Systems, pp.154-161, April 1995

[7] K. Mori, et al., "Autonomous Decentralized Software Structure and Its Application," Proc. of the Fall Joint Computer Conference '86, pp.1056-1063, 1986

[8] F. Kitahara, et al., "Widely-Distributed Train-Traffic Computer Control System and Its Step-by-Step Construction," Proc. of the Second International Symposium on Autonomous Decentralized Systems, pp.93-102, April 1995

[9] Hitachi Ltd., "NX Communication Protocol Specification," ver. 1.3 (in Japanese), 1995

Panel Session P3

Challenges and Future Trends for ADS

Panel Chair

Jürgen Nehmer

Panelists

Hiroyuki Fujita

Michel Gien

Yuji Inoue

Hermann Kopetz

Stephen Yau

Challenges and Future Trends for ADS

Jürgen Nehmer

Panel Chair

University of Kaiserslautern, Germany

Abstract

Future versions of the Internet based on ATM technology and IP/V6 offer novel features for controlling the transmission quality of end-to-end virtual connections to the degree as needed by distributed applications. Ideally, this includes parameters for adjusting packet size, packet transmission rate, packet jitter and the probability for packet losses of a single connection as well as synchronization information for coordinating the packet traffic of related connections as required for multimedia applications. It is claimed that these novel network features will open the Internet for an application domain which were either impossible or had to rely on specific network technologies in the past: globally distributed control systems. Examples of globally distributed control systems are traffic control systems (air, road and railroad), distributed production systems and worldwide energy distribution systems. They share the common property to rely heavily on realtime and reliability constraints which form an important integral part of the systems specification.

This panel aims at discussing the challenges of globally distributed control systems for the Internet and for the next generation computers including operating systems thereby providing some directions for future research and development efforts.

Chair:
Jürgen Nehmer
University of Kaiserslautern, Germany

Panelists:
Hiroyuki Fujita
Tokyo University, Japan

Michel Gien
Chorus Systems, France

Yuji Inoue
NTT, Japan

Hermann Kopetz
Technical University of Vienna, Austria

Stephen Yau
Arizona State University, USA

Author Index

Notes

Notes

Notes

12/12/96